High-Efficiency RF Power Amplifiers and Oscillators

High-Efficiency RF Power Amplifiers and Oscillators

Innovative Engineering Design Approaches

Andrei Grebennikov

Consultant
Cobh, Ireland

WILEY

Published by John Wiley & Sons, Inc., Hoboken, New Jersey.

Library of Congress Cataloging-in-Publication Data:

Name: Grebennikov, Andrei, 1956– author
Title: High-efficiency RF power amplifiers and oscillators : innovative engineering design approaches / Andrei Grebennikov.
Description: Hoboken, New Jersey : Wiley, [2027] | Includes bibliographical references.
Identifiers: LCCN 2026026749 | ISBN 9781394374380 hardback | ISBN 9781394374403 adobe pdf | ISBN 9781394374397 epub | ISBN 9781394374410 obook
Subjects: LCSH: Power amplifiers—Design and construction | Oscillators, Electric—Design and construction | Mechanical efficiency
Classification: LCC TK7871.58.P6 G7 2027
LC record available at https://lccn.loc.gov/2026026749

Cover Design: Wiley
Cover Images: Artisan Graph/stock.adobe.com, Uwe/stock.adobe.com, Andrei Grebennikov

Contents

About the Author

Andrei Grebennikov received his Dipl Eng degree in Radio Electronics from Moscow Institute of Physics and Technology, Moscow, Russia, in 1980, and PhD degree in Radio Engineering from Moscow Technical University of Communications and Informatics, Moscow, Russia, in 1991. He has gained long-term academic and industrial experience working with the Moscow Technical University of Communications and Informatics, Moscow, Russia, the Institute of Microelectronics, Singapore, M/A-COM, Cork, Ireland, Infineon Technologies, Munich, Germany, and Linz, Austria, Bell Labs, Alcatel-Lucent, Dublin, Ireland, Microsemi, Aliso Viejo, California, and Sumitomo Electric, Elstree, UK, as an Engineer, Researcher, Lecturer, and Educator. He has lectured as a Guest Professor at the University of Linz, Linz, Austria, and presented short courses and tutorials as an Invited Speaker at the IEEE Microwave Theory and Techniques Society (MTT-S) International Microwave Symposia (IMS), European and Asia-Pacific Microwave Conferences, the Institute of Microelectronics, Singapore, Motorola Design Centre, Penang, Malaysia, the Tomsk State University of Control Systems and Radioelectronics, Tomsk, Russia, and the Aachen Technical University, Aachen, Germany. Dr. Grebennikov has served as the Guest Editor for the *IEEE Transactions on Microwave Theory and Techniques* and member of editorial board of the *International Journal of RF and Microwave Computer-Aided Engineering*. He is an author and co-author of more than 120 papers, holds 30 European and US patents, and has authored 10 books dedicated to RF and microwave circuit design.

Preface

The main objective of this book is to present in a simplified way the basic information required to design high-efficiency RF and microwave power amplifiers and oscillators, including innovative circuit design approaches to implement simple and effective structures of the load network to maximize efficiency over a wide frequency range with some practical examples. Regardless of the different operation classes like Class F, inverse Class F, or Class E and their combination and subclasses, an efficiency improvement is achieved by providing the nonlinear operation conditions when an active device can subsequently operate in pinch-off, active and saturation regions, resulting in the nonsinusoidal collector (or drain) current and voltage waveforms, for example, symmetrical for Class-F and inverse Class-F and asymmetrical for Class-E operation modes. As a result, the power amplifiers and oscillators operated in a Class F or an inverse Class F can be analyzed explicitly in the frequency domain when the harmonic load impedances are optimized by short-circuit and open-circuit terminations in order to control the voltage and current waveforms at the collector (or drain) to obtain maximum efficiency. However, the power amplifiers and oscillators operated in a Class E can be fully analyzed analytically in the time domain when an efficiency improvement is achieved by realizing close approximation to the ideal on-to-off active device operation in only pinch-off and voltage-saturation modes, so that high voltage and high current at the collector (or drain) do not occur at the same time. Since the transmitting signal is characterized by a high peak-to-average power ratio in modern wireless communication systems with increased bandwidth and high data rate when improving efficiency at significant power backoff has become critical, several circuit design techniques such as different types of the Doherty amplifier and envelope tracking are described to maximize efficiency across the entire backoff output power region of the high-power hybrid power amplifiers for base station applications and integrated power amplifiers for infrastructure applications.

This book is intended for and can be recommended to *university-level professors* as a comprehensive reference material to help in lecturing for graduates and postgraduates students, to *researchers and scientists* to combine the theoretical analysis with practical design and to provide a sufficient basis for innovative ideas and high-efficiency circuit design techniques, and to *practicing designers and engineers* as the book contains numerous well-known and novel practical circuits, architectures, and theoretical approaches with detailed description of their operational principles and applications of the high-efficiency RF and microwave power amplifiers and oscillators. It can be very useful for lecturing to promote the systematic way of thinking with analytical calculations, circuit simulation, and practical verification, thus making a bridge between theory and practice of RF and microwave engineering. As often happens, a new result is the well-forgotten old one. Therefore, the demonstration of not only new results based on new technologies or circuit schematics is given, but some sufficiently old ideas or approaches are also introduced that could be very useful in modern design practice or could contribute to the appearance of new general architectural ideas and specific circuit design techniques.

In Chapter 1, all necessary steps to provide an accurate device modeling procedure starting with the determination of the device small-signal equivalent circuit parameters are described and discussed. A variety of nonlinear models for MOSFET including LDMOSFET, GaAs MESFET, GaN HEMT, and bipolar devices, including GaAs HBTs, are presented. To highlight the advantages or drawbacks of one nonlinear device model over the other, a comparison of the measured and modeled current–voltage (*I–V*) and capacitance–voltage (*C–V*) characteristics and operating frequency ranges are provided. For practical design procedure, simplified equivalent circuits characterizing the device input and output impedances are derived which are necessary to make both impedance matching and harmonic control process clear and sensible.

The concept of impedance matching and the impedance-matching technique, which is very important when designing power amplifiers, are described and presented in Chapter 2. First, the main principles and impedance-matching tools, such as the Smith chart, are described, providing a starting point for the matching design procedure. As an engineering solution in general depends on the different circuit requirements, the designer should choose the optimum solution among a variety of matching networks with only lumped elements, or transmission lines, or both of them together, including a possibility of broadband matching using the stepped transmission line. To simplify and visualize the matching design procedure, an analytical approach, which allows the parameters of the matching circuits to be calculated using simple equations, and the Smith chart traces are discussed and illustrated with several examples of the narrowband and broadband power amplifiers. Finally, lossy matching design technique and broadband matching with prescribed amplitude-frequency response are introduced and explained.

Modern commercial and military communication systems require high-efficiency long-term operating conditions. Chapter 3 describes in detail the possible load-network solutions to provide a high-efficiency power amplifier operation based on using Class-F, inverse Class-F, and Class-E operation modes depending on the technical requirements. In Class-F power amplifiers analyzed in the frequency domain, the fundamental and harmonic load impedances are optimized by short-circuit termination and open-circuit peaking to control the voltage and current waveforms at the drain of the device to obtain maximum efficiency. In Class-E power amplifiers analyzed in the time domain, an efficiency improvement is achieved by realizing the on/off switching operation with special current and voltage waveforms so that high voltage and high current do not exist at the same time. Different configurations of the Class-E load-network structures such as Class E with shunt capacitance and series filter, parallel-circuit Class E, Class E with shunt capacitance and shunt filter, and their broadband applications, including the design example of a two-stage high-efficiency HBT power amplifier using a parallel-circuit Class-E mode, are presented and explained in detail.

Chapter 4 introduces the basic configurations of the power transistor oscillators, with simplified equation-based oscillator analysis using matrix and empirical design approaches. An immittance design approach is introduced and applied to parallel and series feedback oscillators to achieve optimum oscillation conditions, including an example of the optimum MOSFET oscillator design. Different configurations of the high-efficiency Class-F, inverse Class-F, and Class-E power transistor oscillators based on lumped elements and transmission lines for different frequency ranges from RF to millimeter waves, with some examples of practical applications in wireless power transfer (WPT) systems and a design example of a parallel-circuit Class-E power oscillator for compact fluorescent lamps are described and illustrated.

Chapter 5 describes the historical aspect of a Doherty approach to the power amplifier design and modern trends in Doherty amplifier design techniques using multistage and asymmetric architectures. To increase efficiency over the output power backoff range, the Class-F and inverse Class-F operation modes with the second- and third-harmonic control and broadband Class-E operation modes using structures with parallel circuit and shunt filter can be used in the load network. The inverted Doherty architectures, including a three-way Doherty configuration with an enhanced backoff region, are also described and discussed. Examples of the integrated and monolithic Doherty amplifiers and broadband capability of the Doherty amplifier with the design examples of a single-stage high-power asymmetric inverted Doherty amplifier for base station applications and two-stage integrated asymmetric Doherty amplifiers for massive multiple-input multiple-output (MIMO) applications based on GaN technology and operating in a sub-6 GHz frequency band are given.

Chapter 6 describes a variety of alternative circuit design techniques and approaches that can improve the power amplifier and oscillator performances. For example, combining the Class-F and Class-E operation modes allows flattening the drain voltage waveform, thus reducing the voltage peak factor which contributes to more reliable operation of the transistors. In order to increase efficiency over the output power backoff region, an envelope-tracking technique as well as a three-way asymmetric Doherty amplifier with a three-coupled-line combiner and load-modulated balanced power amplifier configurations are discussed and analyzed.

Acknowledgment

The author wishes to thank his wife, Galina Grebennikova, for performing important numerical calculations and figure drawings, as well as for her constant encouragement, inspiration, support, and assistance.

1
Active Devices and Modeling

For accurate power amplifier simulation and its circuit design at different operating frequencies and output power levels, it is important to represent an active device in the form of a nonlinear equivalent circuit, which can adequately describe the small- and large-signal electrical behaviors of the power amplifier up to the device transition frequency f_T and higher to its maximum frequency f_{max} that allows a sufficient number of harmonic components to be taken into account. Better approximation of the final design to the expected performance can only be achieved if the nonlinear device behavior is described accurately. Generally, the power amplifier design based on the nonlinear device models rather than on the blind black-box models enables better understanding and predicting both the active device behavior in particular and power amplifier performance in general, considering that the elements of the device nonlinear equivalent circuit will be the parts of the matching or harmonic-control circuits and directly affect stability and linearity of the power amplifier operation.

1.1 MOSFET Device Modeling

Accurate MOSFET device modeling is an extremely important procedure to develop low-cost silicon integrated circuits using CMOS technology for higher-speed and higher-frequency integrated circuits and subsystems within a shorter design time. The accuracy of MOSFET models to describe their nonlinear behavior is crucial to predict the entire circuit performance. The lumped-parameter equation-based equivalent circuit model provides a clearer physical description and more design flexibility.

The complete small-signal MOSFET equivalent circuit with intrinsic nonlinear model and extrinsic parasitic elements, including the gate–drain substrate coupling network and substrate parameters, is shown in Figure 1.1 [1]. Here, C_{gs} and R_{gs} are the gate–source capacitance and gate channel resistance, C_{gd} and R_{gd} are the gate–drain capacitance and drain ohmic resistance, g_m is the device transconductance, R_{ch} is the differential channel resistance as a result of the channel-length modulation by the drain voltage, C_{ds} is the drain–source capacitance, L_g is the gate lead inductance, R_s and L_s are the source bulk and ohmic resistance and lead inductance, and R_d and L_d are the drain bulk and ohmic resistance and lead inductance, respectively. A coupling path between the gate and the drain through the substrate may significantly affect the transfer performance of the transistor. Under an active bias condition, the conductive channel isolates the gate–bulk coupling and the capacitances C_{gb} and C_{gs} can be merged when the substrate coupling path between the gate and drain terminals can be ignored. However, under pinch-off bias conditions, the conductive channel does not exist, and the signal at the gate terminal is directly coupled to the substrate. This equivalent small-signal device model can accurately characterize the performance of the transistors in nanometer CMOS technology up to 60 GHz and above.

Several degradation mechanisms can significantly affect the performance of CMOS power amplifiers, such as gate-oxide breakdown, hot-carrier degradation, punchthrough, drain–bulk breakdown, or negative bias temperature instability. The gate-oxide breakdown and hot-carrier effect are the two most critical issues

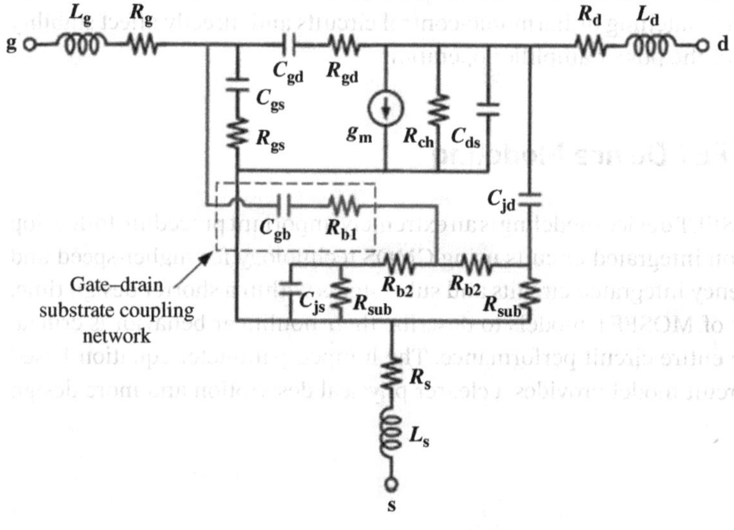

Figure 1.1 Equivalent small-signal circuit of MOSFET device.

of deep submicron CMOS device reliability. The gate-oxide breakdown is the sudden formation of a conductive path in the gate oxide of a MOSFET device as a result of high-voltage application. The allowable gate-oxide voltage becomes smaller as the gate-oxide thickness is reduced. Since the channel voltage varies from the source to the drain, the oxide stress (the voltage difference between the top of the oxide and the bottom of the oxide) is a function of position. The highest stress areas occur at the source and drain oxide edges, and therefore, from a design standpoint, the gate–source and the gate–drain voltages (V_{gs} and V_{gd}) should never exceed prespecified values [2]. There are two basic types of oxide breakdown: hard breakdown and soft breakdown. In hard breakdown, a large gate current increase can be observed, and it has the most disastrous effect on the device performance. This type of breakdown is typically found in thick-oxide devices because of high-voltage stress. In soft breakdown, a small sudden gate current increase and a sudden gate current noise increase are observed [3].

The electrical field near the corner of a silicon–silicon dioxide (Si–SiO_2) interface, where the drain junction is directly under the gate, is the largest in the device and results in the following hot-carrier effects: electron injection into the SiO_2 layer producing a gate current; an avalanche hole–electron pair production process, increasing the substrate potential and resulting in the built-in bipolar transistor and current source between the drain and substrate; or punchthrough for shortened channel lengths of less than 1 μm contributing to a drain–source leakage current [4, 5]. Consequently, for high-efficiency operation modes with a large drain voltage peak factor, the device is liable to a hot-carrier effect when it is turned off.

In short-channel devices, the lateral field can be extremely high, and consequently, carriers can achieve very high energy before losing momentum to a collision with the crystal lattice. Some of the hot energy carriers collide with the lattice before arriving at the drain with sufficient energy to result in impact ionization [2]. This, in addition to potential avalanche multiplication, can result in surface defects, thus resulting in a reduced carrier mobility in the channel. This can also result in a trapped charge in the gate oxide or oxide/silicon interface, shifting the local threshold voltage. Usually, this damage occurs in the drain region where the electric field is very high, resulting in an increase in the device on-resistance and knee voltage.

In a normal operating condition, the drain–source current of a MOSFET device flows close to the surface when a large enough gate bias is applied to create an inversion channel. In the absence of a gate bias, a very little current flows because the drain–bulk and the source–bulk diodes are effectively connected back-to-back with opposite polarities. As larger voltages are applied to the drain, the drain–bulk depletion region extends farther to accommodate the electric potential drop. This depletion region will eventually extend all the way to meet the depletion region of the source–bulk junction, thereby diminishing the potential barrier that stops the direct flow of current between the drain and the source [2]. This punch-through process results in a large current flow that can occur even in the absence of any significant gate bias.

If the gate breakdown is a catastrophic phenomenon, the hot-carrier effect contributes to a long-term power amplifier performance. Because of this hot-carrier effect, the output power of the CMOS power amplifier decreases exponentially, and the slope of this decrease becomes sufficiently small after about 70–80 hours of its operation, suggesting that most of the created trap sites at the Si–SiO_2 interface have been filled by electrons. The recovery of the overall degradation in the output power from its nominal value can be achieved by slightly increasing the gate bias voltage, which indicates that the performance degradation is mainly because of the increase in threshold voltage [6].

1.2 LDMOSFETs

Silicon MOSFET devices have found widespread worldwide applications due to their reliable, low-cost, and high-performance technology. For example, the laterally diffused MOSFET (LDMOSFET) device structures have proven to be highly efficient, high-gain, and linear for high-power RF and microwave base station applications. Initially, several well-known physically based MOSFET models were developed to describe the device's electrical behavior [7, 8]. However, some of them, such as Level 1, Level 2, or Level 3 large-signal models, were very simple and could not describe the current–voltage and voltage–capacitance characteristics with acceptable accuracy. At the same time, the BSIM3 and later BSIM4 models are too complicated and quite formal in general, so, for better learning of the device's basic properties, it is useful to consider its intrinsic nonlinear core circuit only. Moreover, microwave parasitic effects in silicon MOSFETs are not easily physically predictable. The table-based models, such as the Root model, are only accurate for the characterized structures and measurement conditions. An empirical analytical modeling approach is an explicit and valid compromise between physical models and data-based models and provides a clear relationship between the device equivalent circuit parameters and power amplifier performance.

1.2.1 Small-signal Equivalent Circuit

To describe accurately the nonlinear properties of the large-size MOSFET devices, it is necessary to consider the distributed nature of the gate capacitance, because the channel resistance is not equal to zero. In this case, the device channel can be modeled as a bias-dependent distributed transmission line along the channel length, as shown in Figure 1.2, demonstrating the physical structure of an LDMOSFET device. Here, a heavily doped p^+ layer (enhancement and sinker) is inserted between the top source and p^+-substrate (source grounding) for low resistivity to provide high current flow between the drain and source terminals [9]. The lightly doped p-epilayer and n^- layer are required to provide sufficient distance between regions to prevent latchup (forward-biased p–n diodes) and for the drain–source breakdown protection. The parasitic gate–drain capacitance is directly related to the overlap of the gate oxide onto the heavily doped n^+-source region.

This one-dimensional approach assumes a gradual channel approximation when the quantity of charge in the channel is controlled completely by the gate electrode, only fields in the vertical dimension influence the depletion region, and channel current is provided entirely by drift with constant mobility. Despite some drawbacks related to short-channel devices, this approach allows a compromise between accuracy and simplicity of a model derivation.

The *ABCD*-matrix of the given *RC* transmission line can be written as

$$\left[ABCD\right] = \begin{bmatrix} \cosh\gamma L & Z_0\sinh\gamma L \\ \dfrac{\sinh\gamma L}{Z_0} & \cosh\gamma L \end{bmatrix} \tag{1.1}$$

where $\gamma = \sqrt{j\omega R'_{ch}C'_g}$ is the propagation constant, $Z_0 = R'_{ch}/\gamma$ is the characteristic transmission-line impedance, L is the channel length, $R'_{ch} = R_{ch}/L$, $C'_g = C_g/L$, R_{ch} is the channel charging resistance, which is a result of noninstantaneous response to the changes of the gate–source voltage, and C_g is the total gate capacitance.

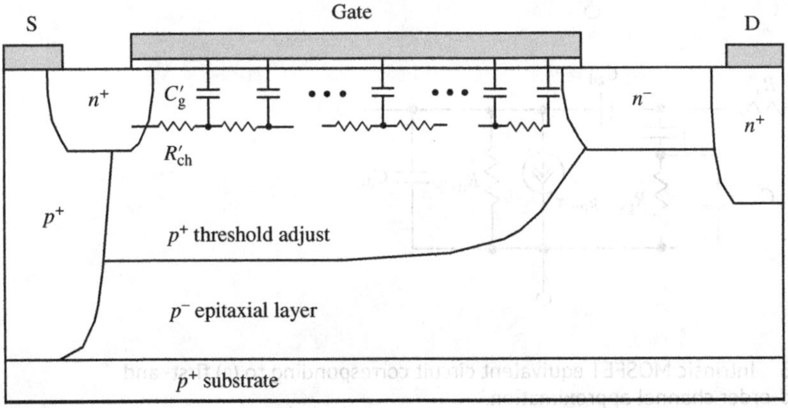

Figure 1.2 Schematic representation of MOSFET distributed channel structure.

In this case, the equivalent gate–source impedance Z_{gs} can be estimated from Eq. (1.1) (matrix) as

$$Z_{gs} = \frac{A}{C} = R_{ch}\frac{\coth \gamma L}{\gamma L} \tag{1.2}$$

The first-order approximation of Z_{gs}, obtained from a power series expansion of Eq. (1.2), yields

$$Z_{gs} = R_{ch}\frac{\coth \gamma L}{\gamma L} \cong \frac{R_{ch}}{\gamma L}\left(\frac{1}{\gamma L} + \frac{\gamma L}{3}\right) = \frac{R_{ch}}{3} + \frac{1}{j\omega C_g} \tag{1.3}$$

From Eq. (1.3), it follows that the MOSFET intrinsic gate–source circuit can be modeled using a simple series circuit with the resistance $R_{gs} = R_{ch}/3$ and the capacitance $C_{gs} = C_g$. Figure 1.3a shows the intrinsic transistor equivalent circuit corresponding to the first-order channel approximation, where τ is the effective channel carrier transit time.

Better accuracy for channel approximation can be obtained from the second-order approximation of Z_{gs}, derived from a power series expansion of Eq. (1.2) as

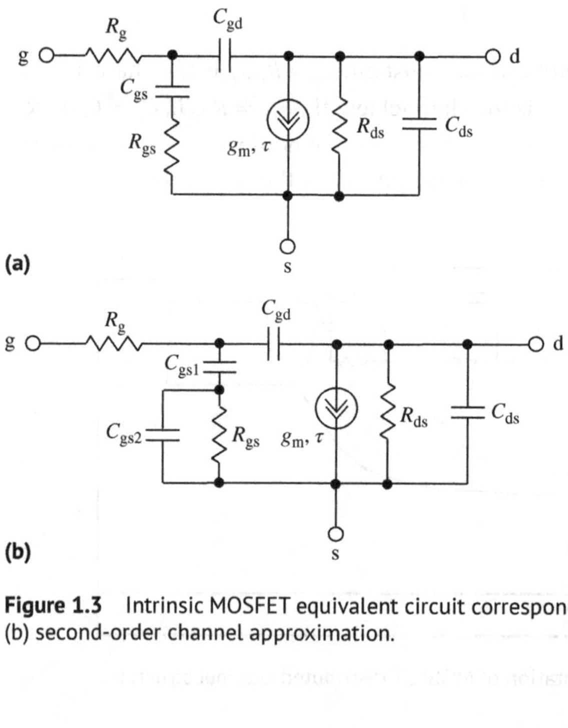

Figure 1.3 Intrinsic MOSFET equivalent circuit corresponding to (a) first- and (b) second-order channel approximation.

$$Z_{gs} = R_{ch}\frac{\coth\gamma L}{\gamma L} \cong \frac{R_{ch}}{\gamma L}\left[\frac{1}{\gamma L} + \frac{\gamma L}{3} - \frac{(\gamma L)^2}{45}\right] = \frac{R_{ch}}{3}\left(1 - \frac{j\omega C_g R_{ch}}{15}\right)$$
$$+ \frac{1}{j\omega C_g} \cong \frac{R_{ch}}{3}/\left(1 + j\omega\frac{R_{ch}}{3}\frac{C_g}{5}\right) + \frac{1}{j\omega C_g} \tag{1.4}$$

As a result, the second-order approximation of the device channel structure can be realized by the series connection of the capacitance $C_{gs1} = C_g$ and the parallel RC circuit, which consists of the resistance $R_{gs} = R_{ch}/3$ and capacitance $C_{gs2} = C_g/5$. The intrinsic transistor equivalent circuit corresponding to the second-order approximation is shown in Figure 1.3b.

For a high-power MOSFET device whose channel width is significantly larger than its channel length, the distributed nature of the total gate resistance $R_{tot} = R_{sh}W/L$ across the width W, where R_{sh} is the sheet resistance of the gate material, must be taken into consideration. In this case, silicon MOSFET can be decomposed into n devices, each with a width of W/n and a gate resistance of R_{tot}/n. For $n \to \infty$, it will be viewed as an array of small transistors distributed along the gate of the device. Commonly, it is necessary to consider a two-dimensional power MOSFET distributed model because it shows a distributed-gate nature along both the channel length and channel width. However, for a short-channel MOSFET, the distributed-gate effect along the channel length can be considered only in the frequency range close to the transition frequency f_T and higher. When $\omega R_{gs}C_{gs} \ll 1$ an analysis of the distributed-gate model along the channel width based on transmission line theory shows that all transistor Y-parameters should be modified $\tanh(\gamma W)/(\gamma W)$ by the term $\tanh(\gamma W)/(\gamma W)$ [10]. However, the linear power series expansion of this term as

$$\frac{\tanh(\gamma W)}{\gamma W} = 1 - j\omega C_g\frac{R_{tot}}{3} \cong \frac{1}{1 + j\omega C_g\frac{R_{tot}}{3}} \tag{1.5}$$

leads to only the additional use of a series lumped gate resistance $R_{tot}/3$ that does not alter the structure of the transistor equivalent circuit. Consequently, the overall gate resistance R_g can generally be divided into two consecutive series resistances as $R_g = R_{ge} + R_{gi}$, where R_{ge} is the extrinsic contact and ohmic gate electrode resistance and $R_{gi} = R_{tot}/3$ is the intrinsic gate resistance due to the distributed-gate structure of the power MOSFET.

Figure 1.4 shows the nonlinear MOSFET equivalent circuit with extrinsic parasitic elements, which can properly describe the nonlinear behavior of both VDMOSFET (vertical double-diffused MOSFET) and LDMOSFET devices [11, 12]. The nonlinear current source $i(v_{gs}, v_{ds}, \tau)$ as a function of the input gate–source

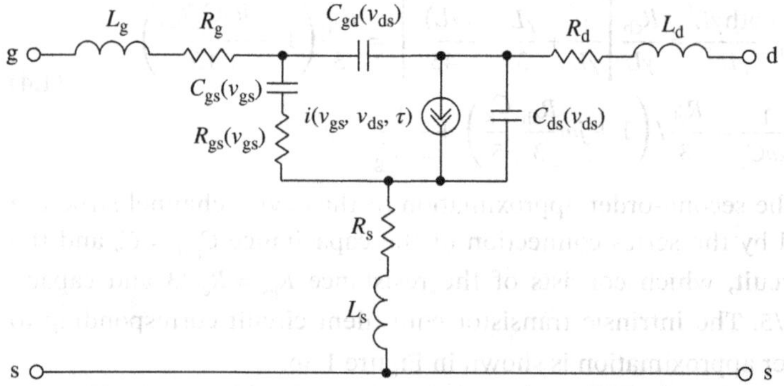

Figure 1.4 Small-signal MOSFET equivalent circuit with extrinsic linear elements.

and output drain–source voltages incorporating self-heating effect can be described sufficiently simply and accurately using hyperbolic functions [12, 13]. Careful analytical description of the transition from quadratic to linear regions of the device transfer characteristic enables more accurate prediction of the intermodulation distortion (IMD) [14]. The overall channel carrier transit time τ also includes an effect of the transcapacitance required for charge conservation. The drain–source capacitance C_{ds} and gate–drain capacitance C_{gd} are considered as junction capacitances that strongly depend on the drain–source voltage. The extrinsic parasitic elements are represented by the gate and drain bondwire inductances L_g and L_d, source inductance L_s, source and drain bulk and ohmic resistances R_s and R_d, and gate contact and ohmic resistance R_g. The effect of the gate–source channel resistance R_{gs} becomes significant at higher frequencies close to transition frequency $f_T = g_m/2\pi C_{gs}$, where g_m is the device transconductance. To account for the self-heating effect and substrate loss, a special four-port thermal circuit and a series combination of the resistance and capacitance between the external drain and source terminals can be included [15].

1.2.2 Nonlinear *I–V* Models

An empirical analytical approach to approximate the nonlinear behavior of the drain current source $I_{ds}(V_{gs}, V_{ds})$ using hyperbolic functions was first applied to a JFET device [16]. In this case, instead of using separate equations for the triode and pinch-off regions, irrespective of the device geometry and material parameters, a general expression based on a hyperbolic tangent function was proposed, written as

$$I_{ds} = I_{dss}\left(1 - \frac{V_{gs}}{V_p}\right)^2 \tanh\alpha\left|\frac{V_{ds}}{V_p - V_{gs}}\right| \tag{1.6}$$

where I_{dss} is the saturation drain current for $V_{gs} = 0$, α is the saturation voltage parameter, and V_p is the pinch-off voltage. As a result, a good agreement was obtained between the predicted and experimental results, showing a promising future of such a simple empirical model.

A similar and sufficiently simple nonlinear model using a hyperbolic tangent function and incorporating the self-heating effect was later proposed to describe the I–V characteristics of a MOSFET device, written as

$$I_{ds} = \beta_{eff} V_{gst}^{VGexp} (1 + \lambda V_{ds}) \tanh\left(\frac{\alpha V_{ds}}{V_{gst}^{SATexp}}\right) \tag{1.7}$$

where

$$\beta_{eff} = \beta / \left(1 + \mu_{crit} V_{gst}^{GMexp}\right)$$

$$V_{gst} = V_{st} \ln\left[1 + \exp\left(\frac{V_{gst1}}{V_{st}}\right)\right]$$

$$V_{gst1} = V_{gs} - V_{th0} - \gamma V_{ds}$$

where VGexp, GMexp, SATexp, and μ_{crit} are the channel current parameters, β and γ are the first-order functions of the channel temperature increase, and α, λ, and V_{st} are the fitting parameters [17, 18]. Similarly, an empirical nonlinear model using both hyperbolic tangent and exponential functions, which is single-piece and continuously differentiable, was developed for silicon LDMOSFET devices [13].

In many applications, it is necessary to take into consideration the MOSFET operation in the weak-inversion region when the gate–source voltage V_{gs} is smaller than the threshold voltage V_{th}. For example, to reduce the IMD of a Class-AB power amplifier when the device is biased around the onset of the strong-inversion region from the weak-inversion region for low drain quiescent current. The drain current in the weak-inversion region is mainly dominated by the diffusion component that increases exponentially with the gate voltage [5]. On the other hand, in the strong-inversion saturation region, when the gate-source voltage is greater than the threshold voltage, the drain current is proportional to the square of $(V_{gs} - V_{th})$.

To obtain continuous behavior from the weak-inversion region to the strong-inversion region for the drain current and a compromise between accurate device modeling and ease of circuit analysis, the following analytical function can be used:

$$I_{ds}(V_{gs}) = A\left\{\ln\left[1 + \exp\left(B\left(V_{gs} - V_{th}\right)\right)\right]\right\}^2 \tag{1.8}$$

where A and B are the approximation parameters. In this case, the drain current is effectively proportional to the square of $(V_{gs} - V_{th})$ when V_{gs} is larger than V_{th} and exponentially decreases with the gate–source voltage when V_{gs} is smaller than V_{th}. The approximation parameters A and B are defined by the following conditions:

$$I_{ds}\big|_{V_{gs}=V_{th}} = I_{th} \qquad \left.\frac{\partial I_{ds}}{\partial V_{gs}}\right|_{V_{gs}=V_{th}} = S_{th} \tag{1.9}$$

where I_{th} is the threshold drain current, as shown in Figure 1.5a, and S_{th} is the slope of the current–voltage transfer characteristic at the threshold point. Then,

$$A = \frac{I_{th}}{(\ln 2)^2} \qquad B = \frac{S_{th}}{I_{th}} \ln 2 \tag{1.10}$$

Consequently, the device transfer characteristic can be defined in terms of only two physical parameters I_{th} and S_{th}, which are easily calculated from the device measurements.

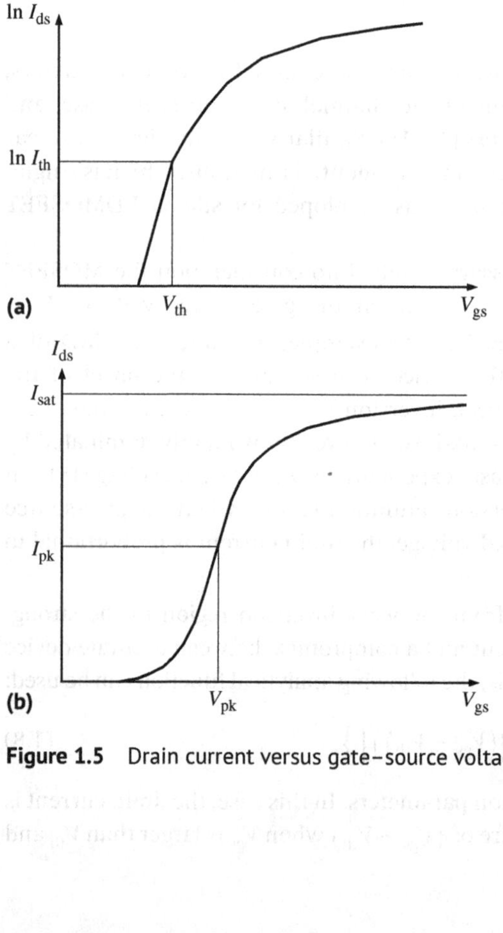

Figure 1.5 Drain current versus gate–source voltage.

To link the linear and saturation regions by a suitable continuous analytical dependence, the following function to describe the I–V curves can be used:

$$I_{ds}\left(V_{gs}, V_{ds}\right) = \left(I_{max}^{-1} + I_{dso}^{-1}\right)^{-1} \tag{1.11}$$

where I_{dso} is the exponential function of V_{gs} and V_{ds}, and I_{max} is the maximum channel current obtained by

$$I_{max}(V_{ds}) = I_{pk}(1 + \lambda V_{ds})\tanh(\alpha V_{ds}) \tag{1.12}$$

where I_{pk} is the drain current, which corresponds to the maximum slope of the $I_{ds}\left(V_{gs}\right)$ curves, as shown in Figure 1.5b [19]. The saturation voltage parameter α affects the slope of the $I_{ds}(V_{ds})$ curves in the linear region, whereas the parameter λ determines the slope of the same drain characteristics in the saturation region.

In view of the monotonous behavior of $I_{ds}(V_{ds})$ curves, the entire drain current–voltage characteristics of a MOSFET device can be described as

$$I_{ds}\left(V_{gs}, V_{ds}\right) = I_0 / \left[1 + \left(\frac{I_0}{I_{max}}\right)^n\right]^{\frac{1}{n}} \tag{1.13}$$

where

$$I_0 = \frac{I_{th}}{(\ln 2)^2\left(1 - \beta V_{gs}\right)}\left\{\ln\left[1 + \exp\left(\frac{S_{th}\ln 2}{I_{th}}\left(V_{gs} - V_{th}\right)\right)\right]\right\}^2$$

$$I_{max} = I_{sat}(1 + \lambda V_{ds})\tanh(\alpha V_{ds})$$

where I_{sat} is the saturated drain current, α is the saturation voltage parameter affecting the slope of the $I_{ds}(V_{ds})$ characteristics in the linear region, λ is the parameter that determines a slope of the same drain characteristics in the saturation region, $V_{th} = V_{tho} - \sigma V_{ds}$, n and β are the fitting parameters that determine the slope of the transfer characteristics under large values of V_{gs}, and σ is the parameter that expresses empirically the dependence of the threshold voltage on V_{ds} [12]. Better accuracy can be achieved by using I_{sat} instead of I_{pk} in Eq. (1.12).

To verify this empirical I–V model, a high-power LDMOSFET with a gate width of 4 cm and a gate length of 1.1 μm (LP801 from Polyfet) was used. The theoretical and experimental output $I_{ds}(V_{ds})$ curves are shown in Figure 1.6, with the resulting current mean-square error of a family of the output I_{ds}– V_{ds} characteristics of 0.5%. The values of the simulated model parameters are indicated in Table 1.1.

The theoretical approximation of $I_{ds}(V_{ds})$ curves is sufficient and accurate for a high-power LDMOSFET and requires only six fitting parameters, four of which

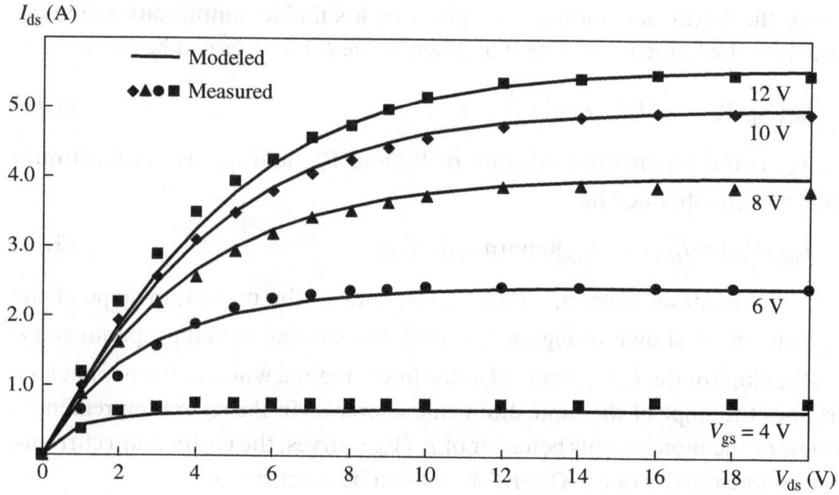

Figure 1.6 Measured and modeled $I_{ds}(V_{ds})$ curves of high-voltage LDMOSFET.

Table 1.1 Simulated $I-V$ model parameters of high-voltage LDMOSFET.

Parameters	α (1/V)	β (1/V)	λ (1/V)	V_{th} (V)	I_{th} (A)	S_{th} (A/V)	I_{sat} (A)	n	σ
Values	0.15	0	0.0005	2.7	0.115	0.2	6.8	1.0	0

are easily determined by the experimental curves. An empirical analytical model allows the description of I_{ds} and g_m as the functions of V_{gs} in the weak-inversion region, as well as in the strong-inversion region of the LDMOSFET operation. Substantially better fitting with $I_{ds}(V_{gs})$ and $g_m(V_{gs})$ experimental curves is achieved compared with the results obtained from two-dimensional simulations, as shown in Figure 1.7 [20].

To verify that this model can be applicable to low-voltage MOSFET devices, the proper low-voltage RF power MOSFET with a gate width of 2 mm was selected [21]. Figure 1.8 shows the theoretical and experimental drain current $I_{ds}(V_{ds})$ curves. In this case, to improve the sensitivity of the drain current I_{ds} under large values of V_{gs}, the term $(1 - \beta V_{gs})$ was added into the denominator for I_0 in Eq. (1.13). The resulting current mean-square error of a family of the output $I_{ds}(V_{ds})$ curves was 0.42% only. The parameters of the simulated $I_{ds}(V_{ds})$ curves are presented in Table 1.2. The comparison of the transfer $I_{ds}(V_{gs})$ curves with the same characteristics calculated by means of the BSIM3v3 model developed for

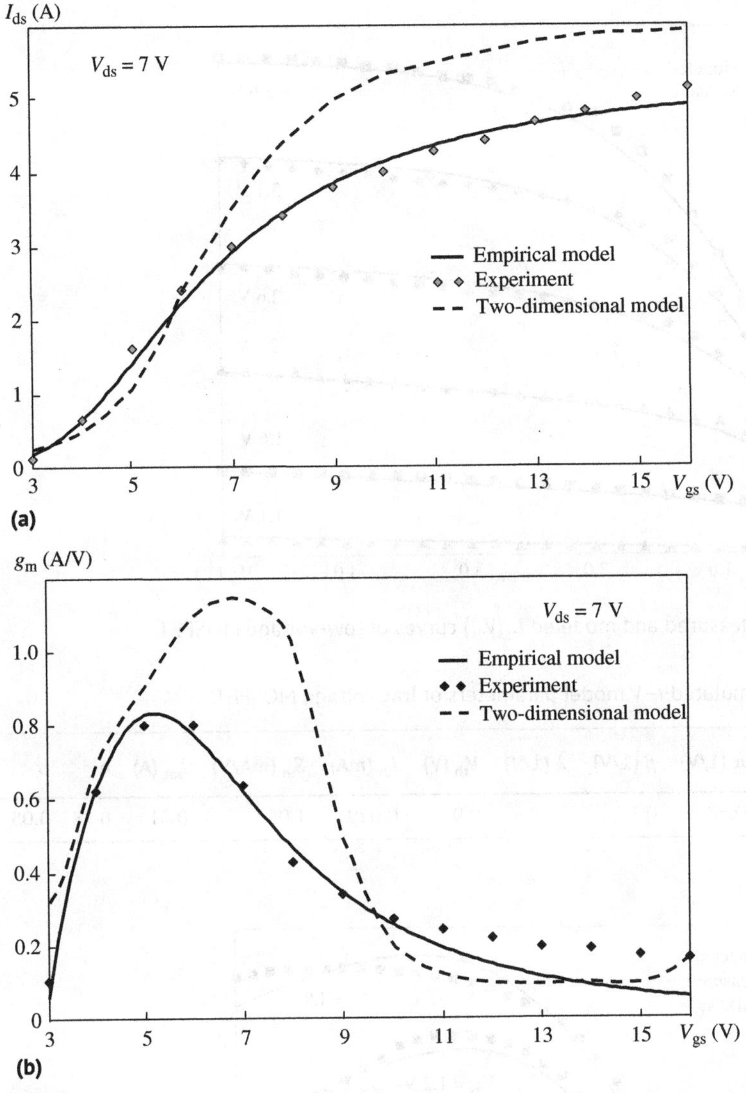

Figure 1.7 Measured and modeled $I_{ds}(V_{gs})$ and $g_m(V_{gs})$ curves of high-voltage LDMOSFET.

modeling of deep submicron devices is shown in Figure 1.9, demonstrating a good agreement with the experimental curves, and they are practically the same as in the case of the BSIM3v3 approximation.

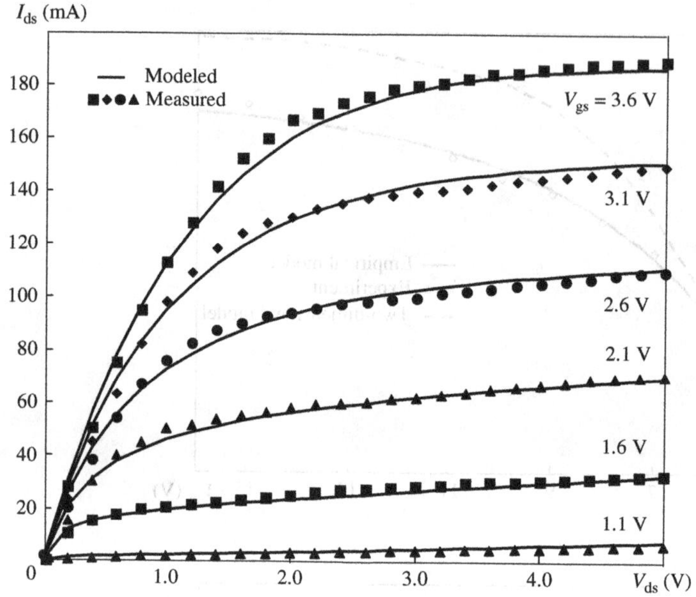

Figure 1.8 Measured and modeled $I_{ds}(V_{ds})$ curves of low-voltage MOSFET.

Table 1.2 Simulated I–V model parameters of low voltage MOSFET.

Parameters	α (1/V)	β (1/V)	λ (1/V)	V_{th} (V)	I_{th} (mA)	S_{th} (mA/V)	I_{sat} (A)	n	σ
Values	0.58	0.17	0	0.9	0.029	1.05	0.31	0.78	0.05

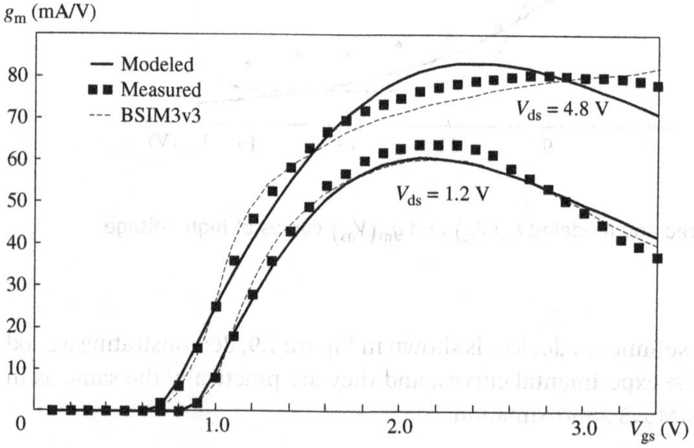

Figure 1.9 Measured and modeled $g_m(V_{gs})$ curves of low-voltage MOSFET.

1.2.3 Nonlinear *C–V* Models

Generally, all capacitances corresponding to the intrinsic equivalent circuit of the LDMOSFET device are nonlinear. However, their nonlinear behavior is different, and the input gate–source capacitance C_{gs} has a dominant effect on the IMD level, especially when the frequency increases in the microwave region [22]. However, the calculation of the gate–source capacitance C_{gs} or the gate–drain capacitance C_{gd} from the charges corresponding to the strong-inversion model only results in a mathematically complicated expression [5]. Therefore, in most cases, when it is necessary to predict efficiency, power gain, or output power of the power amplifier, the capacitances C_{gs} and C_{gd} can be modeled as the fixed capacitances measured at the quiescent bias voltage, and the *p–n* junction diode capacitance model can be applied to the capacitance C_{ds} [20]. The gate–drain capacitance C_{gd} can also be considered as the bias-dependent junction capacitance [23]. With the increase in the drain bias voltage, a depletion region is formed under the oxide in the lightly doped drain region. Therefore, the capacitance C_{gd} can be considered as the junction capacitance, which strongly depends on the drain–source voltage V_{ds}. According to the accurate charge model calculations, the gate–drain capacitance C_{gd} has a strong dependence on V_{gs} only in the moderate-inversion region when $V_{gs} - V_{th} < 1$ V [5]. In this region, the behavior of C_{gd} is similar to C_{gs}, and can be evaluated using the same hyperbolic tangent functions. However, for high-voltage LDMOSFET devices, since the dependence of C_{gd} on V_{gs} is quite small, it seems sufficient to limit its dependence on V_{ds} only. The drain–source capacitance C_{ds} varies due to the change in the depletion region, which is mainly determined by the value of V_{ds}.

To describe the nonlinear gate–source capacitance C_{gs} in a small-signal operation mode as a function of the gate–source voltage V_{gs}, it needs first to consider an appropriate behavior of each of its main composite parts: the intrinsic gate–source capacitance C_{gsi}, including both the gate–source and the source–substrate charge fluctuations, and the gate–substrate capacitance C_{gbi}. The gate–source voltage dependence of these components is substantially different [5]. The intrinsic gate–substrate capacitance C_{gbi} is constant in the accumulation region, where it is equal to the total intrinsic oxide capacitance C_{ox}, slightly decreases in the weak-inversion region, significantly reduces in the moderate-inversion region, and becomes practically constant in the strong-inversion or saturation region. The intrinsic gate–source capacitance C_{gsi} grows rapidly in the moderate-inversion region and equals $2C_{ox}/3$ in the saturation region. The voltage dependence of the total gate–source capacitance C_{gs} as a sum of its components C_{gsi} and C_{gbi} on V_{gs} is shown in Figure 1.10.

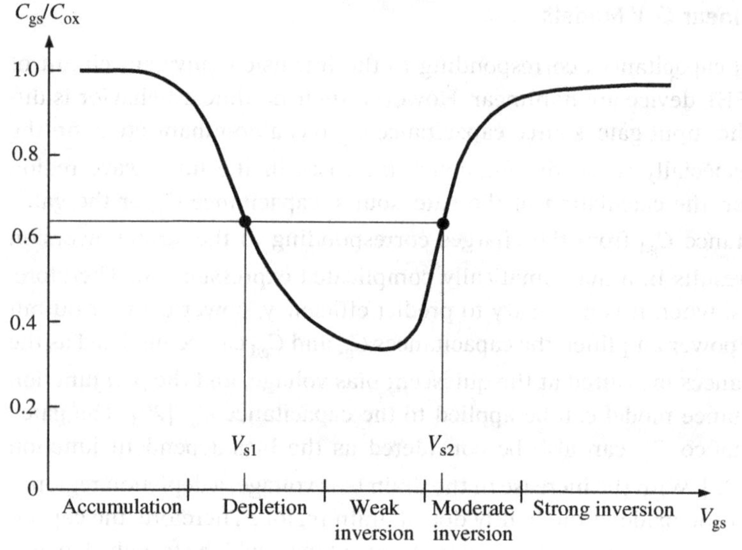

Figure 1.10 Gate–source capacitance versus gate–source voltage.

A hyperbolic tangent function can be used for each of two parts of the dependence $C_{gs}(V_{gs})$, where the gate–source capacitance can be approximated by

$$C_{gs} = C_{gsmin} + C_s\left\{1 + \tanh\left[\frac{S}{C_s}(V_{gs} - V_s)\right]\right\} \tag{1.14}$$

where $C_s = (C_{gsmax} - C_{gsmin})/2$, C_{gsmax} is the maximum gate–source capacitance, C_{gsmin} is the minimum gate–source capacitance, $S = (S_1, S_2)$ is the slope of $C_{gs}(V_{gs})$ at each bend point $V_{gs} = (V_{s1}, V_{s2})$, as shown in Figure 1.10.

$$S_1 = \left.\frac{\partial C_{gs}}{\partial V_{gs}}\right|_{V_{gs}=V_{s1}} \qquad S_2 = \left.\frac{\partial C_{gs}}{\partial V_{gs}}\right|_{V_{gs}=V_{s2}} \tag{1.15}$$

The total gate–source capacitance C_{gs} as a function of V_{gs} can be modeled as

$$C_{gs} = C_{gsmax} - C_{gso}\left\{1 + \tanh\left[\frac{S_1}{C_s}(V_{gs} - V_{s1})\right]\right\} \times \left\{1 + \tanh\left[\frac{S_2}{C_s}(V_{gs} - V_{s2})\right]\right\} \tag{1.16}$$

where $C_{gsmax} = C_{ox}$ and C_{gso} is the model fitting parameter [12].

The approximation function for the gate–source capacitance C_{gs} modeling the dependence on V_{gs} can also be expressed by using the two components, both containing the hyperbolic tangent functions:

$$C_{gs} = C_{gs1} + C_{gs2}\left\{1 + \tanh\left[C_{gs6}\left(V_{gs} + C_{gs3}\right)\right]\right\} + C_{gs4}\left[1 - \tanh\left(C_{gs5}V_{gs}\right)\right] \tag{1.17}$$

where C_{gs1}, C_{gs2}, C_{gs3}, C_{gs4}, C_{gs5}, and C_{gs6} are the approximation parameters [13].

For submicron MOSFET devices, when C_{gs} slightly increases with the increase of V_{ds}, the approximation expression for C_{gs} as a function of both V_{gs} and V_{ds} can be written as

$$C_{gs} = C_{gs0} + C_{gs1}\left\{A_s + B_s\tanh\left[C_s\left(V_{gs} - V_{th}\right)\right]\right\} \times \left\{D_s + E_s\left[1 + \tanh\left(V_{gs} - V_{ds}\right)\right]\tanh\left[F_sV_{ds} - G_sV_{gs}\right]\right\} \tag{1.18}$$

where C_{gs0} is the bias-dependent capacitance, V_{th} is the threshold voltage, C_{gs1} is the scaling factor, and A_s, B_s, C_s, D_s, E_s, F_s, and G_s are the model fitting parameters [24].

On the other hand, for high-voltage MOSFET devices, the dependencies of the gate–drain capacitance C_{gd} and the drain–source capacitance C_{ds} on V_{ds} can be evaluated with sufficient accuracy by the junction capacitance model as

$$C_{gd(ds)} = C_{gdo(dso)}\left(\frac{\phi + V_{dso}}{\phi + V_{ds}}\right)^m \tag{1.19}$$

where m (m_g for C_{gd} or m_d for C_{ds}) is the junction sensitivity depending on a doping profile ($m = 1/3$ for the linearly graded junction, $m = 1/2$ for the abrupt junction, and $m > 1/2$ for the hyperabrupt junction), ϕ is the contact potential, and C_{gdo} and C_{dso} are the junction capacitances when $V_{ds} = V_{dso}$. For practical profiles of the junction, which are neither exactly abrupt nor exactly linearly graded, one often chooses the parameters m and ϕ to obtain the best matching between the theoretical model and the measurements.

The results of the approximation of C_{ds} and C_{gd} as the junction capacitances for high-power LDMOSFET (LP801 from Polyfet) are shown in Figures 1.11 and 1.12, respectively. The fitting parameters of the approximation curves $C_{gd}(V_{ds})$ and $C_{ds}(V_{ds})$ are given in Table 1.3.

For submicron MOSFET devices, to consider the dependence of C_{gd} on both V_{gs} and V_{ds}, the approximation expression for the C_{gd} is written as

$$C_{gd} = C_{gd0} + C_{gd1}\left\{A_d + B_d\tanh\left[C_d\left(D_dV_{gs} - V_{ds}\right) - V_{th}\right]\right\} \tag{1.20}$$

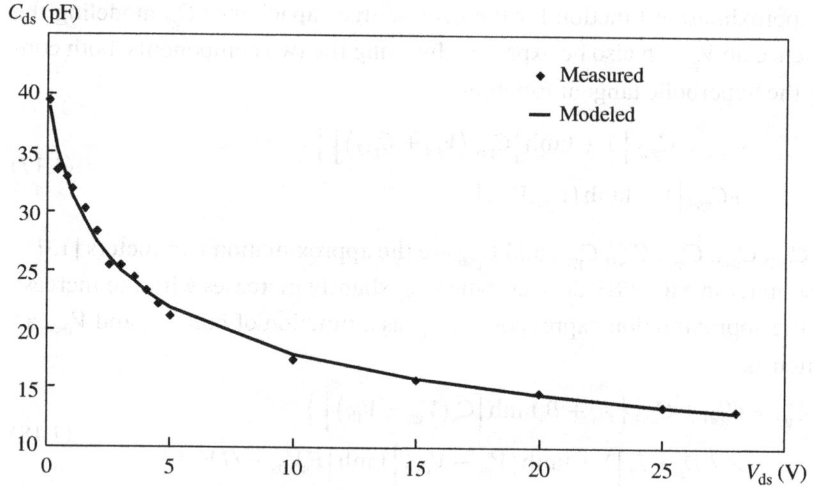

Figure 1.11 Measured and modeled $C_{ds}(V_{ds})$ dependencies of high-voltage LDMOSFET.

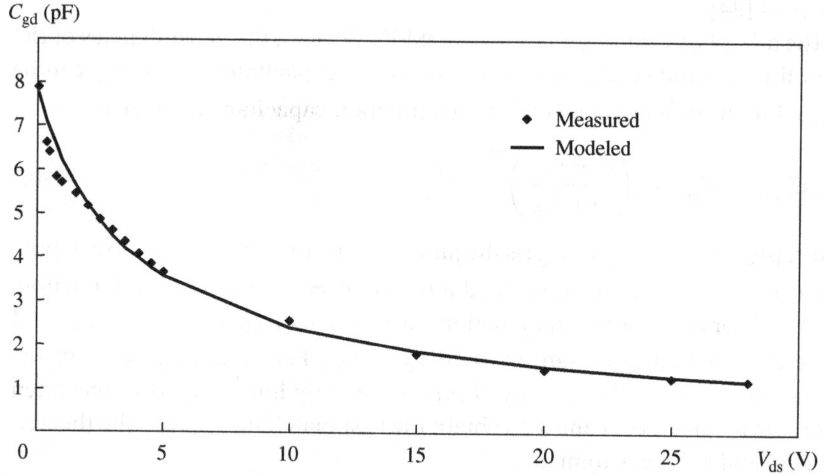

Figure 1.12 Measured and modeled $C_{gd}(V_{ds})$ dependencies of high-voltage LDMOSFET.

Table 1.3 Simulated *C–V* junction model parameters.

Capacitance	$C_{gd(ds)o}$ (pF)	m	ϕ (V)
C_{gd}	7.88	0.8	2.94
C_{ds}	39.42	0.33	1.0

where C_{gd0} is the bias-dependent capacitance, V_{th} is the threshold voltage, C_{gd1} is the scaling factor, whereas A_d, B_d, C_d, and D_d are the model fitting parameters [24].

1.2.4 Charge Conservation

To properly describe the nonlinear device models under the small- and large-signal operation modes, it is necessary to satisfy the charge conservation condition. For a three-terminal MOSFET device, the matrix equation for the small-signal charging circuit in the frequency domain is given by

$$\begin{bmatrix} I_g \\ I_d \\ I_s \end{bmatrix} = j\omega \begin{bmatrix} C_{gg} & -C_{gd} & -C_{gs} \\ -C_{dg} & C_{dd} & -C_{ds} \\ -C_{sg} & -C_{sd} & C_{ss} \end{bmatrix} \begin{bmatrix} V_g \\ V_d \\ V_s \end{bmatrix} \tag{1.21}$$

where I_g, I_d, and I_s are the terminal current amplitudes, V_g, V_d, and V_s are the terminal voltage amplitudes, and the capacitance between any two device terminals (k, l) is described as $C_{kl} = \partial Q_k / \partial V_l$ [5]. To transform a three-terminal device into a two-port network with a common source terminal, the current and voltage terminal conditions should be respectively set as $I_g = I_{gs}$, $I_d = I_{ds}$, $I_s = -(I_{gs} + I_{ds})$, $V_g - V_s = V_{gs}$, $V_d - V_s = V_{ds}$, and the relationships between the terminal capacitances as

$$\begin{aligned} C_{gg} &= C_{gd} + C_{gs} = C_{dg} + C_{sg} \\ C_{dd} &= C_{dg} + C_{ds} = C_{gd} + C_{sd} \\ C_{ss} &= C_{sg} + C_{sd} = C_{gs} + C_{ds} \end{aligned} \tag{1.22}$$

As a result, the admittance Y_c-matrix for such a capacitive two-port network is written as

$$Y_c = \begin{bmatrix} j\omega(C_{gs} + C_{gd}) & -j\omega C_{gd} \\ -j\omega(C_{gd} + C_m) & j\omega(C_{ds} + C_m + C_{gd}) \end{bmatrix} \tag{1.23}$$

where $C_m = C_{dg} - C_{gd}$ is the transcapacitance, C_{gd} represents the effect of the drain on the gate, and C_{dg} represents the effect of the gate on the drain, and these effects are different [8, 25]. Similarly to the I–V characteristics, there is no reason to expect that the effect of the drain voltage on the gate current, which is zero without leakage current, is the same as the effect of the gate voltage on the drain current, which is significantly large.

Therefore, for power MOSFET devices, because the transcapacitance C_m is substantially less than C_{gs}, it can be translated to an additional delay time τ_c

in a frequency range up to f_T by combining it with the transconductance g_m according to

$$g_m - j\omega C_m = g_m\sqrt{1+\left(\frac{\omega}{\omega_T}\frac{C_m}{C_{gs}}\right)^2}\exp\left[-j\tan^{-1}\left(\frac{\omega}{\omega_T}\frac{C_m}{C_{gs}}\right)\right] \cong g_m\exp(-j\omega\tau_c) \tag{1.24}$$

where $\tau_c = C_m/(\omega_T C_{gs})$. To satisfy the charge conservation condition, the total delay time τ shown in the MOSFET equivalent circuit in Figure 1.4 represents both the ideal transit time and delay time due to the transcapacitance. The transcapacitance C_m can be easily added to the drain–source capacitance C_{ds} under parameter extraction procedure.

1.2.5 Gate–Source Resistance

The gate–source resistance R_{gs} is determined by the effect of the channel inertia in responding to rapid changes of the time-varying gate–source voltage, and it varies in such a manner that the charging time $\tau_g = R_{gs}C_{gs}$ remains approximately constant. Thus, the increase of R_{gs} in the velocity saturation region (when the channel conductivity decreases) is partially compensated by the decrease of C_{gs} due to nonuniform channel charge distribution [26]. The effect of R_{gs} becomes significant at higher frequencies close to the transition frequency f_T of the MOSFET and may not be considered when designing RF circuits operating below 2 GHz [21, 27].

1.2.6 Temperature Dependence

Silicon MOSFET devices are very sensitive to the operation temperature T, and their characteristics are strongly temperature-dependent [5]. The main parameters responsible for this are the effective carrier mobility μ and the threshold voltage V_{th}, resulting in the increase in the drain current through $V_{th}(T)$ and the decrease in the drain current through $\mu(T)$ with temperature. Increasing temperature decreases the slope of the $I_{ds}(V_{gs})$ curves. A certain value of V_{gs} can be found, at which the drain current becomes practically temperature-independent over a large temperature range. The variation of V_{th} with temperature in a wide range from −50 to +200 °C represents a nonlinear function, which slowly decreases with temperature and can be approximated by

$$V_{th}(T) = V_{th}(T_{nom}) + V_{T1}\Delta T + V_{T2}\Delta T^2 \tag{1.25}$$

where $\Delta T = T - T_{nom}$, $T_{nom} = 300\,\text{K}(27\,°\text{C})$, and V_{T1} and V_{T2} are the linear and quadratic temperature coefficients for threshold voltage [20].

The variation of μ with temperature is related to the appropriate temperature variation of I_{sat} in Eq. (1.13), representing an almost straight line, which decreases with temperature [20]. The temperature dependence $I_{sat}(T)$ can be approximated by the linear function as

$$I_{sat}(T) = I_{sat}(T_{nom}) + I_T\Delta T \tag{1.26}$$

where I_T is the linear temperature coefficient for the saturation current.

The temperature dependencies of the MOSFET capacitances and series resistances can be described by the following linear equations:

$$C(T) = C(T_{nom}) + C_T\Delta T \tag{1.27}$$

$$R(T) = R(T_{nom}) + R_T\Delta T \tag{1.28}$$

where $C = (C_{gs}, C_{ds}, C_{gd})$, $R = (R_g, R_s, R_d)$, and R_T and C_T are the linear temperature coefficients for the capacitances and resistances, respectively [7, 28].

Figure 1.13 shows the modeled temperature dependencies of $V_{th}(T)$ and $g_m(T)$ for high-power LDMOSFET LP801. The results obtained by using the simple approximations in Eqs. (1.25) and (1.26) with the values $V_{T1} = -2\,\text{mV/°C}$, $V_{T2} = -8.55\,\mu\text{V/(°C)}^2$, and $I_T = -4.5\,\text{mA/°C}$ show a good prediction of the *I–V* characteristics in a wide temperature range.

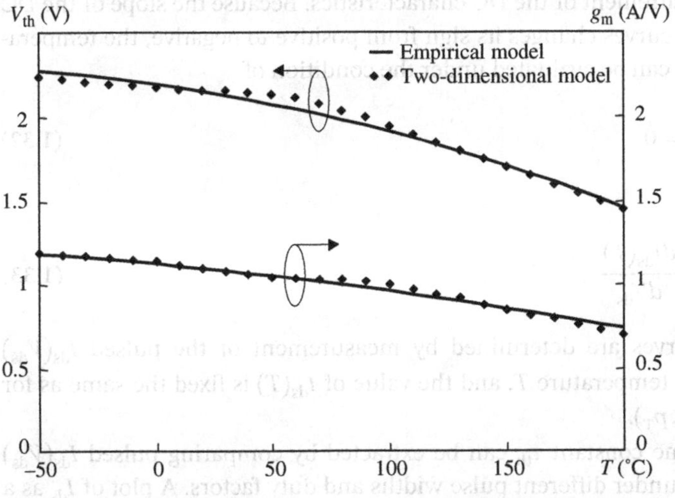

Figure 1.13 Modeled temperature dependencies of V_{th} and g_m.

At high values of V_{gs} and V_{ds} under DC measurement, the slope of $I_{ds}(V_{ds})$ curves can be negative, which occurs due to the self-heating effect in a highly dissipated power region. For the drain current model given by Eq. (1.7), this effect can be considered by adding a linear component describing the temperature dependence as

$$\beta(T) = \beta(T_{nom}) + \beta_T \Delta T_j \tag{1.29}$$

$$\gamma(T) = \gamma(T_{nom}) + \gamma_T \Delta T_j \tag{1.30}$$

where $\Delta T_j = R_{th} P_{dis} + \Delta T$, R_{th}(°C/W) is the thermal resistance, P_{dis} is the DC power consumption in watts caused by DC biasing, and β_T and γ_T are the linear temperature coefficients with negative values, respectively [18].

Another way of considering the effect of the negative conductance at high biasing is to present the nonlinear $I_{ds}(V_{ds})$ model as

$$I_{ds}(T, p_T) = \frac{I_{ds}(T)}{1 + p_T V_d I_{ds}(T)} \tag{1.31}$$

where the drain current source $I_{ds}(T)$ is given by Eq. (1.13), $V_d = V_{ds}/\sqrt{1 + (\omega\tau_{th})^2}$, p_T is the self-heating temperature coefficient, V_{ds} is the drain–source supply voltage, $\tau_{th} = R_{th}C_{th}$ is the thermal time constant, R_{th} is the thermal resistance, C_{th} is the thermal capacitance, and T_j is the function of ambient temperature T and p_T. A thermal equivalent circuit can be added to the nonlinear MOSFET model as a parallel $R_{th}C_{th}$ circuit [13]. The thermal resistance R_{th} can be extracted from the temperature measurement of the DC characteristics. Because the slope of the DC measured $I_{ds}(V_{ds})$ curves changes its sign from positive to negative, the temperature coefficient p_T can be evaluated under the condition of

$$\frac{dI_{ds}(T, p_T)}{dV_{ds}} = 0 \tag{1.32}$$

As a result,

$$p_T = \frac{1}{I_{ds}^2(T)} \frac{dI_{ds}(T)}{dV_{ds}} \tag{1.33}$$

where $I_{ds}(V_{ds})$ curves are determined by measurement of the pulsed $I_{ds}(V_{ds})$ curves at ambient temperature T, and the value of $I_{ds}(T)$ is fixed the same as for zero slope of $I_{ds}(T, p_T)$.

The thermal time constant τ_{th} can be extracted by comparing pulsed $I_{ds}(V_{ds})$ curves calculated under different pulse widths and duty factors. A plot of I_{ds} as a function of pulse width under the fixed gate–source and drain–source bias voltages gives an appropriate value of τ_{th}.

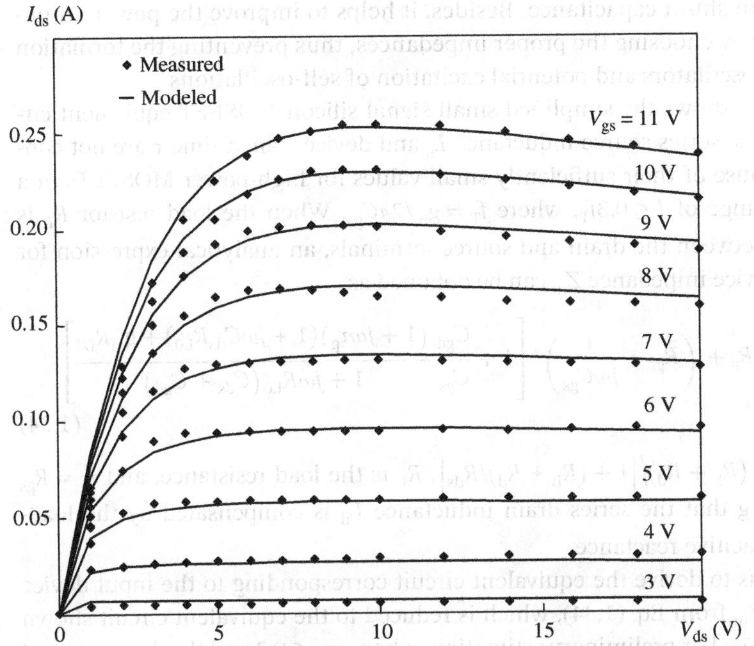

Figure 1.14 Measured and modeled $I_{ds}(V_{ds})$ curves of low-voltage LDMOSFET.

Figure 1.14 shows the comparison between the measured (DC measurement) and modeled $I_{ds}(V_{ds})$ curves for a 12.5-V LDMOSFET cell with the gate geometry of 1.25 μm × 1.44 mm. The effect of self-heating for the model parameters $p_T = 0.035\,1/\text{AV}$, $\alpha = 0.2\,1/\text{V}$, $\beta = \sigma = 0$, $S_{th} = 37\,\text{mA/V}$, $I_{th} = 1.5\,\text{mA}$, $V_{th} = 2.25\,\text{V}$, $I_{sat} = 0.48\,\text{A}$, $n = 1$, and $\lambda = 0.0005\,1/\text{V}$ is described by Eq. (1.31).

1.3 Equivalent Circuits Characterizing Device Input and Output Impedances

To apply the proper matching and harmonic-control techniques to the high-efficiency power amplifier, which is necessary to design, and to make this design clear and comprehensible, it is very important to represent the input and output device impedances separately through the equivalent circuits, the elements of which will be the parts of the corresponding matching and harmonic-control circuits. For example, the series parasitic inductance at the device input may be considered a part of the low-pass matching section, or the shunt capacitance at the device output may represent an important element of the Class-E load network with shunt capacitance. Besides, it helps to improve the power amplifier stability by choosing the proper impedances, thus preventing the formation of parasitic oscillators and potential excitation of self-oscillations.

Figure 1.15 shows the simplified small-signal silicon MOSFET equivalent circuit where the series source inductance L_s and device transit time τ are not considered because of their sufficiently small values for high-power MOSFETs in a frequency range of $f \leq 0.3f_T$, where $f_T = g_m/2\pi C_{gs}$. When the load resistor R_L is connected between the drain and source terminals, an analytical expression for the input device impedance Z_{in} can be obtained as

$$Z_{in} = R_g + \left(R_{gs} + \frac{1}{j\omega C_{gs}}\right) / \left[1 + \frac{C_{gd}}{C_{gs}} \frac{(1 + j\omega\tau_g)(1 + j\omega C_{ds}R_{L0}) + g_m R_{L0}}{1 + j\omega R_{L0}(C_{ds} + C_{gd})}\right] \quad (1.34)$$

where $R_{L0} = (R_L + R_d)/[1 + (R_L + R_d)/R_{ds}]$, R_L is the load resistance, and $\tau_g = R_{gs} C_{gs}$, assuming that the series drain inductance L_d is compensated by the load-network capacitive reactance.

Next step is to derive the equivalent circuit corresponding to the input device impedance Z_{in} from Eq. (1.34), which is reduced to the equivalent circuit shown in Figure 1.16a. For preliminary estimation, when $\omega\tau_g \leq 0.3$ and the device output capacitive reactance is inductively compensated, the input equivalent circuit simplifies significantly and can represent an inductance, a capacitor, and a resistor connected in series, as shown in Figure 1.16b, where

$$L_{in} = L_g \quad (1.35)$$

$$R_{in} \cong R_g + R_{gs} \quad (1.36)$$

$$C_{in} \cong C_{gs} + C_{gd}\left[1 + g_m \frac{R_L + R_d}{1 + R_L/(R_{ds} + R_d)}\right] \quad (1.37)$$

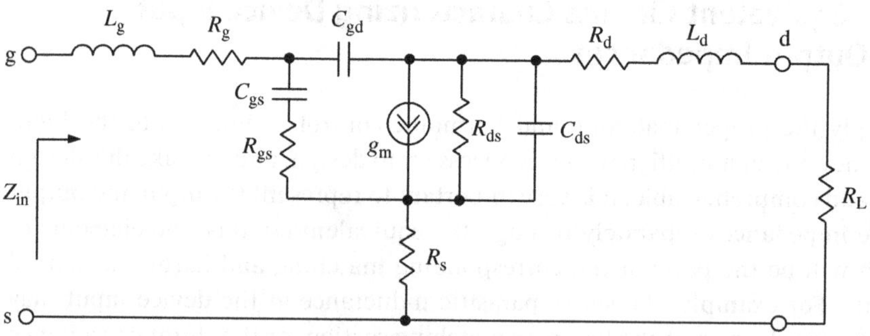

Figure 1.15 Simplified small-signal silicon MOSFET equivalent circuit.

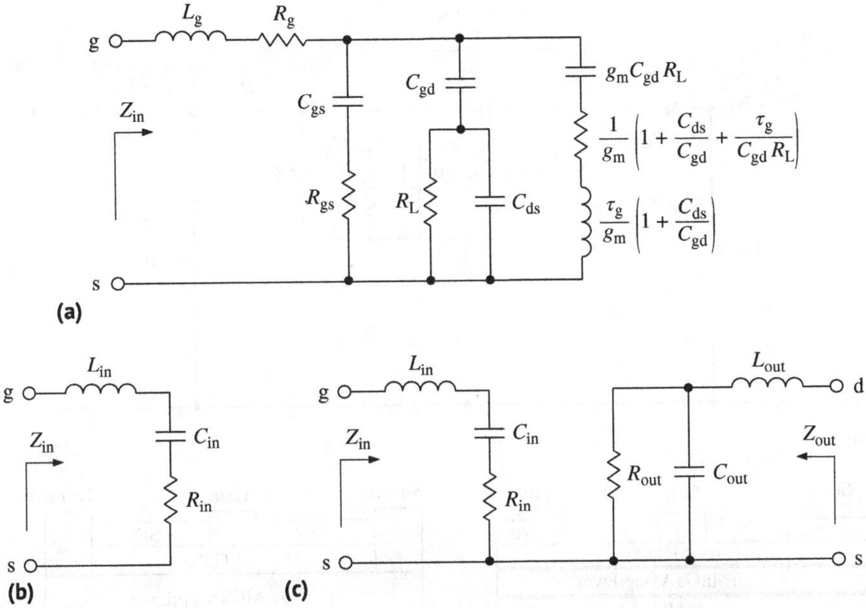

Figure 1.16 Equivalent circuits, characterizing device input and output.

Considering that usually $C_{gd} \ll C_{gs}$, the device equivalent output circuit can be represented by the series inductance $L_{out} = L_d$ and a parallel connection of the equivalent output resistance R_{out} estimated as the ratio of the fundamental voltage and fundamental current Fourier components (presented in Section 3.7 for different operation modes) and the output capacitance $C_{out} = C_{ds} + C_{gd}$, as shown in Figure 1.16c.

1.4 GaAs MESFETs and GaN HEMTs

Adequate representation for MESFETs and HEMTs in a frequency range up to millimeter-wave frequencies can be provided using the nonlinear device model shown in Figure 1.17a, which is very similar to a nonlinear MOSFET model [29, 30]. The intrinsic model is described by the channel charging resistance R_{gs} representing the resistive path for the charging of the gate–source capacitance C_{gs}, the feedback gate–drain capacitance C_{gd}, the drain–source capacitance C_{ds}, the gate–source diode to model the forward conduction current $i_{gs}(v_{gs})$, and the gate–drain diode to account for the gate–drain avalanche current $i_{gd}(v_{gs}, v_{ds})$, which can occur at large-signal operation conditions. The gate–source capacitance

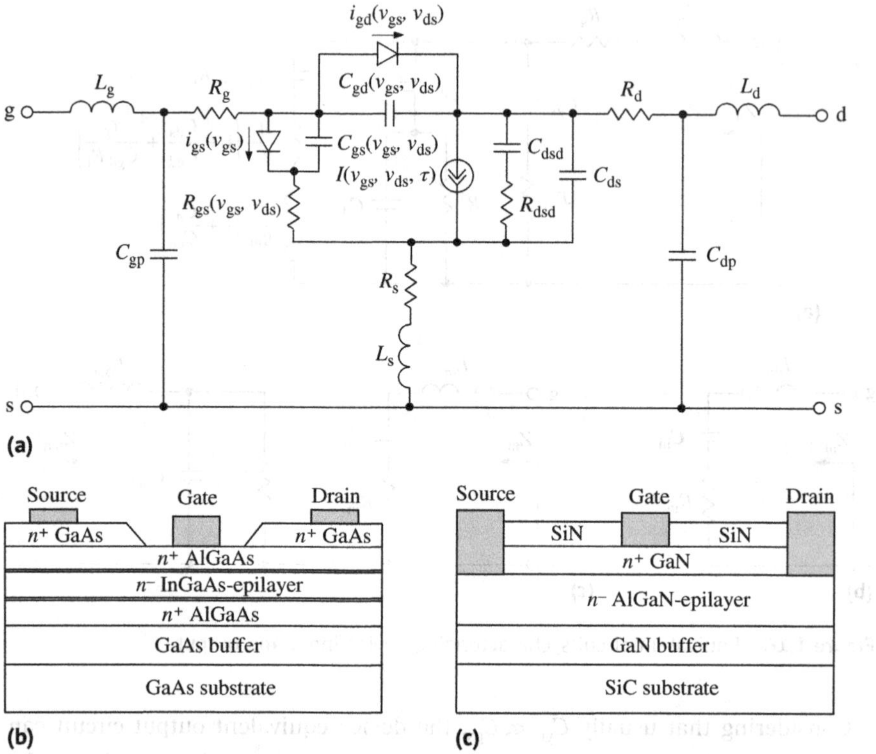

Figure 1.17 Nonlinear MESFET and HEMT model with HEMT physical structures.

C_{gs} and gate–drain capacitance C_{gd} represent the charge depletion region and can be treated as the voltage-dependent Schottky-barrier diode capacitances, being the nonlinear functions of the gate–source voltage v_{gs} and drain–source voltage v_{ds}. For negative gate–source voltage and small drain–source voltage, these capacitances are practically equal. However, when the drain–source voltage is increased beyond the current saturation point, the gate–drain capacitance C_{gd} is much more heavily back-biased than the gate–source capacitance C_{gs}. Therefore, the gate–source capacitance C_{gs} is significantly more important and usually dominates the input impedance of the MESFET or HEMT device. The influence of the drain–source capacitance C_{ds} on the device behavior is insignificant and its value is bias-independent. The capacitance C_{dsd} and resistance R_{dsd} model the dispersion of the MESFET or HEMT current–voltage characteristics due to trapping effect in the device channel, which leads to discrepancy between DC measurement and S-parameter measurements at higher frequencies [31, 32]. A large-signal model for monolithic power amplifier design should be accurate for all operating conditions. In addition, the model parameters should be easily extractable, and the model must be as simple as possible. Various nonlinear MESFET and HEMT models with different complexities are available, and each one can be considered sufficiently accurate for a certain application. However, it is preferable to use the Angelov model for MESFETs and HEMTs, in particular to predict the large-signal behavior of the pHEMT devices using in high-power, high-efficiency millimeter-wave MMICs [33, 34]. By using three additional terms of the gate power-series function in Angelov model, better accuracy can be achieved in a large-signal modeling of AlGaN/GaN HEMT devices on SiC substrate [35]. This model can also be improved by incorporating two additional analytical expressions to model the device behavior in saturation region, which is important to design high-efficiency switchmode power amplifiers, for example, in inverse Class-F mode [36].

Figure 1.17b shows the cross-section of a physical structure of an InGaAs/AlGaAs HEMT device, where an undoped InGaAs *n*-epilayer is used as a channel and two heavily *n*-doped AlGaAs layers with a high energetic barrier for holes are necessary to maximize high electron mobility in the channel. In this case, spacing between AlGaAs layer and InGaAs channel is optimized to achieve a high breakdown voltage. An example of the physical structure of an AlGaN/GaN HEMT device is shown in Figure 1.17c, where an undoped AlGaN *n*-epilayer is used as a channel, *n* -type-doped GaN layer can suppress dispersion in the device current–voltage characteristics, and a SiN passivation layer with optimized parameters contributes to a lower-trap device structure [37]. Thermal conductivity of a GaN HEMT device is improved by using a SiC substrate. Note that the GaN-based technology can provide wider bandwidth and higher efficiency of the power amplifier due to high charge density and the ability to operate at higher voltages for GaN HEMT devices that are characterized by lower output capacitance and on-resistance [38, 39].

The basic electrical properties of the MESFET or HEMT device can be characterized by the admittance *Y*-parameters expressed through the device intrinsic small-signal equivalent circuit as

$$Y_{11} = \frac{j\omega C_{gs}}{1 + j\omega C_{gs} R_{gs}} + j\omega C_{gd} \tag{1.38}$$

$$Y_{12} = -j\omega C_{gd} \tag{1.39}$$

$$Y_{21} = \frac{g_m \exp(-j\omega\tau)}{1 + j\omega C_{gs} R_{gs}} - j\omega C_{gd} \tag{1.40}$$

$$Y_{22} = \frac{1}{R_{ds}} + j\omega \left(C_{ds} + C_{gd} \right) \tag{1.41}$$

where g_m is the device transconductance and R_{ds} is the differential drain resistance [40]. In this case, the dispersion effect, which is important at higher frequencies and modeled by C_{dsd} and R_{dsd}, cannot be considered.

By separating Eqs. (1.38)–(1.41) into their real and imaginary parts, the elements of the small-signal equivalent circuit can be analytically determined as

$$C_{gd} = -\frac{\mathrm{Im}Y_{12}}{\omega} \tag{1.42}$$

$$C_{gs} = -\frac{\mathrm{Im}Y_{11} - \omega C_{gd}}{\omega}\left[1 + \left(\frac{\mathrm{Re}Y_{11}}{\mathrm{Im}Y_{11} - \omega C_{gd}}\right)^2\right] \tag{1.43}$$

$$R_{gs} = \frac{\mathrm{Re}Y_{11}}{(\mathrm{Im}Y_{11} - \omega C_{gd})^2 + (\mathrm{Re}Y_{11})^2} \tag{1.44}$$

$$g_m = \sqrt{(\mathrm{Re}Y_{21})^2 + (\mathrm{Im}Y_{21} + \omega C_{gd})^2}\sqrt{1 + (\omega C_{gs} R_{gs})^2} \tag{1.45}$$

$$\tau = \frac{1}{\omega}\sin^{-1}\left(\frac{-\omega C_{gd} - \mathrm{Im}Y_{21} - \omega C_{gs} R_{gs} \mathrm{Re}Y_{21}}{g_m}\right) \tag{1.46}$$

$$C_{ds} = \frac{\mathrm{Im}Y_{22} - \omega C_{gd}}{\omega} \tag{1.47}$$

$$R_{ds} = \frac{1}{\mathrm{Re}Y_{22}} \tag{1.48}$$

which are valid for a wide frequency range up to the transition frequency f_T [41].

If all extrinsic parasitic elements are known from the data sheet or directly determined from measurements performed at zero drain bias with the gate voltages lower than pinch-off voltage, the only problem is to obtain the admittance Y-parameters of the intrinsic two-port network from on-bias experimental data [42]. Consecutive steps shown in Figure 1.18 can represent such a determination procedure [43]:

- Measurement of the S-parameters of the extrinsic device.
- Transformation of the S-parameters to the impedance Z-parameters with subtraction of the series inductances L_g and L_d.
- Transformation of the impedance Z-parameters to the admittance Y-parameters with subtraction of the parallel capacitances C_{gp} and C_{dp}.
- Transformation of the admittance Y-parameters to the impedance Z-parameters with subtraction of series resistances R_g, R_s, R_d, and inductance L_s.
- Transformation of the impedance Z-parameters to the admittance Y-parameters of the intrinsic device two-port network.

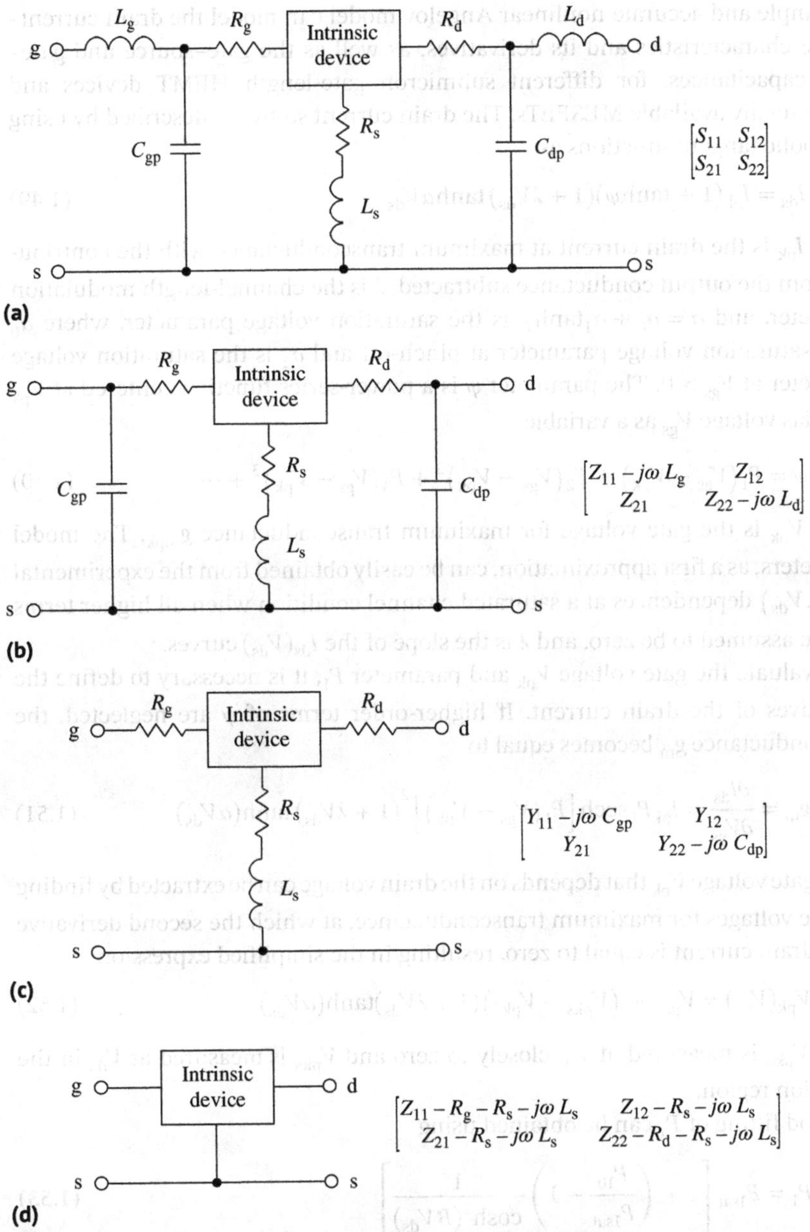

Figure 1.18 Method for extracting device intrinsic *Z*-parameters.

A simple and accurate nonlinear Angelov model can model the drain current–voltage characteristics and its derivatives, as well as the gate–source and gate–drain capacitances, for different submicron gate-length HEMT devices and commercially available MESFETs. The drain current source is described by using hyperbolic tangent functions as

$$I_{ds} = I_{pk}(1 + \tanh\psi)(1 + \lambda V_{ds}) \tanh\alpha V_{ds} \tag{1.49}$$

where I_{pk} is the drain current at maximum transconductance with the contribution from the output conductance subtracted, λ is the channel-length modulation parameter, and $\alpha = \alpha_0 + \alpha_1 \tanh\psi$ is the saturation voltage parameter, where α_0 is the saturation voltage parameter at pinch-off and α_1 is the saturation voltage parameter at $V_{gs} > 0$. The parameter ψ is a power-series function centered at V_{pk} with bias voltage V_{gs} as a variable:

$$\psi = P_1\left(V_{gs} - V_{pk}\right) + P_2\left(V_{gs} - V_{pk}\right)^2 + P_3\left(V_{gs} - V_{pk}\right)^3 + \cdots \tag{1.50}$$

where V_{pk} is the gate voltage for maximum transconductance g_{mpk}. The model parameters, as a first approximation, can be easily obtained from the experimental $I_{ds}\left(V_{gs}, V_{ds}\right)$ dependences at a saturated channel condition when all higher terms in ψ are assumed to be zero, and λ is the slope of the $I_{ds}(V_{ds})$ curves.

To evaluate the gate voltage V_{pk} and parameter P_1, it is necessary to define the derivatives of the drain current. If higher-order terms of ψ are neglected, the transconductance g_m becomes equal to

$$g_m = \frac{\partial I_{ds}}{\partial V_{gs}} = I_{pk} P_1 \operatorname{sech}\left[P_1\left(V_{gs} - V_{pk}\right)\right]^2 (1 + \lambda V_{ds}) \tanh(\alpha V_{ds}) \tag{1.51}$$

The gate voltage V_{pk} that depends on the drain voltage can be extracted by finding the gate voltages for maximum transconductance, at which the second derivative of the drain current is equal to zero, resulting in the simplified expression

$$V_{pk}(V_{ds}) = V_{pk0} + \left(V_{pks} - V_{pk0}\right)(1 + \lambda V_{ds})\tanh(\alpha V_{ds}) \tag{1.52}$$

where V_{pk0} is measured at V_{ds} closely to zero and V_{pks} is measured at V_{ds} in the saturation region.

A good fitting of P_1 can be obtained using

$$P_1 = P_{1sat}\left[1 + \left(\frac{P_{10}}{P_{1sat}} - 1\right)\frac{1}{\cosh^2(B V_{ds})}\right] \tag{1.53}$$

where $P_{10} = g_{m0}/I_{pk0}$ at V_{ds} close to zero and B is the fitting parameter ($B \approx 1.5\alpha$). The parameter P_2 makes the derivative of the drain current asymmetric, whereas the parameter P_3 changes the drain current values at voltages V_{gs} close to pinch-off voltage V_p. Three terms in Eq. (1.50) are usually enough to describe the behavior of the different MESFET or HEMT devices with acceptable accuracy.

The same hyperbolic functions can be used to model the intrinsic device capacitances C_{gs} and C_{gd}. When an accuracy of 5–10% is enough, the gate–source capacitance C_{gs} and gate–drain capacitance C_{gd} can be described by

$$C_{gs} = C_{gs0}\left[1 + \tanh\left(P_{1gsg}V_{gs}\right)\right]\left[1 + \tanh\left(P_{1gsd}V_{ds}\right)\right] \quad (1.54)$$

$$C_{gd} = C_{gd0}\left[1 + \tanh\left(P_{1gdg}V_{gs}\right)\right]\left[1 - \tanh\left(P_{1gdd}V_{ds} + P_{1cc}V_{gs}V_{ds}\right)\right] \quad (1.55)$$

where the product $P_{1cc}V_{gs}V_{ds}$ reflects the cross-coupling of V_{gs} and V_{ds} on C_{gd} and the coefficients P_{1gsg}, P_{1gsd}, P_{1gdg}, and P_{1gdd} are the fitting parameters. The approximate behavior of normalized gate–source capacitance C_{gs}/C_{gs0} (curve 1) and gate–drain capacitance C_{gd}/C_{gd0} (curve 2) as functions of V_{ds} for zero gate–source voltage is shown in Figure 1.19, where C_{gs0} and C_{gd0} are the gate–source and gate–drain capacitances at $V_{ds} = 0$, respectively. The character of the curves is the same for both positive and negative V_{gs}, except that the capacitance range decreases for the same range of V_{ds} with the decrease of V_{gs}.

The drain–source dispersive resistance R_{dsd} as a nonlinear function of the gate–source voltage V_{gs} can be defined by

$$R_{dsd} = R_{dsd0} + \frac{R_{dsdp}}{1 + \tanh\psi} \quad (1.56)$$

where R_{dsd0} is the minimum value of R_{dsd} and R_{dsdp} determines the value of R_{dsd} at the pinch-off [44].

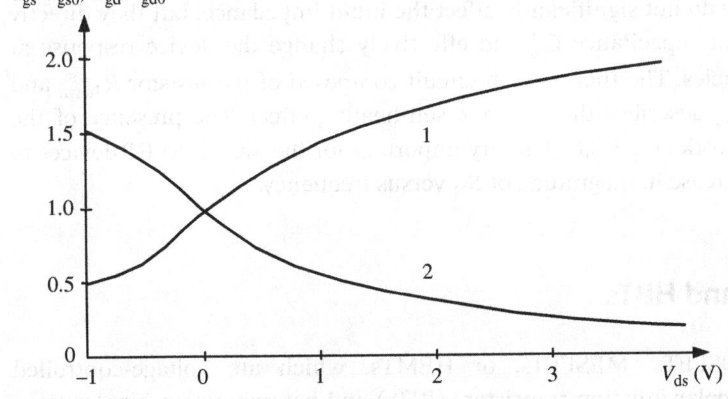

Figure 1.19 Gate–source and gate–drain capacitances versus drain–source voltage.

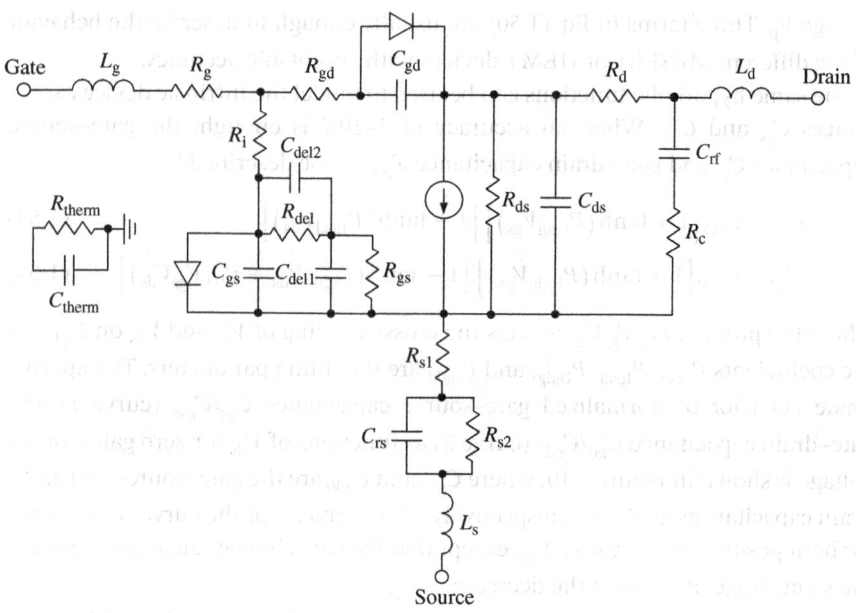

Figure 1.20 Generalized nonlinear equivalent circuit of GaN HEMT and SiC MESFET.

Figure 1.20 shows the generalized nonlinear equivalent circuit, which can describe the nonlinear behavior of GaN HEMT and SiC MESFET devices [35, 45]. Here, the delay network consisting of the resistance R_c and capacitance C_{rf} is introduced to address the dispersion phenomena, whereas the resistance R_{del} and capacitances C_{del1} and C_{del2} are included to more accurately describe the transistor response at high frequencies. The values of C_{del1} and C_{del2} are so small that they do not significantly affect the input impedance, but they directly shunt the input capacitance C_{gs} and effectively change the device response at higher frequencies. The thermal sub-circuit composed of the resistor R_{therm} and capacitor C_{therm} describes the dynamic self-heating effect. The presence of the source RC network (R_{s1}, R_{s2}, C_{rs}) is very important for the SiC MESFET devices to describe an increase in magnitude of S_{21} versus frequency.

1.5 BJTs and HBTs

Unlike LDMOSFETs, MESFETs, or HEMTs, which are voltage-controlled devices, the bipolar junction transistors (BJTs) and heterojunction bipolar transistors (HBTs) are current-controlled devices that control the amount of current flowing through them from the emitter to the collector terminal in proportion to the amount of biasing voltage applied to their base terminal, resulting in a small current flowing into the base terminal to control a much larger collector current forming the basis of transistor action. The principal difference between the BJTs and HBTs is in the use of different semiconductor materials for the emitter–base junction and the base–collector junction, creating a heterojunction that limits the injection of holes from the base into the emitter region. Unlike BJT technology, this allows a high doping density to be used in the base, reducing the base resistance and providing higher gain and maximum operating frequency. As a result, HBTs have become very promising devices for different applications at microwave and millimeter-wave frequencies.

1.5.1 Nonlinear BJT and HBT Models

Figure 1.21a shows the nonlinear model of the bipolar transistor with extrinsic parasitic elements [46, 47]. Such a hybrid-Π equivalent circuit can model the nonlinear electrical behavior of bipolar transistors, particularly HBT devices, with sufficient accuracy up to about 20 GHz or even higher. The intrinsic model is described by the dynamic diode resistance r_π, the total base–emitter junction capacitance and base charging diffusion capacitance C_π, the base–collector diode required to account for the nonlinear effects at the saturation, the internal collector–base junction capacitance C_{ci}, the external distributed collector–base capacitance C_{co}, the collector–emitter capacitance C_{ce}, and the nonlinear current source $i(v_{be},v_{ce})$. The lateral and base semiconductor resistances underneath the base contact and the base semiconductor resistance underneath the emitter are combined into a base-spreading resistance r_b. The extrinsic parasitic elements are represented by the base bondwire inductance L_b, the base ohmic resistance R_b, the emitter ohmic resistance R_e, the emitter inductance L_e, the collector ohmic resistance R_c, and the collector bondwire inductance L_c. To increase the usable operating frequency range of the device up to 50 GHz and higher, it is necessary to include the collector-current delay time τ in the collector current source as $g_m \exp(-j\omega\tau)$.

Figure 1.21b shows the modified version of a bipolar transistor equivalent circuit, where $C_c = C_{co} + C_{ci}$, $r_{b1} = r_b C_{ci}/C_c$, and $r_{b2} = r_b C_{co}/C_c$ [48]. Such an equivalent circuit becomes possible due to an equivalent Π- to T-transformation of the elements r_b, C_{co}, and C_{ci} and a condition $r_b \ll (C_{ci} + C_{co})/\omega C_{ci} C_{co}$, which is usually fulfilled over the frequency range close to the device maximum frequency f_{max}. Then, from a comparison of the transistor nonlinear models, for a bipolar transistor in Figure 1.21b, for a MOSFET device in Figure 1.4, and for a MESFET (or HEMT) device in Figure 1.17a, it is easy to detect the circuit similarity of all these equivalent circuits, which means that the basic circuit design procedure is

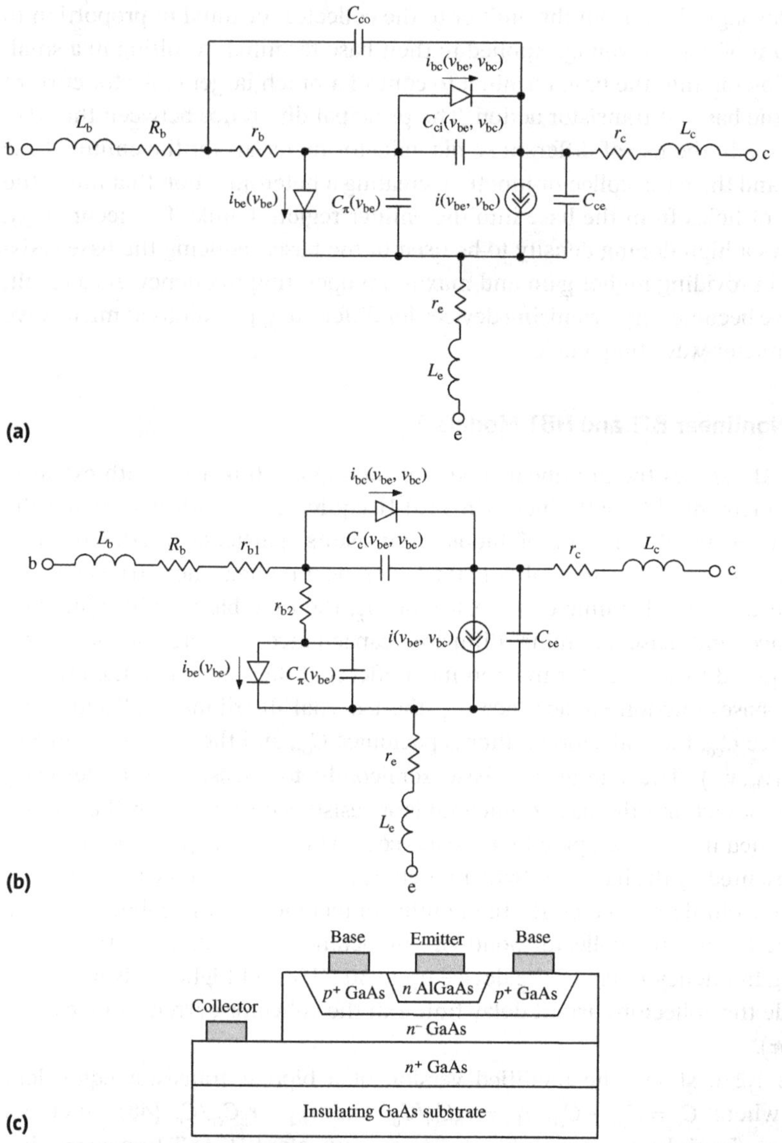

Figure 1.21 Nonlinear BJT and HBT models and HBT physical structure.

very similar for any type of bipolar or field-effect transistors. The difference is in the device physics and values of the model parameters. However, techniques for the representation of the input and output impedances, stability analysis based on feedback effect, and derivation of power gain and efficiency are very similar.

The cross-section of a physical structure of an AlGaAs/GaAs HBT device is shown in Figure 1.21c, with heavily *p*-doped base to reduce base resistance and lightly *n*-doped emitter to minimize emitter capacitance [49]. The lightly *n*-doped collector region allows the collector–base junction to sustain relatively high voltages without breaking down. The forward-bias emitter-injection efficiency is very high because the wider-bandgap AlGaAs emitter injects electrons into the GaAs *p*-base at a lower energy level, but the holes are prevented from flowing into the emitter by a high energy barrier, thus resulting in the ability to decrease the base-length and base-width modulation and to increase the frequency response. By using a wide-bandgap InGaP layer instead of an AlGaAs layer, the device performance over temperature can be improved [50]. The high linearity power performance in a Class-AB condition at the backoff power level, the ruggedness under mismatch and overdrive conditions, and the long lifetime of the InGaP/GaAs HBT technology make it very attractive for the 28-V power amplifier applications [51]. The growth process used for a high-voltage HBT device is identical to the process used for the conventional low-voltage HBT device, which is widely used in handset power amplifiers, except for changes to the collector because of the higher voltage operating requirements. The epitaxial growth process starts with a highly doped *n*-type collector layer and lightly *n*-doped collector drift region, then followed by a heavily doped *p*-type base layer and an InGaP emitter layer and finishes with an InGaAs cap layer [52]. As a result, the high-voltage HBT devices exhibit collector-base breakdown voltages higher than 70 V.

1.5.2 Determination of Equivalent Circuit Elements

The complete bipolar transistor small-signal equivalent circuit with extrinsic parasitic elements is shown in Figure 1.22. Based on this hybrid Π-type representation, the electrical properties of the bipolar transistors, particularly HBT devices, can be described with sufficient accuracy up to 30 GHz [53, 54]. Here, the extrinsic elements R_b, L_b, R_c, L_c, R_e, and L_e are the series resistances and lead inductances associated with the base, collector, and emitter, and C_{pbe}, C_{pbc}, and C_{pce} are the parasitic capacitances associated with the contact pads, respectively. The lateral resistance with the base semiconductor resistance underneath the base contact and the base semiconductor resistance underneath the emitter are combined into a base-spreading resistance r_b. The intrinsic model is described by the dynamic diode resistance r_π, the total base–emitter junction capacitance and base charging capacitance C_π, the transconductance g_m, and the output Early resistance r_{ce}, which model the effect of base-width modulation on the transistor characteristics due to variations in the collector–base depletion region [55]. To increase the usable operating frequency range of the device up to 50 GHz and higher, it is necessary to properly include the collector-current delay time in the current source as $g_m \exp(-j\omega\tau_\pi)$, where $\tau_\pi = r_\pi C_\pi$ is the transit time [56].

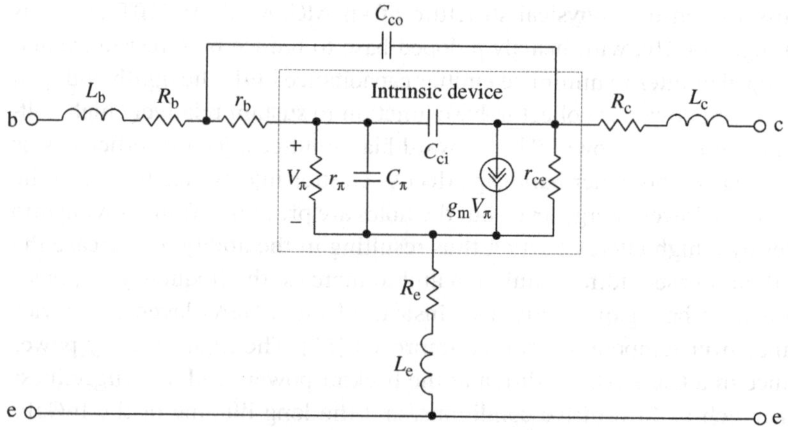

Figure 1.22 Bipolar transistor small-signal equivalent circuit.

If all extrinsic parasitic elements of the device equivalent circuit shown in Figure 1.22 are known, the intrinsic two-port network parameters can be embedded with the following determination procedure [53]:

- Measurement of the S-parameters of the transistor with extrinsic elements.
- Transformation of the S-parameters to the impedance Z-parameters with subtraction of the parasitic series elements L_b, R_b, L_e, R_e, L_c, and R_c.
- Transformation of the new Z-parameters to the Y-parameters with subtraction of the parasitic shunt capacitance C_{co}.
- Transformation of the new Y-parameters to the Z-parameters with subtraction of the parasitic series resistance r_b.
- Transformation of the new Z-parameters to the Y-parameters of the intrinsic device two-port network.

The bipolar transistor intrinsic Y-parameters can be written as

$$Y_{11} = \frac{1}{r_\pi} + j\omega(C_\pi + C_{ci}) \tag{1.57}$$

$$Y_{12} = -j\omega C_{ci} \tag{1.58}$$

$$Y_{21} = g_m \exp(-j\omega\tau_\pi) - j\omega C_{ci} \tag{1.59}$$

$$Y_{22} = \frac{1}{r_{ce}} + j\omega C_{ci} \tag{1.60}$$

After separating Eqs. (1.57)–(1.60) into their real and imaginary parts, the elements of the intrinsic small-signal equivalent circuit can be determined analytically as

$$C_\pi = \frac{\mathrm{Im}(Y_{11} + Y_{12})}{\omega} \tag{1.61}$$

$$r_\pi = \frac{1}{\mathrm{Re}Y_{11}} \tag{1.62}$$

$$C_{\mathrm{ci}} = -\frac{\mathrm{Im}Y_{12}}{\omega} \tag{1.63}$$

$$g_{\mathrm{m}} = \sqrt{(\mathrm{Re}Y_{21})^2 + (\mathrm{Im}Y_{21} + \mathrm{Im}Y_{12})^2} \tag{1.64}$$

$$\tau_\pi = \frac{1}{\omega}\cos^{-1}\frac{\mathrm{Re}Y_{21} + \mathrm{Re}Y_{12}}{\sqrt{(\mathrm{Re}Y_{21})^2 + (\mathrm{Im}Y_{21} + \mathrm{Im}Y_{12})^2}} \tag{1.65}$$

$$r_{\mathrm{ce}} = \frac{1}{\mathrm{Re}Y_{22}} \tag{1.66}$$

The series parasitic resistances r_{b}, R_{b}, R_{e}, and R_{c} can be calculated on the basis of physical parameters of the device or by adding them to the intrinsic device parameters [54]. The external parasitic parallel capacitance C_{co}, as well as the other device capacitances, can be estimated from the device behavior at low frequencies and cutoff operating conditions [57, 58].

1.5.3 Nonlinear Bipolar Device Modeling

The bipolar transistor can be considered as an interacting pair of *p–n* junctions, therefore the approach to describe its nonlinear properties is the same as that used for the diode modeling. The simple large-signal Ebers–Moll model with a single current source between the collector and the emitter is shown in Figure 1.23 [7, 59]. The collector–emitter source current I_{ce} is defined by

$$I_{\mathrm{ce}} = I_{\mathrm{sat}}\left[\exp\left(\frac{V_\pi}{V_{\mathrm{T}}}\right) - \exp\left(\frac{V_{\mathrm{bc}}}{V_{\mathrm{T}}}\right)\right] \tag{1.67}$$

where I_{sat} is the bipolar transistor reverse saturation current, and V_{T} is the thermal voltage obtained by

$$V_{\mathrm{T}} = \frac{kT}{q} \tag{1.68}$$

where q is the electron charge, k is Boltzmann's constant, and T is the temperature in kelvin.

The device terminal currents are defined as $I_{\mathrm{c}} = I_{\mathrm{ce}} - I_{\mathrm{bc}}$, $I_{\mathrm{e}} = -I_{\mathrm{ce}} - I_{\mathrm{be}}$, and $I_{\mathrm{b}} = I_{\mathrm{be}} + I_{\mathrm{bc}}$, where the diode currents are given by

$$I_{\mathrm{be}} = \frac{I_{\mathrm{sat}}}{\beta_{\mathrm{F}}}\left[\exp\left(\frac{V_\pi}{V_{\mathrm{T}}}\right) - 1\right] \tag{1.69}$$

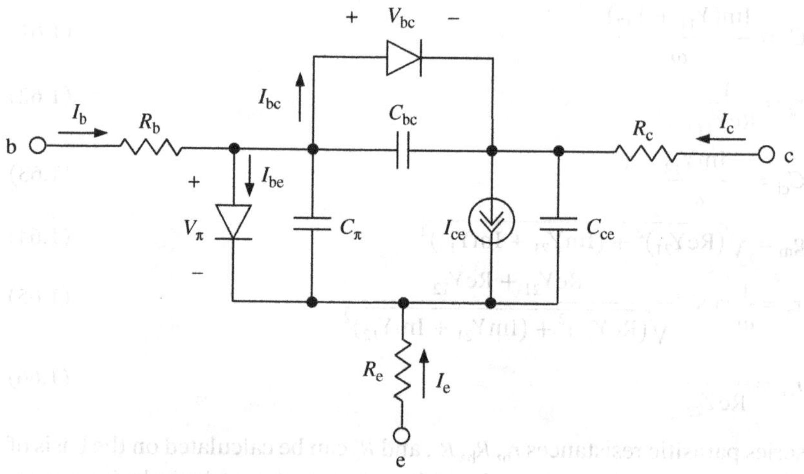

Figure 1.23 Large-signal Ebers–Moll model.

$$I_{bc} = \frac{I_{sat}}{\beta_R}\left[\exp\left(\frac{V_{bc}}{V_T}\right) - 1\right] \tag{1.70}$$

where β_F and β_R are the large-signal forward current gain and reverse current gain of a common emitter bipolar transistor, respectively.

The device capacitances C_π and C_{bc} consist of two components each and are modeled by the diffusion capacitance and junction capacitance, respectively, as

$$C_\pi = \tau_F \frac{dI_{be}}{dV_\pi} + C_{jeo}\left(1 - \frac{V_\pi}{\phi_e}\right)^{-m_e} \tag{1.71}$$

$$C_{bc} = \tau_R \frac{dI_{bc}}{dV_{bc}} + C_{jco}\left(1 - \frac{V_{bc}}{\phi_c}\right)^{-m_c} \tag{1.72}$$

where τ_F and τ_R are the ideal total forward time and reverse transit time, C_{jeo} and C_{jco} are the base–emitter and base–collector zero-bias junction capacitances, and m_e and m_c are the base–emitter and base–collector junction grading factors, respectively.

The collector–emitter substrate capacitance C_{ce} should be considered when designing the integrated circuits. Its representation is adequate for many cases, because the epitaxial-layer–substrate junction is reversed-biased for isolation purposes, and usually it is modeled as a capacitance with constant value.

The Ebers–Moll model cannot describe the second-order effects that can influence the accuracy of Eq. (1.67). One deals with the width of the neutral base region when the widths of the depletion regions change with V_π and V_{bc}, thus providing a base-width modulation (Early effect [55]). Others involve the device behavior at low and high currents, resulting in a variation of the large-signal forward current gain β_F with collector current I_{ce}. For example, as the number of electrons injected into the base approaches the background doping concentration, the number of holes will increase to maintain space-charge neutrality, and the gain will drop (Webster effect [60]). At the same time, if the electron concentration overwhelms the doping in the collector, the base will push out, increasing the neutral base region and again reducing the gain (Kirk effect [61]). In addition, to find a better approximation of the distributed structure of the base–collector junction at microwave frequencies, the junction capacitance C_{bc} should be divided into two separate capacitances: internal C_{ci} and external C_{co}. The lateral resistance and the base semiconductor resistance underneath the base contact and the base semiconductor resistance underneath the emitter are combined into a base-spreading resistance r_b.

Figure 1.24 shows the modified Π-type Gummel–Poon nonlinear model of the bipolar transistor, which can describe the nonlinear electrical behavior of bipolar transistors, particularly HBT devices, with sufficient accuracy up to microwave frequencies [7, 62]. For the Gummel–Poon large-signal model, the collector source current I_{ce} is determined by

$$I_{ce} = \frac{I_{ss}}{q_b}\left[\exp\left(\frac{V_\pi}{n_F V_T}\right) - \exp\left(\frac{V_{bc}}{n_R V_T}\right)\right] \tag{1.73}$$

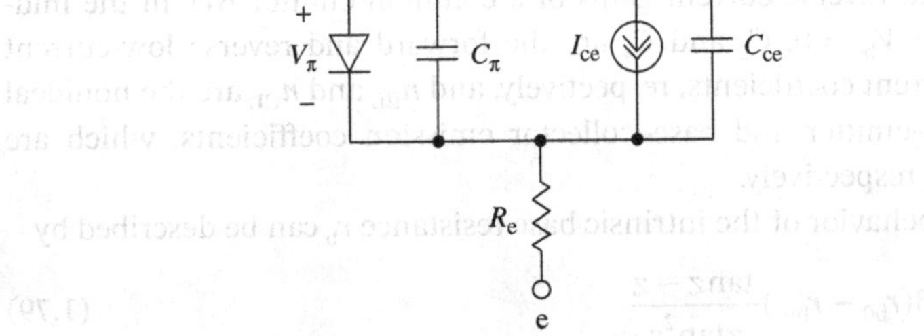

Figure 1.24 Large-signal Gummel–Poon model.

where I_{ss} is the BJT fundamental constant defined at zero-bias condition, n_F is the forward current emission coefficient, n_R is the reverse current emission coefficient, and q_b is the variable parameter defined by

$$q_b = \frac{q_1}{2} + \sqrt{\left(\frac{q_1}{2}\right)^2 + q_2} \tag{1.74}$$

where

$$q_1 = 1 + \frac{V_\pi}{V_B} + \frac{V_{bc}}{V_A} \tag{1.75}$$

$$q_2 = \frac{I_{ss}}{I_{KF}}\left[\exp\left(\frac{V_\pi}{n_F V_T}\right) - 1\right] + \frac{I_{ss}}{I_{KR}}\left[\exp\left(\frac{V_{bc}}{n_R V_T}\right) - 1\right] \tag{1.76}$$

where V_A is the forward Early voltage, V_B is the reverse Early voltage, I_{KF} is the forward knee current for high-level injection in the normal active region, and I_{KR} is the reverse knee current for low-level injection in inverse region. Although the physical mechanism of the collector current behavior given by Eqs. (1.74)–(1.76) is related to silicon BJTs, the highly doped base found in HBTs makes it unlikely that their base–collector potential will significantly change the base width, or that the injected electron concentration will approach the background doping at typical current levels [63]. However, mainly for historical reasons, and so that the model can still be used for silicon BJTs, they are usually retained for HBTs, although their physical behaviors are not exactly relevant.

The currents through model diodes are defined by

$$I_{be} = \frac{I_{sat}(0)}{\beta_{FM}(0)}\left[\exp\left(\frac{V_\pi}{n_F V_T}\right) - 1\right] + C_2 I_{sat}(0)\left[\exp\left(\frac{V_\pi}{n_{EL} V_T}\right) - 1\right] \tag{1.77}$$

$$I_{bc} = \frac{I_{sat}(0)}{\beta_{RM}(0)}\left[\exp\left(\frac{V_{bc}}{n_R V_T}\right) - 1\right] + C_4 I_{sat}(0)\left[\exp\left(\frac{V_{bc}}{n_{CL} V_T}\right) - 1\right] \tag{1.78}$$

where $I_{sat}(0)$ is the saturation current for $V_{bc} = 0$, $\beta_{FM}(0)$ and $\beta_{RM}(0)$ are the large-signal forward and reverse current gains of a common-emitter BJT in the mid-current region for $V_{bc} = 0$, C_2 and C_4 are the forward and reverse low-current nonideal base current coefficients, respectively, and n_{EL} and n_{CL} are the nonideal low-current base–emitter and base–collector emission coefficients, which are usually close to 2, respectively.

The nonlinear behavior of the intrinsic base resistance r_b can be described by

$$r_b = r_{bm} + 3(r_{b0} - r_{bm})\frac{\tan z - z}{z\tan^2 z} \tag{1.79}$$

where

$$z = \frac{\sqrt{1 + \frac{144}{\pi^2}\frac{I_b}{I_{rb}}} - 1}{\frac{24}{\pi^2}\sqrt{\frac{I_b}{I_{rb}}}} \tag{1.80}$$

where r_{bm} is the minimum base resistance that occurs at a high current level, r_{b0} is the base resistance at zero bias with a small base current level, and I_{rb} is the current where the base resistance falls halfway to its minimum value [7, 62].

The intrinsic device capacitances C_π, C_{ci}, and C_{co} are modeled by the diffusion capacitance and junction capacitance, each being

$$C_\pi = \frac{d}{dV_\pi}\left(\tau_{FF}\frac{I_{cc}}{q_b}\right) + C_{jeo}\left(1 - \frac{V_\pi}{\phi_e}\right)^{-m_e} \tag{1.81}$$

$$C_{ci} = \tau_R \frac{dI_{bc}}{dV_{bc}} + k_c C_{jco}\left(1 - \frac{V_{bc}}{\phi_c}\right)^{-m_c} \tag{1.82}$$

$$C_{co} = C_{jco}(1 - k_c)\left(1 - \frac{V_{bco}}{\phi_c}\right)^{-m_c} \tag{1.83}$$

where k_c is the fraction of the base–collector junction capacitance connected to the base resistance r_b, V_{bco} is the voltage through the capacitance C_{co}, and τ_{FF} is the modulated transit time defined by

$$\tau_{FF} = \tau_F \left[1 + X_{\tau F}\left(\frac{I_{cc}}{I_{cc} + I_{\tau F}}\right)^2 \exp\left(\frac{V_{bc}}{1.44V_{\tau F}}\right)\right] \tag{1.84}$$

where $X_{\tau F}$ is the transit time bias dependence coefficient, $I_{\tau F}$ is the high-current parameter for effect on τ_F, $V_{\tau F}$ is the value of V_{bc} where the exponential equals 0.5, and

$$I_{cc} = \frac{I_{ss}}{q_b}\left[\exp\left(\frac{V_\pi}{n_F V_T}\right) - 1\right] \tag{1.85}$$

From Eq. (1.81), the nonlinear behavior of the capacitance C_π strongly depends on the effect of the transit time modulation characterized by τ_{FF}. This transit charge variation results in significant changes of the transition frequency f_T at various operation conditions. For example, at medium currents, f_T reaches its peak value and is practically constant. Here, the ideal transit time is defined by $\tau_F = 1/(2\pi f_T)$ and the dominated base–emitter diffusion capacitance increases linearly with collector current. At low currents, f_T is dominated by junction capacitance and increases with the increase in collector current. At high currents, the widening of the charge-neutral base region and pushing of the entire space-charge region toward the heavily doped collector region (Kirk effect) degrades the frequency response of the transistor by increasing the transit time and decreasing f_T. In this case, the transit time is modeled by τ_{FF}.

The extended nonlinear HBT equivalent circuit developed for computer-aided simulations, which is shown in Figure 1.25, is based on the representation of the collector current source through the power series and diffusion capacitances through the hyperbolic functions [64]. Here, to fully describe the losses related to the base–emitter and base–collector junctions, additional model resistances R_{be1}, R_{be2}, R_{bc1}, and R_{bc2} are included, as well as the delay time $\tau_{del} = R_{del}C_{del}$ is taken into account for better accuracy of the collector current frequency response.

The more complicated large-signal models, such as VBIC, HICUM, or MEXTRAM, include the effects of self-heating of a bipolar transistor, take into account the parasitic *p–n–p* transistor formed by the base, collector, and substrate regions, provide an improved description of depletion capacitances at large forward bias, and consider the avalanche and tunneling currents and other nonlinear effects corresponding to distributed high-frequency effects [63, 65]. The most common implementation of a lumped equivalent thermal network, describing the HBT self-heating behavior, represents a single- or two-section low-pass *RC* network with thermal resistances and capacitances.

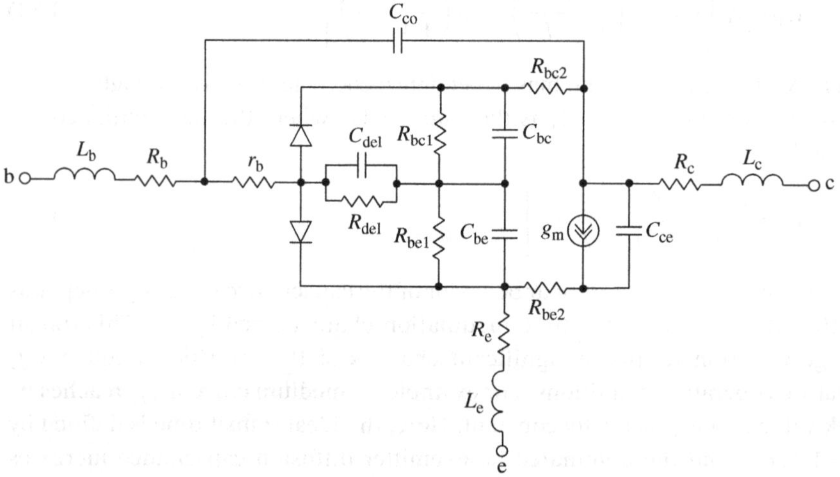

Figure 1.25 Extended nonlinear HBT equivalent circuit.

1.5.4 Large-signal Collector Capacitance

To equivalently estimate the input and output impedances of a bipolar transistor under large-signal operation, it makes sense to simplify the bipolar transistor nonlinear equivalent circuit, by limiting to the main nonlinear elements only. In this case, the external elements can be omitted as their effect is not so significant, especially at lower frequencies, and they can be included within the input matching circuit and load network, respectively, when, for example, the series parasitic base inductance can be considered as a part of a low-pass L-type matching circuit. For instance, for a bipolar transistor, the simplified equivalent circuit of which is shown in Figure 1.26, all elements of its equivalent circuit are nonlinear, depending significantly on operation mode, especially the transconductance g_m and base–emitter capacitance C_π. The base–emitter capacitance C_π consists of the diffusion and junction capacitances, and, at high frequencies, its reactance is sufficiently low, enabling the shunting of the base–emitter forward-biased diode. Because the transition frequency is defined by $\omega_T = g_m/C_\pi$, it is sufficient to consider only the nonlinear elements g_m, ω_T, and collector capacitance C_c, considering that the base resistance r_b poorly depends on the bias mode. As a result, the equivalent output capacitance can be defined as $C_{out} \cong C_c$ because $C_\pi \gg C_c$. The real part of the input impedance R_{in} can approximately be represented by the base resistance r_b, whereas the input equivalent capacitance can be defined as $C_{in} \cong C_\pi + C_c$. The feedback effect of the collector capacitance C_c through C_{c0} and C_{ci} is sufficiently high when load variations are directly transferred to the device input to a significant extent. At higher frequencies, since the input impedance of the high-power bipolar transistor is sufficiently low, it becomes inductive in view of the significant effect of the series parasitic base inductance.

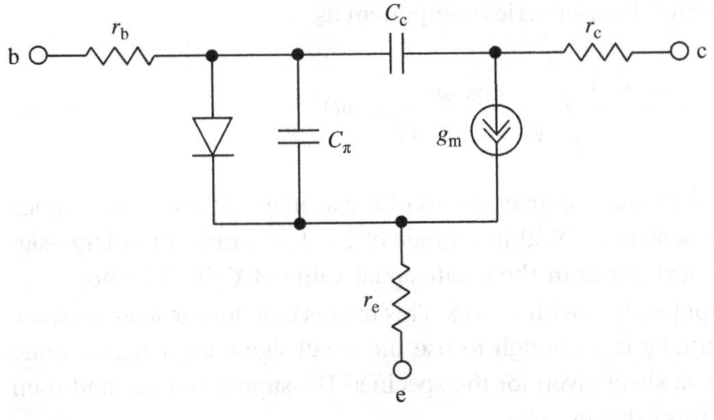

Figure 1.26 Bipolar transistor simplified equivalent circuit.

The collector capacitance C_c represents a junction capacitance and can be approximated by

$$C_c = \frac{C_{c0}}{\left(1 + \frac{v_c}{\phi}\right)^{\gamma}} \tag{1.86}$$

where ϕ is the built-in junction potential, γ is the junction sensitivity corresponding to the base–emitter and base–collector junction grading factors (m_e and m_c), and C_{c0} is the initial capacitance when $v_c = 0$. Note that similar equation is used to describe the behavior of the gate–drain and drain–source capacitances of high-power LDMOSFETs where the junction sensitivity γ corresponds to the gate–drain and drain–source junction grading factors (m_g and m_d) given by Eq. (1.19).

In a quasilinear case when considering the fundamental frequency only, the voltage across the capacitance is written as $v_c = V_{cc} + V_c \sin\omega t$, where V_{cc} is the collector DC-supply voltage and V_c is the fundamental voltage amplitude. Then, the current flowing through the collector capacitance can be defined as

$$\begin{aligned} i_c = C_c(v_c)\frac{dv_c}{dt} &= \frac{\omega C_{c0} V_c \cos\omega t}{\left(1 + \frac{V_{cc}}{\phi} + \frac{V_c}{\phi}\sin\omega t\right)^{\gamma}} \\ &= \frac{\omega C_c(V_{cc})\, V_c \cos\omega t}{(1 + \xi \sin\omega t)^{\gamma}} \end{aligned} \tag{1.87}$$

where $C_c(V_{cc})$ is the small-signal capacitance in the operating point and $\xi = V_c/(V_{cc} + \varphi)$.

As a result, the average large-signal collector capacitance C_{c1} can be calculated through the fundamental Fourier series component as

$$C_{c1}(V_c) = \frac{I_{c1}}{\omega V_c} = \frac{C_c(V_{cc})}{\pi} \int_0^{2\pi} \frac{\cos^2\omega t}{(1 + \xi \sin\omega t)^{\gamma}} d(\omega t) \tag{1.88}$$

Figure 1.27 shows the voltage dependencies of the average collector capacitance for different junction sensitivity. Within a range of $\xi < 1$, the maximum large-signal value of $C_{c1}(V_c)$ deviates from the small-signal value of $C_c(V_{cc})$ by not more than 20% for an abrupt junction with $\gamma = 0.5$. This means that, for the load-network design procedure, initially it is enough to use the small-signal capacitance value from the transistor data sheet given for the specified DC-supply voltage, and then just need to increase this value by 20%.

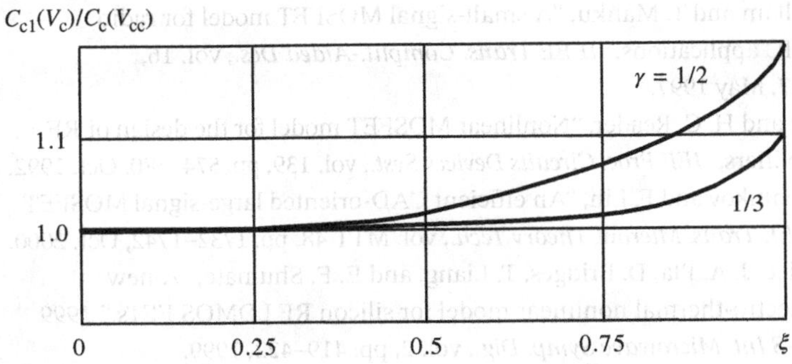

Figure 1.27 Large-signal behavior of collector capacitance.

References

1 Y. Wu, Q. Xu, C. Zhao, J. Liu, Y. Yu, W. Yin, and K. Kang, "An improved RF MOSFET model accounting substrate coupling among terminals," *IEEE Microw. Wirel. Comp. Lett.*, vol. 28, pp. 138–140, Feb. 2018.

2 I. Aoki, S. Kee, R. Magoon, R. Aparicio, F. Bohn, J. Zachan, G. Hatcher, D. McClymont, and A. Hajimiri, "A fully integrated quad-band GSM/GPRS CMOS power amplifier," *IEEE J. Solid-State Circuits*, vol. SC-43, pp. 2747–2758, Dec. 2008.

3 M. Depas, T. Nigam, and M. M. Heyns, "Soft breakdown of ultra-thin gate oxide layers," *IEEE Trans. Electron Devices*, vol. ED-43, pp. 1499–1504, Aug. 1996.

4 R. S. Muller and T. I. Kamins, *Device Electronics for Integrated Circuits*, Wiley, 1986.

5 Y. Tsividis, *Operation and Modeling of the MOS Transistors*, McGraw-Hill, 1999.

6 V. R. Vathulya, T. Sowlati, and D. Leenaerts, "Class 1 bluetooth power amplifier with 24 dBm output power and 48% PAE at 2.4 GHz in 0.25 μm CMOS," *Proc. 27th European Solid-State Circuits Conf.*, pp. 57–60, 2001.

7 G. Massobrio and P. Antognetti, *Semiconductor Device Modeling with SPICE*, McGraw-Hill, 1993.

8 Y. P. Tsividis and K. Suyama, "MOSFET modeling for analog circuit CAD: Problems and prospects," *IEEE J. Solid-State Circuits*, vol. SC-29, pp. 210–216, Mar. 1994.

9 N. Chevaux and M. M. De Souza, "Comparative analysis of VDMOS/LDMOS power transistors for RF amplifiers," *IEEE Trans. Microw. Theory Tech.*, vol. MTT-57, pp. 2643–2651, Nov. 2009.

10 E. Abou-Allam and T. Manku, "A small-signal MOSFET model for radio frequency IC applications," *IEEE Trans. Comput.-Aided Des.*, vol. 16, pp. 437–447, May 1997.
11 G. A. Holle and H. C. Reader, "Nonlinear MOSFET model for the design of RF power amplifiers," *IEE Proc. Circuits Devices Syst.*, vol. 139, pp. 574–580, Oct. 1992.
12 A. V. Grebennikov and F. Lin, "An efficient CAD-oriented large-signal MOSFET model," *IEEE Trans. Microw. Theory Tech.*, vol. MTT-48, pp. 1732–1742, Oct. 2000.
13 W. R. Curtice, J. A. Pla, D. Bridges, T. Liang, and E. E. Shumate, "A new dynamic electro-thermal nonlinear model for silicon RF LDMOS FETs," *1999 IEEE MTT-S Int. Microwave Symp. Dig.*, vol. 2, pp. 419–422, 1999.
14 C. Fager, J. C. Pedro, N. B. Carvalho, and H. Zirath, "Prediction of IMD in LDMOS transistor amplifiers using a new large-signal model," *IEEE Trans. Microw. Theory Tech.*, vol. MTT-50, pp. 2834–2842, Dec. 2002.
15 W. R. Curtice, L. Dunleavy, W. Claussen, and R. Pengelly, "New LDMOS model delivers powerful transistor library: The CMC model," *High Freq. Electron.*, vol. 3, pp. 18–25, Oct. 2004.
16 T. Taki, "Approximation of junction field-effect transistor characteristics by a hyperbolic function," *IEEE J. Solid-State Circuits*, vol. SC-13, pp. 724–726, Oct. 1978.
17 M. Miller, T. Dinh, and E. Shumate, "A new empirical large signal model for silicon RF LDMOS FETs," *1997 IEEE MTT-S Technol. Wireless Appl. Symp. Dig.*, pp. 19–22, 1997.
18 D. Heo, E. Chen, E. Gebara, S. Yoo, J. Laskar, and T. Anderson, "Temperature dependent MOSFET RF large signal model incorporating self-heating effects," *1999 IEEE MTT-S Int. Microwave Symp. Dig.*, pp. 415–418, 1999.
19 Y. C. Chen, D. L. Ingram, H. C. Yen, R. Lai, and D. C. Streit, "A new empirical *I–V* model for HEMT devices," *IEEE Microw. Guided Wave Lett.*, vol. 8, pp. 342–344, Oct. 1998.
20 P. Perugupalli, M. Trivedi, K. Shenai, and S. K. Leong, "Modeling and characterization of an 80 V silicon LDMOSFET for emerging RFIC applications," *IEEE Trans. Electron Devices*, vol. ED-45, pp. 1468–1478, July 1998.
21 M. C. Ho, K. Green, R. Culbertson, J. Y. Yang, D. Ladwig, and P. Ehnis, "A physical large signal Si MOSFET model for RF circuit design," *1997 IEEE MTT-S Int. Microwave Symp. Dig.*, vol. 2, pp. 391–394, 1997.
22 J. Angel Garcia, A. Mediavilla, J. Carlos Pedro, N. B. Carvalho, A. Tazon, and J. L. Garcia, "Characterizing the gate to source nonlinear capacitor role on FET IMD performance," *1998 IEEE MTT-S Int. Microwave Symp. Dig.*, vol. 3, pp. 1635–1637, 1998.
23 D. Moncoqut, D. Farenc, P. Rossel, G. Charitat, H. Tranduc, J. Victory, and I. Pages, "LDMOS transistor for smart power circuits: Modeling and design," *Proc. 1996 IEEE Bipolar/BiCMOS Circuits and Technology Meeting (BCTM)*, pp. 216–219, 1996.

24 C. E. Biber, M. L. Schmatz, T. Morf, U. Lott, and W. Bachtold, "A nonlinear microwave MOSFET model for SPICE simulators," *IEEE Trans. Microw. Theory Tech.*, vol. MTT-46, pp. 604–610, May 1998.

25 S. Lai, C. Fager, D. Kuylenstierna, and I. Angelov, "LDMOS modeling," *IEEE Microw. Mag.*, vol. 14, pp. 108–116, Jan. 2013.

26 R. Sung, P. Bendix, and M. B. Das, "Extraction of high-frequency equivalent circuit parameters of submicron gate-length MOSFET's," *IEEE Trans. Electron Devices*, vol. ED-45, pp. 1769–1775, Aug. 1998.

27 B. J. Cheu and P. K. Ko, "Measurement and modeling of short-channel MOS transistor gate capacitances," *IEEE J. Solid-State Circuits*, vol. SC-22, pp. 464–472, June 1987.

28 R. E. Anholt and S. E. Swirhun, "Experimental investigations of the temperature dependence of GaAs FET equivalent circuits," *IEEE Trans. Electron Devices*, vol. ED-39, pp. 2029–2036, Sept. 1992.

29 J. M. Dortu, J. E. Muller, M. Pirola, and G. Ghione, "Accurate large-signal GaAs MESFET and HEMT modeling for power MMIC amplifier design," *Int. J. Microw. Mill.-Wave Comput.-Aided Eng.*, vol. 5, pp. 195–208, Sep. 1995.

30 L. S. Liu, J. G. Ma, and G. I. Ng, "Electrothermal large-signal model for III-V FETs including frequency dispersion and charge conservation," *IEEE Trans. Microw. Theory Tech.*, vol. MTT-38, pp. 822–824, June 1990.

31 C. J. Wei, Y. Tkachenko, and D. Bartle, "An accurate large-signal model of GaAs MESFET which accounts for charge conservation, dispersion, and self-heating," *IEEE Trans. Microw. Theory Tech.*, vol. MTT-46, pp. 1638–1644, Nov. 1998.

32 A. Jarndal and G. Kompa, "Large-signal model for AlGaN/GaN HEMTs accurately predicts trapping- and self-heating-induced dispersion and intermodulation distortion," *IEEE Trans. Electron Devices*, vol. ED-54, pp. 2830–2836, Nov. 2007.

33 I. Angelov, H. Zirath, and N. A. Rorsman, "A new empirical nonlinear model for HEMT and MESFET devices," *IEEE Trans. Microw. Theory Tech.*, vol. MTT-40, pp. 2258–2266, Dec. 1992.

34 O. S. A. Tang, K. H. G. Duh, S. M. J. Liu, P. M. Smith, W. F. Kopp, T. J. Rogers, and D. J. Pritchard, "Design of high-power, high-efficiency 60-GHz MMIC's using an improved nonlinear PHEMT model," *IEEE J. Solid-State Circuits*, vol. SC-32, pp. 1326–1333, Sep. 1997.

35 I. Angelov, V. Desmaris, K. Dynefors, P. A. Nilsson, N. Rorsman, and H. Zirath, "On the large-signal modelling of AlGaN/GaN HEMTs and SiC MESFETs," *13th European GAAS Symp. Dig.*, pp. 309–312, 2005.

36 A. Garcia-Osorio, J. R. Loo-Yau, J. A. Reynoso-Hernandez, S. Ortega, and J. L. del Valle-Padilla, "An empirical I-V nonlinear model suitable for GaN FET Class F PA design," *Microw. Opt. Technol. Lett.*, vol. 53, pp. 1256–1259, June 2011.

37 K. Joshin and T. Kikkawa, "High-power and high-efficiency GaN HEMT amplifiers," *2008 IEEE Radio and Wireless Symp. Dig.*, pp. 65–68, 2008.

38 T. Ishida, "GaN HEMT technologies for space and radio applications," *Microw. J.*, vol. 54, pp. 56–66, Aug. 2011.

39 U. K. Mishra, P. Parikh, Y. F. Wu, "AlGan/GaN HEMTs – An overview of device operation and applications," *Proc. IEEE*, vol. 90, pp. 1022–1031, June 2002.

40 A. Grebennikov, *RF and Microwave Power Amplifier Design*, McGraw-Hill, 2015.

41 M. Berroth and R. Bosch, "Broad-band determination of the FET small-signal equivalent circuit," *IEEE Trans. Microw. Theory Tech.*, vol. MTT-38, pp. 891–895, July 1990.

42 Q. Fan, J. H. Leach, and H. Morkoc, "Small signal equivalent circuit modeling for AlGaN/GaN HFET: Hybrid extraction method for determining circuit elements of AlGaN/GaN HFET," *Proc. IEEE*, vol. 98, pp. 1140–1150, July 2010.

43 G, Dambrine, A. Cappy, F. Heliodore, and E. Playez, "A new method for determining the FET small-signal equivalent circuit," *IEEE Trans. Microw. Theory Tech.* vol. MTT-36, pp. 11512–11259, July 1988.

44 I. Angelov, L. Bengtsson, and M. Garcia, "Extensions of the Chalmers nonlinear HEMT and MESFET model," *IEEE Trans. Microw. Theory Tech.*, vol. MTT-44, pp. 1664–1674, Oct. 1996.

45 I. Angelov, K. Andersson, D. Schreurs, D. Xiao, N. Rorsman, V. Desmaris, M. Sudow, and H. Zirath, "Large-signal modelling and comparison of AlGaN/GaN HEMTs and SiC MESFETs." *Proc. 2006 Asia-Pacific Microwave Conf.*, pp. 1–4, 2006.

46 N. M. Rohringer and P. Kreuzgruber, "Parameter extraction for large-signal modeling of bipolar junction transistors," *Int. J. Microw. Mill.-Wave Comput.-Aided Eng.*, vol. 5, pp. 161–172, Sep. 1995.

47 J. P. Fraysee, D. Floriot, P. Auxemery, M. Campovecchio, R. Quere, and J. Obregon, "A non-quasi-static model of GaInP/AlGaAs HBT for power applications," *1997 IEEE MTT-S Int. Microwave Symp. Dig.*, pp. 377–382, 1997.

48 V. M. Bogachev and V. V. Nikiforov, *Transistor Power Amplifiers* (in Russian), Energiya, 1978.

49 P. M. Asbeck, M. F. Chang, K. C. Wang, D. L. Miller, G. J. Sullivan, N. H. Sheng, E. Sovero, and J. A. Higgins, "Heterojunction bipolar transistors for microwave and millimeter-wave integrated circuits," *IEEE Trans. Microw. Theory Tech.*, vol. MTT-35, pp. 1462–1468, Dec. 1987.

50 Y. S. Lin and J. J. Jiang, "Temperature dependence of current gain, ideality factor, and offset voltage of AlGaAs/GaAs and InGaP/GaAs HBTs," *IEEE Trans. Electron Devices*, vol. ED-56, pp. 2945–2951, Dec. 2009.

51 N. L. Wang, W. Ma, S. Xu, E. Camargo, X. Sun, and P. Hu, "28-V high-linearity and rugged InGaP/GaAs HBT," *2006 IEEE MTT-S Int. Microwave Symp. Dig.*, pp. 881–884, 2006.

52 C. Steinberger, T. Landon, C. Suckling, J. Nelson, J. Delaney, and J. Hitt, "250 W HVHBT Doherty with 57% WCDMA efficiency linearized to 255 dBc for 2c11 6.5 dB PAR," *IEEE J. Solid-State Circuits*, vol. SC-43, pp. 2218–2228, Oct. 2008.

53 D. Costa, W. U. Liu, and J. S. Harris, "Direct extraction of the AlGaAs/GaAs heterojunction bipolar transistor small-signal equivalent circuit" *IEEE Trans. Electron Devices*, vol. ED-38, pp. 2018–2024, Sep. 1991.

54 H. Ghaddab, F. M. Ghannouchi, and F. Bouallegue, "Small-signal modeling of HBT's using a hybrid optimization/statistical technique," *IEEE Trans. Microw. Theory Tech.*, vol. MTT-46, pp. 292–298, Mar. 1998.

55 J. M. Early, "Effects of space-charge layer widening in junction transistors," *Proc. IRE*, vol. 40, pp. 1401–1406, Nov. 1952.

56 D. A. Teeter and W. R. Curtice, "Comparison of hybrid Pi and Tee HBT circuit topologies and their relationship to large signal modeling," *1997 IEEE MTT-S Int. Microwave Symp. Dig.*, vol. 2, pp. 375–378, 1997.

57 A. Samelis and D. Pavlidis, "DC to high-frequency HBT-model parameter evaluation using impedance block conditioned optimization," *IEEE Trans. Microw. Theory Tech.*, vol. MTT-45, pp. 886–897, June 1997.

58 B. Li, F. S. Prasad, L. W. Yang, and S. C. Wang, "A semianalytical parameter-extraction procedure for HBT equivalent circuit," *IEEE Trans. Microw. Theory Tech.*, vol. MTT-46, pp. 1427–1435, Oct. 1998.

59 J. J. Ebers and J. L. Moll, "Large-signal behavior of junction transistors," *Proc. IRE*, vol. 42, pp. 1761–1772, Dec. 1954.

60 W. M. Webster, "On the variation of junction-transistor current amplification factor with emitter current," *Proc. IRE*, vol. 42, pp. 914–920, June 1954.

61 C. T. Kirk, "A theory of transistor cutoff frequency (f_T) falloff at high current densities," *IEEE Trans. Electron Devices*, vol. ED-9, pp. 164–174, Mar. 1962.

62 H. K. Gummel and H. C. Poon, "An integral charge control model of bipolar transistors," *Bell Syst. Tech. J.*, vol. 49, pp. 827–852, May–June 1970.

63 J. McMacken, S. Nedeljkovich, J. Gering, and D. Halchin, "HBT modeling," *IEEE Microw. Mag.*, vol. 9, pp. 48–72, Apr. 2008.

64 I. Angelov, K. Choumei, and A. Inoue, "An empirical HBT large-signal model for CAD," *Int. J. RF Microw. Comput.-Aided Eng.*, vol. 13, pp. 518–533, Nov. 2003.

65 M. Reisch, *High-Frequency Bipolar Transistors*, Springer, 2003.

2

Impedance Matching

Generally, an optimum impedance-matching solution depends on many factors, such as the impedances to be matched, simplicity in practical realization, frequency bandwidth and its flatness, design implementation and adjustability, stable operation conditions for the entire power amplifier, possibility for harmonic termination to maximize efficiency, and sufficient higher-order harmonic suppression. As a result, many types of matching networks are available, including low-pass, high-pass, or bandpass structures which are based on lumped elements and different types of transmission lines. To simplify and visualize the matching design procedure, an analytical approach, which allows calculating the parameters of the matching circuits using simple equations, and consideration of the impedance Smith chart traces according to the design equations are very helpful.

2.1 Conjugate-matching Condition

Impedance matching is necessary to provide maximum delivery of the RF power available from the source to the load because, generally, when the electrical signal propagates in the circuit, a portion of this signal will be reflected at the interface between the sections with different impedances. Therefore, it is necessary to establish the conditions that allow fully transmitting the entire electrical signal without any reflection. To determine the optimum value of the load impedance Z_L, at which the power delivered to the load is maximal, consider the equivalent circuit shown in Figure 2.1a.

The power delivered to the load can be defined as

$$P = \frac{1}{2} V_{in}^2 \operatorname{Re}\left(\frac{1}{Z_L}\right) = \frac{1}{2} V_S^2 \left|\frac{Z_L}{Z_S + Z_L}\right|^2 \operatorname{Re}\left(\frac{1}{Z_L}\right) \tag{2.1}$$

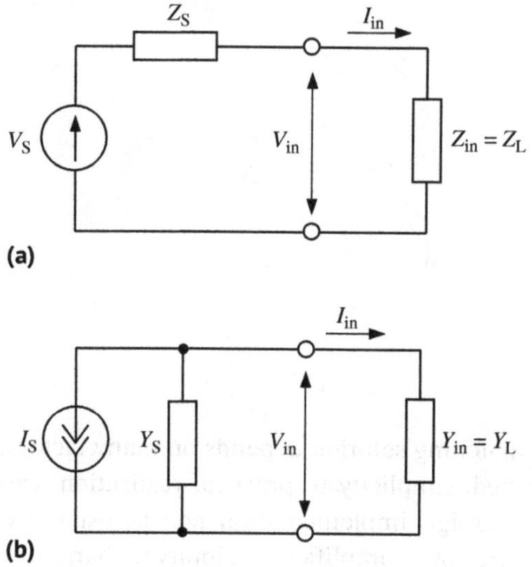

Figure 2.1 Equivalent circuits with (*a*) voltage and (*b*) current sources.

where $Z_S = R_S + jX_S$ is the source impedance, $Z_L = R_L + jX_L$ is the load impedance, V_S is the source voltage amplitude, and V_{in} is the load voltage amplitude. Substituting the real and imaginary parts of the source and load impedances Z_S and Z_L into Eq. (2.1) yields

$$P = \frac{1}{2} V_S^2 \frac{R_L}{(R_S + R_L)^2 + (X_S + X_L)^2} \tag{2.2}$$

From Eq. (2.2), it follows that, if the source impedance Z_S is fixed, it is necessary to vary the real and imaginary parts of the load impedance Z_L until maximum power is delivered to the load. To maximize the output power, the following analytical conditions in the form of derivatives with respect to the output power are necessary to be applied:

$$\frac{\partial P}{\partial R_L} = 0 \qquad \frac{\partial P}{\partial X_L} = 0 \tag{2.3}$$

Then, applying these conditions to Eq. (2.2), the following system of two equations can be obtained:

$$\frac{1}{(R_L + R_S)^2 + (X_L + X_S)^2} - \frac{2R_L(R_L + R_S)}{\left[(R_L + R_S)^2 + (X_L + X_S)^2\right]^2} = 0 \tag{2.4}$$

$$\frac{2X_L(X_L + X_S)}{\left[(R_L + R_S)^2 + (X_L + X_S)^2\right]^2} = 0 \tag{2.5}$$

Simplifying Eqs. (2.4) and (2.5) results in

$$R_S^2 - R_L^2 + (X_L + X_S)^2 = 0 \tag{2.6}$$

$$X_L(X_L + X_S) = 0 \tag{2.7}$$

Then, solving Eqs. (2.6) and (2.7) simultaneously for R_S and X_S results in

$$R_S = R_L \tag{2.8}$$

$$X_L = -X_S \tag{2.9}$$

or, in an impedance form,

$$Z_L = Z_S^* \tag{2.10}$$

where * denotes the complex conjugate value [1, 2].

Equation (2.10) is called the impedance conjugate-matching condition, and its fulfillment results in maximum power delivered to the load for fixed source impedance. It should be noted that the term "impedance" was introduced by Oliver Lodge in 1889, and it meant the ratio *V/I* in the special circuit comprised of a resistance and an inductance, where *I* and *V* are the amplitudes of an alternating current and the driving force which produced it [3].

The admittance conjugate-matching condition, applied to the equivalent circuit shown in Figure 2.1b, is written as

$$Y_L = Y_S^* \tag{2.11}$$

which can be readily obtained in the same way.

Thus, the conjugate-matching condition in a common case can be determined through the immittance *W*-parameters, which represent any system of the impedance *Z*-parameters or admittance *Y*-parameters, in the form of

$$W_L = W_S^* \tag{2.12}$$

where W_S is the source immittance and W_L is the load immittance. The term "immittance" was introduced by Bode to refer to a complex number, which may be either the impedance or the admittance of a system [4].

For a single-stage high-efficiency power amplifier, the matching circuit is connected between the source and the input of an active device, as shown in Figure 2.2a, and between the harmonic-control circuit at the output of an active device and the load, as shown in Figure 2.2b. In some cases, the input matching circuit can include the resonant circuit tuned to the second or third harmonic to improve the power amplifier efficiency by providing faster switching time of the transistor. For a multistage power amplifier, the load represents an input circuit of the next stage. Therefore, the matching circuit is connected between the output of the active device of the preceding amplifier stage and the input

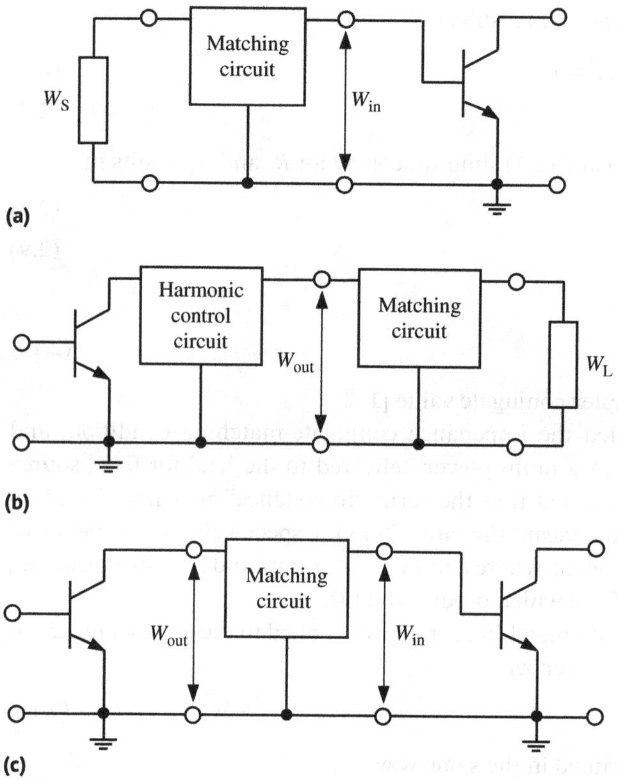

Figure 2.2 Matching circuit arrangements.

of the active device of the succeeding stage of the power amplifier, as shown in Figure 2.2c. The main objective is to properly transform the load immittance W_L to the optimum device output immittance W_{out}, the value of which is properly determined by the supply voltage, the output power, the saturation voltage of the active device, and the selected class of the active device operation to maximize the operating efficiency and output power of the power amplifier. It should be noted that Eq. (2.12) is given in a general immittance form without indication of whether it is used in a small-signal or large-signal application. In the latter case, this only means that the device immittance W-parameters are fundamentally averaged over large-signal swing across the device equivalent circuit parameters and that the conjugate-matching principle is valid in both small-signal and large-signal applications where the optimum equivalent device output resistance (or conductance) at the fundamental is matched to the load resistance (or conductance) and the effect of the device output reactive elements is eliminated by the conjugate reactance of the load network. In addition, the matching circuits should be designed to realize the required voltage and current waveforms at the device output to provide the stability of operation conditions and to satisfy the requirements for the power amplifier amplitude and phase characteristics. The losses in the matching circuits must be as small as possible to deliver the output power to the load with maximum efficiency. Finally, it is desirable that the matching circuit be easy to tune.

2.2 Matching with Lumped Elements

Generally, there is a variety of configurations for matching networks to efficiently deliver the signal from the source to the load, and applying any of these matching networks to the power amplifier structure depends on its class of operation, level of output power, operating frequency, frequency bandwidth, or level of harmonic suppression. The lumped matching networks in the form of (*a*) L-transformer, (*b*) Π-transformer, or (*c*) T-transformer shown in Figure 2.3, respectively, have proved for a long time to be effective for power amplifier design [5]. The simplest and most popular matching network is the circuit in the form of the L-transformer. The transforming properties of this matching circuit can be analyzed by using the equivalent transformation of a parallel into a series representation of the *RX* network.

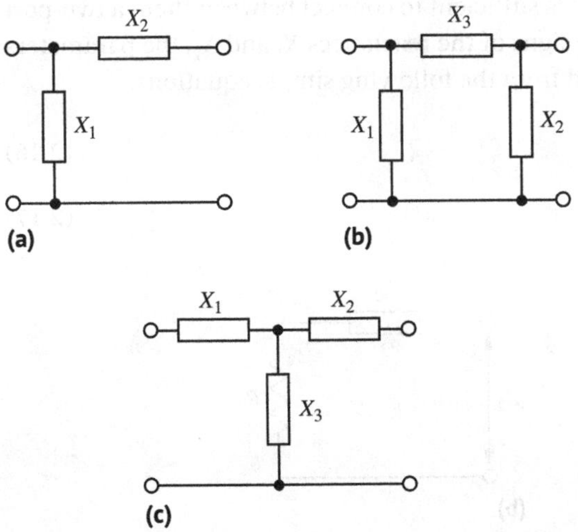

Figure 2.3 Matching networks in the form of L-, Π-, and T-transformers.

2.2.1 Analytic Design Technique

Consider the parallel *RX* circuit shown in Figure 2.4a, where R_1 is the real (resistive) part and X_1 is the imaginary (reactive) part of the network impedance $Z_1 = jX_1R_1/(R_1 + jX_1)$, and the series *RX* network shown in Figure 2.4b, where R_2 is the resistive part and X_2 is the reactive part of the circuit impedance $Z_2 = R_2 + jX_2$. These two impedance circuits (series and parallel) can be considered equivalent at some frequency if $Z_1 = Z_2$, resulting in

$$R_2 + jX_2 = \frac{R_1X_1^2}{R_1^2 + X_1^2} + j\frac{R_1^2X_1}{R_1^2 + X_1^2} \tag{2.13}$$

Equation (2.13) can be rearranged into two separate equations for real and imaginary parts according to

$$R_1 = R_2\left(1 + Q^2\right) \tag{2.14}$$

$$X_1 = X_2\left(1 + Q^{-2}\right) \tag{2.15}$$

where $Q = R_1/|X_1| = |X_2|/R_2$ is the circuit quality factor, which is equal for both the series and parallel *RX* networks.

Consequently, if the reactive impedance $X_1 = -X_2\left(1 + Q^{-2}\right)$ with an opposite sign for the reactance X_1 of the parallel circuit in Eq. (2.15) is connected in parallel to the series R_2X_2 circuit, it allows the reactive impedance (or reactance) X_2 of the series circuit to be compensated. In this case, the input impedance Z_{in} of such a two-port network, which is shown in Figure 2.5, will be only resistive and equal to R_1. Consequently, to transform the resistance R_1 into the other resistance R_2 at a certain frequency, it is sufficient to connect between them a two-port L-transformer with the opposite signs of the reactances X_1 and X_2, the parameters of which can be easily calculated from the following simple equations:

$$|X_1| = \frac{R_1}{Q} \tag{2.16}$$

$$|X_2| = R_2Q \tag{2.17}$$

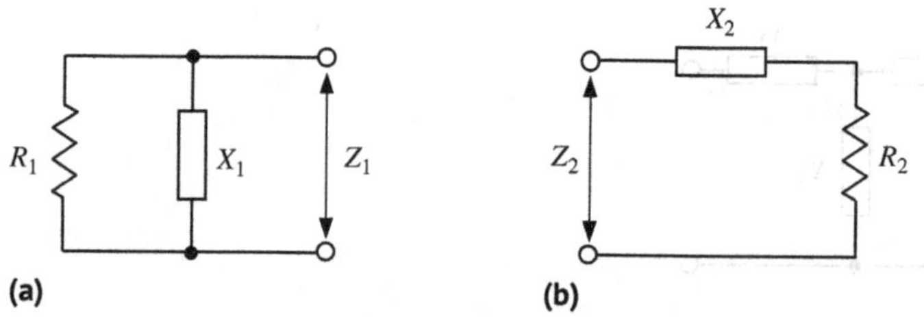

Figure 2.4 Impedance parallel and series equivalent circuits.

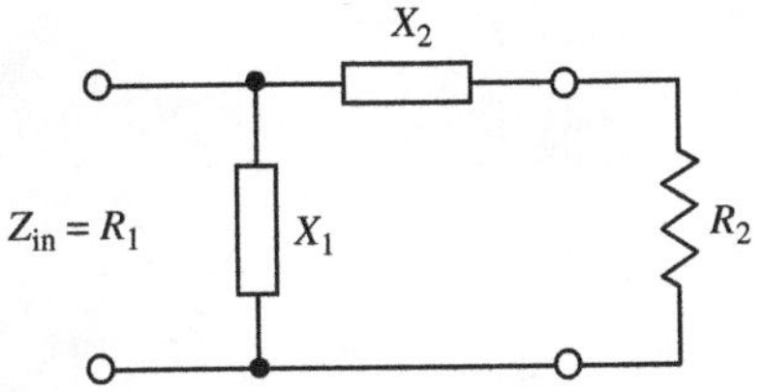

Figure 2.5 Input impedance of two-port network.

where

$$Q = \sqrt{\frac{R_1}{R_2} - 1} \tag{2.18}$$

is the circuit or loaded quality factor (also can be indicated as Q_L) expressed through the resistances to be matched. Thus, to design a matching circuit with fixed resistances to be matched, it needs first to calculate the loaded quality factor Q (or Q_L) according to Eq. (2.18) and then to define the reactive elements according to Eqs. (2.16) and (2.17). Note that, for such a low-pass or high-pass L-type impedance transformer, the source impedance R_1 should always be higher than the load impedance R_2 according to Eq. (2.18).

2.2.2 Low-pass and High-pass L-type Transformers

Because of the opposite signs of the reactances X_1 and X_2 of the L-type impedance transformer, shown in Figure 2.5, the two possible circuit configurations (one in the form of a low-pass filter section and another in the form of a high-pass filter section) with the same transforming properties can be realized, as shown in Figure 2.6 together with the design equations.

If this matching circuit has small values of Q, a wider frequency bandwidth but poor out-of-band suppression can be achieved. However, with large values of Q, the frequency bandwidth is substantially reduced. Therefore, there is always a compromise between the frequency bandwidth, out-of-band suppression, and the number of impedance-matching sections to be used. For the case of $R_1/R_2 \geq 10$, which corresponds to the condition of $Q \geq 3$, the out-of-band suppression factor F_n of such a low-pass L-transformer can be approximately evaluated by the same formulas as for a parallel resonant circuit, for which the loaded quality factor Q is defined as $f_0/\Delta f$, where f_0 is the center bandwidth frequency and Δf is the frequency 3-dB bandwidth (level at −3 dB from maximum level), as

$$F_n = Q(n^2 - 1) \tag{2.19}$$

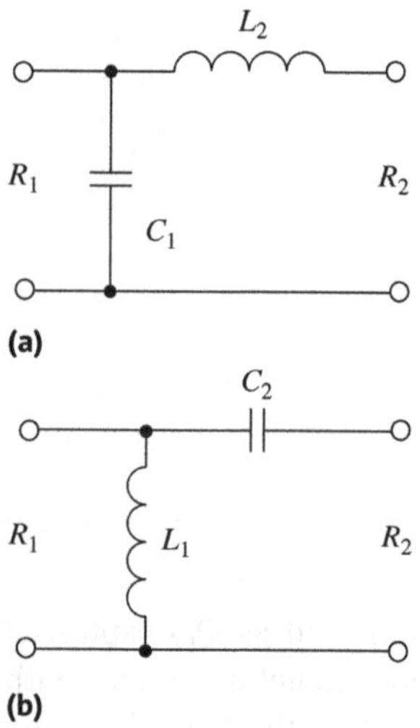

$$C_1 = \frac{Q}{\omega R_1}$$

$$L_2 = \frac{QR_2}{\omega}$$

$$L_1 = \frac{R_1}{\omega Q}$$

$$C_2 = \frac{1}{Q\omega R_2}$$

Figure 2.6 The L-type matching circuits and relevant equations.

where n is the harmonic number [6]. The out-of-band suppression factor F_n of the matching circuit represents the ratio of the selected harmonic component in the input (collector or drain) current to the same harmonic component in the output (load) current.

The transformer efficiency η_T is determined by the ratio of P_L/P_{in}, where P_{in} is the power at the input of the transformer and P_L is the transformer load power. Such a transformer efficiency for the L-type impedance transformer with negligible losses in the capacitor can be calculated by

$$\eta_T = 1/\left(1 + \frac{Q}{Q_{ind}}\right) \tag{2.20}$$

where Q_{ind} is the inductor quality factor [7]. From Eq. (2.20), it follows that the efficiency of the impedance transformer decreases with the increase of Q. This means that, for the same R_1 and series parasitic resistance of the circuit inductor, a lower resistance R_2 provides a higher current flowing through this inductor, which leads to an additional power dissipation. An analysis of Eqs. (2.16) and (2.17) shows that, for the given resistances R_1 and R_2, each element of the low-pass or high-pass L-transformer can have only one value at a fixed frequency. Consequently, it is difficult to satisfy simultaneously such contradictory requirements as efficiency, frequency bandwidth, and out-of-band suppression.

To avoid the parasitic low-frequency oscillations in the power amplifier and to increase the level of the harmonic suppression, it may be necessary to connect

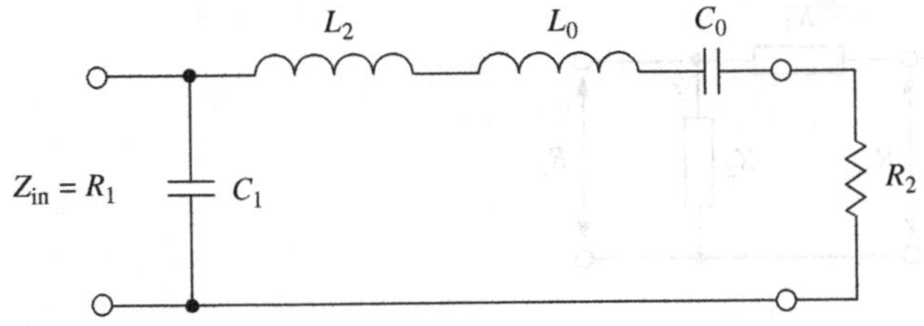

Figure 2.7 Low-pass L-transformer with an additional series filter.

an additional L_0C_0 series circuit, the resonant frequency of which is equal to the operating frequency of the power amplifier, as shown in Figure 2.7. In this case, the out-of-band suppression factor F_n defined by Eq. (2.19) is written as

$$F_n = Q_\Sigma (n^2 - 1) \tag{2.21}$$

where $Q_\Sigma = Q + Q_0$ and $Q_0 = \omega L_0/R_2$. Better harmonic suppression is achieved at the expense of the frequency bandwidth narrowing.

In practice, it makes sense to use the single-section L-transformers in power amplifiers as the input and interstage matching circuits, where the requirements for out-of-band suppression and efficiency are not as high as for the output matching circuits. In this case, the main advantage of such an L-transformer is its simplicity when only two reactive elements with fast tuning are needed. For larger values of $Q \geq 10$, it is possible to use a cascade connection of the L-transformers, which allows wider frequency bandwidth and higher transformer efficiency to be realized.

2.2.3 The Π- and T-type Transformers

An appropriate connection of two or more low-pass and high-pass impedance L-transformers allows improving the power amplifier performance by satisfying to some extent such contradictory requirements as frequency bandwidth and out-of-band suppression. In the simplest case, by connecting the two L-transformers, the matching circuits in the form of a Π-transformer shown in Figure 2.8a or a T-transformer shown in Figure 2.8b can be realized. For each L-transformer, the resistances R_1 and R_2 are transformed to some intermediate resistance R_0 with the value of $R_0 < (R_1, R_2)$ for a Π-transformer and the value of $R_0 > (R_1, R_2)$ for a T-transformer. The value of R_0 is not fixed and can be chosen arbitrarily, depending on the frequency bandwidth. This means that, compared to the simple L-transformer with fixed parameters for the same ratio of R_2/R_1, the parameters of a Π-transformer or a T-transformer can be different. However, they provide

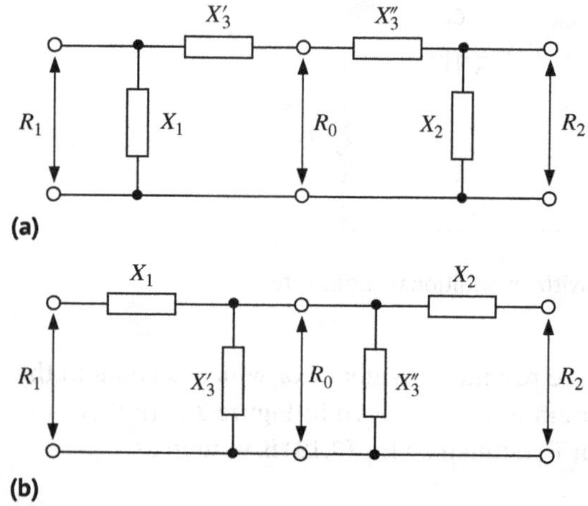

Figure 2.8 Matching circuits, developed by connecting two L-transformers.

narrower frequency bandwidths due to higher loaded quality factors because the intermediate resistance R_0 is either greater or smaller than each of the resistances R_1 and R_2. By considering the two possible circuit configurations of the L-transformer, which are shown in Figure 2.6, there is a possibility to design different circuit configurations of such two-port impedance transformers, as shown in Figure 2.8a with a series reactance $X_3 = X_3' + X_3''$ and in Figure 2.8b with a shunt reactance $X_3 = X_3' X_3'' / (X_3' + X_3'')$.

Several of the most widely used two-port Π-transformers, together with the design formulas, are shown in Figure 2.9 [7, 8]. The low-pass Π-transformer shown in Figure 2.9a is usually used as an output matching circuit of high-power amplifiers in a Class-B operation mode when it is necessary to achieve a sinusoidal drain (or collector) voltage waveform by appropriate harmonic suppression. In addition, it is convenient to use some of them as interstage matching circuits in low-power and medium-power amplifiers when it is necessary to provide sinusoidal voltage waveforms both at the drain (or collector) of the preceding-stage transistor and at the gate (or base) of the succeeding-stage transistor. In this case, for the Π-transformer with shunt capacitors, the input and output capacitances of these transistors can be easily included into the matching circuit elements C_1 and C_2, respectively. Besides, a Π-transformer can be directly used as the load network for a high-efficiency Class-E mode with proper calculation of its design parameters, especially when using an additional series resonant circuit to provide high impedance for the high-order harmonic components to prevent their flowing into the load.

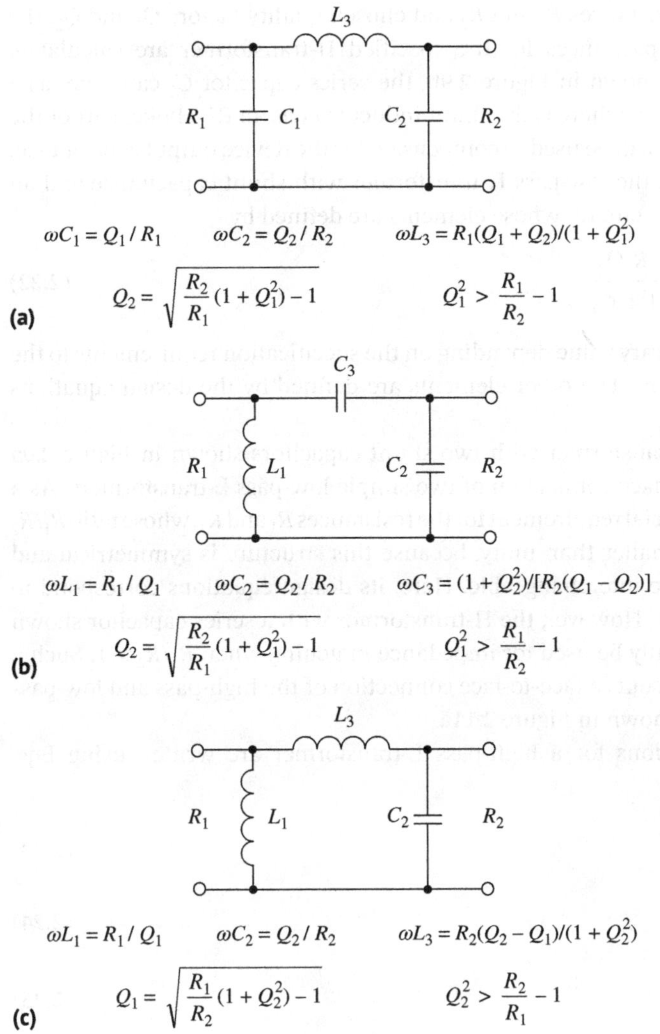

Figure 2.9 The Π-transformers and relevant equations.

To calculate the matching circuit parameters, it needs first to estimate the quality factor Q_1 for the Π-transformers shown in Figure 2.9a,b or the quality factor Q_2 for the Π-transformer shown in Figure 2.9c, which should not be chosen too low. Then, calculate the quality factor Q_2 for the Π-transformers shown in Figure 2.9a,b for a chosen Q_1 or the quality factor Q_1 for the Π-transformer shown in Figure 2.9c for a chosen Q_2. Note that the chosen factors Q_1 or Q_2 should not be too low, otherwise the values of the calculated Q_1 or Q_2 will be very small.

Finally, for given resistances R_1 and R_2 and chosen quality factors Q_1 and Q_2, the inductances and capacitances for the specified Π-transformer are calculated. In a Π-transformer shown in Figure 2.9b, the series capacitor C_2 can serve as a DC-blocking capacitor, whereas the shunt inductor L_1 as an RF choke, part of the inductance of which can be used to compensate for the device output capacitance.

Figure 2.10 shows the low-pass Π-transformer with shunt capacitance and an additional L_0C_0 series circuit, whose elements are defined by

$$\omega L_0 = \frac{1}{\omega C_0} = \frac{R_1 Q_0}{1 + Q_1^2} \tag{2.22}$$

where Q_0 is the arbitrary value depending on the specification requirements to the harmonic suppression. The other elements are defined by the design equations given in Figure 2.9a.

The low-pass Π-transformer with two shunt capacitors shown in Figure 2.9a represents a face-to-face connection of two simple low-pass L-transformers. As a result, there is no special requirement for the resistances R_1 and R_2, whose ratio R_1/R_2 can be greater or smaller than unity, because this structure is symmetrical and input and output are interchangeable. Here, its design equations correspond to the case of $R_1/R_2 > 1$. However, the Π-transformer with a series capacitor shown in Figure 2.9b can only be used for impedance matching when $R_1/R_2 > 1$. Such a Π-transformer represents a face-to-face connection of the high-pass and low-pass L-transformers, as shown in Figure 2.11a.

The design equations for a high-pass L-transformer are written using Eqs. (2.16)–(2.18) as

$$\omega L_1 = \frac{R_1}{Q_1} \tag{2.23}$$

$$\omega C_3' = \frac{1}{Q_1 R_0} \tag{2.24}$$

$$Q_1^2 = \frac{R_1}{R_0} - 1 \tag{2.25}$$

where Q_1 is the circuit quality factor and R_0 is the intermediate resistance.

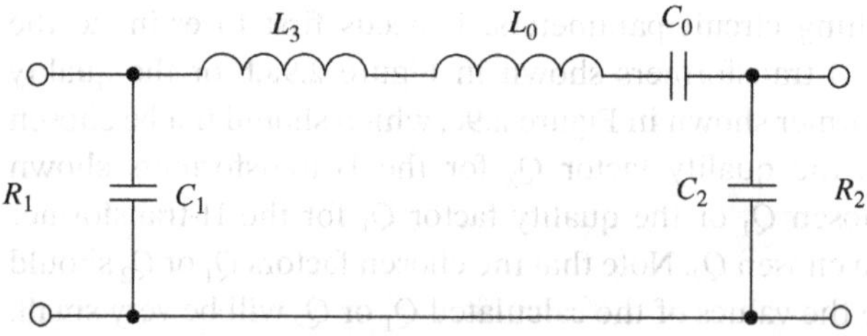

Figure 2.10 Low-pass Π-transformer with an additional series filter.

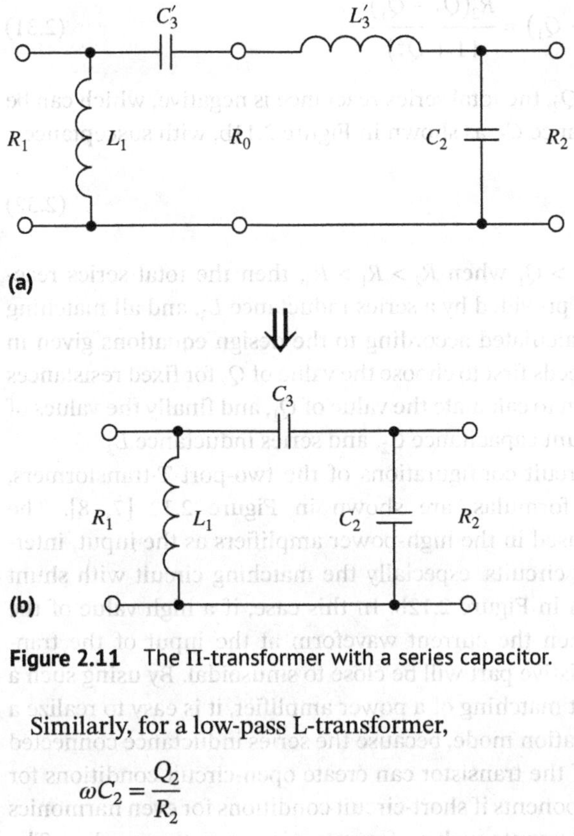

Figure 2.11 The Π-transformer with a series capacitor.

Similarly, for a low-pass L-transformer,

$$\omega C_2 = \frac{Q_2}{R_2} \tag{2.26}$$

$$\omega L_3' = Q_2 R_0 \tag{2.27}$$

$$Q_2^2 = \frac{R_2}{R_0} - 1 \tag{2.28}$$

Because it is assumed that $R_1 > R_2 > R_0$, from Eq. (2.25), it follows that the quality factor Q_1 of a high-pass L-transformer can be chosen from

$$Q_1^2 > \frac{R_1}{R_2} - 1 \tag{2.29}$$

Substituting Eq. (2.25) into Eq. (2.28) results in

$$Q_2 = \sqrt{\frac{R_2}{R_1}\left(1 + Q_1^2\right) - 1} \tag{2.30}$$

Combining the reactances of two series elements (capacitor C_3' and inductor C_3') given by Eqs. (2.24) and (2.27) yields

$$\omega L_3' - \frac{1}{\omega C_3'} = R_0(Q_2 - Q_1) = \frac{R_2(Q_2 - Q_1)}{(1 + Q_2^2)} \tag{2.31}$$

As a result, because $Q_1 > Q_2$, the total series reactance is negative, which can be provided by a series capacitance C_3, as shown in Figure 2.11b, with susceptance

$$\omega C_3 = \frac{1 + Q_2^2}{R_2(Q_1 - Q_2)} \tag{2.32}$$

On the other hand, if $Q_2 > Q_1$ when $R_2 > R_1 > R_0$, then the total series reactance is positive that can be provided by a series inductance L_3, and all matching circuit parameters can be calculated according to the design equations given in Figure 2.9c. In this case, it needs first to choose the value of Q_2 for fixed resistances R_1 and R_2 to be matched, then to calculate the value of Q_1, and finally the values of the shunt inductance L_1, shunt capacitance C_2, and series inductance L_3.

Some of the matching circuit configurations of the two-port T-transformers, together with the design formulas, are shown in Figure 2.12 [7, 8]. The T-transformers are usually used in the high-power amplifiers as the input, interstage, and output matching circuits, especially the matching circuit with shunt and series capacitors shown in Figure 2.12b. In this case, if a high value of the inductance L_2 is chosen, then the current waveform at the input of the transistor with a small input resistive part will be close to sinusoidal. By using such a T-transformer for the output matching of a power amplifier, it is easy to realize a high-efficiency Class-F operation mode, because the series inductance connected to the drain (or collector) of the transistor can create open-circuit conditions for higher-order harmonic components if short-circuit conditions for even harmonics are provided by additional elements such as, for example, a quarterwave line. The calculation of the matching circuit parameters of the T-transformers is similar to that for the Π-transformers, beginning from the estimation of the circuit quality factors Q_1 and Q_2.

The low-pass T-transformer with two series inductors, shown in Figure 2.12a, represents a back-to-back connection of two simple low-pass L-transformers. In this case, the resistance ratio R_1/R_2 can be greater or smaller than unity, similar to a Π-transformer with two shunt capacitors shown in Figure 2.9a, where the design equations also correspond to the case of $R_1/R_2 > 1$. However, the T-transformer with series and shunt capacitors shown in Figure 2.12b can only be used for impedance matching when $R_1/R_2 > 1$. Such a T-transformer represents a back-to-back connection of the high-pass and low-pass L-transformers, as shown in Figure 2.13a.

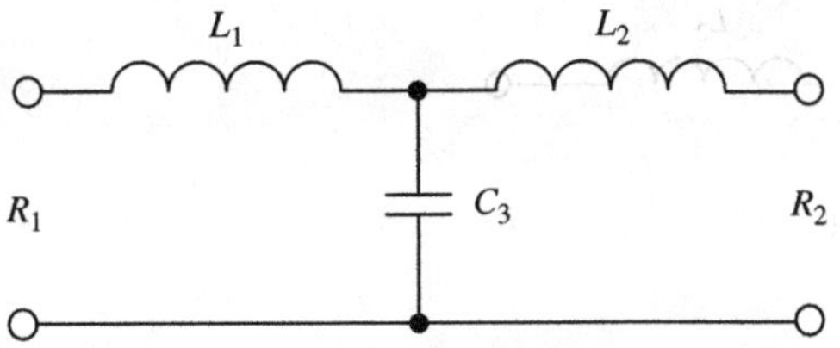

$\omega L_1 = Q_1 R_1$ $\quad \omega L_2 = Q_2 R_2$ $\quad \omega C_3 = (Q_1 + Q_2)/[R_2(1 + Q_2^2)]$

$$Q_1 = \sqrt{\frac{R_2}{R_1}(1 + Q_2^2) - 1} \qquad Q_2^2 > \frac{R_1}{R_2} - 1$$

(a)

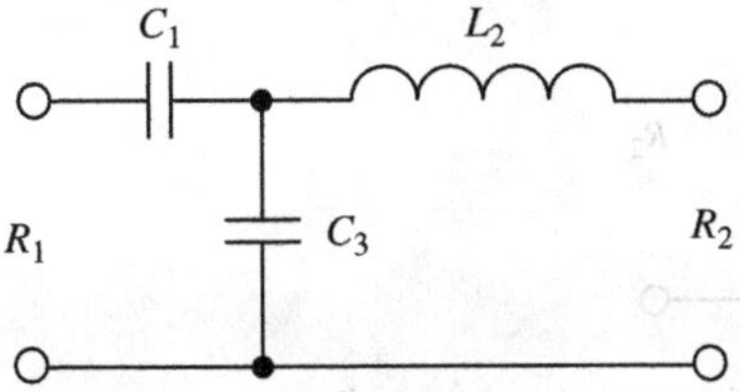

$\omega C_1 = 1 / R_1 Q_1$ $\quad \omega L_2 = Q_2 R_2$ $\quad \omega C_3 - (Q_2 - Q_1)/[R_2(1 + Q_2^2)]$

$$Q_1 = \sqrt{\frac{R_2}{R_1}(1 + Q_2^2) - 1} \qquad Q_2^2 > \frac{R_1}{R_2} - 1$$

(b)

$\omega C_1 = 1 / R_1 Q_1$ $\quad \omega L_2 = Q_2 R_2$ $\quad \omega L_3 = R_2(1 + Q_2^2)/(Q_1 - Q_2)$

$$Q_2 = \sqrt{\frac{R_1}{R_2}(1 + Q_1^2) - 1} \qquad Q_1^2 > \frac{R_2}{R_1} - 1$$

(c)

Figure 2.12 The T-transformers and relevant equations.

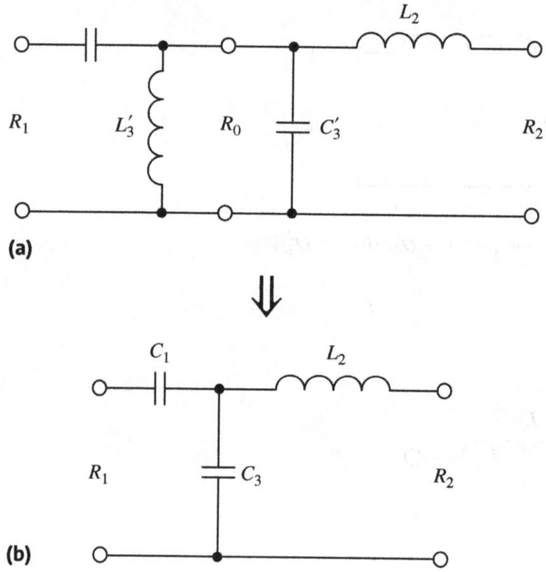

Figure 2.13 The T-transformer with series and shunt capacitors.

The design equations for a high-pass section of such a T-transformer are written using Eqs. (2.16)–(2.18) as

$$\omega C_1 = \frac{1}{R_1 Q_1} \tag{2.33}$$

$$\omega L_3' = \frac{R_0}{Q_1} \tag{2.34}$$

$$Q_1^2 = \frac{R_0}{R_1} - 1 \tag{2.35}$$

where Q_1 is the circuit quality factor and R_0 is the intermediate resistance.

Similarly, for a low-pass section,

$$\omega L_2 = Q_2 R_2 \tag{2.36}$$

$$\omega C_3' = Q_2 / R_0 \tag{2.37}$$

$$Q_2^2 = \frac{R_0}{R_2} - 1 \tag{2.38}$$

Because it is assumed that $R_0 > R_1 > R_2$, from Eq. (2.38), it follows that the quality factor Q_2 of a low-pass L-transformer can be chosen from the condition of

$$Q_2^2 > \frac{R_1}{R_2} - 1 \tag{2.39}$$

Substituting Eq. (2.38) into Eq. (2.35) results in

$$Q_1 = \sqrt{\frac{R_2}{R_1}\left(1 + Q_2^2\right) - 1} \tag{2.40}$$

Combining the susceptances of two shunt elements (inductor L_3' and capacitor C_3') given by Eqs. (2.34) and (2.37) yields

$$\omega C_3' - \frac{1}{\omega L_3'} = \frac{Q_2 - Q_1}{R_0} = \frac{Q_2 - Q_1}{R_2\left(1 + Q_2^2\right)} \tag{2.41}$$

As a result, because $Q_2 > Q_1$, the total shunt susceptance is positive, which can be provided by a shunt capacitance C_3, as shown in Figure 2.13b, with susceptance

$$\omega C_3 = \frac{Q_2 - Q_1}{R_2\left(1 + Q_2^2\right)} \tag{2.42}$$

On the other hand, if $Q_1 > Q_2$ when $R_0 > R_2 > R_1$, then the total shunt susceptance is negative that can be provided by a shunt inductance L_3, and all matching circuit parameters can be calculated according to the design equations given in Figure 2.12c. In this case, it needs first to choose the value of Q_1 for fixed resistances R_1 and R_2 to be matched, then to calculate the value of Q_2, and finally the values of the series capacitance C_1, series inductance L_2, and shunt inductance L_3.

If the elements of the Π- and T-transformers are chosen according to the condition $X_1 = X_2 = -X_3$, then the input impedance Z_{in} of the transformer loaded by the resistance R_L (from any side) is equal to

$$Z_{in} = R_{in} = \frac{X^2}{R_L} \tag{2.43}$$

where $X = |X_i|$, $i = 1, 2, 3$. As a result, the input impedance Z_{in} will be resistive, regardless of the value of the load resistance R_L. For example, setting $R_L = R_2$ for the transformers shown in Figures 2.9a and 2.12a results in

$$R_1 = \frac{X^2}{R_2} \tag{2.44}$$

When $X_1 \neq X_2 \neq -X_3$, the input impedances of the Π- and T-transformers will be resistive for only one particular value of R_L.

2.2.4 Equal-Q Broadband Matching

Application of each matching circuit in the form of an L-, Π-, or T-transformer significantly depends on the value and character (inductive or capacitive) of the device input and output impedances, level of the out-of-band intermodulation and harmonic suppression, and frequency bandwidth, which should be covered. Despite the Π- and T-transformers representing simple combinations of the low-pass and high-pass L-transformers, they are characterized by narrower frequency bandwidths than the L-transformers. However, there is a possibility to increase the operating frequency bandwidth by providing the ladder-type matching networks, representing the low-pass, high-pass, or a combination of them. In this case, the low-pass L-sections can improve the harmonic and high-frequency intermodulation suppression, whereas the high-pass L-section can provide better suppression of the low-frequency intermodulation components, thus improving the stability of the power amplifier operation.

Consider the design example of a 10-W 300-MHz bipolar power amplifier with a supply voltage of 12.5 V, providing a power gain of at least 10 dB. The first design step is to select an appropriate transistor that allows both simplifying the circuit design procedure and satisfying the specified requirements. Usually, the manufacturer states the values of the input and output impedances (or admittances) at the nominal operation point in the datasheet for the device. For example, the above requirements can be realized by the bipolar transistor operating at 300 MHz with the input impedance $Z_{in} = (1.3 + j0.9)\Omega$, which is intended for power amplification in a Class-AB mode.

In this case, the input impedance Z_{in} is expressed as a series combination of the transistor input resistance and inductive reactance. To match the series input inductive impedance to the standard 50-Ω source impedance at a certain frequency, it is convenient to use the matching circuit in the form of a T-transformer shown in Figure 2.12b, where the series capacitor C_1 can also serve as a blocking capacitor. Figure 2.14 shows the complete input network, including the input device elements and a low-pass T-type matching circuit. From the reactive part of the input impedance Z_{in}, it follows that, at the operating frequency of 300 MHz, the input inductance will be equal approximately to 0.5 nH.

It is necessary first to calculate the circuit quality factor Q_2, which is required to determine the parameters of a low-pass T-type matching circuit, resulting in

$$Q_2 > \sqrt{\frac{R_1}{R_{in}} - 1} = 6.1.$$

Consequently, the value of Q_2 must be larger than 6.1. For example, $Q_2 = 6.5$ provides a 3-dB bandwidth of 300 MHz/6.5 = 46 MHz. In this case, the value of Q_1 will be equal to 0.35. As a result, the values of the input matching circuit elements are

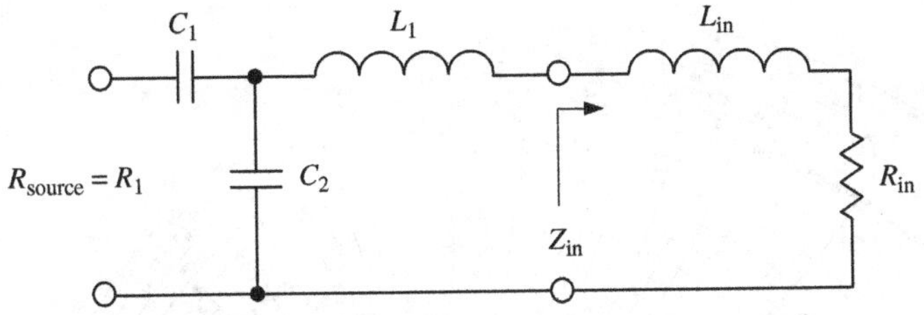

Figure 2.14 Complete input network circuit.

$$C_1 = \frac{1}{\omega Q_1 R_1} = 30\,\text{pF}$$

$$L_1 + L_{\text{in}} = \frac{Q_2 R_{\text{in}}}{\omega} = 4.5\,\text{nH} \Rightarrow L_1 = 4.0\,\text{nH}$$

$$C_2 = \frac{Q_2 - Q_1}{\omega R_{\text{in}}\left(1 + Q_2^2\right)} = 59\,\text{pF}.$$

This type of low-pass T-transformer is widely used in practical matching circuit design because of its design simplicity and implementation convenience. In addition, the sufficiently small value of the series capacitance C_1 can contribute to the elimination of the low-frequency parasitic oscillations in the case of a multistage power amplifier. For visualization, the function of each element can be graphically traced on the Smith chart as shown in Figure 2.15.

The easiest and most convenient way to plot the traces of the matching circuit elements is by initially plotting the traces of $Q_2 = 6.5$ and $Q_1 = 0.35$. The semicircle of equal Q is plotted by considering that, for each point located on this circle, a ratio of X/R or B/G must be the same. The trace of the series inductance L_1 must be plotted as far as the intersection point with Q_2-semicircle. This means that, beginning at Z_{in}, a curve of increasing inductive reactance must be plotted as far as Q_2-semicircle. The value of L_1 is determined from the normalized inductive impedance at this intersection point. Then, because of normalization to 50 Ω, the chart value must be multiplied by the factor of 50. The trace of the parallel capacitance C_2 must be plotted using admittance semicircles. The previous impedance point located at Q_2-semicircle is converted to its appropriate admittance one. This point is symmetrical to the impedance point relative to the center point and is located on a straight line from the intersection point drawn through the center of the Smith chart into its lower half at the same distance from the center point. A curve from this point with constant conductance and increasing capacitive susceptance is plotted. These points are transformed to appropriate impedances using a line through the center point extended at equal distance on the other side and stop when the transformed curve reaches the Q_1-semicircle. In other words, it

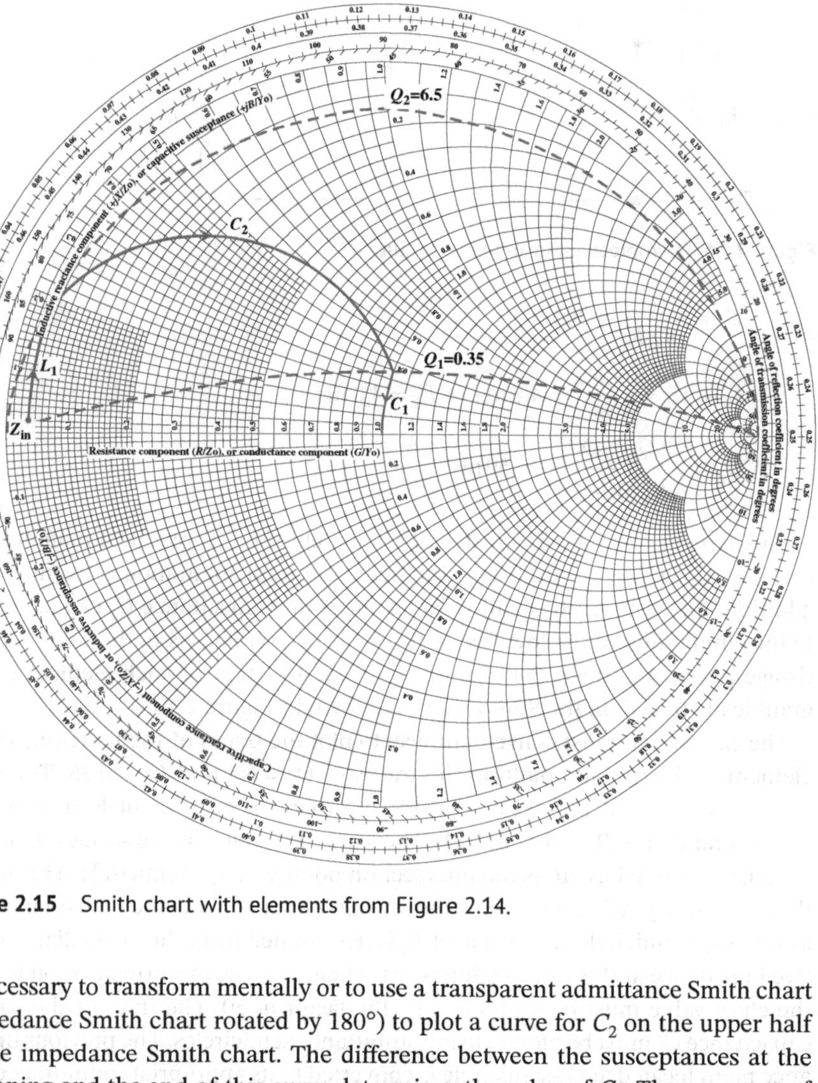

Figure 2.15 Smith chart with elements from Figure 2.14.

is necessary to transform mentally or to use a transparent admittance Smith chart (impedance Smith chart rotated by 180°) to plot a curve for C_2 on the upper half of the impedance Smith chart. The difference between the susceptances at the beginning and the end of this curve determines the value of C_2. Then, a curve of reducing inductive reactance is plotted down to the center point in order to determine the value of the series capacitor C_1.

Now let us demonstrate a lumped matching circuit technique to design a 150-W MOSFET power amplifier with the supply voltage of 50 V, operating in a frequency bandwidth of 132–174 MHz and providing a power gain greater than 10 dB. These requirements can be satisfied using a silicon MOSFET device designed for power amplification in the VHF frequency range. In this case, the center bandwidth frequency is equal to $f_c = \sqrt{132 \times 174} = 152\,\text{MHz}$. For this frequency, the manufacturer states the following values of the input and output impedances: $Z_{in} = (0.9 - j1.2)\Omega$ and $Z_{out} = (1.8 + j2.1)\Omega$. Both Z_{in} and Z_{out} represent the series combination of an input or output resistance with a capacitive or inductive reactance, respectively. To cover the required frequency bandwidth, the low-Q matching circuits should be used that allows reduction of the in-band amplitude ripple and improvement of the input voltage standing wave ratio (VSWR). The value of a quality factor for 3-dB bandwidth level must be less than $Q = 152/(174 - 132) = 3.6$. As a result, it is very convenient to design the input and output matching circuits using the simple L-transformers in the form of the low-pass and high-pass matching sections with a constant value of Q [9].

To match the input series capacitive impedance to the standard 50-Ω source impedance in a sufficiently wide frequency bandwidth, it is preferable to use the three L-type matching sections (high-pass section as the first one followed by two low-pass sections), as shown in Figure 2.16. From the negative reactive part of the input impedance Z_{in}, it follows that the input capacitance at the operating frequency of 152 MHz is equal to approximately 873 pF. To compensate for this capacitive reactance at the center bandwidth frequency, it is sufficient to connect an inductance of 1.3 nH in series to it. Now, when the device input capacitive reactance is compensated, to simplify the matching design procedure, it is best to cascade the L-transformers with equal value of Q. Although equal-Q values are not absolutely necessary, this provides a convenient guide for both analytical calculation of the matching circuit parameters and the Smith chart graphical design.

In this case, the following ratio can be written for the input matching circuit:

$$\frac{R_1}{R_2} = \frac{R_2}{R_3} = \frac{R_3}{R_{in}} \tag{2.45}$$

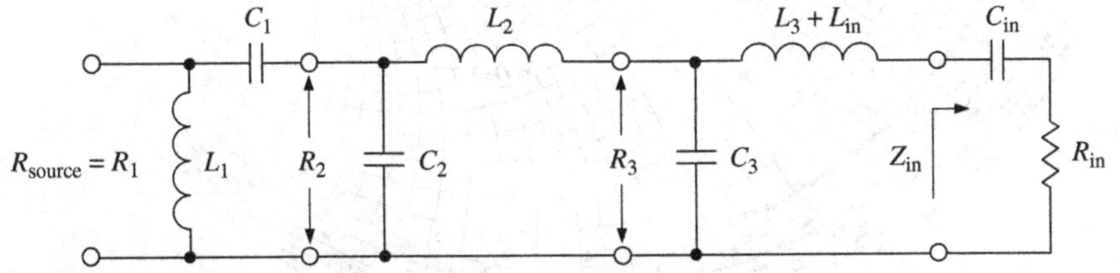

Figure 2.16 Complete broadband input matching circuit.

resulting in $R_2 = 13\,\Omega$ and $R_3 = 3.5\,\Omega$ for $R_{source} = R_1 = 50\,\Omega$ and $R_{in} = 0.9\,\Omega$. Consequently, the quality factor of each L-transformer is equal to $Q = 1.7$ according to Eq. (2.18). The elements of the input matching circuit using the formulas given in Figure 2.6 can be calculated as $L_1 = 31$ nH, $C_1 = 47$ pF, $L_2 = 6.2$ nH, $C_2 = 137$ pF, $L_3 = 1.6$ nH, and $C_3 = 509$ pF.

This equal-Q approach significantly simplifies the matching circuit design when using the Smith chart. After calculating the value of Q, it is necessary to plot a semi-circle of equal Q values on the Smith chart. Then, each element of the input matching circuit can be readily determined, as shown in Figure 2.17. Each trace for the series inductance must be plotted until the intersection point with Q-semicircle,

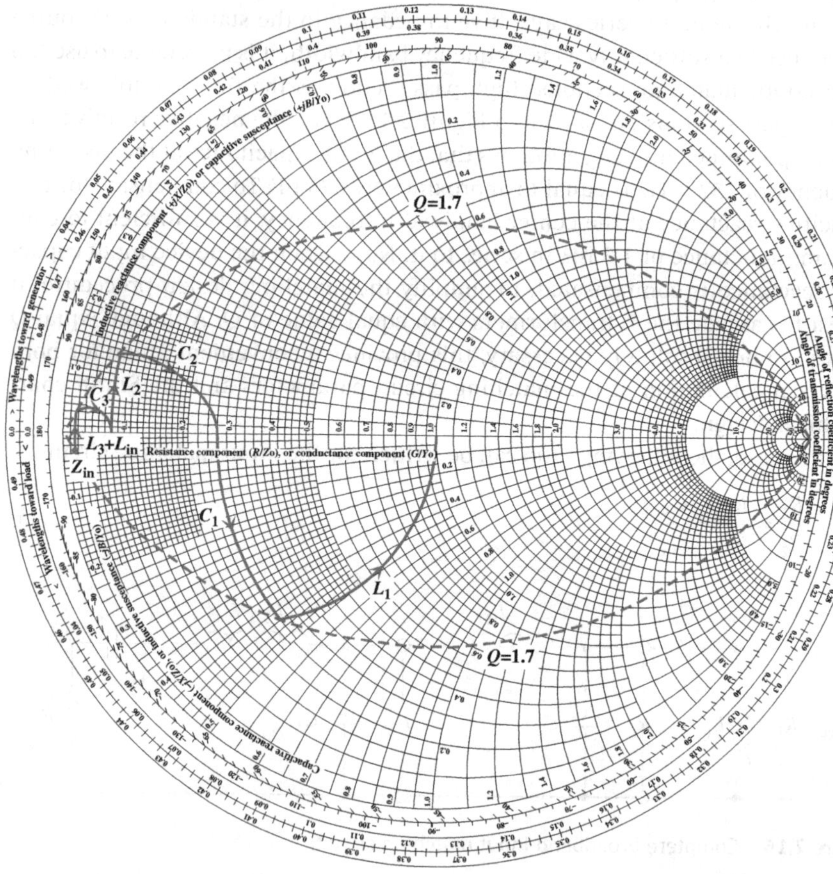

Figure 2.17 Smith chart with elements from Figure 2.16.

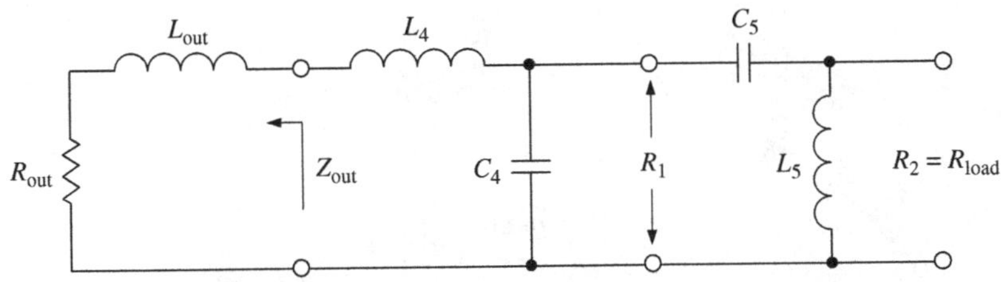

Figure 2.18 Complete broadband output network circuit.

whereas each trace for the parallel capacitance should be plotted until intersection with the horizontal real axis.

To match the output series inductive impedance, which is usually the case for the packaged high-power device due to its output lead, to the standard 50-Ω load impedance, it is sufficient to use only two L-type matching sections (low-pass and high-pass), as shown in Figure 2.18. At the operating frequency of 152 MHz, the transistor series output inductance is equal to approximately 2.2 nH. This inductance can be used as part of the low-pass L-type matching section. For an output matching circuit, the condition of equal-Q values results in

$$\frac{R_2}{R_1} = \frac{R_1}{R_{out}} \tag{2.46}$$

with the value $R_1 = 9.5\,\Omega$ for $R_{load} = R_2 = 50\,\Omega$ and $R_{out} = 1.8\,\Omega$. Consequently, a quality factor of each L-transformer is equal to $Q = 2.1$, which is substantially smaller than the value of Q for 3-dB bandwidth level. Now it is necessary to check the value of a series inductance of the low-pass section, which must exceed the value of 2.2 nH for correct matching procedure. The appropriate calculation gives a value of the total series inductance $L_4 + L_{out}$ of approximately 4 nH. As a result, the values of the output matching circuit elements are $L_4 = 1.8$ nH, $C_4 = 231$ pF, $C_5 = 52$ pF, and $L_5 = 25$ nH.

The output matching circuit design using the Smith chart with constant Q-semicircles is shown in Figure 2.19. For the final high-pass L-type matching section, a trace for the series capacitance C_5 must be plotted until the intersection with $Q = 2.1$ semicircle, whereas a trace for the shunt inductance L_5 should be plotted until the intersection with the center point of the Smith chart.

2.3 Matching with Transmission Lines

At very high frequencies, it is very difficult to implement lumped elements with predefined accuracy in view of a significant effect of their parasitic parameters, for example, the parasitic interturn and direct-to-ground capacitances for lumped

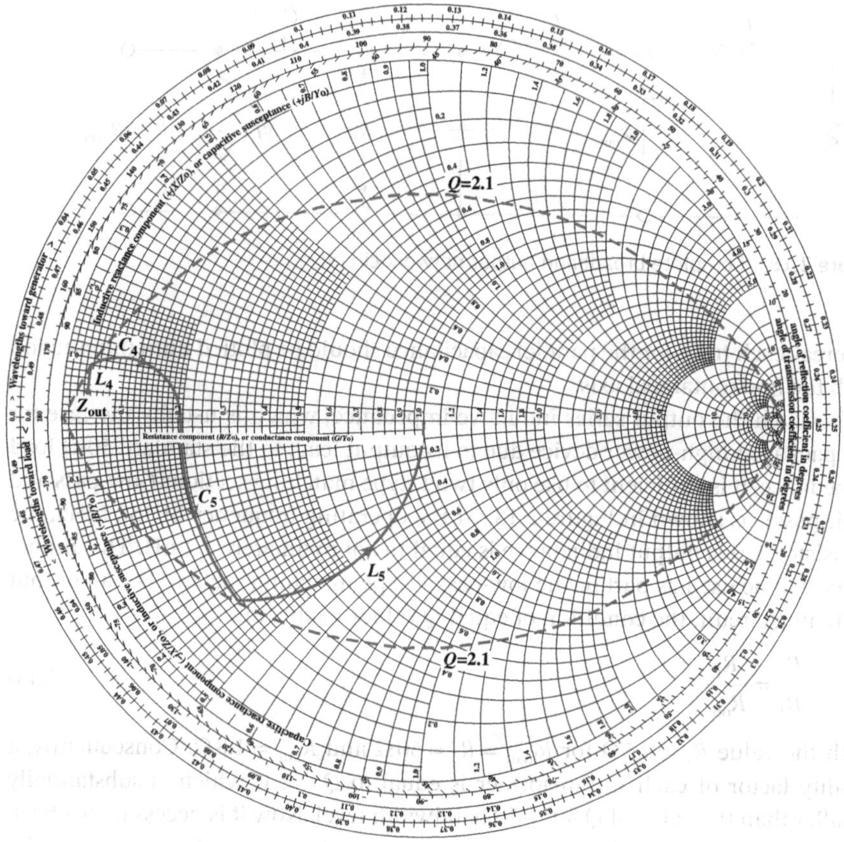

Figure 2.19 Smith chart with elements from Figure 2.18.

inductors and the stray inductance for lumped capacitors. However, these parasitic parameters become a part of a distributed *LC* structure such as a transmission line [3]. In this case, for a microstrip line, the series inductance is associated with the flow of current in the conductor, and the shunt capacitance is associated with the strip separated from the ground by the dielectric substrate. If the microstrip line is wide, the inductance is reduced, but the capacitance is large. However, for a narrow microstrip line, the inductance is increased, but the capacitance is small.

2.3.1 Transmission-line Transformer

Figure 2.20 shows the impedance-matching circuit in the form of a transmission-line transformer connected between the source impedance Z_S and load

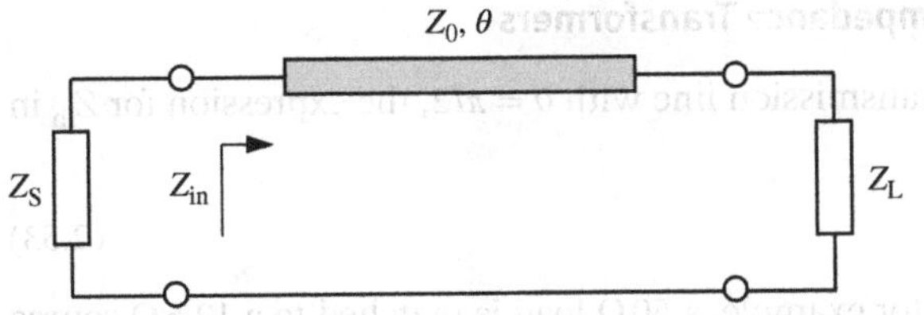

Figure 2.20 Transmission-line impedance transformer.

impedance Z_L. The input impedance as a function of the length of the transmission line with arbitrary load impedance is

$$Z_{in} = Z_0 \frac{Z_L + jZ_0 \tan\theta}{Z_0 + jZ_L \tan\theta} \tag{2.47}$$

where Z_0 is the characteristic impedance, $\theta = \beta l$ is the electrical length of the transmission line, $\beta = \omega\sqrt{\mu_r \varepsilon_r}/c$ is the phase constant, c is the speed of light in free space, μ_r is the substrate permeability, ε_r is the substrate permittivity, ω is the angular frequency, and l is the geometrical length of the transmission line [10, 11].

To provide a complex conjugate matching of the input transmission-line impedance Z_{in} with the source impedance $Z_S = R_S + jX_S$ when $R_S = \text{Re}Z_{in}$ and $X_S = -\text{Im}Z_{in}$, Eq. (2.47) can be rewritten as

$$R_S - jX_S = Z_0 \frac{R_L + j(X_L + Z_0 \tan\theta)}{Z_0 - X_L \tan\theta + jR_L \tan\theta} \tag{2.48}$$

Generally, Eq. (2.48) can be divided into two equations representing the real and imaginary parts, respectively,

$$R_S(Z_0 - X_L \tan\theta) - R_L(Z_0 - X_S \tan\theta) = 0 \tag{2.49}$$

$$X_S(X_L \tan\theta - Z_0) - Z_0(X_L + Z_0 \tan\theta) + R_S R_L \tan\theta = 0 \tag{2.50}$$

Solving Eqs. (2.49) and (2.50) for the two independent variables Z_0 and θ yields

$$Z_0 = \sqrt{\frac{R_S\left(R_L^2 + X_L^2\right) - R_L\left(R_S^2 + X_S^2\right)}{R_L - R_S}} \tag{2.51}$$

$$\theta = \tan^{-1}\left(Z_0 \frac{R_S - R_L}{R_S X_L - X_S R_L}\right) \tag{2.52}$$

As a result, the transmission line with the characteristic impedance Z_0, determined by Eq. (2.51), and the electrical length θ, determined by Eq. (2.52), can match any source and load impedances when the impedance ratio gives a positive value inside the square root in Eq. (2.51).

2.3.2 The $\lambda/4$ and $\lambda/8$ Impedance Transformers

For a quarter-wavelength transmission line with $\theta = \pi/2$, the expression for Z_{in} in Eq. (2.47) is simplified to

$$Z_{in} = Z_0^2/Z_L \tag{2.53}$$

from which it follows that, for example, a 50 Ω load is matched to a 12.5 Ω source with the characteristic impedance of 25 Ω. Usually, such a quarterwave impedance transformer is used for impedance matching in a narrow frequency bandwidth of 10–20%, and its length is chosen at the bandwidth center frequency. However, using a multi-section quarterwave transformer widens the bandwidth and expands the choice of the substrate to include materials with high dielectric permittivity, which reduces the transformer's size.

In the case of a quarterwave transformer, Eq. (2.48) can be divided into two separate equations representing the real and imaginary parts of the source impedance Z_S as

$$R_S = Z_0^2 \frac{R_L}{R_L^2 + X_L^2} \tag{2.54}$$

$$X_S = -Z_0^2 \frac{X_L}{R_L^2 + X_L^2} \tag{2.55}$$

For a purely real load impedance with $X_L = 0$, a quarterwave transmission line with the characteristic impedance Z_0 can provide impedance matching for a purely active source and load only when

$$Z_0 = \sqrt{R_S R_L} \tag{2.56}$$

However, for a particular case of a purely active source when $Z_S = R_S$, the ratio between the load and transmission-line parameters can be expressed by

$$X_L Z_0 (1 - \tan^2\theta) + (Z_0^2 - X_L^2 - R_L^2)\tan\theta = 0 \tag{2.57}$$

Then, for the electrical length of the transmission line $\theta = \pi/4$ or 45°, the expression for the characteristic impedance Z_0 given by Eq. (2.51) can be simplified to

$$Z_0 = |Z_L| = \sqrt{R_L^2 + X_L^2} \tag{2.58}$$

whereas the required real source impedance R_S should be equal to

$$R_S = R_L \frac{Z_0}{Z_0 - X_L} \tag{2.59}$$

Consequently, any load impedance can be transformed to a real source impedance defined by Eq. (2.59) using a $\lambda/8$ transformer, the characteristic impedance of which is equal to the magnitude of the load impedance [12].

Applying the same approach to match purely resistive load with the source impedance, the total matching circuit that includes two $\lambda/8$ transformers and a $\lambda/4$ transformer can provide impedance matching between any complex source impedance Z_S and load impedance Z_L, as shown in Figure 2.21.

2.3.3 Low-pass L- and Π-transformers with Series Transmission Line

In practice, to simplify the power amplifier design at microwave frequencies, the simple matching circuits are very often used, including a low-pass L-transformer with a series transmission line as the basic matching section. It is convenient to analyze the transforming properties of this matching circuit by substituting the equivalent transformation of the parallel *RX* circuit into the series one. For example, R_1 is the resistance and $X_1 = -1/\omega C$ is the reactance of the impedance $Z_1 = jR_1X_1/(R_1 + jX_1)$ for a parallel *RC* circuit, and $R_{in} = \text{Re}Z_{in}$ is the resistance and $X_{in} = \text{Im}Z_{in}$ is the reactance of the impedance $Z_{in} = R_{in} + jX_{in}$ for the series transmission-line circuit loaded by the resistance R_2, as shown in Figure 2.22.

In this case, for a conjugate matching when $Z_1 = Z_{in}^*$,

$$\frac{R_1X_1^2}{R_1^2 + X_1^2} + j\frac{R_1^2X_1}{R_1^2 + X_1^2} = R_{in} - jX_{in} \tag{2.60}$$

The solution of Eq. (2.60) can be obtained in the form of the two expressions for real and imaginary impedance parts as

$$R_1 = R_{in}\left(1 + Q^2\right) \tag{2.61}$$

$$X_1 = -X_{in}\left(1 + Q^{-2}\right) \tag{2.62}$$

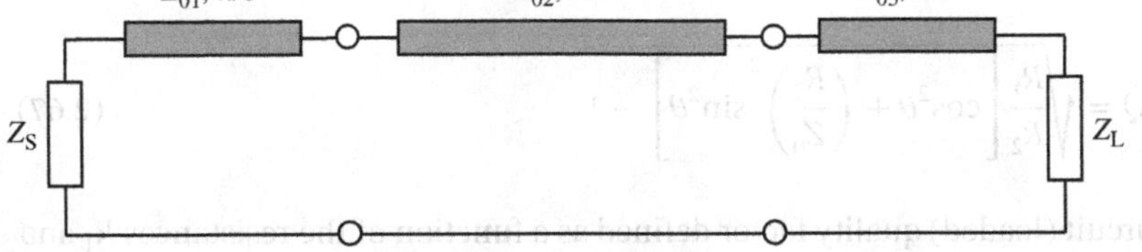

Figure 2.21 Transmission-line transformer for any source and load impedances.

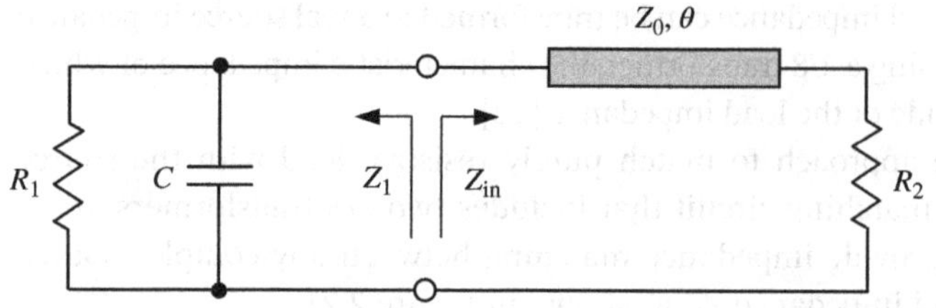

Figure 2.22 Low-pass L-transformer with series transmission line.

where $Q = R_1/|X_1| = X_{in}/R_{in}$ is the quality factor equal for both parallel capacitive and series transmission-line circuits. By using Eq. (2.47), the real and imaginary parts of the input impedance Z_{in} can be written as

$$R_{in} = Z_0^2 R_2 \frac{1 + \tan^2\theta}{Z_0^2 + (R_2 \tan\theta)^2} \tag{2.63}$$

$$X_{in} = Z_0 \tan\theta \frac{Z_0^2 - R_2^2}{Z_0^2 + (R_2 \tan\theta)^2} \tag{2.64}$$

From Eq. (2.64), it follows that an inductive input impedance, which is necessary to compensate for the capacitive parallel component, is provided when $Z_0 > R_2$ for $\theta < \pi/2$ and $Z_0 < R_2$ for $\pi/2 < \theta < \pi$. As a result, to transform the resistance R_1 into the other resistance R_2 at a certain frequency, it is necessary to connect a low-pass L-transformer (including a parallel capacitor and a series transmission line) between them. When one parameter (usually the characteristic impedance Z_0) is known in advance, the remaining matching circuit parameters can be calculated from the following two equations:

$$C = \frac{Q}{\omega R_1} \tag{2.65}$$

$$\sin 2\theta = \frac{2Q}{\frac{Z_0}{R_2} - \frac{R_2}{Z_0}} \tag{2.66}$$

where

$$Q = \sqrt{\frac{R_1}{R_2}\left[\cos^2\theta + \left(\frac{R_2}{Z_0}\right)^2 \sin^2\theta\right] - 1} \tag{2.67}$$

is the circuit (loaded) quality factor defined as a function of the resistances R_1 and R_2 and the parameters of the transmission line (characteristic impedance Z_0 and electrical length θ).

It follows from Eqs. (2.66) and (2.67) that the electrical length θ can be calculated as a result of the numerical solution of a transcendental equation with one unknown parameter. However, it is more convenient to combine these two equations and to rewrite them in the implicit form as

$$\frac{R_1}{R_2} = \frac{1 + \left(\frac{Z_0}{R_2} - \frac{R_2}{Z_0}\right)^2 \sin^2\theta\cos^2\theta}{\cos^2\theta + \left(\frac{R_2}{Z_0}\right)^2 \sin^2\theta} \tag{2.68}$$

Figure 2.23 shows the resistance ratio of R_1/R_2 as a function of the normalized parameter Z_0/R_2 and electrical length θ in the form of two nomographs: for the case of $Z_0/R_2 > 0$ shown in Figure 2.23a and for the case of $Z_0/R_2 < 0$ shown in Figure 2.23b. When the input resistance R_1 and load resistance R_2 are known in advance, it is easy to evaluate the required value of θ for a fixed transmission-line characteristic impedance Z_0 using these two nomographs. The graphical results show that, in contrast to the lumped low-pass L-transformer, the transmission-line low-pass L-transformer can match purely resistive source and load impedances with any ratio of R_1/R_2.

A low-pass transmission-line Π-transformer can be realized by back-to-back connection of the two low-pass transmission-line L-transformers, as shown in Figure 2.24a, where the resistances R_1 and R_2 are transformed into some intermediate resistance R_0. In this case, to minimize the length of a transmission line, the value of R_0 should be smaller than that of both R_1 and R_2, $R_0 < (R_1, R_2)$. The same procedure for a low-pass transmission-line T-transformer shown in Figure 2.24b gives a value of R_0 that is larger than that of both R_1 and R_2, $R_0 > (R_1, R_2)$. Then, for a low-pass T-transformer, two shunt adjacent capacitors are combined, whereas for a low-pass Π-transformer, two adjacent series transmission lines are combined into a single transmission line with total electrical length.

For a Π-transformer, the lengths of each part of the combined transmission line can be calculated by equating the imaginary parts of the impedances from both sides at the reference plane A–A' to zero, which means that the intermediate impedance R_0 is real. This leads to two quadratic equations to calculate the electrical lengths θ_1 and θ_2 of the combined series transmission line written as

$$\tan^2\theta_1 - \frac{R_1}{Z_0 Q_1}\left[1 - (1 + Q_1^2)\left(\frac{Z_0}{R_1}\right)^2\right]\tan\theta_1 - 1 = 0 \tag{2.69}$$

$$\tan^2\theta_2 - \frac{R_2}{Z_0 Q_2}\left[1 - (1 + Q_2^2)\left(\frac{Z_0}{R_2}\right)^2\right]\tan\theta_2 - 1 = 0 \tag{2.70}$$

where $Q_1 = \omega C_1 R_1$ and $Q_2 = \omega C_2 R_2$.

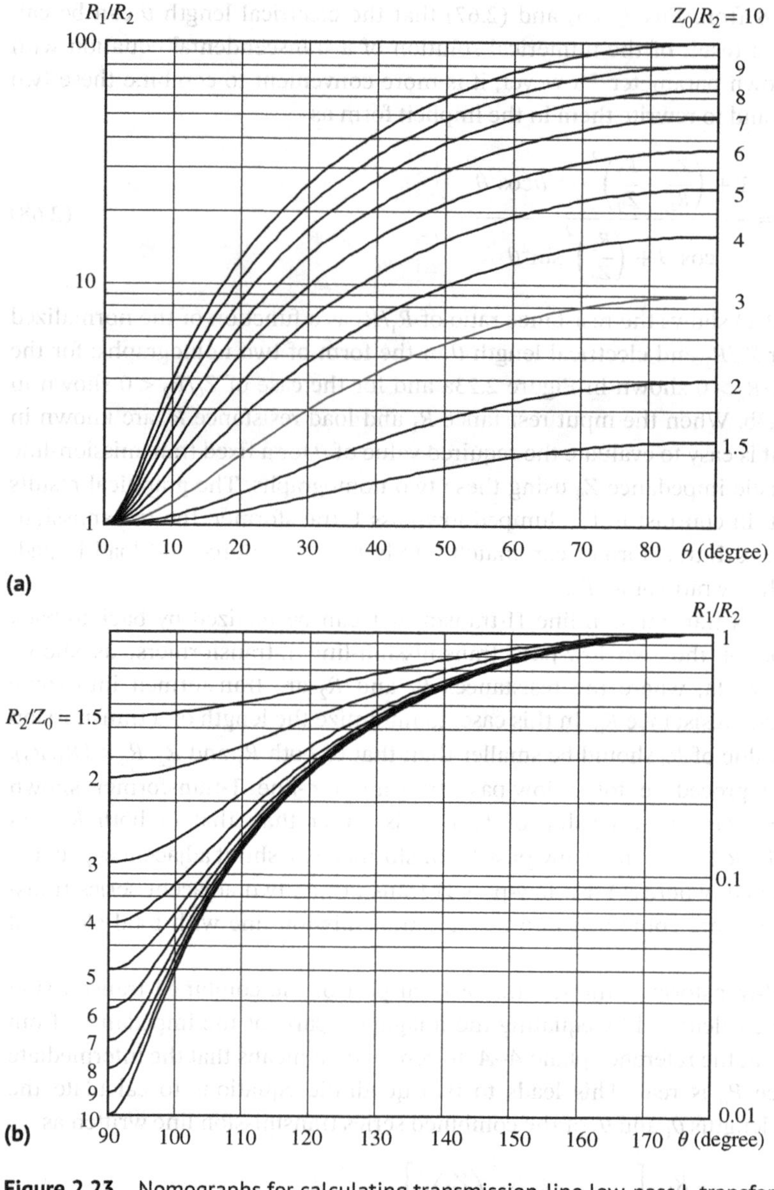

Figure 2.23 Nomographs for calculating transmission-line low-pass L-transformer.

However, to simplify the matching circuit design procedure, it is very helpful to use the nomographs shown in Figure 2.23. If the values of R_1 and R_2 are selected in advance to set the intermediate resistance R_0 and the characteristic impedance of

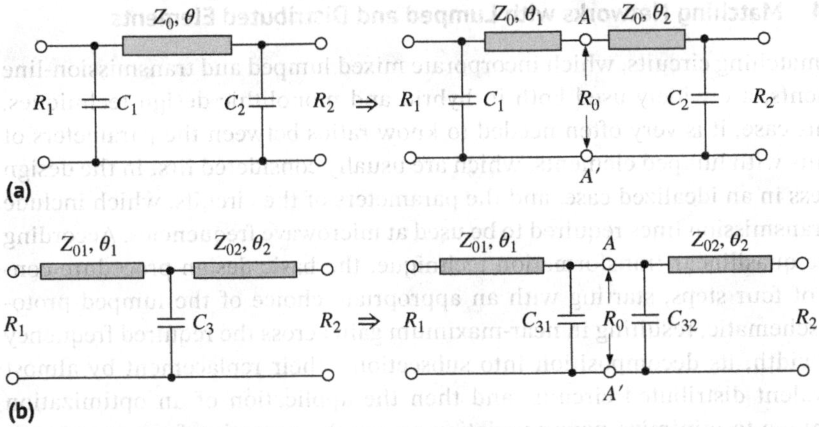

Figure 2.24 Low-pass Π- and T-transformers with transmission lines.

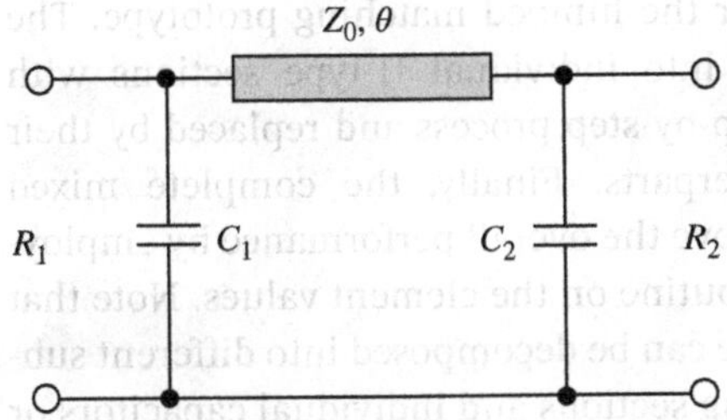

$$\omega C_1 = \frac{Q}{R_1} \qquad \omega C_2 = \frac{Q}{R_2} \qquad \theta = \theta_1 + \theta_2$$

$$Q_2 = \sqrt{\frac{R_2}{R_1}\left(1 + Q_1^2\right) - 1} \qquad Q_1 > \sqrt{\frac{R_1}{R_2} - 1}$$

Figure 2.25 Low-pass transmission-line Π-transformer and relevant equations.

the transmission line Z_0 is known, the values of θ_1 and θ_2 can be easily determined from one of these nomographs.

A widely used low-pass Π-transformer with two shunt capacitors along with its design formulas is shown in Figure 2.25 [13]. Such a transformer can be conveniently used as an output matching circuit when the device collector (or drain–source) capacitance can be considered as a first shunt capacitance and the parasitic series lead inductance can easily be added in series to the transmission line. In addition, it is convenient to use this transformer as the matching circuits in balanced power amplifiers where the shunt capacitors can be connected between the two series transmission lines due to the effect of virtual grounding.

2.3.4 Matching Networks with Lumped and Distributed Elements

The matching circuits, which incorporate mixed lumped and transmission-line elements, are widely used both in hybrid and monolithic design techniques. In this case, it is very often needed to know ratios between the parameters of circuits with lumped elements, which are usually considered first in the design process in an idealized case, and the parameters of the circuits, which include the transmission lines required to be used at microwave frequencies. According to the quasilinear transformation technique, the basic design procedure consists of four steps, starting with an appropriate choice of the lumped prototype schematic, resulting in near-maximum gain across the required frequency bandwidth, its decomposition into subsections, their replacement by almost equivalent distributed circuits, and then the application of an optimization technique to minimize power variation across the operating frequency bandwidth [14].

Consider a periodic lumped *LC* structure in the form of a low-pass ladder Π-network, which is used as a basis for the lumped matching prototype. The lumped prototype should be split up into individual Π-type sections with equal capacitances by a consecutive step-by-step process and replaced by their equivalent distributed network counterparts. Finally, the complete mixed matching structure is optimized to improve the overall performance by employing a standard nonlinear optimization routine on the element values. Note that generally the lumped prototype structure can be decomposed into different subnetworks, including also L-type matching sections and individual capacitors or inductors.

To define a single-frequency equivalence between low-pass Π-transformers using both lumped and distributed elements, it is convenient to use two-port transmission *ABCD*-parameters. The *ABCD*-matrices of these low-pass Π-transformers shown in Figure 2.26a can respectively be given by

$$\begin{aligned}[ABCD]_{\mathrm{L}} &= \begin{bmatrix} 1 & 0 \\ j\omega C & 1 \end{bmatrix} \begin{bmatrix} 1 & j\omega L \\ 0 & 1 \end{bmatrix} \begin{bmatrix} 1 & 0 \\ j\omega C & 1 \end{bmatrix} \\ &= \begin{bmatrix} 1-\omega^2 LC & j\omega L \\ j\omega C(2-\omega^2 LC) & 1-\omega^2 LC \end{bmatrix}\end{aligned} \tag{2.71}$$

$$\begin{aligned}[ABCD]_{\mathrm{T}} &= \begin{bmatrix} 1 & 0 \\ j\omega C & 1 \end{bmatrix} \begin{bmatrix} \cos\theta & jZ_0\sin\theta \\ j\dfrac{\sin\theta}{Z_0} & \cos\theta \end{bmatrix} \begin{bmatrix} 1 & 0 \\ j\omega C & 1 \end{bmatrix} \\ &= \begin{bmatrix} \cos\theta - \omega C_{\mathrm{T}} Z_0 \sin\theta & jZ_0\sin\theta \\ \dfrac{j}{Z_0}\left(2\omega C_{\mathrm{T}} Z_0\cos\theta + \sin\theta - \omega^2 C_{\mathrm{T}}^2 Z_0^2 \sin\theta\right) & \cos\theta - \omega C_{\mathrm{T}} Z_0 \sin\theta \end{bmatrix}\end{aligned} \tag{2.72}$$

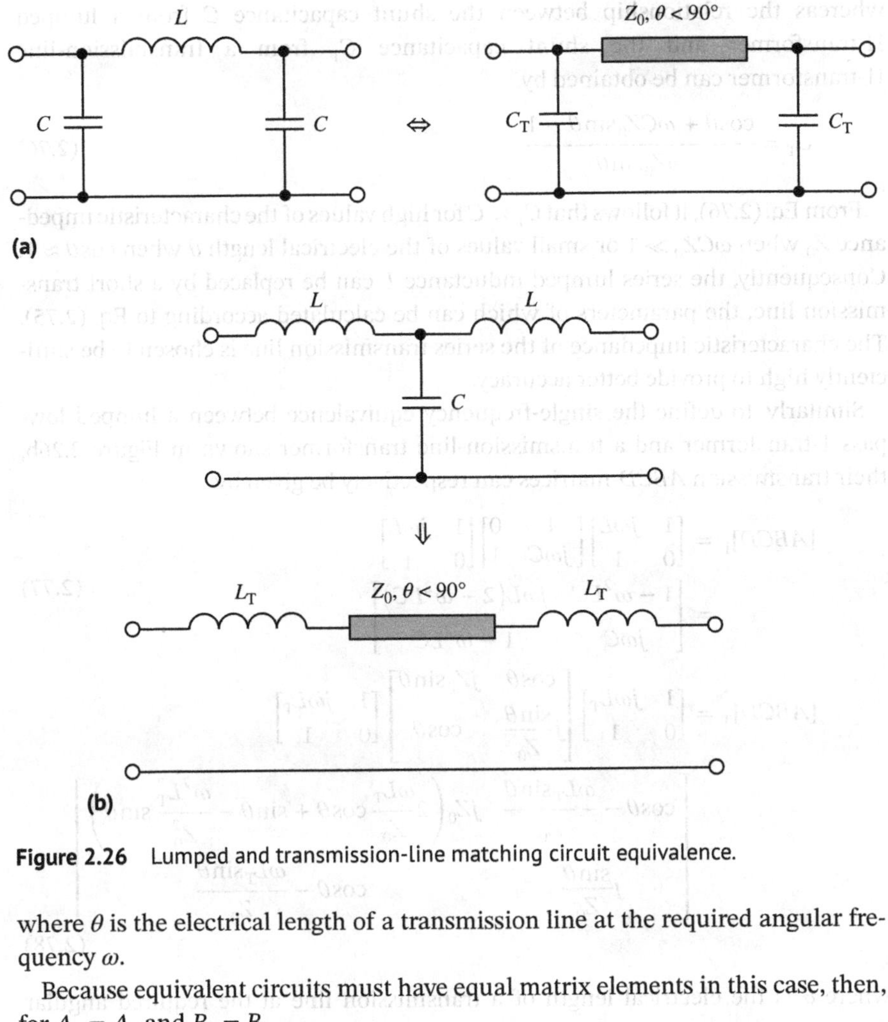

Figure 2.26 Lumped and transmission-line matching circuit equivalence.

where θ is the electrical length of a transmission line at the required angular frequency ω.

Because equivalent circuits must have equal matrix elements in this case, then, for $A_L = A_T$ and $B_L = B_T$,

$$1 - \omega^2 LC = \cos\theta - \omega C_T Z_0 \sin\theta \tag{2.73}$$

$$j\omega L = jZ_0 \sin\theta \tag{2.74}$$

As a result, the equivalent series inductance L can be expressed through the parameters of the transmission line as

$$L = \frac{Z_0 \sin\theta}{\omega} \tag{2.75}$$

whereas the relationship between the shunt capacitance C from a lumped Π-transformer and the shunt capacitance C_T from a transmission-line Π-transformer can be obtained by

$$C_T = \frac{\cos\theta + \omega C Z_0 \sin\theta - 1}{\omega Z_0 \sin\theta} \tag{2.76}$$

From Eq. (2.76), it follows that $C_T \approx C$ for high values of the characteristic impedance Z_0 when $\omega C Z_0 \gg 1$ or small values of the electrical length θ when $\cos\theta \approx 1$. Consequently, the series lumped inductance L can be replaced by a short transmission line, the parameters of which can be calculated according to Eq. (2.75). The characteristic impedance of the series transmission line is chosen to be sufficiently high to provide better accuracy.

Similarly, to define the single-frequency equivalence between a lumped low-pass T-transformer and a transmission-line transformer shown in Figure 2.26b, their transmission *ABCD*-matrices can respectively be given by

$$\begin{aligned}[ABCD]_L &= \begin{bmatrix} 1 & j\omega L \\ 0 & 1 \end{bmatrix} \begin{bmatrix} 1 & 0 \\ j\omega C & 1 \end{bmatrix} \begin{bmatrix} 1 & j\omega L \\ 0 & 1 \end{bmatrix} \\ &= \begin{bmatrix} 1 - \omega^2 LC & j\omega L\left(2 - \omega^2 LC\right) \\ j\omega C & 1 - \omega^2 LC \end{bmatrix} \end{aligned} \tag{2.77}$$

$$\begin{aligned}[ABCD]_T &= \begin{bmatrix} 1 & j\omega L_T \\ 0 & 1 \end{bmatrix} \begin{bmatrix} \cos\theta & jZ_0 \sin\theta \\ j\dfrac{\sin\theta}{Z_0} & \cos\theta \end{bmatrix} \begin{bmatrix} 1 & j\omega L_T \\ 0 & 1 \end{bmatrix} \\ &= \begin{bmatrix} \cos\theta - \dfrac{\omega L_T \sin\theta}{Z_0} & jZ_0\left(2\dfrac{\omega L_T}{Z_0}\cos\theta + \sin\theta - \dfrac{\omega^2 L_T^2}{Z_0^2}\sin\theta\right) \\ j\dfrac{\sin\theta}{Z_0} & \cos\theta - \dfrac{\omega L_T \sin\theta}{Z_0} \end{bmatrix} \end{aligned} \tag{2.78}$$

where θ is the electrical length of a transmission line at the required angular frequency ω.

For equivalent matrix elements $A_L = A_T$ and $C_L = C_T$,

$$1 - \omega^2 LC = \cos\theta - \frac{\omega L_T}{Z_0}\sin\theta \tag{2.79}$$

$$j\omega C = j\frac{\sin\theta}{Z_0} \tag{2.80}$$

As a result, the equivalent shunt capacitance C can be expressed through the parameters of the transmission line as

$$C = \frac{\sin\theta}{\omega Z_0} \tag{2.81}$$

whereas the relationship between the series inductance L from a lumped low-pass T-transformer and the series inductance L_T from a transmission-line transformer can be obtained by

$$L_T = \frac{\cos\theta + \frac{\omega L}{Z_0}\sin\theta - 1}{\frac{\omega}{Z_0}\sin\theta} \tag{2.82}$$

From Eq. (2.82), it follows that $L_T \approx L$ for small values of the characteristic impedance Z_0 when $\omega L/Z_0 \gg 1$ or small values of the electrical length θ when $\cos\theta \approx 1$. Consequently, the shunt lumped capacitor C can be replaced by a short transmission line, the parameters of which can be calculated according to Eq. (2.81). In this case, the characteristic impedance of the series transmission line is chosen to be sufficiently low to provide better accuracy.

In order to provide the design method using a single-frequency equivalent technique, the following consecutive design steps applied to the low-pass lumped Π-type transformer can be performed:

- Designate the lumped low-pass Π-type C_1–L_1–C_2 section to be replaced.
- From a chosen Π-type C_1–L_1–C_2 section, form the symmetrical C–L–C ladder section with equal capacitances C, as shown in Figure 2.27a. The choice of capacitances is arbitrary but their values cannot exceed the minimum of (C_1, C_2).
- Calculate the parameters of the symmetrical C_T–TL–C_T section using the parameters of the lumped equivalent Π-section by setting the electrical length θ of the transmission line according to Eqs. (2.75) and (2.76). Here, it is assumed that the minimum of the capacitances C_1 and C_2 should be equal to C_T or greater so that C_T can be readily embedded in the new C_T–TL–C_T section.
- Finally, replace the Π-type C_1–L_1–C_2 ladder section by the equivalent symmetrical C_T–TL–C_T section and combine adjacent shunt capacitors, as shown in Figure 2.27b, where the loaded shunt capacitances C_A and C_B are given as: $C_A = C_1' + C_T$ and $C_B = C_2' + C_T$.

2.3.5 Equal-*Q* Broadband Matching

A similar equal-Q broadband matching approach used for a lumped-element matching circuit design can be applied to the design of the matching circuit with mixed lumped and transmission-line elements. Consider the design example of a broadband 150-W power amplifier that operates over a frequency bandwidth of 470–860 MHz, uses a 28-V supply voltage, and provides a power gain of more than 10 dB. In this case, it is convenient to use a high-power balanced LDMOSFET device

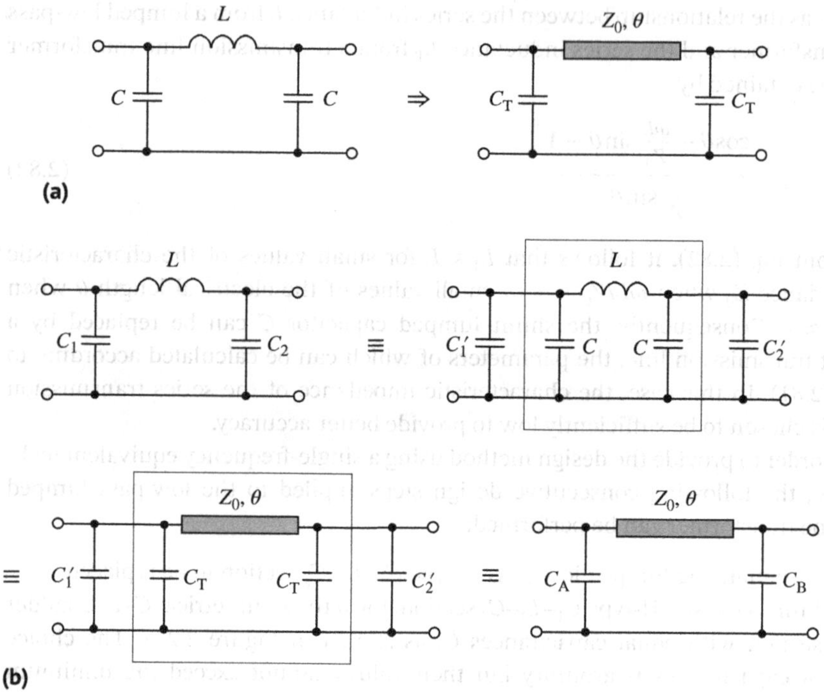

Figure 2.27 Transforming design procedure for lumped and distributed Π-type matching circuits.

specially designed for UHF TV transmitters. The manufacturer states the value of the input impedance for each balanced transistor part as $Z_{in} = (1.7 + j1.3)\,\Omega$ at the center bandwidth frequency $f_c = \sqrt{470 \times 860} = 635\,\text{MHz}$. The input impedance Z_{in} represents a series combination of the input resistance and inductive reactance. To cover the required frequency bandwidth, several low-Q matching circuits should be used to reduce in-band amplitude flatness and improve input VSWR.

In order to achieve a 3-dB frequency bandwidth, the value of a quality factor must be less than $Q = 635/(860–470) = 1.63$. Based on this value of Q, the next step is to define a number of matching sections. For example, for a single-stage input lumped-element matching circuit, the value of a quality factor Q is

$$Q > \sqrt{\frac{50}{1.7} - 1} = 5.33$$

which means that the entire frequency range can be appropriately covered using a multistage matching circuit only.

In this case, it is important that the device input quality factor is smaller than 1.63, being equal to $Q_{in} = 1.3/1.7 = 0.76$. It is very convenient to design the input matching circuit (as well as the output matching circuit) by using simple low-pass L-transformers, composed of the series transmission line and shunt capacitor each, with a constant value of Q for each balanced part of the active device. Then, these two input matching circuits are combined by inserting shunt capacitors, the values of which are reduced by twice, between the two series transmission lines.

To match the series input inductive impedance to the standard 50 Ω source, three low-pass L-transformers connected in series should be used, as shown in Figure 2.28. In this case, the input resistance R_{in} can be assumed to be constant over the entire frequency range. At the center bandwidth frequency of 635 MHz, the input inductance is approximately equal to 0.3 nH. By taking this inductance into account, it is necessary to subtract the appropriate value of the electrical length θ_{in} from the total electrical length θ_3. Because of the short size of this transmission line when $\tan\theta_{in} \approx \theta_{in}$, the value of θ_{in} can be easily calculated by

$$\theta_{in} \cong \frac{X_{in}}{Z_0} = \frac{\omega L_{in}}{Z_0} \tag{2.83}$$

According to Eq. (2.67), there are two simple possibilities to provide input matching using a technique with equal quality factors of the L-transformers. One option is to use the same values of characteristic impedance for all transmission lines; the other one is to use the same electrical lengths for all transmission lines. Consider the first approach, which also allows direct use of the Smith chart, and choose the value of the characteristic impedance $Z_0 = Z_{01} = Z_{02} = Z_{03} = 50\,\Omega$. The ratio of the input and output resistances can be written as

$$\frac{R_1}{R_2} = \frac{R_2}{R_3} = \frac{R_3}{R_{in}} \tag{2.84}$$

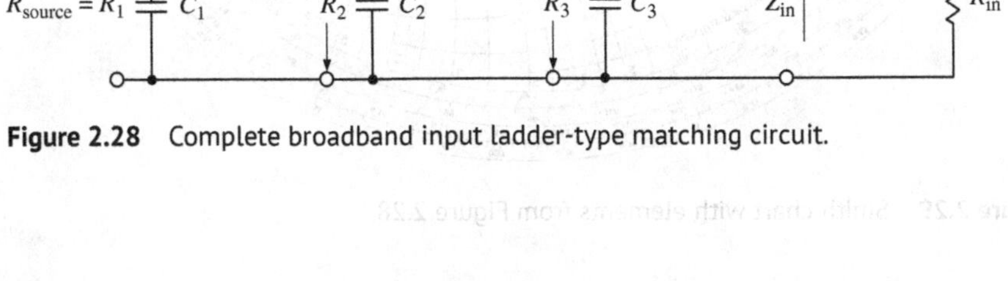

Figure 2.28 Complete broadband input ladder-type matching circuit.

which gives the values $R_2 = 16.2\,\Omega$ and $R_3 = 5.25\,\Omega$ for $R_{source} = R_1 = 50\,\Omega$ and $R_{in} = 1.7\,\Omega$. The values of electrical lengths are determined from the nomograph shown in Figure 2.23a as $\theta_1 = 30°$, $\theta_2 = 7.5°$, and $\theta_3 = 2.4°$.

To calculate the quality factor Q from Eq. (2.67), which must be equal for each L-transformer, it is enough to know the electrical length θ_1 of the first L-transformer. The remaining two electrical lengths can be directly obtained from Eq. (2.66). As a result, the quality factor of each L-transformer is equal to $Q = 1.2$. The values of the shunt capacitances using Eq. (2.65) are $C_1 = 6\,\text{pF}$, $C_2 = 19\,\text{pF}$, and $C_3 = 57\,\text{pF}$.

For a constant Q, the design of the matching circuit can be significantly simplified by using the Smith chart. After calculating the value of Q, it is necessary to plot a constant Q-semicircle on the Smith chart. Figure 2.29 shows the input

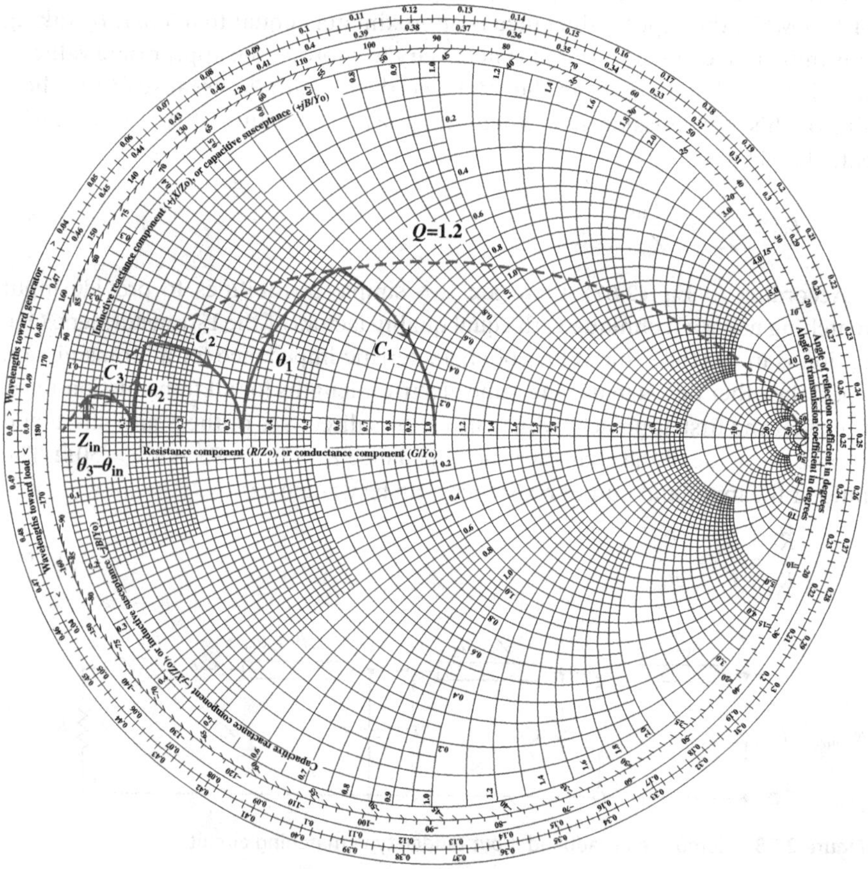

Figure 2.29 Smith chart with elements from Figure 2.28.

matching circuit design using the Smith chart with a constant Q-semicircle, where the curves for the series transmission lines represent the arcs of the circles with center point at the center of the Smith chart. The capacitive traces are moved along the circles with the increasing susceptances and constant conductances.

Another approach assumes the same values of electrical lengths $\theta = \theta_1 = \theta_2 = \theta_3$ and calculates the characteristic impedances of the series transmission lines from Eq. (2.67) at equal ratios of the input and output resistances according to Eq. (2.84). Such an approach is more convenient in practical design, because, when using the microstrip lines with standard characteristic impedance $Z_0 = 50\Omega$, the electrical length of the transmission line adjacent to the transistor input terminal is usually too short. In this case, it makes sense to set the characteristic impedance of the first transmission line to $Z_{01} = 50\Omega$. Then, the value of $\theta = 30°$ is determined directly from the nomograph shown in Figure 2.23a. Further calculation of Q from Eq. (2.65) for fixed θ and Z_0/R_2 yields $Q = 1.2$. The characteristic impedances of the remaining two transmission lines are then calculated directly from Eq. (2.65) or Eq. (2.67). Their values are $Z_{02} = 15.7\Omega$ and $Z_{03} = 5.1\Omega$ with the same values of the shunt capacitances. Generally, the characteristic impedances and electrical lengths of the transmission lines can be optimized for the convenience of practical implementation.

2.3.6 Broadband Matching with Stepped Transmission Lines

The lumped-element or mixed matching networks generally work well at sufficiently low frequencies (up to one or several gigahertz). However, the lumped elements, such as inductors and capacitors, are difficult to implement at microwave frequencies where they can be treated as distributed elements. In addition, the quality factors at higher frequencies for inductors are sufficiently small, resulting in additional power losses. An alternative impedance-matching technique is based on the multi-section matching transformers consisting of the stepped transmission-line sections with different characteristic impedances and electrical lengths [15]. These transformers, in contrast to the continuously tapered transmission-line transformers, are significantly shorter and provide broader performance. Figure 2.30 shows the schematic structure of a stepped transmission-line transformer, which consists of a cascaded connection of n uniform sections of different characteristic impedances $Z_1, Z_2, \ldots, Z_n$ and lengths $l_1, l_2, \ldots, l_n$. In the simplest case, all n uniform sections have equal quarter-wavelength $l = \lambda_0/4$, where λ_0 is the wavelength corresponding to the center bandwidth frequency. Such a stepped transmission-line transformer represents an antimetric structure, for which the ratio between the characteristic impedances of its transmission-line sections can be written in the general form as

$$Z_i Z_{n+1-i} = Z_S Z_L \tag{2.85}$$

where $i = 1, 2, \ldots, n$, Z_S is the source impedance, and Z_L is the load impedance [16].

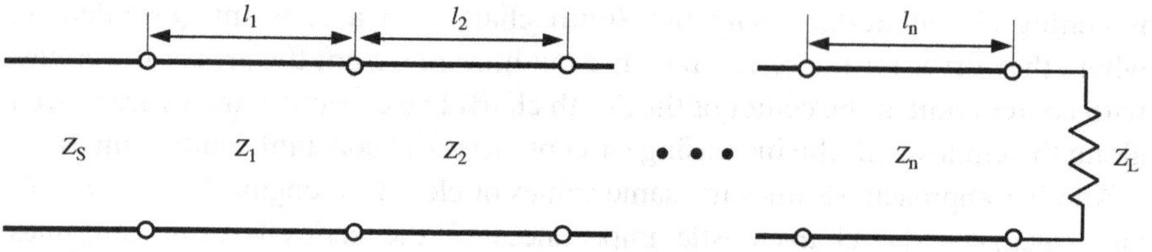

Figure 2.30 Schematic structure of stepped transmission-line transformer.

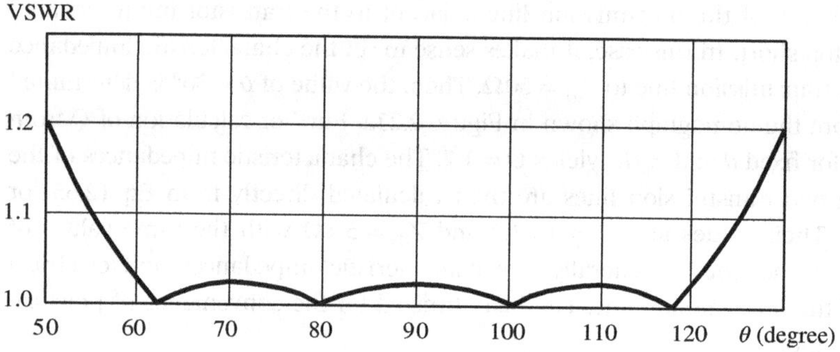

Figure 2.31 Theoretical frequency bandwidth of the five-step transformer.

In Figure 2.31, as a practical example, the minimum possible VSWR is plotted for the five-step transmission-line impedance transformer with a total characteristic-impedance variation of 8:1, which was designed for maximum VSWR of 1.021 in an octave frequency bandwidth, and where each section is of a quarterwave electrical length [17].

The main drawback of the stepped quarterwave transformers is their significant total length $L = n\lambda_0/4$. However, it is possible to reduce the overall transformer length by applying other profiles of its structure. The stepped transformers using n cascaded uniform transmission-line sections of various lengths with alternating impedances are shorter by 1.5–2 times. In this case, the number of sections n is always an even number and the section impedances can be equal to the source and load impedances to be matched, as shown in Figure 2.30 for $Z_S = Z_{2n}$ and $Z_L = Z_{2n-1}$. For example, the input and output matching circuits of a microwave GaAs MESFET power amplifier, which was designed to operate in a frequency bandwidth of 4–8 GHz, were composed of the stepped microstrip lines where all high-impedance sections were made 50 Ω and all low-impedance sections were made 10 Ω [18].

To define the unknown section lengths, the optimization approach to achieve the global minimum of the objective function $|\Gamma(\theta,A)|$ can be used according to

$$\min_{A} \max_{\theta \in [\theta_1,\theta_2]} |\Gamma(\theta,A)| \tag{2.86}$$

where θ_1 and θ_2 are the electrical lengths at the low- and high-frequency bandwidth edges, respectively, and the vector $A = (A_1, A_2, \ldots, A_n)$ consists of the normalized section lengths l_i/λ_0 as components [19]. By solving Eq. (2.86) numerically, the optimum Chebyshev characteristics can be provided by the stepped transmission-line structure with

$$l_i = l_{n+1-i} \tag{2.87}$$

where $i = 1, 2, \ldots, n/2$.

The total length of such a stepped transmission-line transformer can be further reduced by using the structure representing the cascade connection of n transmission-line sections of the same length $l < \lambda_0/4$ with

$$\begin{aligned} &Z_1 < Z_3 < \cdots < Z_{n-1} \\ &Z_2 < Z_4 < \cdots < Z_n \end{aligned} \tag{2.88}$$

where n is an even number and $Z_1 > Z_n$ when $Z_S < Z_L$ [20]. An example of the stepped transmission-line transformer to match the source impedance of 25 Ω with the load impedance of 50 Ω is shown in Figure 2.32a, where the electrical length of each section is equal to $\lambda_0/12$. In this case, the total transformer length is shorter by three times compared to the basic structure with the quarterwave sections, and an octave passband from 2 to 4 GHz for the lossless ideal transmission-line sections is provided, as shown in Figure 2.32b, with an input return loss better than 25 dB. However, it requires the use of a high-impedance ratio for its sections, reaching 30–50 when the source and load impedances differ significantly.

To reduce the high-impedance ratio of the stepped transformers with a short total length, their generalized structure, representing the cascaded even n sections of different lengths l_i and impedances Z_i, as shown in Figure 2.30, can be optimized. The optimum Chebyshev characteristics for this structure can be provided with the ratios between the lengths and characteristic impedances of its sections according to

$$\begin{aligned} l_i &= l_{n+1-i} \\ Z_i Z_{n+1-i} &= Z_S Z_L \end{aligned} \tag{2.89}$$

where $i = 1, 2, \ldots, n/2$ and

$$Z_{n-1} > Z_{n-3} > \cdots > Z_1 > Z_n > Z_{n-2} > \cdots > Z_2 \tag{2.90}$$

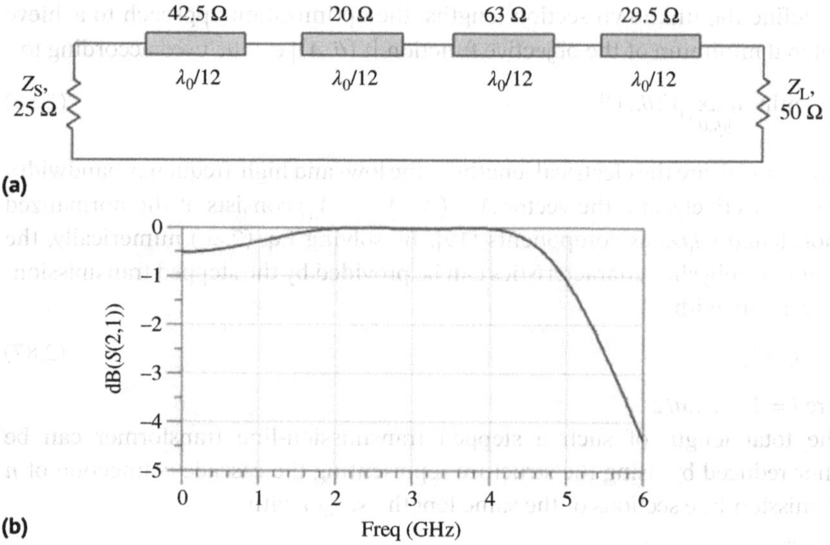

Figure 2.32 Stepped transmission-line transformer with equal-length sections.

where the impedances of both even and odd sections decrease in the direction from higher impedance Z_L to lower impedance Z_S, and the impedance of any odd section is always larger than that of any even section [19]. The lengths of even sections decrease in the direction from the transmission line of a smaller impedance, whereas those of odd sections increase in the same direction.

For a new structure of the stepped transmission-line transformer with the reduced total electrical length, the lengths of each section are defined by Eq. (2.87), and equal characteristic impedances for both odd and even sections differ from the source and load impedances according to

$$\begin{aligned} Z_1 &= Z_3 = \cdots = Z_{n-1} \\ Z_2 &= Z_4 = \cdots = Z_n \end{aligned} \tag{2.91}$$

where $Z_1Z_2 = Z_SZ_L$, $Z_n < Z_S$, and $Z_{n-1} > Z_L$ [19]. In particular situations of a high impedance-matching ratio at microwave frequencies, when it is necessary to match the standard source and load impedance of 50 Ω with the device input and output impedance of 1 Ω and smaller, respectively, both the length and width of the microstrip-line sections can be optimized.

Table 2.1 gives the optimum parameters for different four-section transformers ($n = 4$) designed to match the transmission lines with impedances $Z_S = 25\,\Omega$ and $Z_L = 50\,\Omega$ in an octave frequency range, where the section lengths l_i and total length L are normalized to λ_0 [19].

Table 2.1 Optimum parameters for different four-section transformers.

Equal-length structure

$\|\Gamma\|_{max}$	$l_{1,2,3,4}$	$Z_1(\Omega)$	$Z_2(\Omega)$	$Z_3(\Omega)$	$Z_4(\Omega)$	L
0.065	0.0833	42.38	19.80	63.13	29.49	0.3330
0.071	0.0625	55.75	13.58	92.03	22.42	0.2500
0.074	0.0418	82.58	8.45	148.01	12.14	0.1670
0.076	0.0313	109.64	6.17	202.48	11.40	0.1250

Generalized structure

	$l_{1,2}$	$l_{3,4}$	$Z_1(\Omega)$	$Z_2(\Omega)$	$Z_3(\Omega)$	$Z_4(\Omega)$	L
0.068	0.0625	0.0833	51.75	18.20	68.73	24.15	0.2916
0.070	0.0525	0.0725	62.00	15.28	81.85	20.15	0.2500
0.075	0.0320	0.0510	103.00	10.13	123.39	12.14	0.1660
0.076	0.0205	0.0420	152.90	7.78	160.95	8.18	0.1250

New structure

	$l_{1,2}$	$l_{3,4}$	$Z_1(\Omega)$	$Z_2(\Omega)$	L
0.064	0.0479	0.1171	52.38	23.86	0.3330
0.070	0.0405	0.0841	72.91	17.14	0.2500
0.074	0.0282	0.0553	114.55	10.91	0.1670
0.075	0.0213	0.0412	155.67	8.03	0.1250

2.4 Lossy Match Design Technique

In many practical cases, especially at lower frequencies, to provide broadband matching with minimum gain flatness and input reflection coefficient across the entire operating frequency bandwidth, it is sufficient to use the resistive shunt element at the transistor input. An additional matching improvement with reference to upper frequencies can be achieved by using the inductive reactive element connected in series to the resistor. The resistive nature of this type of network may also improve amplifier stability and intermodulation distortion. To provide a broadband performance for microwave GaAs MESFET power amplifiers, a resistively loaded shunt network, where the resistor is connected in series with a short-circuited quarterwave microstrip line to decrease the loaded quality factor without greatly reducing the maximum available gain, was used in the input matching circuit to cover a frequency bandwidth of 2–6.2 GHz [21]. For ultra-broadband high-gain multistage amplifiers, using a simple lossy compensation shunt circuit with a resistor connected in series with an inductor placed at the input and output of each transistor in parallel with the second-order *LC* circuits had resulted in a power gain of 12 ± 1.5 dB with a VSWR of less than 2.5 from 150 MHz to 16 GHz for a three-stage GaAs MESFET amplifier [22].

For a broadband lossy match silicon MOSFET high-power amplifier, when the internal device input circuit can be equivalently presented by the series $R_{in}C_{in}$ circuit (see Section 1.3), as shown in Figure 2.33a, it is sufficient to use a simple gain-compensation network with a resistor connected in series with a lumped inductor when the operating frequencies are low enough compared to the device transition frequency f_T [23]. In this case, it is very important to optimize the elements of a lossy matching circuit to achieve minimum gain flatness over the maximum frequency bandwidth. To provide the constant real part of the input impedance Z_{in} in a frequency range up to (0.1–0.2) f_T, it is enough to use a simple lossy compensation circuit consisting of an inductor L_{corr} and a resistor R_{corr} connected in series, as shown in Figure 2.33b.

The total input impedance of both the lossy match gain-compensation circuit and device input circuit is written as

$$Z_{in} = \frac{R_{corr} - \omega^2 C_{in} R_{in} L_{corr} + j\omega(L_{corr} + C_{in} R_{in} R_{corr})}{1 - \omega^2 L_{corr} C_{in} + j\omega C_{in}(R_{corr} + R_{in})} \tag{2.92}$$

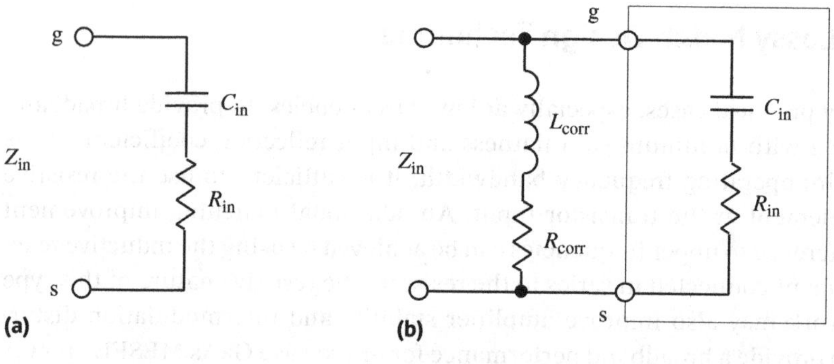

Figure 2.33 Broadband lossy matching for MOSFET power amplifier.

whose real and imaginary parts can be expressed through the circuit parameters by

$$\mathrm{Re}Z_{\mathrm{in}} = \frac{(1-\omega^2 L_{\mathrm{corr}}C_{\mathrm{in}})(R_{\mathrm{corr}} - \omega^2 C_{\mathrm{in}}R_{\mathrm{in}}L_{\mathrm{corr}}) + \omega^2 C_{\mathrm{in}}(R_{\mathrm{corr}}+R_{\mathrm{in}})(L_{\mathrm{corr}}+C_{\mathrm{in}}R_{\mathrm{in}}R_{\mathrm{corr}})}{(1-\omega^2 L_{\mathrm{corr}}C_{\mathrm{in}})^2 + (\omega C_{\mathrm{in}})^2(R_{\mathrm{corr}}+R_{\mathrm{in}})^2} \tag{2.93}$$

$$\mathrm{Im}Z_{\mathrm{in}} = \omega\frac{L_{\mathrm{corr}}(1-\omega^2 L_{\mathrm{corr}}C_{\mathrm{in}}) - C_{\mathrm{in}}(R_{\mathrm{corr}}^2 - \omega^2 L_{\mathrm{corr}}C_{\mathrm{in}}R_{\mathrm{in}}^2)}{(1-\omega^2 L_{\mathrm{corr}}C_{\mathrm{in}})^2 + (\omega C_{\mathrm{in}})^2(R_{\mathrm{corr}}+R_{\mathrm{in}})^2} \tag{2.94}$$

Under the condition $R = R_{\mathrm{corr}} = R_{\mathrm{in}}$, the equations for $\mathrm{Re}Z_{\mathrm{in}}$ and $\mathrm{Im}Z_{\mathrm{in}}$ can be reduced to

$$\mathrm{Re}Z_{\mathrm{in}} = R\frac{(1-\omega^2 L_{\mathrm{corr}}C_{\mathrm{in}})^2 + 2\omega^2 C_{\mathrm{in}}(L_{\mathrm{corr}} + C_{\mathrm{in}}R^2)}{(1-\omega^2 L_{\mathrm{corr}}C_{\mathrm{in}})^2 + (2\omega C_{\mathrm{in}}R)^2} \tag{2.95}$$

$$\mathrm{Im}Z_{\mathrm{in}} = \omega(1-\omega^2 L_{\mathrm{corr}}C_{\mathrm{in}})\frac{L_{\mathrm{corr}} - C_{\mathrm{in}}R^2}{(1-\omega^2 L_{\mathrm{corr}}C_{\mathrm{in}})^2 + (2\omega C_{\mathrm{in}}R)^2} \tag{2.96}$$

From Eq. (2.96), it follows that the reactive part of the input impedance Z_{in} becomes zero, that is, $\mathrm{Im}Z_{\mathrm{in}} = X_{\mathrm{in}} = 0$, when

$$L_{\mathrm{corr}} = C_{\mathrm{in}}R^2 \tag{2.97}$$

which leads to a pure active input impedance Z_{in} obtained as

$$Z_{\mathrm{in}} = R = R_{\mathrm{in}} \tag{2.98}$$

However, when the frequency increases, the voltage amplitude applied to the input capacitance C_{in} decreases. This leads to an appropriate decrease in the power gain at higher bandwidth frequencies. Because of the small values of R_{in} for high-power MOSFETs, the power gain may not be high enough. Therefore, it is necessary to provide an additional impedance matching with lossless matching circuits in order to match with the source impedance of 50 Ω or the high output impedance of the active device of the previous power amplifier stage. The requirements for a lossy *RL* matching circuit can be simplified by imposing some limitation on maximum values of magnitude and phase of the input impedance or reflection coefficient.

Consider the design solution applied to the broadband input lossy matching circuits using optimization procedure where the optimization goal is to maximize the voltage *V* across the input capacitance C_{in} at higher bandwidth frequencies with minimum voltage flatness within the operating broadband frequency bandwidth $\omega_{\mathrm{low}} \leq \omega \leq \omega_{\mathrm{high}}$ on the example of two lumped impedance correction circuits, a simple $R_{\mathrm{corr}}L_{\mathrm{corr}}$ circuit shown in Figure 2.34a and the $R_{\mathrm{corr}}L_{\mathrm{corr}}C_{\mathrm{corr}}$ circuit

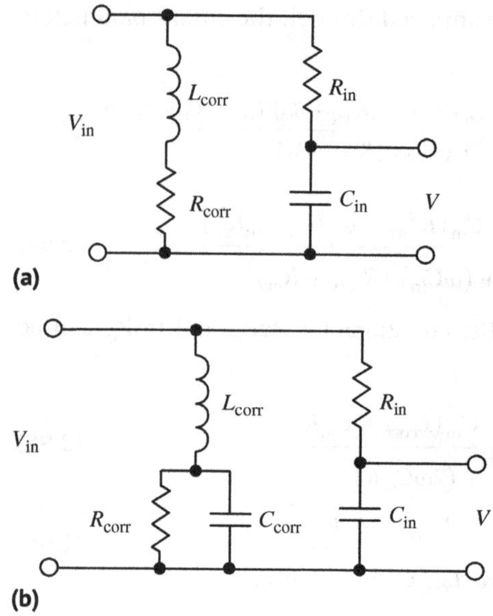

Figure 2.34 Broadband lumped lossy matching circuits.

with an additional capacitor connected in parallel to the resistor R_{corr} shown in Figure 2.34b [24].

From the optimization process, it follows that a simple two-element $R_{corr}L_{corr}$ compensating circuit allows increasing the voltage across the input capacitance C_{in} by 5 dB, compared to the circuit with a shunt resistor R_{corr} only, whereas a three-element $R_{corr}L_{corr}C_{corr}$ circuit allows further voltage increasing by 2 dB. Figure 2.35 shows the broadband performance using a three-element lossy matching circuit shown in Figure 2.34b for 1-dB (curve 1) and 3-dB (curve 2) gain flatness requirements, where V_{in0} is the initial input voltage amplitude before optimization. In Table 2.2, the optimum parameters of these lossy matching circuits are given for different voltage flatness, which can be used to calculate the lossy matching circuit elements according to $L_{corr} = L^{o}_{corr}/\omega^{2}_{high}C_{in}$, $R_{corr} = R^{o}_{corr}/\omega_{high}C_{in}$, and $C_{corr} = C_{in}/C^{o}_{corr}$, where ω_{high} is the high bandwidth frequency, L^{o}_{corr}, R^{o}_{corr}, and C^{o}_{corr} are the optimum corrections coefficients.

Figure 2.36a shows the circuit schematic of a broadband LDMOSFET power amplifier with the device geometry of 1.25 μm × 40 mm. The optimized input three-element lossy matching circuit allows a very broadband operation to be provided with minimum power gain flatness, and a 1 : 2 output transformer contributes to an increase in the output power level. The capacitor of 20 pF connected in parallel with the resistor of 27 Ω provides an additional increase

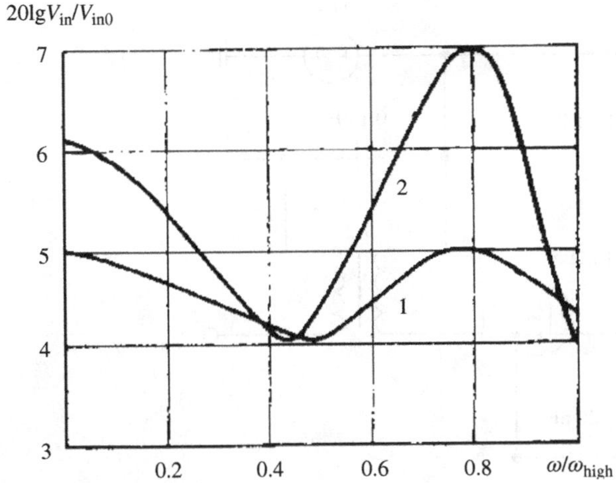

Figure 2.35 Frequency response of a three-element lossy matching circuit.

Table 2.2 Optimum normalized lossy matching circuit parameters.

	1-dB flatness			3-dB flatness		
Circuit type	L^{o}_{corr}	R^{o}_{corr}	C^{o}_{corr}	L^{o}_{corr}	R^{o}_{corr}	C^{o}_{corr}
Figure 2.34a	0.983	1.30		1.88	1.41	
Figure 2.34b	2.56	1.77	1.99	3.17	2.03	1.46

of power gain at higher bandwidth frequencies. The simulation results are shown in Figure 2.36b, where an output power of 22–25 W with a power gain of 13.7 ± 0.3 dB in a frequency range of 5–300 MHz was achieved (curve 1). In this case, the input return loss is greater than 8 dB up to 225 MHz (curve 2). However, when a 50 Ω load is directly connected to the device drain through the blocking capacitor, this results in output power levels in the range of 6–7 W.

2.5 Broadband Matching with Prescribed Amplitude-frequency Response

An additional difficulty in designing the broadband transistor power amplifier apart from the broadband impedance transformation with large impedance ratio and prescribed reactive constraints is that the matching networks must be designed with prescribed tapered magnitude characteristics to compensate for the transistor

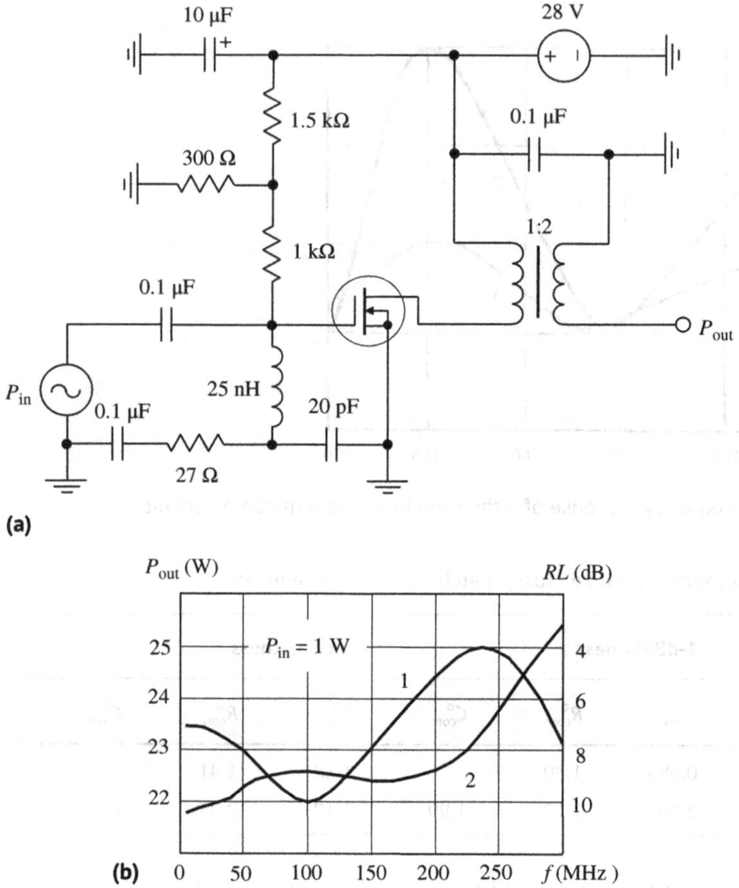

Figure 2.36 Schematic and performance of broadband LDMOSFET power amplifier.

gain roll-off at higher frequencies that is inherent with the high-frequency bipolar and field-effect transistors. The gain tapering is a direct consequence of the fact that the transistor's maximum available gain decreases with increasing frequency. Therefore, for a flat overall gain response, matching networks with tapered magnitude characteristics must be used. The exact frequency dependence of the taper or gain slope varies with specific types of transistors, as well as with output power levels (transition from a linear small-signal to a large-signal operation mode).

For example, for common-emitter bipolar transistor, the input matching network must be designed to simultaneously provide broadband matching to compensate for the series reactive impedance represented by the transistor series *RL* equivalent circuit (base resistance r_b and lead inductance L_b) and broadband impedance transformation of 50 Ω source to the real part of the input impedance (usually r_b of the order of 1 Ω or less for a high-power Class-C biased transistor). However, for the MOSFET or MESFET device, its input circuit can be modeled more closely by a series *RC* circuit (gate–source resistance R_{gs} and gate–source capacitance C_{gs}). At the same time, the reactance of the output matching network is represented by a shunt collector capacitance C_c for the bipolar device and mainly by a drain–source capacitance for the MOSFET or MESFET device, and the equivalent load impedance at the fundamental frequency depends on the DC-supply voltage and output power.

The input matching circuit is responsible for achieving the corresponding gain level over the operating frequency bandwidth. The collector or drain load network, on the other hand, can be designed independently to satisfy the output power and efficiency requirements. A pragmatic approach to the design of a power amplifier stage is to first determine a suitable output matching network and then to design the input matching section to satisfy gain and input return loss requirements [25]. The most promising topology for the input matching consists of a two-section low-pass impedance-transforming structure, which easily absorbs the series *RL* nature of the packaged bipolar transistor input. The low-pass structure, which employs shunt capacitors, is also more readily constructed with microstrip lines than a high-pass or bandpass structure, which would require series capacitors in their realization. Although, in some cases, when it is necessary to improve operation stability, the high-pass matching section can be used together with a low-pass one, especially in the intermediate matching stages. The design philosophy for the input matching circuit is based on the gain response of the transistor versus frequency. Therefore, it is important to find a matching network that closely matches the transistor at the highest bandwidth frequency, where the transistor power gain will be the lowest. At lower frequencies, the power gain is greater but must be sacrificed by selective mismatching, so that the overall gain of the power amplifier will be flat. Thus, the matching networks must exhibit some prescribed tapered magnitude characteristics to compensate for the intrinsic transistor gain roll-off. Note that the best gain-bandwidth performance can be realized with a lumped-element equalizer, whereas simple transmission-line networks can usually be expected to provide near-optimum gain [26].

Figure 2.37a shows the general low-pass ladder-type matching network configuration used for broadband matching network design to approximate a prescribed tapered magnitude gain characteristic [25, 27]. This ladder network must provide not only the prescribed gain slope but also a broadband impedance transformation. Let the gain function for an *n*-element low-pass *LC* ladder network be given by

$$|S_{21}(\omega)|^2 = \frac{1}{1 + \left[Q(\omega^2)\right]^2} \tag{2.99}$$

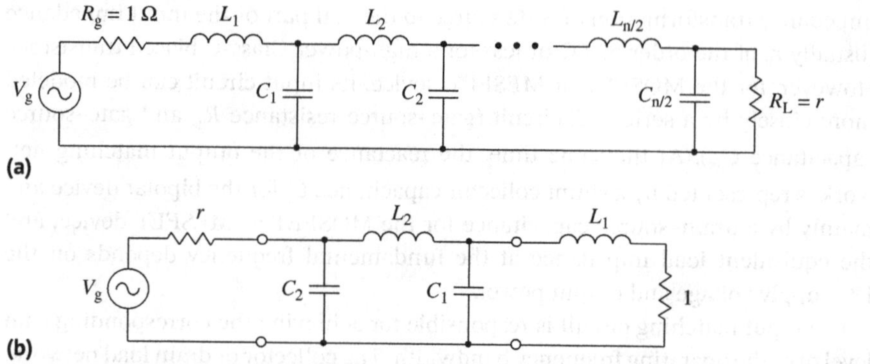

Figure 2.37 Low-pass tapered magnitude marching networks.

where

$$Q(\omega^2) = a_0 + a_1\omega^2 + \cdots + a_{n/2}\omega^n \tag{2.100}$$

and n must be an even number. Once an approximation is found from Eq. (2.99), an optimization procedure can be used to adjust the actual element values [28].

To choose first the polynomial $Q(\omega^2)$ in Eq. (2.99), let $G(\omega)$ be the desired realizable gain function with a prescribed taper. Then, it needs to be set

$$G(\omega_k) = \frac{1}{1 + \left(\sum_{i=0}^{n/2} a_i \omega_k^{2i}\right)^2} \qquad k = 0, 1, \ldots, \frac{n}{2} \tag{2.101}$$

The condition given in Eq. (2.101) can be rewritten as

$$\sum_{i=0}^{n/2} a_i \omega_k^{2i} = C_k \qquad \text{for}\, k = 0, 1, \ldots, \frac{n}{2} \tag{2.102}$$

where

$$C_k = \sqrt{\frac{1}{G(\omega_k)} - 1} \tag{2.103}$$

with a convenient way to choose the values ω_k to space them evenly over the band of interest using either a linear or a logarithmic frequency scale [28]. Figure 2.37b shows the low-pass tapered magnitude *LC* matching network configuration with $n = 4$.

For a two-stage microwave amplifier, whose block schematic is shown in Figure 2.38b, the typical specification of a good input and output matching

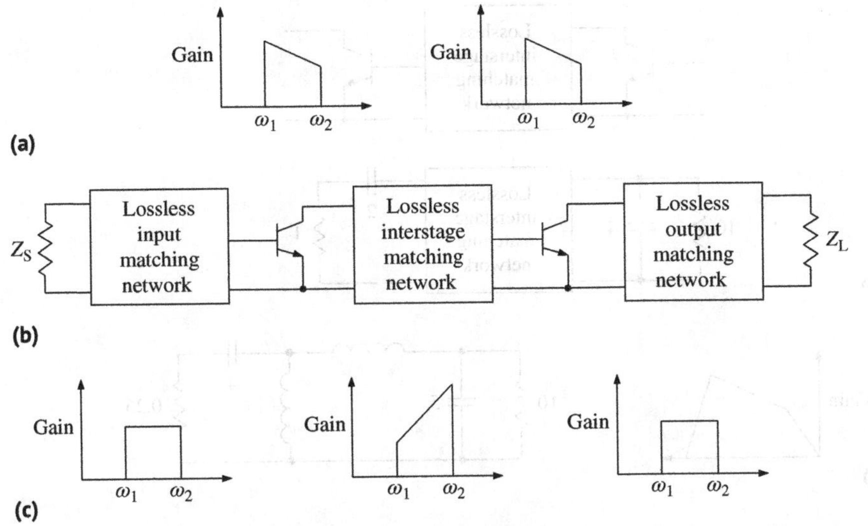

Figure 2.38 Block schematic and frequency responses of a two-stage power amplifier.

implies a frequency response of the input and output matching networks that is flat over the power amplifier passband, and the interstage matching network must provide a positive-sloped gain with increasing frequency, as shown in Figure 2.38c, to compensate for the gain roll-off of each transistor, as shown in Figure 2.38a, thus resulting in an overall flat power gain of a two-stage microwave amplifier [29].

The design example steps for the synthesis of the interstage matching network of an arbitrary specified gain versus frequency are outlined in Figure 2.39, where the output impedance of the first-stage device is represented by a shunt *RC* circuit and the input impedance of the second-stage device is approximated by a series *RC* circuit, as shown in Figure 2.39a with normalized parameters for each *RC* circuit [29]. The insertion loss *IL* of the lossless matching network shown in Figure 2.39b, and having all transmission zeroes at zero and infinite frequencies, can be expressed as

$$IL = \frac{a_0 + a_2\omega^2 + \cdots + a_{2n}\omega^{2n}}{\omega^{2j}} \tag{2.104}$$

where $n = 3$ is the number of natural frequencies of the network, $j = 2$ is the number of transmission zeros at zero frequency, and $n - j = 1$ is the number of transmission zeros at infinite frequency.

The overall synthesis procedure consists of two basic steps: an insertion-loss function is first obtained that approximates a flat passband response in an

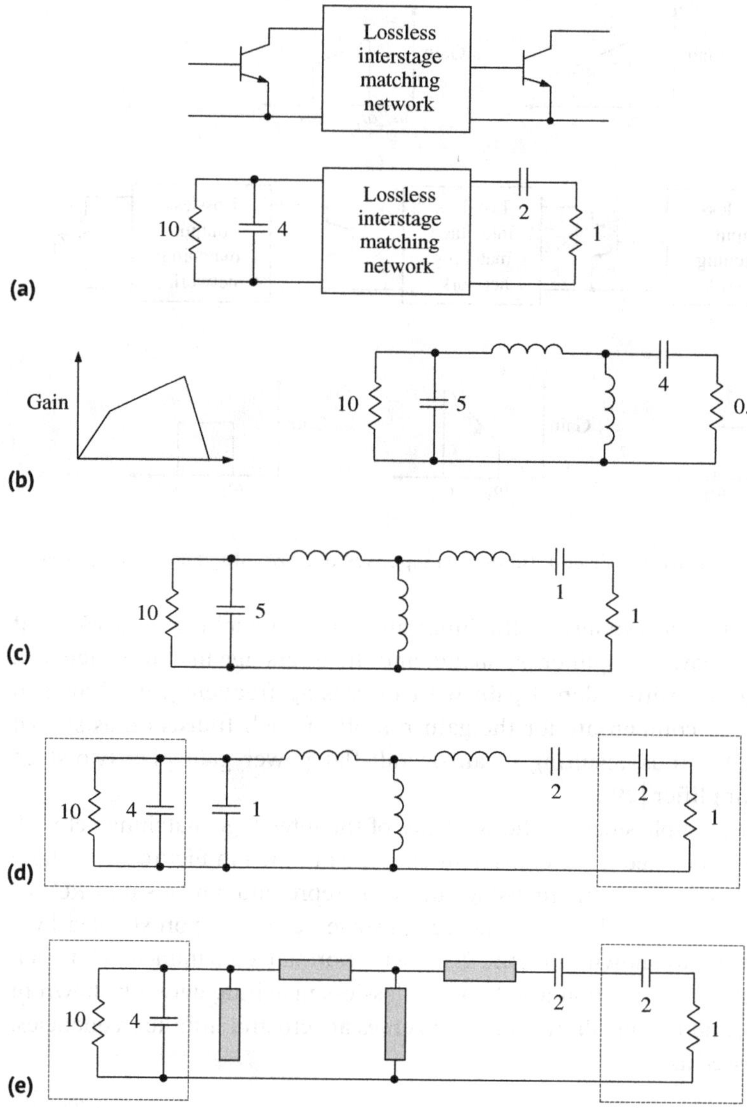

Figure 2.39 Design synthesis process of interstage matching circuit for microwave power amplifier.

equiripple or maximally flat manner, and then a straightforward computational procedure yields a network as prescribed by the insertion-loss function. At the same time, the synthesis of interstage matching networks requires that insertion-loss functions of specified gain versus frequency slope, bandwidth, and ripple be

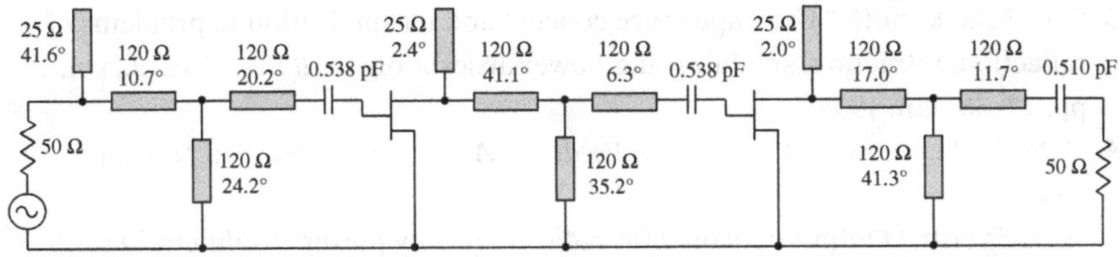

Figure 2.40 Design example of a two-stage microstrip GaAs MESFET amplifier.

found. In this case, the insertion-loss functions approximating $6k$-dB/octave gain slope, where k is an integer, are easily obtained from flat insertion-loss functions. The flat insertion-loss function is first normalized to an upper cutoff frequency of 1 rad/s and then divided by ω^{2k}. Once the sloped insertion-loss function is obtained, the computation procedure yields a network having the specified frequency response.

The practical steps in the interstage matching network design process shown in Figure 2.39 can be listed as follows: (i) model the input and output impedances of the active devices to be used in the microwave power amplifier; (ii) select a topology consistent with device parasitics; (iii) adjust the gain-bandwidth performance to ensure an inclusion of parasitics; (iv) select the reflection coefficient zeroes consistent with inclusion of parasitics; (v) transform impedances to desired levels; (vi) transform the lumped design to a transmission-line realization; (vii) analyze the resultant design by itself and/or as a part of the complete amplifier design; and (viii) optimize the amplifier design if needed [30]. Given that many topologies of the matching networks are available, a good topology would accommodate the existing parasitics and provide a wide range of impedance transformation capabilities. Figure 2.40 shows the design example of a two-stage microwave GaAs MESFET amplifier with three matching networks (input, interstage, and output) having a similar structure, where the input and interstage matching networks provide a gain slope of 6 dB/octave and the output matching network has a flat frequency response, resulting in an overall gain of 15 ± 2 dB over the frequency bandwidth of 6–12 GHz [30].

References

1 H. W. Bode, "A method of impedance correction," *Bell Syst. Tech. J.*, vol. 9, pp. 794–838, Oct. 1930.

2 S. Roberts, "Conjugate-image impedances," *Proc. IRE*, vol. 34, pp. 198–204, Apr. 1946.

3 S. A. Schelkunoff, "The impedance concept and its application to problems of reflection, refraction, shielding and power absorption," *Bell Syst. Tech. J.*, vol. 17, pp. 17–48, Jan. 1938.

4 H. W. Bode, *Network Analysis and Feedback Amplifier Design*, Van Nostrand, 1945.

5 W. L. Everitt, "Output networks for radio-frequency power amplifiers," *Proc. IRE*, vol. 19, pp. 725–737, May 1931.

6 L. F. Gray and R. Graham, *Radio Transmitters*, McGraw-Hill, 1961.

7 V. M. Bogachev and V. V. Nikiforov, *Transistor Power Amplifiers* (in Russian), Energiya, 1978.

8 A. V. Grebennikov, "A simplified CAD approach to analyzing lumped elements," *RF Des.*, pp. 22–38, June 2000.

9 A. Tam, "Network building blocks balance power amp parameters," *Microw. RF*, vol. 23, pp. 81–87, July 1984.

10 P. H. Smith, *Electronic Applications of the Smith Chart*, Noble Publishing, 2000.

11 D. M. Pozar, *Microwave Engineering*, Wiley, 2004.

12 D. H. Steinbrecher, "An interesting impedance matching network," *IEEE Trans. Microw. Theory Tech.*, vol. MTT-15, p. 382, June 1967.

13 A. V. Grebennikov, "Create transmission-line matching circuits for power amplifiers," *Microw. RF*, vol. 39, pp. 113–122, Oct. 2000.

14 B. S. Yarman and A. Aksen, "An integrated design tool to construct lossless matching networks with mixed lumped and distributed elements," *IEEE Trans. Circuits and Systems – I: Fundam. Theory Appl.*, vol. CAS-I-39, pp. 713–723, Sep. 1992.

15 H. Q. Tserng, V. Sokolov, H. M. Macksey, and W. R. Wisseman, "Microwave power GaAs FET amplifiers," *IEEE Trans. Microw. Theory Tech.*, vol. MTT-24, pp. 936–943, Dec. 1976.

16 H. J. Riblet, "General synthesis of quarter-wave impedance transformers," *IRE Trans. Microw. Theory Tech.*, vol. MTT-5, pp. 36–43, Apr. 1957.

17 S. B. Cohn, "Optimum design of stepped transmission-line transformers," *IRE Trans. Microw. Theory Tech.*, vol. MTT-3, pp. 16–21, Apr. 1955.

18 R. E. Neidert and H. A. Willing, "Wide-band gallium arsenide power MESFET amplifiers," *IEEE Trans. Microw. Theory Tech.*, vol. MTT-24, pp. 342–350, June 1976.

19 V. P. Meschanov, I. A. Rasukova, and V. D. Tupikin, "Stepped transformers on TEM-transmission lines," *IEEE Trans. Microw. Theory Tech.*, vol. MTT-44, pp. 793–798, June 1996.

20 G. L. Matthaei, "Short-step Chebyshev impedance transformers," *IEEE Trans. Microw. Theory Tech.*, vol. MTT-14, pp. 372–383, Aug. 1966.

21 D. P. Hornbuckle and L. J. Kuhlman, Jr., "Broad-band medium-power amplification in the 2–12.4-GHz range with GaAs MESFET's," *IEEE Trans. Microw. Theory Tech.*, vol. MTT-24, pp. 338–342, June 1976.

22 J. Obregon, R. Funck, and S. Barvet, "A 150MHz–16GHz FET amplifier," *1981 IEEE Int. Solid-State Circuits Conf. Dig.*, pp. 66–67, 1981.

23 V. V. Nikiforov, T. T. Kulish, and I. V. Shevnin, "Broadband HF-VHF MOSFET power amplifier design (in Russian)," *Poluprovodnikovaya Elektronika v Tekhnike Svyazi*, vol. 23, pp. 27–36, 1983.

24 V. V. Nikiforov and S. Y. Terentiev, "Synthesis of correction circuits of broadband power amplifiers using nonlinear programming methods (in Russian)," *Poluprovodnikovaya Elektronika v Tekhnike Svyazi*, vol. 26, pp. 27–36, 1986.

25 O. Pitzalis, Jr. and R. A. Gilson, "Broad-band microwave class-C transistor amplifiers," *IEEE Trans. Microw. Theory Tech.*, vol. MTT-21, pp. 660–668, Nov. 1973.

26 R. S. Tucker, "Gain-bandwidth limitations of microwave transistor amplifiers," *IEEE Trans. Microw. Theory Tech.*, vol. MTT-21, pp. 322–327, May 1973.

27 O. Pitzalis, Jr. and R. A. Gilson, "Tables of impedance matching networks which approximate prescribed attenuation versus frequency slopes," *IEEE Trans. Microw. Theory Tech.*, vol. MTT-19, pp. 381–386, Apr. 1971.

28 W. H. Ku and W. C. Petersen, "Optimum gain-bandwidth limitations of transistor amplifiers as reactively constrained active two-port networks," *IEEE Trans. Circuits Syst.*, vol. CAS-22, pp. 523–533, June 1975.

29 D. J. Mellor and J. G. Linvill, "Synthesis of interstage networks with prescribed gain versus frequency slopes," *IEEE Trans. Microw. Theory Tech.*, vol. MTT-23, pp. 1013–1020, Dec. 1975.

30 D. J. Mellor, "Improved computer-aided synthesis tools for the design of matching networks for wide-band microwave amplifiers," *IEEE Trans. Microw. Theory Tech.*, vol. MTT-34, pp. 1276–1281, Dec. 1986.

3
High-efficiency Power Amplifier Design

Efficiency as a key parameter of the power amplifier can be maximized by using biharmonic or polyharmonic modes using additional single-resonant or multi-resonant circuits tuned to the odd or even harmonic components in the load network when flattening of the voltage or current waveforms at the transistor output is provided. As an alternative, an efficiency improvement in practical implementation is achieved by providing the proper phasing for each harmonic component, resulting in asymmetrical voltage and current waveforms that do not overlap each other. In both cases, ideally, the transistor should operate as a lossless switch that contributes to the complete transformation of the DC-supply power to the power at the fundamental frequency delivered to the load. However, their load network implementations are different, and in the latter case, it is much simpler due to not requiring the harmonically tuned multi-resonant circuits.

3.1 Class-F Circuit Design

The practical solution of improving efficiency by approximating the anode voltage waveform to a square wave in vacuum-tube amplifiers to minimize the power losses in the vacuum tubes was proposed in the early 1920s by using a multi-resonant load network with third- and higher-order harmonic traps connected in series to the anode [1]. However, the effect of the inclusion of a third-harmonic resonator in the load network, as shown in Figure 3.1a, was described and analyzed in detail only 1.5 decades later [2, 3]. It was shown that the symmetrical anode voltage waveform and level of its depression can be provided with opposite phase conditions at the waveform midpoints between the fundamental and third harmonics with their optimum voltage amplitudes. Besides, to maximize efficiency of the vacuum-tube amplifier with a better approximation of a square

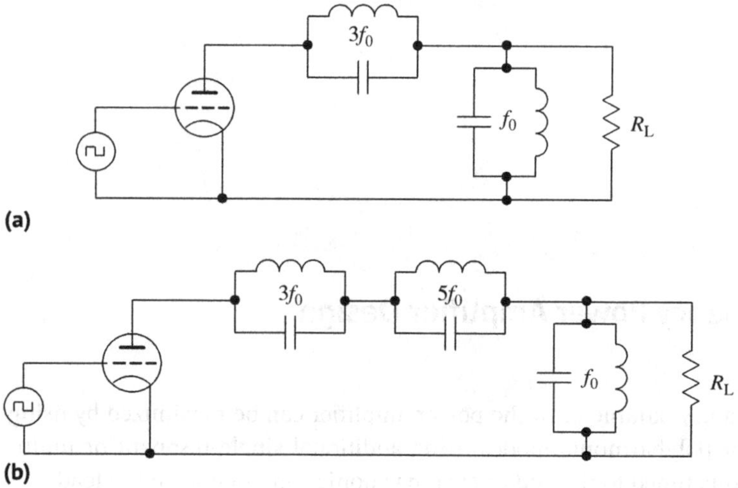

Figure 3.1 High-efficiency vacuum-tube amplifiers with multi-resonant circuits.

voltage anode waveform, it was also suggested to use in the load network both the third- and fifth-harmonic resonators connected in series, shown in Figure 3.1b [4].

In practical power amplifier design, especially at very high frequencies, when it becomes difficult to control the higher-order harmonics, efficiency may significantly degrade. To understand the basics of a multi-resonant approach, a Fourier-series analysis can be used to evaluate the contribution of each harmonic component to the ideal half-sinusoidal current waveform and square voltage waveform. In this case, the partial Fourier series of normalized current $i(\omega t)$ and voltage $v(\omega t)$ should be calculated according to

$$\frac{i(\omega t)}{I_0} = 1 - \frac{\pi}{2}\sin\omega t - 2\sum_{n=2,4,6,\ldots}^{N} \frac{\cos n\omega t}{n^2 - 1} \tag{3.1}$$

$$\frac{v(\omega t)}{V_0} = 1 + \frac{4}{\pi}\sin\omega t + \frac{4}{\pi}\sum_{n=3,5,7,\ldots}^{N} \frac{\sin n\omega t}{n} \tag{3.2}$$

where I_0 is the DC current and V_0 is the DC-supply voltage.

Figure 3.2 shows that the shapes of the voltage and current waveforms can be significantly changed with increasing fundamental voltage amplitude by adding even one additional harmonic component being properly phased. For example, the proper ratio between the amplitudes of the fundamental and third-harmonic components can provide the flattened voltage waveform with minimum depression and maximum difference between its peak amplitude and amplitude of the fundamental component, as shown in Figure 3.2a. Similarly, the combination of

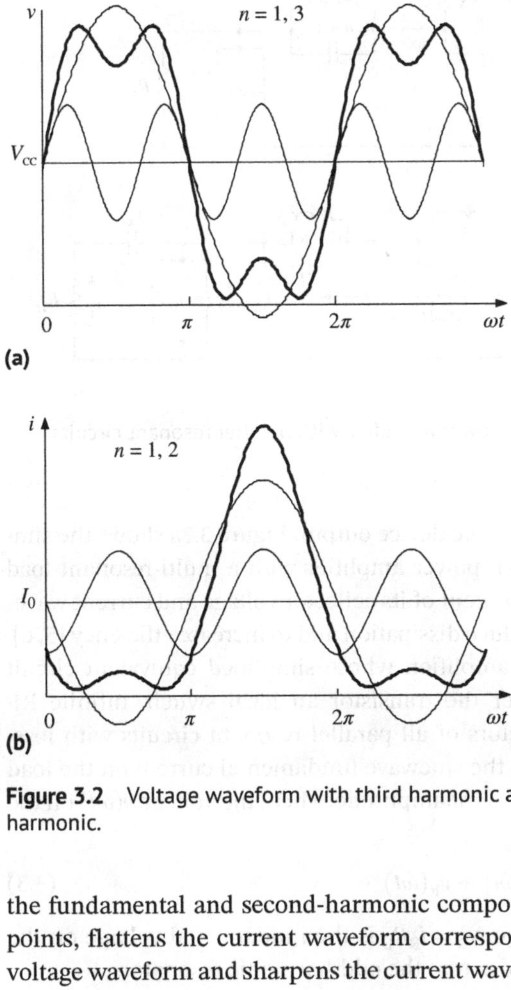

Figure 3.2 Voltage waveform with third harmonic and current waveform with second harmonic.

the fundamental and second-harmonic components, being in phase at the midpoints, flattens the current waveform corresponding to maximum values of the voltage waveform and sharpens the current waveform corresponding to minimum values of the voltage waveform, as shown in Figure 3.2b. The optimum ratio between the amplitudes of the fundamental-frequency and second-harmonic components can maximize a peak value of the current waveform, with its minimum value determined by the device saturation resistance in a practical circuit. Thus, power loss due to the active device can be minimized because the results of the integration over a period when the minimum voltage corresponds to the maximum current will give a small value compared with the power delivered to the load.

3.1.1 Idealized Class-F Mode

Generally, an infinite number of the odd-harmonic tank resonators can maintain a square collector (anode or drain) voltage waveform, also providing a

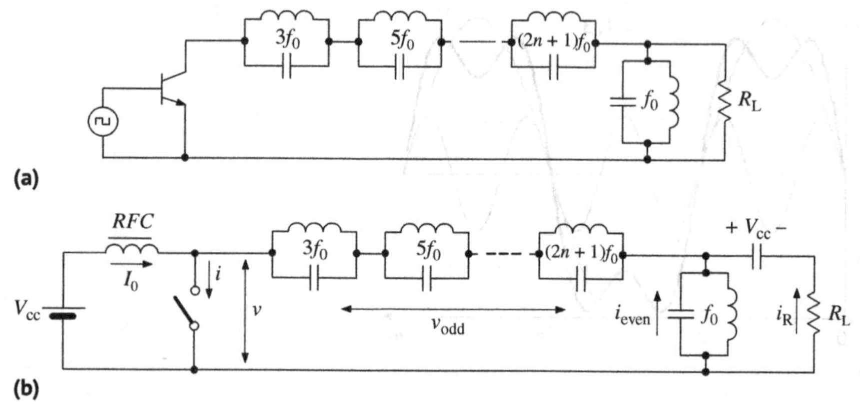

Figure 3.3 Load networks of Class-F power amplifier with parallel resonant circuits.

half-sinusoidal current waveform at the device output. Figure 3.3a shows the simplified schematic of such a Class-F power amplifier with a multi-resonant load network to control the harmonic content of its collector voltage and current waveforms, thereby shaping them to reduce dissipation and to increase efficiency [5, 6].

To analyze the Class-F power amplifier, whose simplified equivalent circuit is shown in Figure 3.3b, consider the transistor an ideal switch, infinite RF choke inductance and quality factors of all parallel resonant circuits with lossless elements, 50% duty ratio, and the sinewave fundamental current on the load $i_R(\omega t) = I_R \sin \omega t$, where I_R is the current amplitude. Then, the voltage $v(\omega t)$ on the switch can be written as

$$v(\omega t) = V_{cc} + v_{odd}\left[(2n+1)\omega t\right] + v_R(\omega t) \tag{3.3}$$

where V_{cc} is the DC-supply voltage, $v_R = i_R R_L$ is the fundamental voltage on the load resistor, and v_{odd} is the voltage across the odd-harmonic resonators.

Adding a phase shift of π in both parts of Eq. (3.3) results in

$$v(\omega t + \pi) = V_{cc} - v_{odd}\left[(2n+1)\omega t\right] - v_R(\omega t) \tag{3.4}$$

As a result, combining Eqs. (3.3) and (3.4) yields

$$v(\omega t) = 2V_{cc} - v(\omega t + \pi) \tag{3.5}$$

As it is seen from Eq. (3.5), the only possible waveform for the collector voltage is a square wave consisting of only DC, fundamental-frequency, and odd-harmonic components and having a maximum value of $2V_{cc}$.

When the switch is turned on in the interval $0 < \omega t \leq \pi$, the current $i(\omega t)$ through the switch is obtained as

$$i(\omega t) = I_0 + i_{\text{even}}(2n\omega t) + i_{\text{R}}(\omega t) \tag{3.6}$$

while the current $i(\omega t + \pi)$ is equal to zero when the switch is turned off in the interval $\pi < \omega t \leq 2\pi$ written by

$$0 = I_0 + i_{\text{even}}(2n\omega t) - i_{\text{R}}(\omega t) \tag{3.7}$$

Then, substituting Eq. (3.7) into Eq. (3.6) results in

$$i(\omega t) = 2i_{\text{R}}(\omega t) = 2I_{\text{R}}\sin(\omega t) \tag{3.8}$$

where the amplitude of current on the switch during the interval $0 < \omega t \leq \pi$ is two times greater than the amplitude of the fundamental current. As a result, generally Eq. (3.6) can be written as

$$i(\omega t) = I_{\text{R}}(\sin\omega t + |\sin\omega t|) \tag{3.9}$$

representing the switch current in the form of half-sine pulses with the amplitude equal to two times the load current amplitude.

Figure 3.4a shows the sinewave current on the load, the ideal collector square-wave voltage waveform consisting of the fundamental and odd harmonics is shown in Figure 3.4b, and the ideal collector half-sine current waveform composed of the fundamental and even harmonics is shown in Figure 3.4c. Since the collector (or drain) current and voltage waveforms do not overlap simultaneously, such a condition corresponds to an idealized Class-F mode with 100% collector efficiency.

From the Fourier-series analysis of the collector (or drain) current and voltage waveforms, the expressions for the DC-supply current I_0, the fundamental current amplitude I_1, and the fundamental voltage amplitude V_1 can be properly derived using Eqs. (3.5) and (3.9) as

$$I_0 = \frac{1}{2\pi}\int_0^{\pi} 2I_{\text{R}}\sin\omega t\, d\omega t = \frac{2I_{\text{R}}}{\pi} \tag{3.10}$$

$$I_1 = \frac{1}{\pi}\int_0^{\pi} 2I_{\text{R}}\sin^2\omega t\, d\omega t = I_{\text{R}} \tag{3.11}$$

$$V_1 = V_{\text{R}} = \frac{1}{\pi}\int_{\pi}^{2\pi} 2V_{\text{cc}}\sin(\omega t + \pi) d\omega t = \frac{4V_{\text{cc}}}{\pi} \tag{3.12}$$

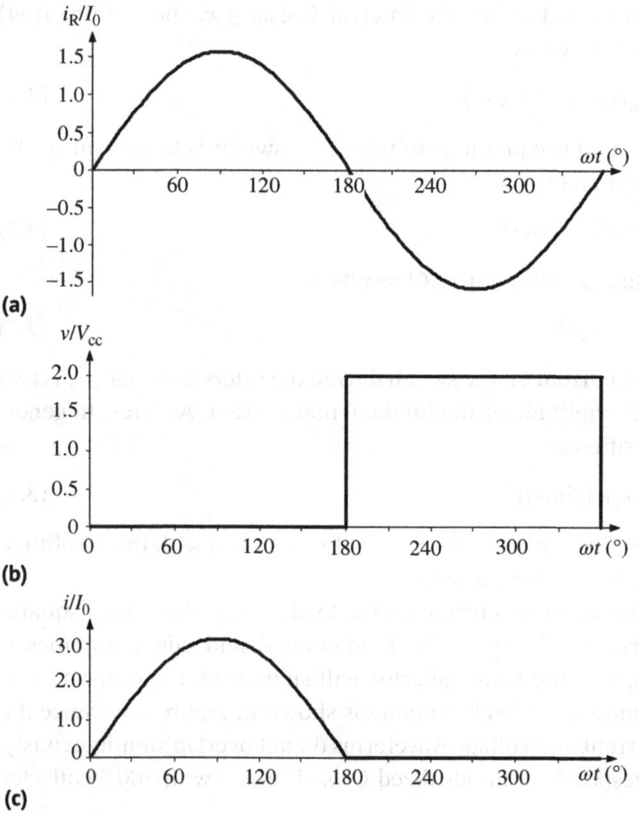

Figure 3.4 Voltage and current waveforms of an ideal Class-F power amplifier.

where $V_R = I_R R_L$ is the fundamental voltage amplitude across the load resistor R_L.

Hence, the DC-supply power and the fundamental output power are obtained by

$$P_0 = V_{cc}\ I_0 = \frac{2V_{cc}I_R}{\pi} \tag{3.13}$$

$$P_1 = \frac{V_1 I_1}{2} = \frac{2V_{cc}\ I_R}{\pi} \tag{3.14}$$

respectively, with a maximum ideal collector (drain) efficiency of

$$\eta = \frac{P_1}{P_0} = 100\% \tag{3.15}$$

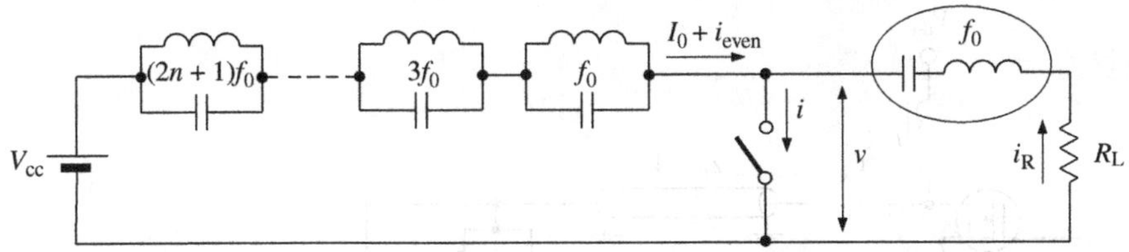

Figure 3.5 Multi-resonant load network of Class-F power amplifier with series filter.

The impedance conditions for the transistor output (collector or drain) for an idealized Class-F mode using Eqs. (3.10)–(3.12) must be equal to

$$Z_1 = R_1 = \frac{V_1}{I_1} = \frac{8}{\pi^2}\frac{V_{cc}}{I_0} \tag{3.16}$$

$$Z_{2n} = 0 \quad \text{for even harmonics} \tag{3.17}$$

$$Z_{2n+1} = \infty \quad \text{for odd harmonics} \tag{3.18}$$

which coincide with the required impedances in the limiting case of the optimum efficiency Class-B power amplifier [7].

The square-wave voltage and half-sine current waveforms can be obtained in a similar way for a multi-resonant load network shown in Figure 3.5, where the parallel resonators are connected in series between the DC supply and the switch being properly tuned to the fundamental frequency, third, and higher-order odd harmonics, whereas the sinusoidal current flows to the load through the fundamentally tuned series resonant circuit.

3.1.2 Class F with Quarterwave Transmission Line

A series quarterwave transmission line loaded by a parallel resonant circuit ideally can control an infinite number of harmonics by providing a square voltage waveform and a half-sinusoidal current waveform at the device output, as shown in Figure 3.6. In this case, the short circuit provided by an ideal parallel resonant circuit at the output of the transmission line for even harmonics is repeated, thus producing the corresponding short circuits at the drain. At the same time, for odd harmonics, the short circuits at the load produce an open circuit at the drain with a resistive load at the fundamental. Such type of a Class-F power amplifier with a series quarterwave transmission line was proposed to operate at a sufficiently high frequency, where the load network with lumped elements only is difficult to implement, and the device parasitics are sufficiently small [5, 6]. To approximate the Class-F voltage and current waveforms, the transistor should act as a

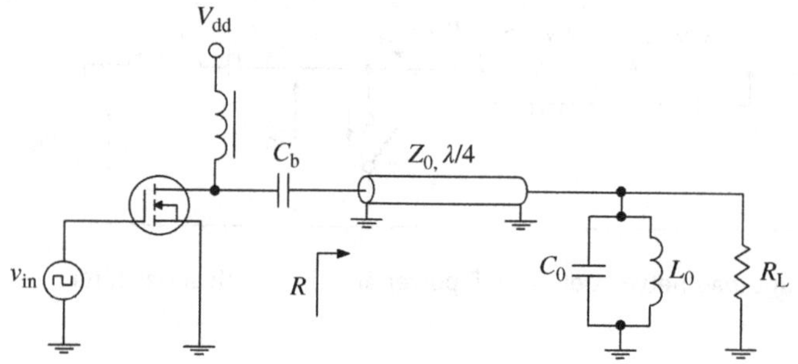

Figure 3.6 Class-F power amplifier with series quarterwave transmission line.

switch rather than a saturating current source under high drive levels. In this case, the quarterwave transmission line is operated as a quarterwave impedance transformer according to

$$R = \frac{Z_0^2}{R_L} \tag{3.19}$$

where Z_0 is the transmission-line characteristic impedance [8].

Figure 3.7a shows another circuit structure of the Class-F power amplifier with a shunt quarterwave line connected between the DC power supply and the transistor output (collector or drain). Such a location of the quarterwave transmission line does not require an RF choke and a series DC-blocking capacitor in view of a series resonant L_0C_0 circuit tuned to the fundamental frequency and connected between the transistor collector (or drain) and the load. However, in this case, an impedance transformation of the Class-F load resistance R to the standard load resistance R_L if required should be provided separately by additional lumped elements or a transmission line.

In order to provide an analytical insight into the Class-F power amplifier with a quarterwave transmission line, the main fundamental properties of a quarterwave transmission line should be derived when the transmission line is considered an element with finite delay time in the time domain. Figure 3.7b shows a simplified load network of the Class-F power amplifier consisting of a parallel quarterwave transmission line grounded at its end through the DC power supply, a series fundamentally tuned L_0C_0 circuit, and a load resistance R. In an idealized case, the intrinsic device output capacitance is assumed to be negligible to affect the power amplifier RF performance, and the sinusoidal output current i_R flows to the load R due to the sufficiently high value of the loaded quality factor Q_L of the series resonant L_0C_0 circuit.

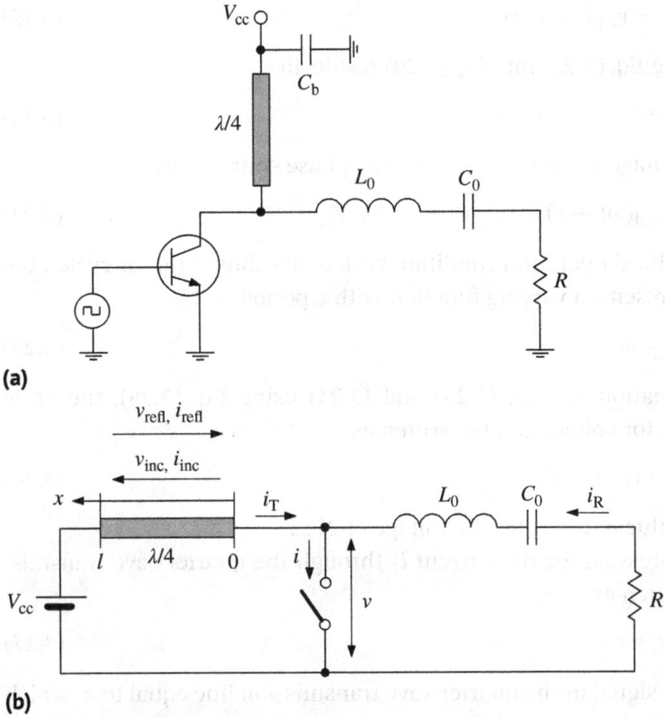

Figure 3.7 Class-F power amplifier with shunt quarterwave transmission line.

The electrical behavior of a homogeneous lossless quarterwave transmission line connected to the DC voltage supply with RF grounding can be described considering the distribution of the voltage $v(t,x)$ along the transmission line where it can be represented as a sum of the incident voltage $v_{inc}(\omega t - 2\pi x/\lambda)$ and the reflected voltage $v_{refl}(\omega t + 2\pi x/\lambda)$ in any cross-section of the transmission line, as shown in Figure 3.7b [9]. For $x = 0$, the voltage $v(t,x)$ is equal to the collector voltage,

$$v(\omega t) = v(t,0) = v_{inc}(\omega t) + v_{refl}(\omega t) \tag{3.20}$$

At the other end of the transmission line, when $x = \lambda/4$, the voltage is equal to the DC-supply voltage V_{cc},

$$V_{cc} = v(t, \pi/2) = v_{inc}(\omega t - \pi/2) + v_{refl}(\omega t + \pi/2) \tag{3.21}$$

At the same time, since the time moment t was chosen arbitrarily, Eq. (3.21) can be rewritten introducing a phase shift of $\pi/2$ for each voltage as

$$v_{\rm inc}(\omega t) = V_{\rm cc} - v_{\rm refl}(\omega t + \pi) \tag{3.22}$$

Then, substituting Eq. (3.22) into Eq. (3.20) results in

$$v(\omega t) = v_{\rm refl}(\omega t) - v_{\rm refl}(\omega t + \pi) + V_{\rm cc} \tag{3.23}$$

and the collector voltage can be obtained for the phase shift of π by

$$v(\omega t + \pi) = v_{\rm refl}(\omega t + \pi) - v_{\rm refl}(\omega t + 2\pi) + V_{\rm cc} \tag{3.24}$$

Thus, for an idealized operation condition with a 50% duty ratio (or cycle), the voltage $v_{\rm refl}(\omega t)$ represents a varying function with a period of 2π,

$$v_{\rm refl}(\omega t) = v_{\rm refl}(\omega t + 2\pi) \tag{3.25}$$

Finally, by summation of Eqs. (3.23) and (3.24) using Eq. (3.25), the basic expression for collector voltage can be written as

$$v(\omega t) = 2V_{\rm cc} - v(\omega t + \pi) \tag{3.26}$$

with a maximum value of the collector voltage of $2V_{\rm cc}$.

Similarly, the expression for the current $i_{\rm T}$ through the quarterwave transmission line can be written as

$$i_{\rm T}(\omega t) = i_{\rm T}(\omega t + \pi) \tag{3.27}$$

with the period of a signal in the quarterwave transmission line equal to π, which means that it contains only even harmonics, because a short-circuited quarterwave transmission line at its end results in an infinite impedance at odd harmonics at its input.

If the transistor is operated as an ideal switch, it is turned on during the interval $0 < \omega t \leq \pi$ where $v = 0$ and turned off during the interval $\pi < \omega t \leq 2\pi$, where $v = 2V_{\rm cc}$ according to Eq. (3.26). In this case, during the interval $\pi < \omega t \leq 2\pi$ when the switch is turned off, the load is directly connected to the transmission line and $i_{\rm T} = -i_{\rm R} = -I_{\rm R}\sin\omega t$. Hence, during the interval $0 < \omega t \leq \pi$ when the switch is turned on, $i_{\rm T} = I_{\rm R}\sin\omega t$ according to Eq. (3.27). Thus, the current flowing into the quarterwave transmission line at any ωt can be represented by

$$i_{\rm T}(\omega t) = I_{\rm R}|\sin\omega t| \tag{3.28}$$

where $I_{\rm R}$ is the amplitude of current flowing into the load.

Since the collector current is defined as $i = i_{\rm T} + i_{\rm R}$, then it can be written as

$$i(\omega t) = I_{\rm R}(\sin\omega t + |\sin\omega t|) \tag{3.29}$$

representing the half-sine pulses with the amplitude equal to two times the load current amplitude.

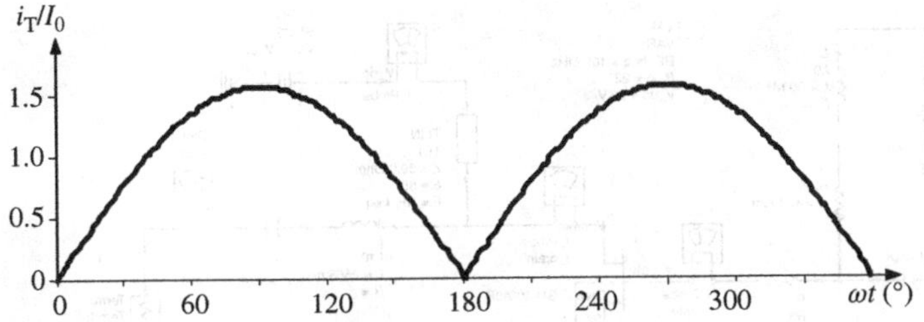

Figure 3.8 Current waveform in an ideal quarterwave transmission line.

Consequently, for a sinewave current flowing into the load due to the infinite loaded quality factor of the series fundamentally tuned L_0C_0 circuit shown in Figure 3.4a, the ideal collector voltage and current waveforms can be represented by the corresponding normalized square and half-sine waveforms shown in Figure 3.4b,c, respectively, where I_0 is the DC current. Here, a sum of the fundamental and odd harmonics approximates a square voltage collector waveform, and a sum of the fundamental and even harmonics approximates a half-sine collector current waveform. The waveform corresponding to the normalized current flowing into the quarterwave transmission line shown in Figure 3.8 represents a sum of even harmonics representing a full-wave rectified sine wave. As a result, the shapes of the collector current and voltage waveforms provide a condition when the current and voltage do not overlap simultaneously.

Figure 3.9 shows the simulated circuit schematic of a Class-F power amplifier with a shunt quarterwave line using a 28-V 10-W Cree GaN HEMT power transistor CGH40010, similar to that presented in [10]. To better illustrate the drain voltage and current waveforms with minimum effect of the device parasitic output parameters, a sufficiently low operating frequency of 100 MHz was chosen. In this case, a simple lossy *RL* input shunt circuit is used to match the device input impedance to a 50-Ω source and to compensate for the device input gate–source capacitance C_{in} of about 5 pF at the fundamental frequency that resulted in a small-signal S_{11} magnitude better than −25 dB. The series 50-Ω resistor connected to the device gate is additionally included to provide an unconditional operation stability. The simulated drain voltage, which is close to a square waveform, and the drain current, which is close to a half-sinusoidal waveform, are shown in Figure 3.10a, where small waveform ripples (minimized by load-network parameter optimization) can be explained because of some effect of the device output drain–source capacitance C_{out} of about 1.3 pF and package parasitics at higher-order harmonic components. As a result, the maximum drain efficiency of 84.7% with a power gain of 20 dB and an output power of 40 dBm at a DC-supply voltage of 28 V was obtained with a sinewave driving signal, as shown in Figure 3.10b. In this case, to better approximate

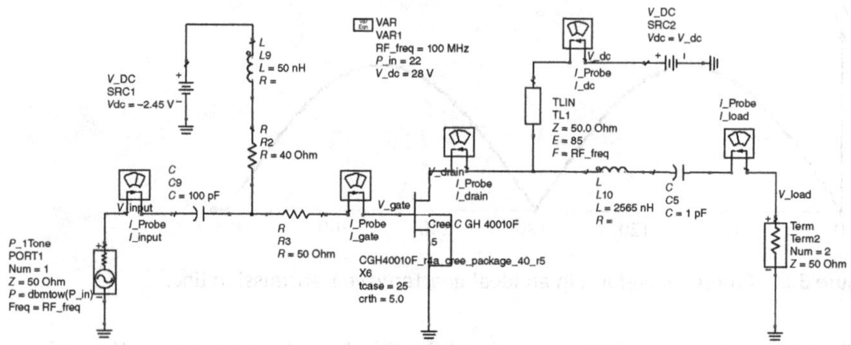

Figure 3.9 Simulation schematic of Class-F power amplifier with parallel quarterwave line.

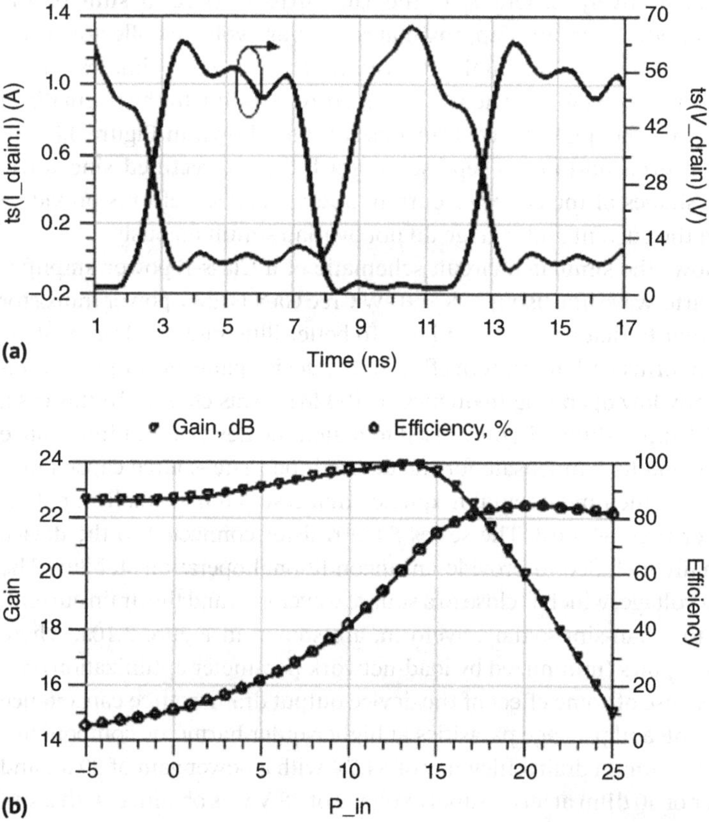

Figure 3.10 Simulated performance of Class-F power amplifier with parallel quarterwave line.

the switchmode operation of a 10-W GaN HEMT device, it is necessary to slightly increase the input signal amplitude to operate mostly in a saturated mode characterized by more than 3-dB gain compression point.

3.1.3 Effect of Saturation Resistance

In order to minimize power losses, it is important to analytically evaluate how significant the effect of the transistor saturation resistance r_{sat} on the Class-F power amplifier performance when the transistor is turned on. Figure 3.11 shows the simplified equivalent circuit of a Class-F power amplifier with a quarterwave transmission line where a nonideal switch represents the transistor with saturation resistance r_{sat} and parasitic output capacitance C_{out}. When the switch is turned on within the interval $0 < \omega t \leq \pi$, the saturation voltage v_{sat} due to the current $i(\omega t)$ flowing through the switch is written as

$$v_{sat}(\omega t) = V_{sat} \sin \omega t = 2I_R i_{sat} \sin \omega t \tag{3.30}$$

where the saturation voltage amplitude V_{sat} is obtained using Eq. (3.12) as

$$V_{sat} = 2V_R \frac{r_{sat}}{R} = \frac{8V_{cc}}{\pi} \frac{r_{sat}}{R} \tag{3.31}$$

Figure 3.12 shows the corresponding idealized collector current and voltage waveforms with nonzero saturation resistance, where the half-sine current flowing through the saturation resistance r_{sat} causes some deviation of the voltage waveform from the ideal square wave. In this case, the bottom part of the voltage waveform becomes sinusoidal with the amplitude V_{sat} within the interval $0 < \omega t \leq \pi$. As follows from Eq. (3.5), similar sinewave behavior will also correspond to the top part of the voltage waveform within the interval $\pi < \omega t \leq 2\pi$.

The power losses, considering the saturation resistance r_{sat} can be estimated using Eqs. (3.8), (3.10), and (3.12) as

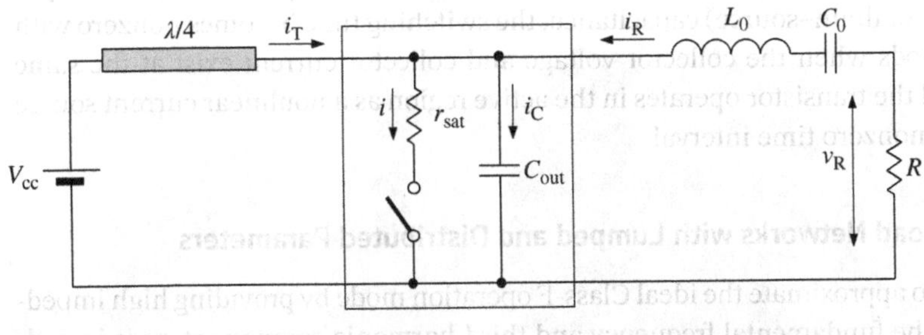

Figure 3.11 Schematic with saturation resistance and shunt capacitance.

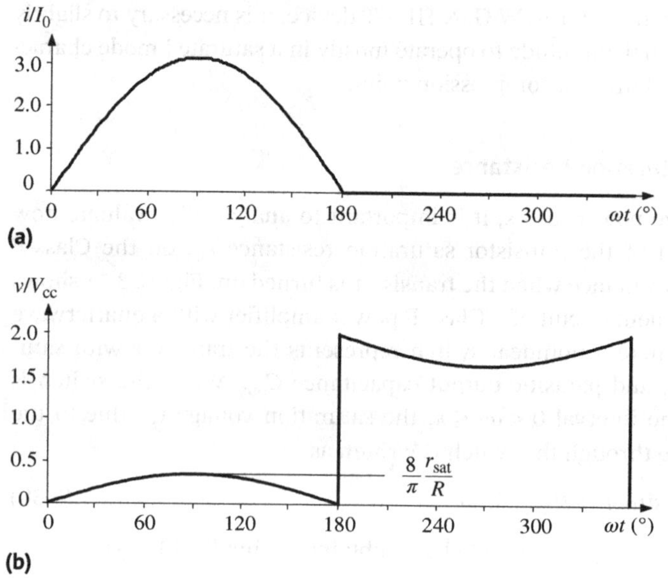

Figure 3.12 Current and voltage waveforms with nonzero saturation resistance.

$$\frac{P_{sat}}{P_0} = \frac{1}{2\pi}\int_0^{2\pi} \frac{i^2(\omega t) r_{sat}}{I_0 V_{cc}} d\omega t = \frac{r_{sat}}{2\pi I_0 V_{cc}} \int_0^{2\pi} (2I_R)^2 \sin^2\omega t \, d\omega t$$
$$= \frac{r_{sat} I_R}{V_{cc}} \frac{I_R}{I_0} = \frac{r_{sat}}{R} \frac{I_R}{I_0} \frac{V_R}{V_{cc}} = \frac{2r_{sat}}{R} \tag{3.32}$$

Consequently, the collector (or drain) efficiency can be evaluated by

$$\eta = 1 - \frac{P_{sat}}{P_0} = 1 - \frac{2r_{sat}}{R} \tag{3.33}$$

It should be noted that, at higher frequencies, due to the presence of the output (collector or drain–source) capacitance, the switching time becomes nonzero with time periods when the collector voltage and collector current exist at the same time, and the transistor operates in the active region as a nonlinear current source during a nonzero time interval.

3.1.4 Load Networks with Lumped and Distributed Parameters

In order to approximate the ideal Class-F operation mode by providing high impedances at the fundamental frequency and third-harmonic component, zero impedance at the second-harmonic component at the collector (or drain) of the transistor,

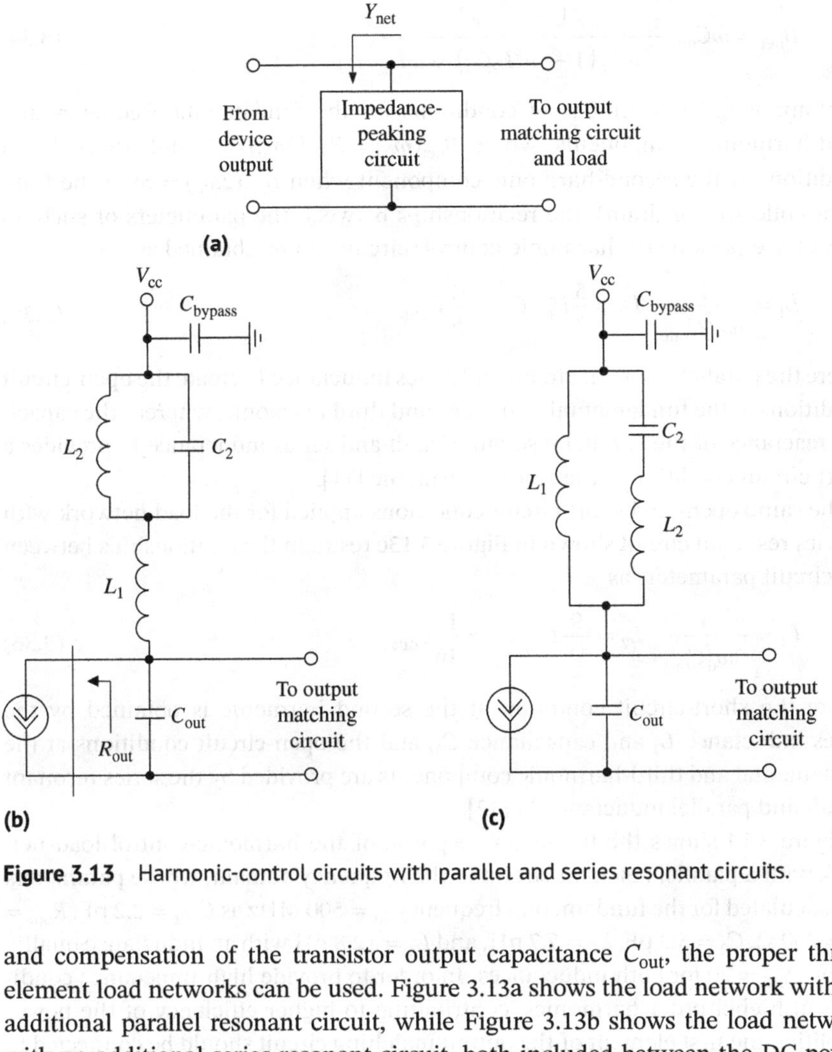

Figure 3.13 Harmonic-control circuits with parallel and series resonant circuits.

and compensation of the transistor output capacitance C_{out}, the proper three-element load networks can be used. Figure 3.13a shows the load network with an additional parallel resonant circuit, while Figure 3.13b shows the load network with an additional series resonant circuit, both included between the DC power supply and transistor output [11, 12]. Here, the output circuit of the transistor represents a multi-harmonic current source with an equivalent output resistance at the fundamental frequency R_{out} and a transistor output (collector or drain–source) capacitance C_{out}, where R_{out} is the defined as a ratio of the fundamental voltage amplitude to the fundamental current amplitude on the transistor.

The reactive part of the output admittance (or susceptance) $B_{net} = \text{Im}(Y_{net})$ of the load network, which includes a parallel resonant circuit and the transistor output capacitance C_{out}, shown in Figure 3.13b, can be written as

$$B_{\text{net}} = \omega C_{\text{out}} - \frac{1 - \omega^2 L_2 C_2}{\omega L_1 \left(1 - \omega^2 L_2 C_2\right) + \omega L_2} \tag{3.34}$$

By applying the open-circuit conditions for the fundamental-frequency and third-harmonic components when $B_{\text{net}}(\omega_0) = B_{\text{net}}(3\omega_0) = 0$ and short-circuit condition for the second-harmonic component when $B_{\text{net}}(2\omega_0) = \infty$ at the transistor collector (or drain), the relationships between the parameters of such an impedance-peaking (or harmonic-control) circuit can be obtained as

$$L_1 = \frac{1}{6\omega_0^2 C_{\text{out}}} \quad L_2 = \frac{5}{3} L_1 \quad C_2 = \frac{12}{5} C_{\text{out}} \tag{3.35}$$

where the parallel resonant circuit and series inductance L_1 create the open-circuit conditions at the fundamental frequency and third harmonic, whereas the capacitive reactance of the parallel resonant circuit and series inductance L_1 provides a short-circuit condition at the second harmonic [11].

The same open- and short-circuit conditions applied for the load network with a series resonant circuit shown in Figure 3.13c result in the relationships between the circuit parameters as

$$L_1 = \frac{4}{9\omega_0^2 C_{\text{out}}} \quad L_2 = \frac{9}{15} L_1 \quad C_2 = \frac{15}{16} C_{\text{out}} \tag{3.36}$$

where the short-circuit condition at the second harmonic is obtained by the series inductance L_2 and capacitance C_2, and the open-circuit conditions at the fundamental and third-harmonic components are provided by the series resonant circuit and parallel inductance L_1 [12].

Figure 3.14 shows the frequency response of the harmonic-control load network with a parallel resonant circuit in the frequency domain, whose parameters are calculated for the fundamental frequency f_0 = 500 MHz as C_{out} = 2.2 pF, $R_{\text{out}} = R_{\text{L}}$ = 200 Ω, C_2 = 5.3 pF, L_1 = 7.7 nH, and L_2 = 12.8 nH with an inductance quality factor Q_{ind} = 20 for both inductances. In order to provide high-impedance conditions at higher-order harmonics, contributing to higher efficiency of the power amplifier, the first element of the output matching circuit should be connected to the transistor output in series, being inductive.

At higher frequencies or in the case of a high-power mode when the transistor output resistance at the fundamental frequency is sufficiently small, effect of the output series inductance, including the bondwire and lead inductances of the transistor package, becomes significant. Figure 3.15a shows the simplified circuit schematic of a two-stage GaAs HBT power amplifier designed to use in cellphone transmitters in a frequency range of 698–716 MHz, with the final stage operated in a Class-F mode with second-harmonic control. Here, in order to realize the second-harmonic short-circuit condition at the collector, the series circuit composing of a

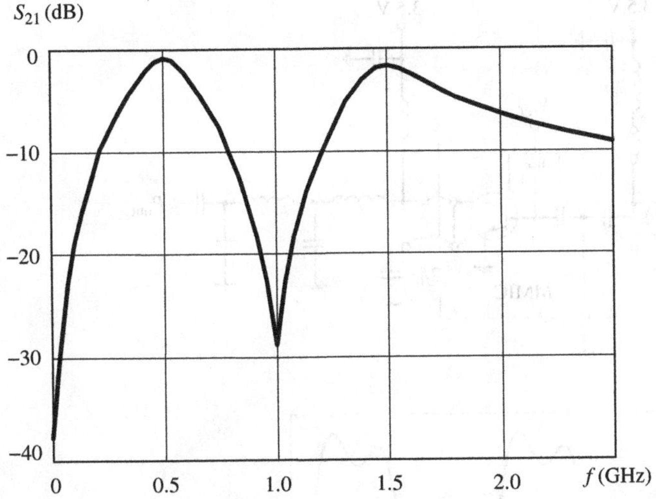

Figure 3.14 Frequency response of harmonic-control load network with parallel resonant circuit.

bondwire and grounded external shunt capacitor connected directly to the device collector is used. At the same time, the second-harmonic short circuit together with the DC-feed and series output matching inductances implemented as narrow high-impedance microstrip lines provides a sufficiently high inductive impedance at the third harmonic. As a result, the collector efficiency of 73% was achieved at 710 MHz, with the corresponding collector voltage and current waveforms shown in Figure 3.15b.

The transistor output resistance R_{out} at the fundamental frequency is evaluated as the equivalent resistance defined at the fundamental frequency for an ideal Class-F operation mode with zero saturation voltage, which is written as, $R_{out} = R_1^{(F)} = V_1/I_1$, where V_1 and I_1 are the fundamental-frequency voltage and current amplitudes at the transistor output, respectively. Using Eqs. (3.10) and (3.14) results in

$$R_1^{(F)} = \frac{8}{\pi^2}\frac{V_{cc}}{I_0} = \frac{8}{\pi^2}\frac{V_{cc}^2}{P_1} = \left(\frac{4}{\pi}\right)^2 R_1^{(B)} \tag{3.37}$$

where $R_1^{(B)} = V_{cc}/I_1 = V_{cc}^2/2P_1$ is the transistor output resistance at the fundamental frequency for an ideal Class-B mode.

For short-circuit conditions at all even harmonics and open-circuit conditions at the third harmonic in an idealized Class-F power amplifier, the maximum

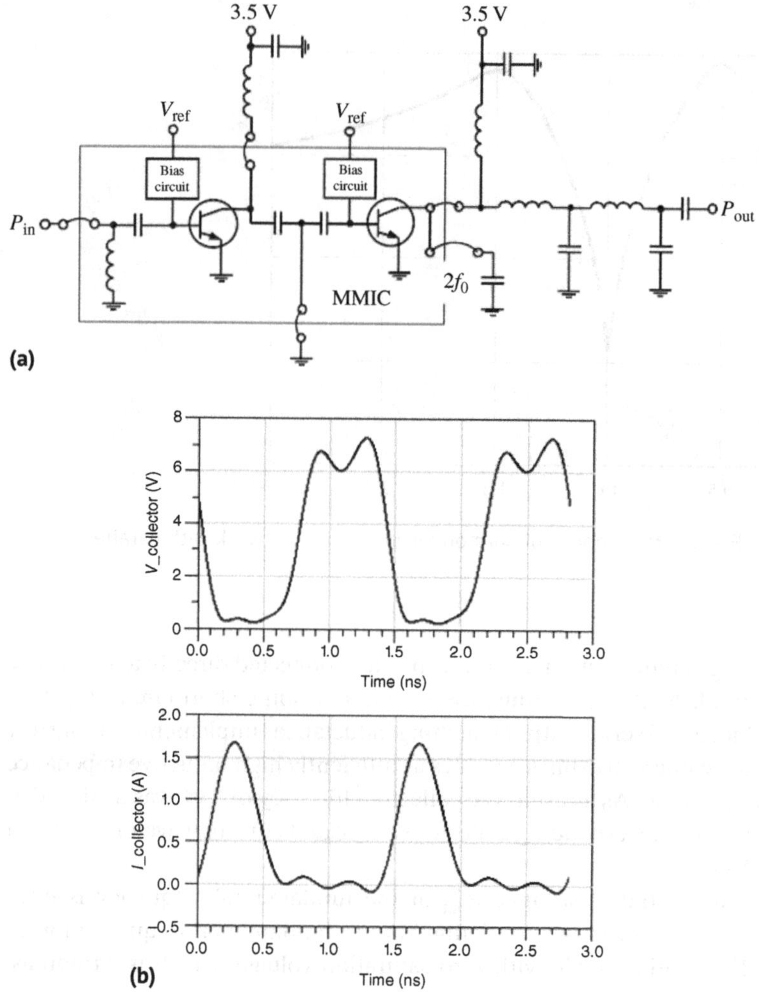

Figure 3.15 Schematic and collector voltage and current waveforms of a two-stage GaAs HBT power amplifier with second-harmonic control for handset application.

drain efficiency can be increased up to 90% [13]. At microwave frequencies, it is convenient to realize such an operation mode using the microstrip lines of different characteristic impedances and electrical lengths in the load network, including the impedance matching at the fundamental frequency. Figure 3.16 shows the simplified circuit schematic of the load network based on microstrip lines corresponding to the Class-F mode with a shunt short-circuited quarterwave

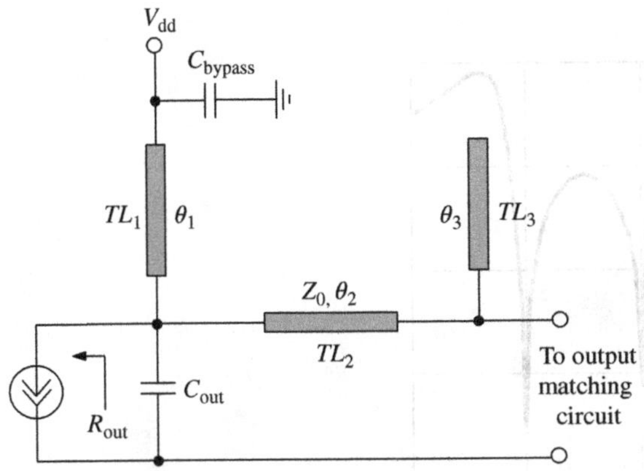

Figure 3.16 Transmission-line load network for Class-F power amplifier.

microstrip line TL_1 with the electrical length $\theta_1 = 90°$ that provides short-circuit conditions at even harmonics [11].

In this case, the electrical length θ_3 of an open-circuit stub TL_3 is equal to 30° to provide a quarter wavelength at the third harmonic for a short-circuit condition at the right-hand side of the series transmission line TL_2, the electrical length θ_2 of which should provide an inductive reactance to create a parallel resonant circuit with the transistor output capacitance C_{out} at the third harmonic. The quarterwave transmission line TL_1 located between the DC power supply and the collector (drain) provides a short-circuit condition at even harmonics. Such a simple impedance-control load network is very convenient for practical implementation [14]. The electrical lengths of the transmission lines in a Class-F load network with the third- and even-harmonic control are obtained at the fundamental frequency as

$$\theta_1 = \frac{\pi}{2} \quad \theta_2 = \frac{1}{3}\tan^{-1}\left(\frac{1}{3Z_0 \quad \omega_0 C_{out}}\right) \quad \theta_3 = \frac{\pi}{6} \tag{3.38}$$

where Z_0 is the characteristic impedance of the series transmission line TL_2 and ω_0 is the fundamental frequency. Note that the proper adjustment of the characteristic impedance of the transmission line TL_3 may contribute to the conjugate impedance matching at the fundamental frequency.

The frequency response of the microstrip impedance-control Class-F load network to be implemented on alumina substrate is shown in Figure 3.17, with the device output resistance R_{out} of 50 Ω, output capacitance C_{out} of 2.2 pF, characteristic impedance of microstrip lines Z_0 of 50 Ω, and electrical length θ_2 of 15°. In this case, an additional output impedance matching with the standard

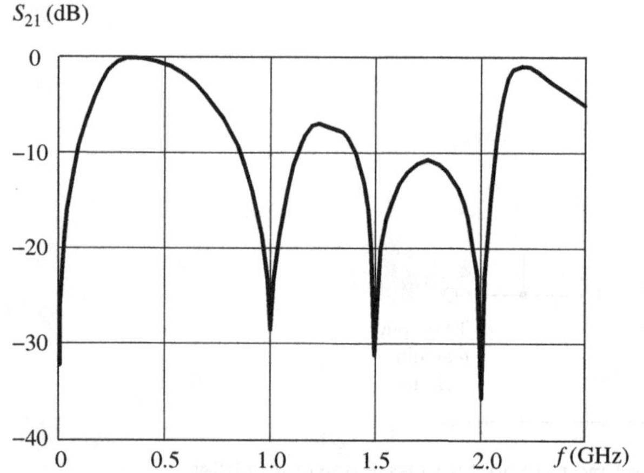

Figure 3.17 Frequency response of microstrip Class-F load network.

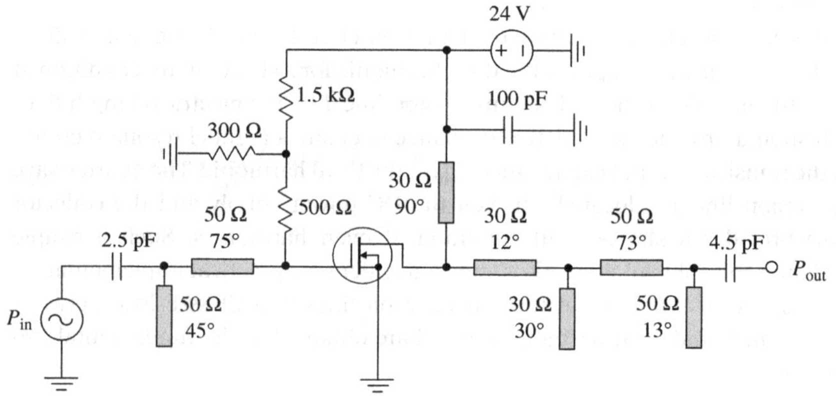

Figure 3.18 Simulation schematic of transmission-line LDMOSFET Class-F power amplifier.

50-Ω load at the fundamental frequency $f_0 = 500$ MHz is required, which can be achieved simply by optimizing the characteristic impedances of the series microstrip line TL_2 and open-circuit stub TL_3 instead of using a separate matching section.

Figure 3.18 shows the equivalent circuit of a simulated 500-MHz transmission-line Class-F power amplifier where an LDMOSFET device with a gate length of 1.25 μm and a gate width of 7 × 1.44 mm and T-type low-pass input and output matching circuits, consisting of the series microstrip lines, shunt open-circuit

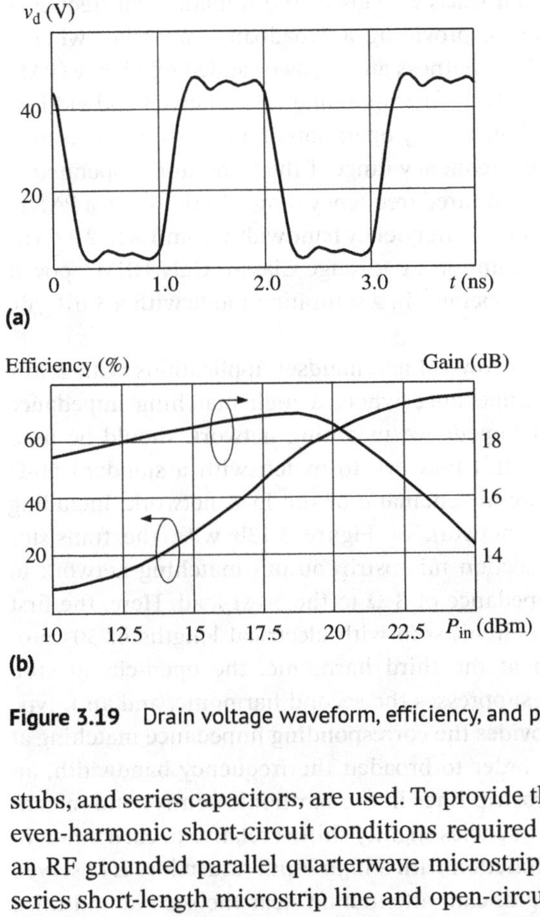

Figure 3.19 Drain voltage waveform, efficiency, and power gain.

stubs, and series capacitors, are used. To provide third-harmonic open-circuit and even-harmonic short-circuit conditions required for a Class-F operation mode, an RF grounded parallel quarterwave microstrip line and a combination of the series short-length microstrip line and open-circuit stub with electrical length of 30° at the fundamental frequency are used, resulting in a close to the square drain voltage waveform, as shown in Figure 3.19a, and providing the drain efficiency over 75% with a maximum output power of 8 W, as shown in Figure 3.19b.

3.1.5 Broadband Capability of Class-F Power Amplifiers

Generally, in an idealized Class-F operation mode, the reactances in the load network, including the transistor output capacitance, should be compensated to provide the conjugate matching at the fundamental frequency, short-circuit conditions at even harmonics, and open-circuit conditions at the third- and higher-order odd harmonics, resulting in symmetrical collector voltage and current waveforms. This means that the high-efficiency performance realized in a Class-F operation mode inherently is narrowband. However, the load network can be designed to achieve quasi-optimal Class-F loads at the fundamental-frequency and second-harmonic components, providing a broadband capability when a 21-dBm output power with a 0.5-dB flatness and a power-added efficiency (PAE) better than 65% from 575 to 915 MHz, corresponding to a relative bandwidth of 45%, was achieved [15]. The high-efficiency operation of a Class-F power amplifier can be extended over a wider frequency range if the transistor is operated in a deep saturation mode within the desired frequency range. In this case, a PAE of more than 60% was measured over 20% frequency bandwidth from 1.8 to 2.2 GHz, with an output power of 40–42 dBm for a two-stage Class-F GaN HEMT power amplifier, where the second stage operates in a saturation mode with a 3-dB gain compression point or more [16].

For high-power base station or low-voltage handset applications with a low value of the transistor output impedance where a high matching impedance ratio is required, an additional impedance matching network should be used in addition to the harmonic-control network to match with a standard 50-Ω load. Figure 3.20a shows the circuit schematic of the load network, including a third-order harmonic-control network of Figure 3.13b with the transistor output capacitance and a two-section microstrip output matching network to match the transistor output impedance of 3 Ω to the 50-Ω load. Here, the first series microstrip line and open-circuit stub with electrical lengths of 30° provide an open-circuit condition at the third harmonic, the open-circuit stub with an electrical length of 45° suppresses the second harmonic, and an L-type microstrip matching section provides the corresponding impedance matching at the fundamental frequency. In order to broaden the frequency bandwidth, an electrical length of 30° for microstrip lines in the harmonic-control and matching networks is chosen at the frequencies slightly shifted relative to each other. In this case, the flattened transfer function with a high input return loss is observed from 700 to 900 MHz with a significant attenuation near the second and third harmonics, as shown in Figure 3.20b.

Figure 3.21a shows the impedance behavior at the fundamental frequency from 700 to 900 MHz when the reactive part of the input impedance Z_{net} changes from capacitive at 700 MHz to inductive close to 900 MHz. At the same time, the second-harmonic reactance varies from inductive at 1.4 GHz to capacitive at 1.8 GHz, as shown in Figure 3.21b, similar to the third-harmonic reactance varying from inductive to capacitive as well, as shown in Figure 3.21c. The flattening of the real and imaginary parts of the load-network impedance is provided by the reactance compensation when the opposite reactances are provided by the harmonic-control and matching networks at the frequency bandwidth edges, resulting in the variation of $\text{Re}(Z_{net})$ from 2.2 to 3.8 Ω, as shown in Figure 3.22a, and $\text{Im}(Z_{net})$ from −2.5 to 2.5 Ω, as shown in Figure 3.22b.

In the idealized Class-F operation mode with symmetrical drain voltage and current waveforms, the reactances at the fundamental, second, and third harmonics are equal to zero, corresponding to the middle points of the impedance curves shown in Figure 3.21. However, when moving to higher frequencies, the impedance distribution becomes different, being inductive at the fundamental and purely capacitive at the second and third harmonics, corresponding to the Class-E mode with asymmetrical drain voltage and current waveforms. But when moving to lower frequencies, the impedance becomes capacitive at the fundamental and purely inductive at the second and third harmonics, corresponding to the inverse Class-E mode with asymmetrical drain voltage and current waveforms as well. This means that such a broadband performance in reality corresponds to a mixed-mode Class-E^{-1}FE operation mode, which is called in some publications as a continuous Class-F mode where the high-efficiency broadband performance is achieved at the expense of significant complications of the load network with multiple open-circuit stubs [17, 18]. Similarly, the broadband performance can be achieved using an inverse Class-F mode at the center bandwidth frequency and the Class-E mode with an inductive impedance at the fundamental and the

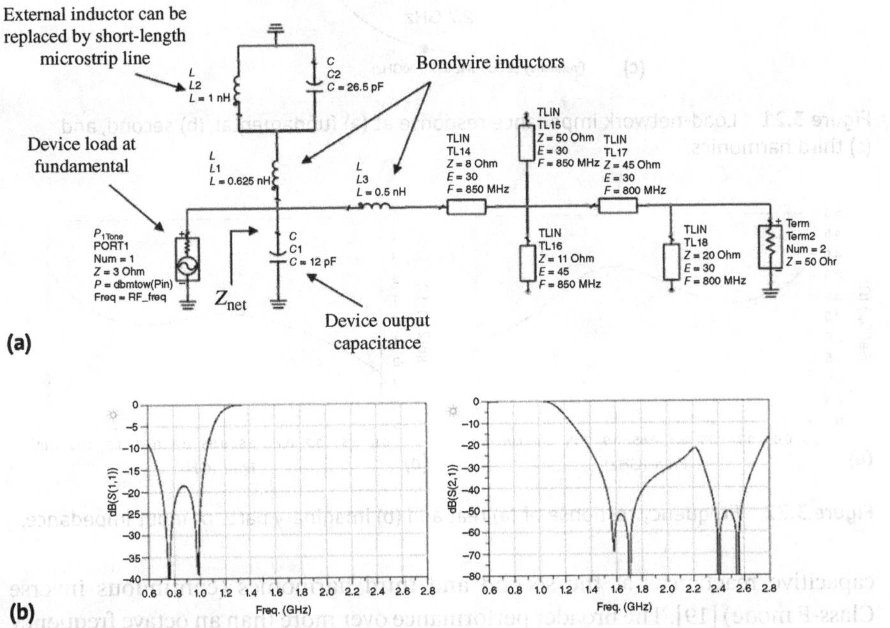

Figure 3.20 Schematic of (a) Class-F load network with broadband capability and (b) its frequency response.

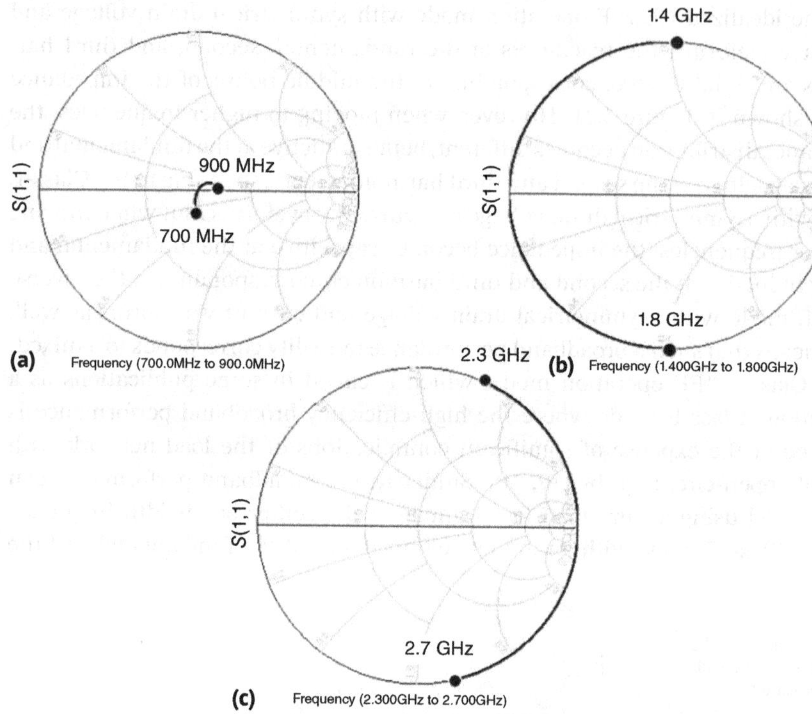

Figure 3.21 Load-network impedance response at (a) fundamental, (b) second, and (c) third harmonics.

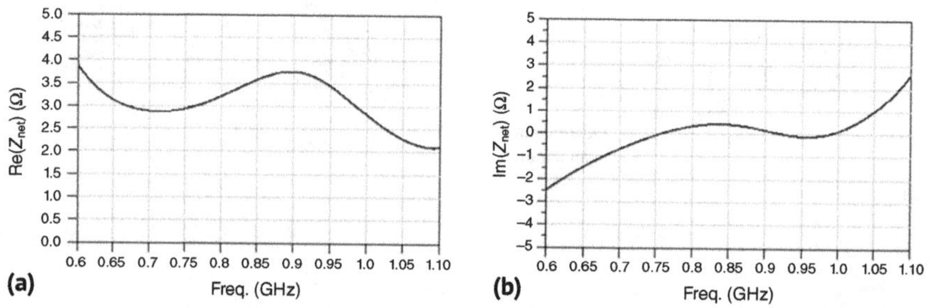

Figure 3.22 Frequency response of (a) real and (b) imaginary parts of input impedance.

capacitive reactances at the second and third harmonics (continuous inverse Class-F mode) [19]. The broader performance over more than an octave frequency bandwidth can be obtained using both continuous Class-F and continuous inverse Class-F modes simultaneously [20]. In all these implementations using so-called continuous Class-F and continuous inverse Class-F operation modes, the load networks contain a multiple number of open-circuit stubs, and the matching impedance ratio at the fundamental is not so high. It should be noted that, for any set of harmonic reactances, the maximum efficiency can be achieved by proper adjustment of the voltage and current waveforms and the fundamental-frequency load reactance [21].

3.2 Inverse Class F

A high-efficiency biharmonic operation mode can also be realized with an additional parallel resonant circuit tuned to the second harmonic of the fundamental frequency, being included in series with the load network, creating a high impedance at the second harmonic and resulting in an efficiency improvement. In this case, the partial Fourier series of the voltage $v(t)$ and current $i(t)$ for a biharmonic operation condition are written in normalized forms as

$$\frac{v(\omega t)}{V_0} = 1 - \frac{\pi}{2}\sin\omega t - \frac{2}{3}\cos 2\omega t \tag{3.39}$$

$$\frac{i(\omega t)}{I_0} = 1 + \frac{4}{\pi}\sin\omega t + \frac{4}{3\pi}\sin 3\omega t \tag{3.40}$$

where V_0 is the DC-supply voltage and I_0 is the DC current component. However, the ideal half-sine voltage and square current waveforms can be obtained by considering an infinite number of voltage and current harmonics in a Fourier-series expansion.

Figure 3.23 shows that the shapes of the voltage and current waveforms can be significantly changed with increasing fundamental current amplitude by adding even one additional harmonic component being properly phased. For example, the combination of the fundamental-frequency and third-harmonic components with an out-of-phase shift at the center of symmetry results in a flattened current waveform with depression in its center. It is clearly seen from Figure 3.23a that the proper ratio between the amplitudes of the fundamental and third-harmonic components can provide the flattened current waveform with minimum depression and maximum difference between its peak amplitude and amplitude of the fundamental component. Similarly, the combination of the fundamental-frequency and second-harmonic components, being in phase at the midpoints, sharpens the current waveform, corresponding to minimum values of the voltage waveform, as shown in Figure 3.23b. The optimum ratio between the amplitudes of the fundamental and second harmonics can maximize a peak value of the voltage waveform, with its minimum value determined by the device saturation

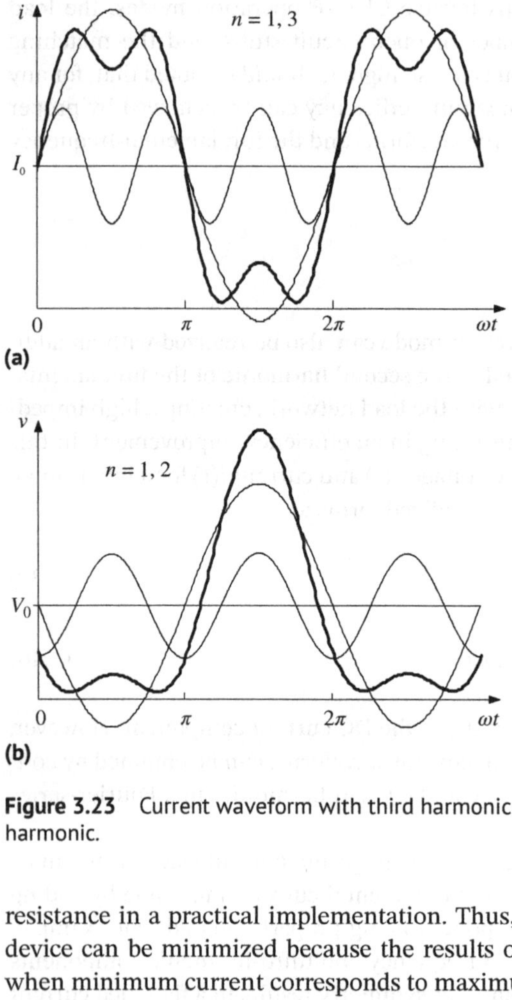

Figure 3.23 Current waveform with third harmonic and voltage waveform with second harmonic.

resistance in a practical implementation. Thus, the power loss due to the active device can be minimized because the results of the integration over the period when minimum current corresponds to maximum voltage will give a small value compared with the power delivered to the load.

The biharmonic operation mode using a second-harmonic parallel resonant circuit connected in series to the anode of the vacuum tube, as shown in Figure 3.24a, was first described and analyzed in the early 1940s [3, 22]. It was shown that the symmetrical anode current waveform and level of its depression can be provided with the opposite phase conditions between the fundamental and second-harmonic components and an optimum value of the ratio between their voltage amplitudes. In practical vacuum-tube power amplifiers intended to operate at very high frequencies, the peak output power and anode efficiency can

(a)

(b)

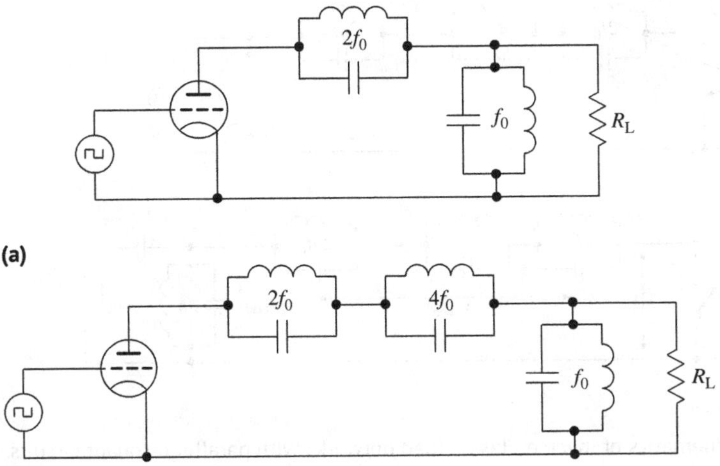

Figure 3.24 Schematics of multi-harmonic vacuum-tube power amplifiers.

therefore be increased by 1.15–1.2 times [23]. In addition, it was later suggested that using an additional resonator tuned to the fourth harmonic and connected in series with the second-harmonic resonator, as shown in Figure 3.24b, contributes to further increase of anode efficiency with an approximate square voltage-driving waveform [24].

3.2.1 Idealized Inverse Class-F Mode

Figure 3.25 shows the simplified schematics of an idealized inverse Class-F power amplifier with a multi-resonant load network to control the harmonic content of its collector voltage and current waveforms, thereby shaping them to reduce dissipation and to increase efficiency, where the voltage $v_{\mathrm{even}} = v_{\mathrm{even}}(2n\omega t)$ is the voltage across the even-harmonic parallel resonators connected in series and current $i_{\mathrm{odd}} = i_{\mathrm{odd}}\left[(2n+1)\,\omega t\right]$ is the current through the fundamental-frequency parallel resonator which includes odd harmonics, $n = 1,2,\ldots$.

The term "inverse" implies the interchanged collector voltage and current waveforms compared to a conventional case of Class F under the same idealized assumptions, with a sinewave load current shown in Figure 3.26a, the ideal collector current square waveform, consisting of the fundamental and odd harmonics, shown in Figure 3.26b, and the collector voltage half-sine waveform,

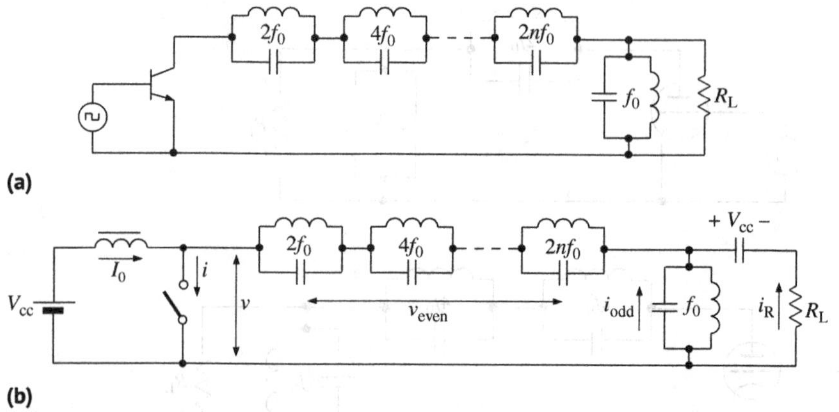

Figure 3.25 Schematics of inverse Class-F load networks with parallel resonant circuits.

consisting of the fundamental and even harmonics, shown in Figure 3.26c, which correspond to an inverse Class F.

The current and voltage waveforms corresponding to an inverse Class-F operation mode can be written using Eqs. (3.5) and (3.9) for a conventional Class-F mode as

$$i(\omega t) = 2I_0 - i(\omega t + \pi) \tag{3.41}$$

$$v(\omega t) = V_R(\sin\omega t + |\sin\omega t|) \tag{3.42}$$

where I_0 is the DC current amplitude and V_R is the fundamental-frequency amplitude, with the maximum value of the collector current $2I_0$.

The expressions for the DC-supply voltage V_{cc}, fundamental voltage amplitude V_1, and fundamental current amplitude I_1 in the collector (or drain) voltage and current waveforms using a Fourier analysis and Eqs. (3.41) and (3.42) can be obtained as

$$V_{cc} = \frac{1}{2\pi}\int_0^{\pi} 2V_R \sin\omega t\, d\omega t = \frac{2V_R}{\pi} \tag{3.43}$$

$$V_1 = \frac{1}{\pi}\int_0^{\pi} 2V_R \sin^2\omega t\, d\omega t = V_R \tag{3.44}$$

$$I_1 = I_R = \frac{1}{\pi}\int_{\pi}^{2\pi} 2I_0 \sin(\omega t + \pi) d\omega t = \frac{4I_0}{\pi} \tag{3.45}$$

where I_R is sinewave load current amplitude.

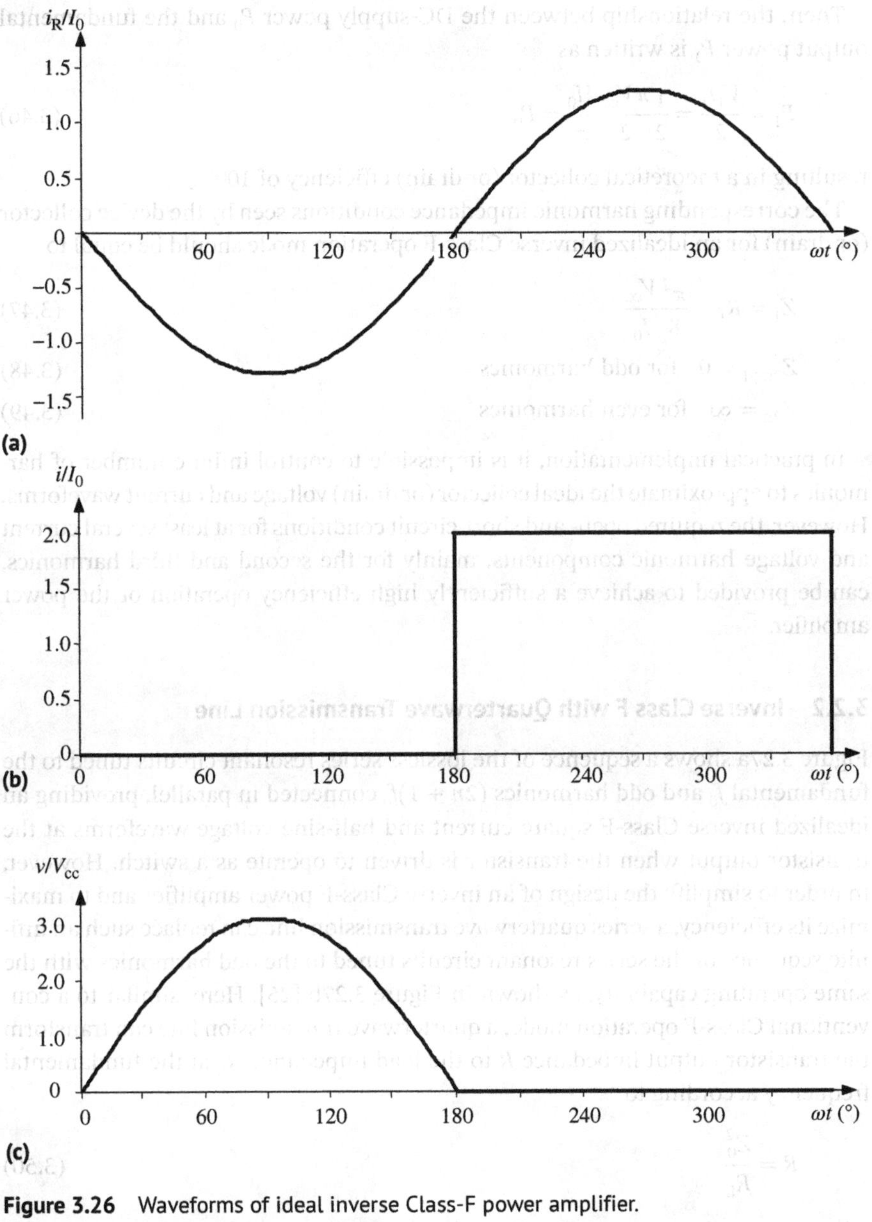

Figure 3.26 Waveforms of ideal inverse Class-F power amplifier.

Then, the relationship between the DC-supply power P_0 and the fundamental output power P_1 is written as

$$P_1 = \frac{V_1 I_1}{2} = \frac{1}{2}\frac{\pi V_{cc}}{2}\frac{4I_0}{\pi} = P_0 \tag{3.46}$$

resulting in a theoretical collector (or drain) efficiency of 100%.

The corresponding harmonic impedance conditions seen by the device collector (or drain) for an idealized inverse Class-F operation mode should be equal to

$$Z_1 = R_1 = \frac{\pi^2}{8}\frac{V_{cc}}{I_0} \tag{3.47}$$

$$Z_{2n+1} = 0 \quad \text{for odd harmonics} \tag{3.48}$$

$$Z_{2n} = \infty \quad \text{for even harmonics} \tag{3.49}$$

In practical implementation, it is impossible to control infinite number of harmonics to approximate the ideal collector (or drain) voltage and current waveforms. However, the required open- and short-circuit conditions for at least several current and voltage harmonic components, mainly for the second and third harmonics, can be provided to achieve a sufficiently high-efficiency operation of the power amplifier.

3.2.2 Inverse Class F with Quarterwave Transmission Line

Figure 3.27a shows a sequence of the lossless series resonant circuits tuned to the fundamental f_0 and odd harmonics $(2n + 1)f_0$ connected in parallel, providing an idealized inverse Class-F square current and half-sine voltage waveforms at the transistor output when the transistor is driven to operate as a switch. However, in order to simplify the design of an inverse Class-F power amplifier and to maximize its efficiency, a series quarterwave transmission line can replace such an infinite sequence of the series resonant circuits tuned to the odd harmonics with the same operating capability, as shown in Figure 3.27b [25]. Here, similar to a conventional Class-F operation mode, a quarterwave transmission line can transform the transistor output impedance R to the load impedance R_L at the fundamental frequency according to

$$R = \frac{Z_0^2}{R_L} \tag{3.50}$$

where Z_0 is the transmission-line characteristic impedance. Because of the presence of the quarterwave transmission line, the output sinewave voltage on the load resistor R_L is shifted by 90° relative to the fundamental frequency of the drain voltage.

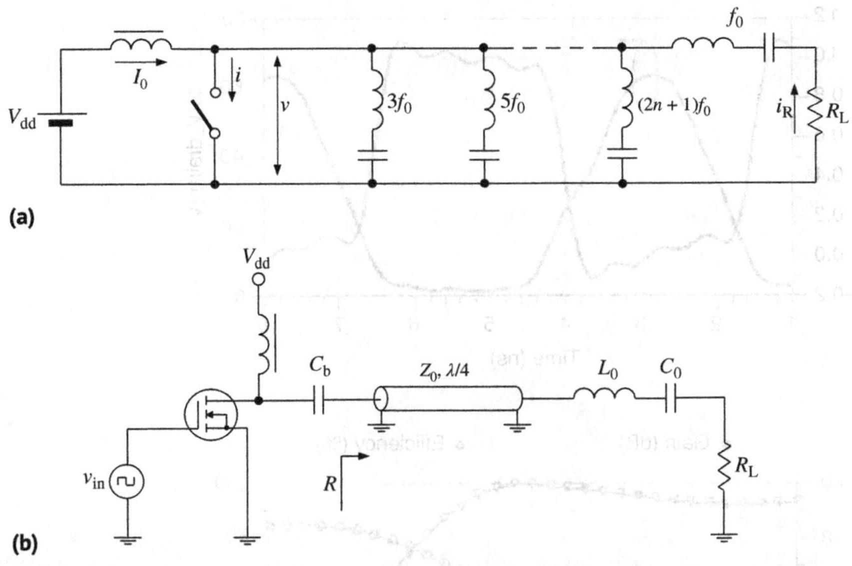

Figure 3.27 Schematics of ideal inverse Class-F load networks.

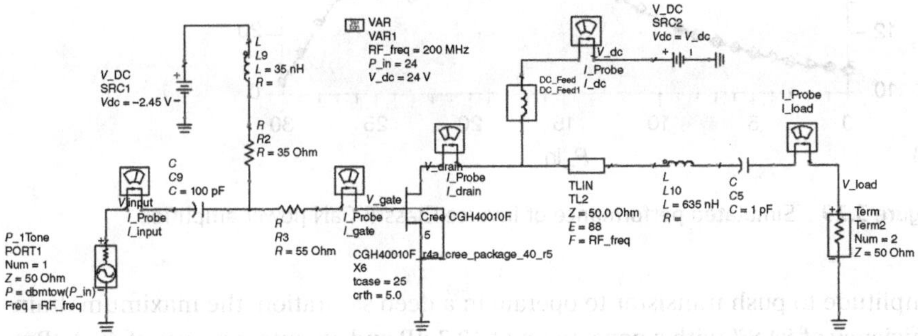

Figure 3.28 Simulation setup of inverse Class-F GaN power amplifier with series quarterwave line.

The simulation setup of the inverse Class-F power amplifier with a series quarterwave transmission line using a 10-W GaN HEMT CGH40010 transistor is shown in Figure 3.28 [10]. Here, the lossy *RL* shunt circuit and series 55-Ω resistor are used to match to the source impedance and to provide unconditional operation at the operating frequency of 200 MHz. Figure 3.29a shows the simulated drain voltage waveform, which is close to a half-sine wave, and current waveform, which is close to a square wave. For a sinewave driving signal with increased

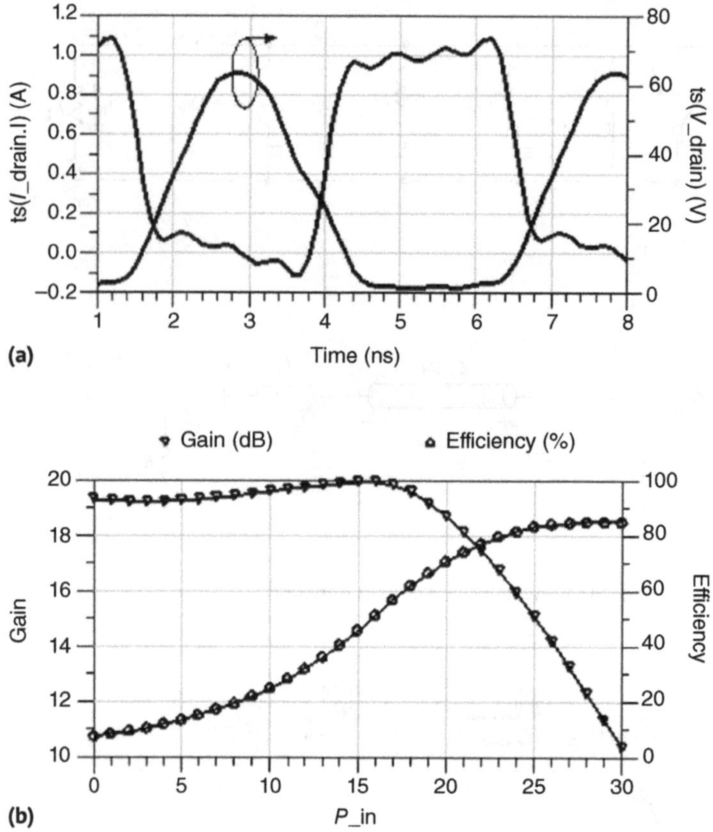

Figure 3.29 Simulated performance of inverse Class-F GaN power amplifier.

amplitude to push transistor to operate in a deep saturation, the maximum drain efficiency of 84.8% with a power gain of 12.3 dB and an output power of 40.3 dBm was simulated with an input power of 28 dBm, as shown in Figure 3.29b.

3.2.3 Load Networks with Lumped and Distributed Parameters

Practical realization of an inverse Class-F load network with multiple series resonant circuits can cause a significant implementation problem. Therefore, in order to compromise efficiency and circuit complexity, it is sufficient to use the three- or four-element resonant circuit in the load network of the power amplifier. In this case, the operation with the second-harmonic open circuit and third-harmonic short circuit is a promising concept for low-voltage power amplifiers [26]. In

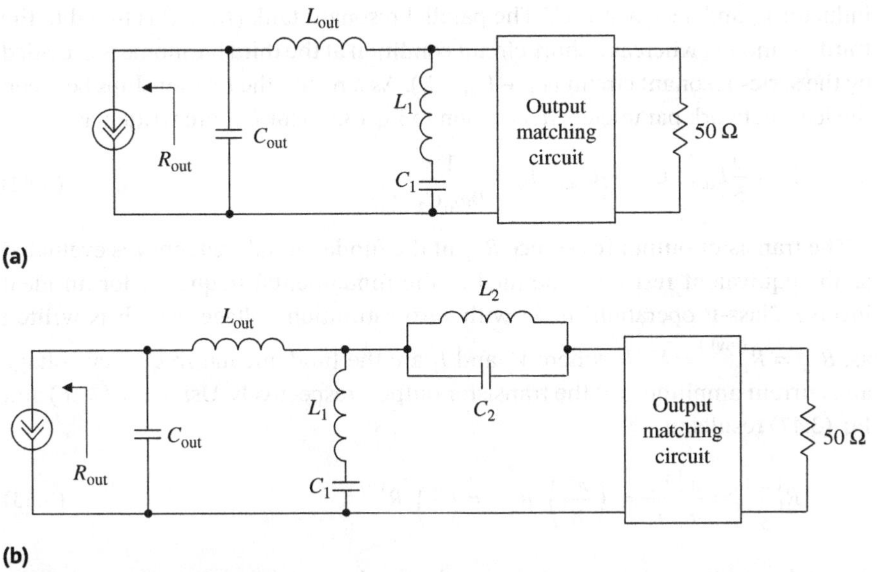

Figure 3.30 Lumped load networks with second- and third-harmonic control.

addition, it is necessary to take into account that, in practice, both extrinsic and intrinsic transistor parasitic elements, such as the output shunt capacitance (collector or drain–source) or series inductance (bondwire and package lead), have a substantial effect on the efficiency.

Figure 3.30a shows the equivalent circuit of the second-harmonic impedance-control load network, where the series circuit consisting of an inductor L_1 and a capacitor C_1 creates a resonance at the second harmonic. Because the device output inductance L_{out} and capacitance C_{out} are tuned to create an open-circuit condition at the second harmonic, the transistor collector (or drain) sees the resultant high impedance at the second harmonic. If necessary, an external inductance can be added to the transistor output inductance L_{out} directly at the transistor output (collector or drain). As a result, the values of the load-network parameters for given C_{out} and an arbitrarily chosen C_1 are

$$L_{out} = \frac{1}{4\omega_0^2 C_{out}} \quad L_1 = \frac{1}{4\omega_0^2 C_1} \tag{3.51}$$

In order to control both the second and third harmonics, a parallel resonant circuit can be connected in series to the output in addition to the shunt series resonant circuit, as shown in Figure 3.30b. Here, the required short-circuit condition at the second harmonic is provided by the series resonant circuit consisting of an inductor L_1 and a capacitor C_1. The parallel resonant tank (L_2, C_2) is tuned to the third harmonic, whereas a short-circuit condition at the third harmonic is provided by the series resonant circuit ($L_1 + L_{out}$, C_1). As a result, the relationships between the load-network parameters in addition to Eq. (3.51) for L_{out} are written as

$$L_1 = \frac{9}{5}L_{out} \quad C_1 = \frac{5}{9}C_{out} \quad L_2 = \frac{1}{9\omega_0^2 C_2} \tag{3.52}$$

The transistor output resistance R_{out} at the fundamental frequency is evaluated as the equivalent resistance defined at the fundamental frequency for an ideal inverse Class-F operation mode with zero saturation voltage, which is written as, $R_{out} = R_1^{(invF)} = V_1/I_1$, where V_1 and I_1 are the fundamental-frequency voltage and current amplitudes at the transistor output, respectively. Using Eq. (3.16) and Eq. (3.47) results in

$$R_1^{(invF)} = \frac{\pi^2}{8}\frac{V_{cc}}{I_0} = \left(\frac{\pi^2}{8}\right)^2 R_1^{(F)} = \left(\frac{\pi}{2}\right)^2 R_1^{(B)} \tag{3.53}$$

where $R_1^{(F)}$ is the output resistance at the fundamental frequency for a conventional Class-F operation mode, and $R_1^{(B)} = V_{cc}/I_1 = V_{cc}^2/2P_1$ is the output resistance at the fundamental frequency for an ideal Class-B mode. From Eq. (3.53), it follows that the equivalent output resistance for an ideal inverse Class-F mode is more than 2.4 times higher compared to that in the Class-B mode and 1.5 times higher compared to that in a Class-F mode, which simplifies the load-network design due to a lower impedance transformation ratio.

Unlike an idealized Class-F power amplifier where the short-circuit conditions for even harmonics can be provided by a single parallel transmission line, an inverse Class-F power amplifier cannot simultaneously provide all voltage third- and higher-order odd harmonic short-circuit conditions. Nevertheless, applying the current second-harmonic peaking and voltage third-harmonic shorting with a sufficiently simple circuit schematic convenient for practical implementation can result in a maximum drain efficiency of more than 80% [27]. Figure 3.31 shows the output transmission-line network with the second- and third-harmonic control, whose structure is similar to that used to provide a Class-F operation mode.

Here, the electrical lengths of the transmission lines at the fundamental frequency are equal to

$$\theta_1 = \frac{\pi}{3} \quad \theta_2 = \frac{1}{2}\tan^{-1}\left[\left(2Z_0\omega_0 C_{out} + \frac{1}{\sqrt{3}}\right)^{-1}\right] \quad \theta_3 = \frac{\pi}{4} \tag{3.54}$$

where Z_0 is the transmission-line characteristic impedance. The transmission line TL_1 with electrical length $\theta_1 = 60°$ at the fundamental frequency provides a

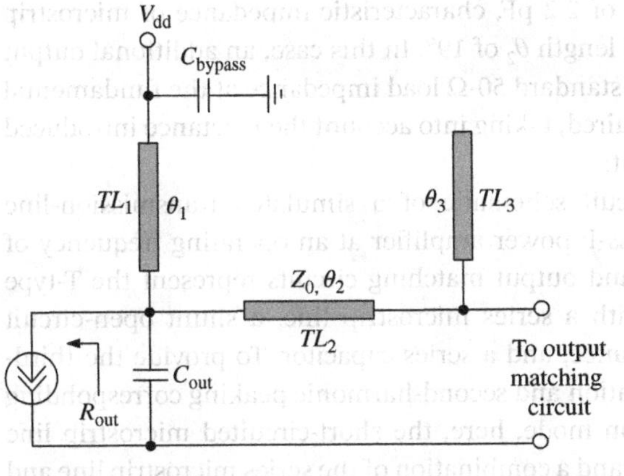

Figure 3.31 Transmission-line load network for inverse Class-F power amplifier.

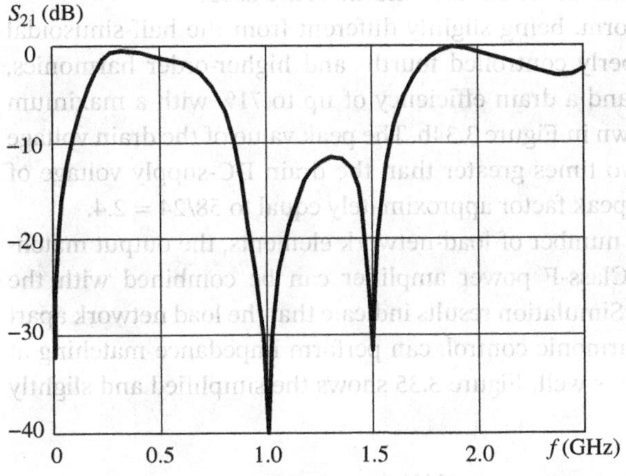

Figure 3.32 Frequency response of the transmission-line network of Figure 3.31.

short-circuit condition for the third harmonic and introduces a capacitive reactance at the second harmonic. The open-circuit stub TL_3, with electrical length $\theta_3 = 45°$, creates a short-circuit condition at the end of the transmission line TL_2 at the second harmonic. Thus, the series transmission line TL_2 having an inductive reactance is tuned to the parallel resonance condition at the second harmonic with the device output capacitance C_{out} and short-circuited transmission line TL_1.

The frequency response of the microstrip impedance-control inverse Class-F load network with second- and third-harmonic control to be implemented on alumina substrate is shown in Figure 3.32, with the device output resistance R_{out} of

50 Ω, output capacitance C_{out} of 2.2 pF, characteristic impedance of microstrip lines Z_0 of 50 Ω, and electrical length θ_2 of 19°. In this case, an additional output impedance matching with the standard 50-Ω load impedance at the fundamental frequency $f_0 = 500$ MHz is required, taking into account the reactance introduced by the harmonic-control circuit.

Figure 3.33 shows the circuit schematic of a simulated transmission-line 28-V LDMOSFET inverse Class-F power amplifier at an operating frequency of 500 MHz, where both input and output matching circuits represent the T-type low-pass matching circuits with a series microstrip line, a shunt open-circuit stub having a capacitive reactance, and a series capacitor. To provide the third-harmonic short-circuit termination and second-harmonic peaking corresponding to an inverse Class-F operation mode, here, the short-circuited microstrip line with an electrical length of 60° and a combination of the series microstrip line and open-circuit stub with an electrical length of 45° for the second-harmonic open-circuit and the third-harmonic short-circuit conditions are used.

The drain voltage waveform, being slightly different from the half-sinusoidal shape in view of not properly controlled fourth- and higher-order harmonics, is shown in Figure 3.34a, and a drain efficiency of up to 71% with a maximum output power of 8 W is shown in Figure 3.34b. The peak value of the drain voltage amplitude is more than two times greater than the drain DC-supply voltage of 24 V, resulting in a voltage peak factor approximately equal to 58/24 = 2.4.

In order to minimize the number of load-network elements, the output matching circuit in an inverse Class-F power amplifier can be combined with the harmonic-control network. Simulation results indicate that the load network apart of the second- and third-harmonic control, can perform impedance matching at the fundamental frequency as well. Figure 3.35 shows the simplified and slightly

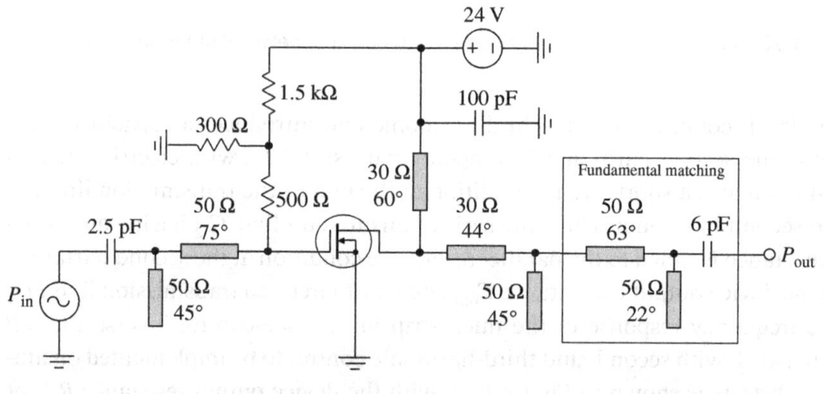

Figure 3.33 Simulation schematic of 500-MHz transmission-line inverse Class-F power amplifier.

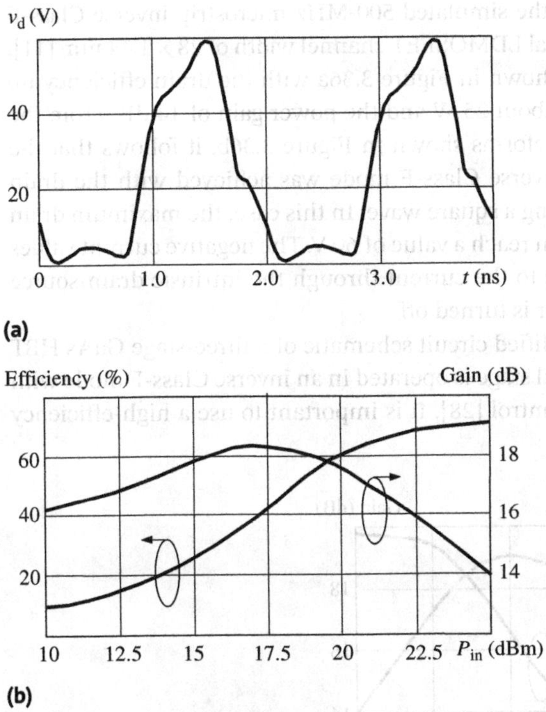

Figure 3.34 Simulated voltage waveform, efficiency, and power gain.

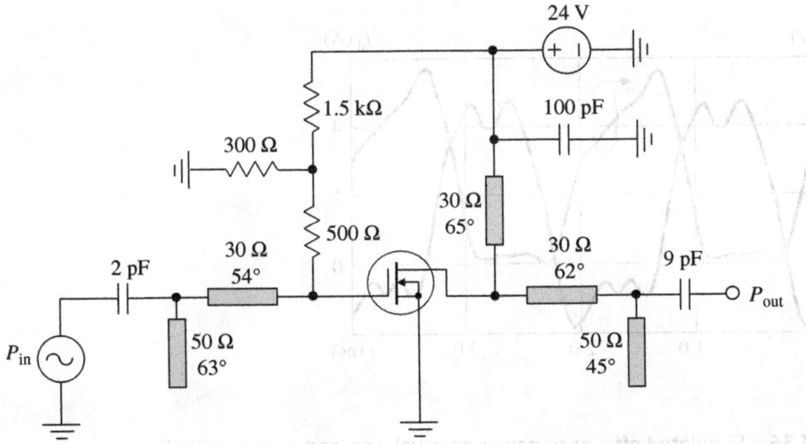

Figure 3.35 Simulated modified 500-MHz microstrip inverse Class-F power amplifier.

modified circuit schematic of the simulated 500-MHz microstrip inverse Class-F high-power amplifier with a total LDMOSFET channel width of 28 × 1.44 mm [11].

The simulation results are shown in Figure 3.36a with the drain efficiency up to 78% at an output power of about 25 W and the power gain of 14 dB. From the drain voltage and current waveforms shown in Figure 3.36b, it follows that the operation mode close to an inverse Class-F mode was achieved with the drain current waveform approximating a square wave. In this case, the maximum drain voltage amplitude does not even reach a value of 60 V. The negative current values in a current waveform are due to the current through the intrinsic drain-source capacitance when the transistor is turned off.

Figure 3.37a shows the simplified circuit schematic of a three-stage GaAs HBT power amplifier, where the final stage is operated in an inverse Class-F mode with second- and third-harmonic control [28]. It is important to use a high-efficiency

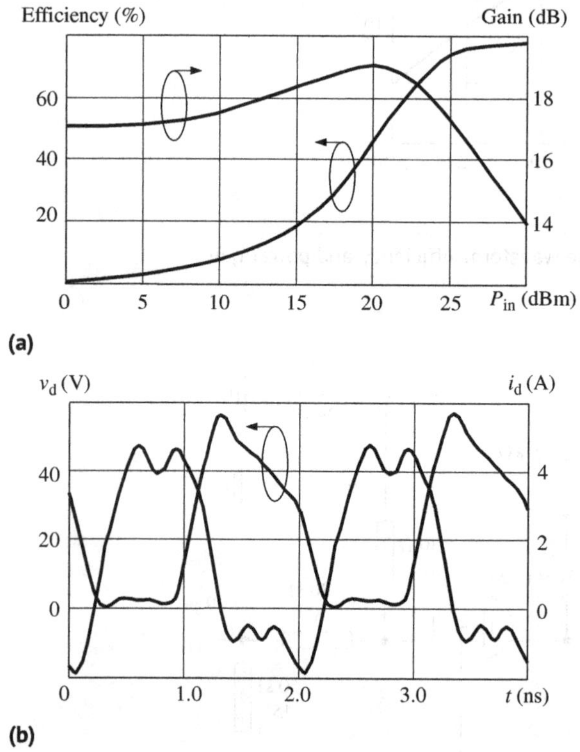

Figure 3.36 Simulated efficiency, power gain, voltage, and current waveforms.

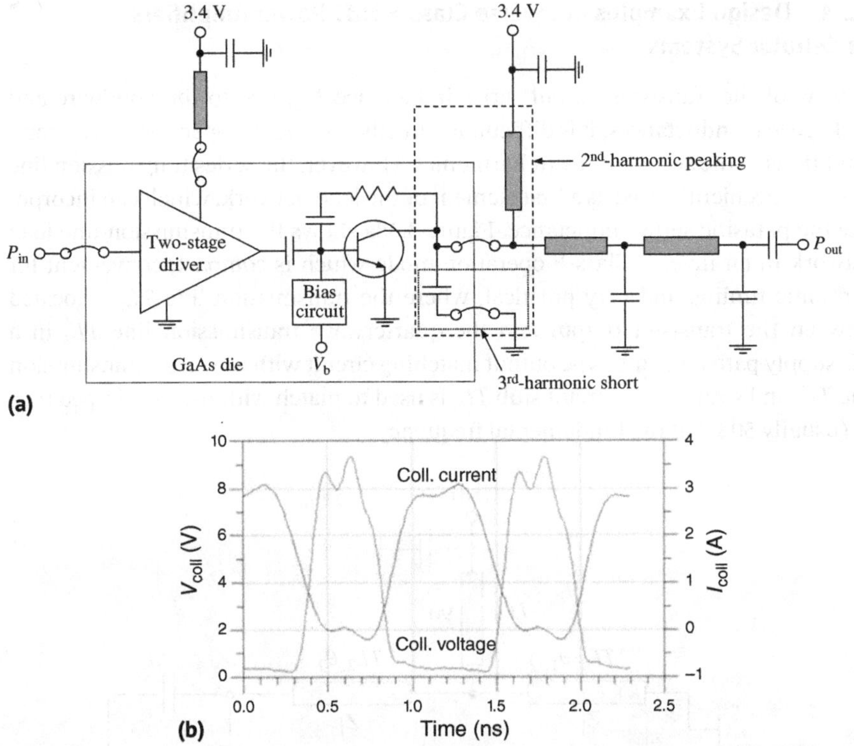

Figure 3.37 Schematic and waveforms of GaAs HBT power amplifier.

final stage in cellular handsets to reduce power consumption and cost. In this case, a two-section output matching network is used to match the device output impedance of $2.2 + j1.1\ \Omega$ at the fundamental frequency to achieve a flat output power response in a frequency bandwidth of 824–915 MHz. The series resonant circuit with an on-die capacitor and a bondwire is tuned to resonate at the third harmonic to provide the third-harmonic short-circuit condition. At the same time, the second-harmonic high-impedance condition is realized by combining the reactances in the load network, using a parallel microstrip line with an electrical length of about 45°. The collector current and voltage waveforms, approximating an inverse Class-F operation mode with a close to rectangular current waveform is shown in Figure 3.37b. As a result, the output power of 35 dBm with a flatness of ±0.125 dB and a PAE of about 60% in the frequency band of 824–915 MHz were measured for the module, occupying an area of 12 mm^2.

3.2.4 Design Examples of Inverse Class-F HBT Power Amplifiers for Cellular Systems

In view of the transistor output series inductance L_{out} due to the bondwire and package lead inductances, it is difficult to directly provide the open- or short-circuit conditions at the second or third harmonics. However, the series transmission line can be conveniently used as a first element of the load network, which can incorporate the parasitic series inductance. Figure 3.38a shows the transmission-line load network in an inverse Class-F operation mode which is compact, convenient for harmonic tuning, and very practical, where the transmission line TL_1 is located between the transistor output and the quarterwave transmission line TL_3 in a DC-supply path and an L-type output matching circuit with the series transmission line TL_2 and shunt open-circuit stub TL_4 is used to match with the load impedance R_L (usually 50 Ω) at the fundamental frequency.

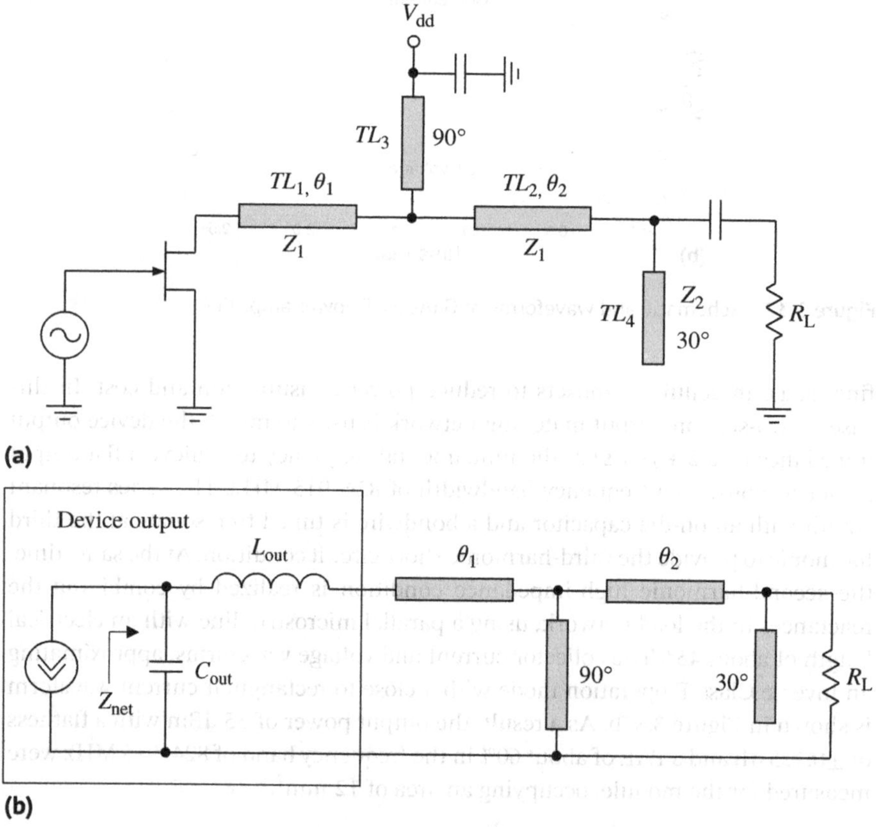

Figure 3.38 Equivalent circuits of transmission-line inverse Class-F power amplifier.

The equivalent circuit of a transmission-line inverse Class-F load network is shown in Figure 3.38b, with a conjugate load matching at the fundamental, a high reactance at the second harmonic, and a low reactance at the third harmonic created at the transistor output providing by the two series transmission lines TL_1 and TL_2, whose electrical lengths depend on the values of the transistor output shunt capacitance C_{out} and series inductance L_{out}, a quarterwave short-circuit stub TL_3, and an open-circuit stub TL_4 with electrical length of 30° [29, 30]. The output capacitance C_{out} can represent both intrinsic drain–source capacitance C_{ds} and extrinsic drain pad-contact capacitance C_{dp} of the nonlinear large-signal GaN HEMT equivalent circuit, while the output inductance L_{out} model the combined effect of the bondwire and package lead inductances [31].

The impedance conditions at the fundamental, second, and third harmonics seen by the transistor current source for an inverse Class-F load network can be written using Eqs. (3.47)–(3.49) as

$$\mathrm{Re}Z_{net}(\omega_0) = R \tag{3.55}$$

$$\mathrm{Im}Z_{net}(2\omega_0) = \infty \tag{3.56}$$

$$\mathrm{Im}Z_{net}(\omega_0) = \mathrm{Im}\ Z_{net}(3\omega_0) = 0 \tag{3.57}$$

where the transistor output resistance R at the fundamental frequency is defined in an inverse Class-F operation mode as

$$R = \frac{\pi^2}{4} R^{(B)} = \frac{\pi^2}{8} \frac{(V_{dd} - V_{sat})^2}{P_{out}} \tag{3.58}$$

where $R^{(B)} = (V_{dd} - V_{sat})^2/2P_{out}$ is the transistor output resistance in a Class-B operation mode, V_{dd} is the DC-supply voltage, V_{sat} is the saturation voltage, and P_{out} is the output power at the fundamental frequency.

Figure 3.39a shows the transmission-line load network shown in Figure 3.38b at the fundamental frequency, where an L-type transmission-line matching network provides an impedance matching of the transistor output impedance R with the load impedance R_L at the fundamental frequency by properly setting the transmission-line characteristic impedances Z_1 and Z_2. The load network at the second harmonic is shown in Figure 3.39b where an open-circuit condition for the second harmonic by the second-harmonic tank with the shunt capacitor C_{out} and series inductive reactance due to series inductance L_{out} and transmission line TL_1 considering the short-circuit condition of the quarterwave transmission line TL_3. Figure 3.39c shows the load network at the third harmonic when an open-circuit condition at the input of the quarterwave transmission line TL_3 and zero reactance of the series resonant circuit composed of a series inductance L_{out} and series transmission lines TL_1 and TL_2 are created.

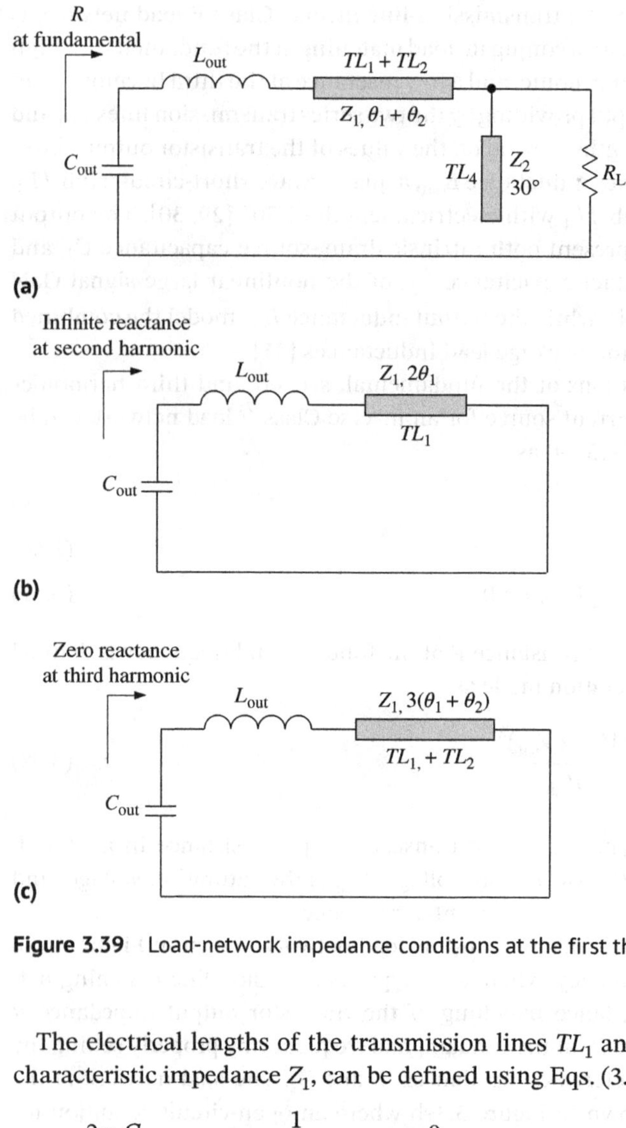

Figure 3.39 Load-network impedance conditions at the first three harmonics.

The electrical lengths of the transmission lines TL_1 and TL_2, having the same characteristic impedance Z_1, can be defined using Eqs. (3.56) and (3.57) as

$$2\omega_0 C_{out} - \frac{1}{2\omega_0 L_{out} + Z_1 \tan 2\theta_1} = 0 \tag{3.59}$$

$$3\omega_0 L_{out} + Z_1 \tan 3(\theta_1 + \theta_2) = 0 \tag{3.60}$$

resulting in

$$\theta_1 = \frac{1}{2}\tan^{-1}\frac{1 - (2\omega_0)^2 L_{out} C_{out}}{2Z_1 \omega_0 C_{out}} \tag{3.61}$$

$$\theta_2 = \frac{\pi}{3} - \frac{1}{3}\tan^{-1}\frac{3\omega_0 L_{out}}{Z_1} - \theta_1 \tag{3.62}$$

where the characteristic impedance Z_1 should be initially chosen.

The transmission-line load network shown in Figure 3.40a can be equivalently represented by a lumped low-pass Π-type matching circuit, where $C = \tan 30°/\omega_0 Z_2$ and $L \approx (Z_1/\omega_0)\sin(\theta_1 + \theta_2) + L_{out}$ due to the sufficiently short length of the combined transmission line $TL_1 + TL_2$, typically much less than 60° at the fundamental frequency, as shown in Figure 3.40b. As a result, the characteristic impedances Z_1 and Z_2 can be derived as

$$Z_1 \cong \frac{\omega_0(L - L_{out})}{\sin(\theta_1 + \theta_2)} \tag{3.63}$$

$$Z_2 = \frac{1}{\omega_0 C\sqrt{3}} \tag{3.64}$$

For a lumped low-pass Π-type matching circuit shown in Figure 2.9a in Chapter 2, when $R_L > R$ and $Q = \omega_0 C_{out} R$,

$$Q_L = \sqrt{\frac{R_L}{R}(1 + Q^2) - 1} \tag{3.65}$$

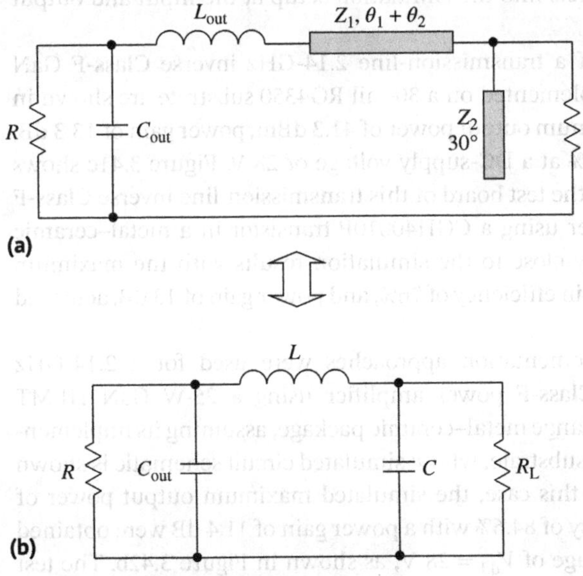

Figure 3.40 Load network at fundamental frequency.

$$C = \frac{Q_L}{\omega_0 R_L} \tag{3.66}$$

$$L = \frac{Q + Q_L}{1 + Q_L^2} \frac{R_L}{\omega_0} \tag{3.67}$$

from where, for the fixed parameters L_{out}, C_{out}, R, and R_L, the loaded quality factor Q_L can be calculated from Eq. (3.65). Then, the capacitance C is obtained from Eq. (3.66) following by the calculation of the characteristic impedance Z_2 from Eq. (3.64). When the inductance L is obtained from Eq. (3.67), the other parameters, such as the electrical lengths θ_1 and θ_2 and the characteristic impedance Z_1, can be calculated from a system of three equations given by Eqs. (3.61)–(3.63) for particular values of a shunt capacitance C_{out} and a series inductance L_{out}.

Figure 3.41a shows the simulated circuit schematic of a transmission-line inverse Class-F power amplifier using a 28-V 10-W GaN HEMT CGH40010 transistor and transmission-line load network with second- and third-harmonic tuning operating at 2.14 GHz, as shown in Figure 3.38a. The input matching circuit provides the conjugate matching with the standard 50-Ω source at the fundamental frequency. The parameters of the transmission lines in the load network are optimized to maximize the power amplifier performance and for implementation convenience. It is very important to properly model the transistor input and output package leads as external elements to minimize the effect of their inductances by adding these models into the simulation setup at the input and output of the transistor.

The simulation results of a transmission-line 2.14-GHz inverse Class-F GaN HEMT power amplifier implemented on a 30-mil RO4350 substrate are shown in Figure 3.41b, with the maximum output power of 41.3 dBm, power gain of 13.3 dB, and drain efficiency of 80.3% at a DC-supply voltage of 28 V. Figure 3.41c shows the measurement results of the test board of this transmission-line inverse Class-F GaN HEMT power amplifier using a CGH40010P transistor in a metal–ceramic pill-type package being very close to the simulation results with the maximum output power of 41 dBm, drain efficiency of 76%, and power gain of 13 dB, achieved without any tuning [29].

Similar design and implementation approaches were used for a 2.14-GHz transmission-line inverse Class-F power amplifier using a 25-W GaN HEMT CGH40025F transistor in a flange metal–ceramic package, assuming its implementation into a 30-mil RO4350 substrate, whose simulated circuit schematic is shown in Figure 3.42a [32, 33]. In this case, the simulated maximum output power of 43.4 dBm and drain efficiency of 84.6% with a power gain of 11.4 dB were obtained at the drain DC-supply voltage of $V_{dd} = 28$ V, as shown in Figure 3.42b. The test board of a 25-W inverse Class-F GaN HEMT power amplifier implemented on a

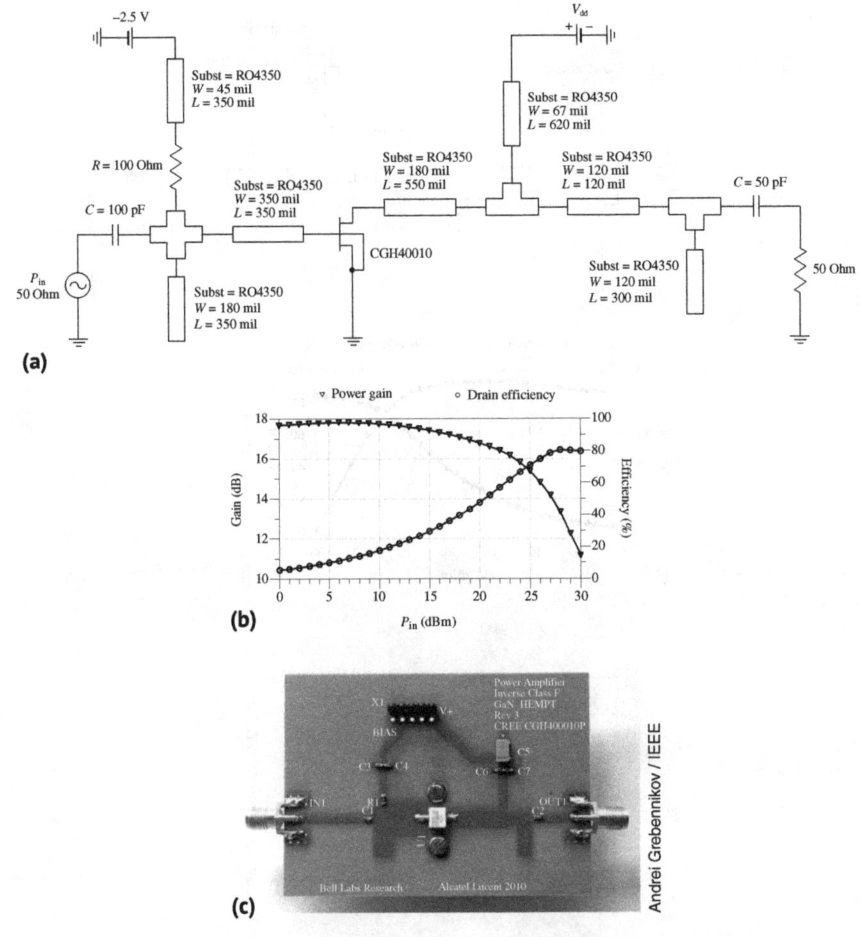

Figure 3.41 Microstrip-line 10-W inverse Class-F GaN HEMT power amplifier.

30-mil RO4350 substrate is shown in Figure 3.42c, with the measured maximum output power of 43.5 dBm, drain efficiency of 82.1%, and power gain of 12.2 dB at the operating frequency of 2.14 GHz with the slightly increased drain DC-supply voltage $V_{dd} = 32$ V were also achieved without any additional tuning.

If the output capacitance C_{out} is enough small and output inductance L_{out} is sufficiently large, the electrical length θ_2 of the transmission line TL_2 can be set to zero for certain values of C_{out} and L_{out}, which corresponds to the special case of the simplified inverse Class-F load network shown in Figure 3.43a when Eqs. (3.59) and (3.60) can be written in a normalized form as

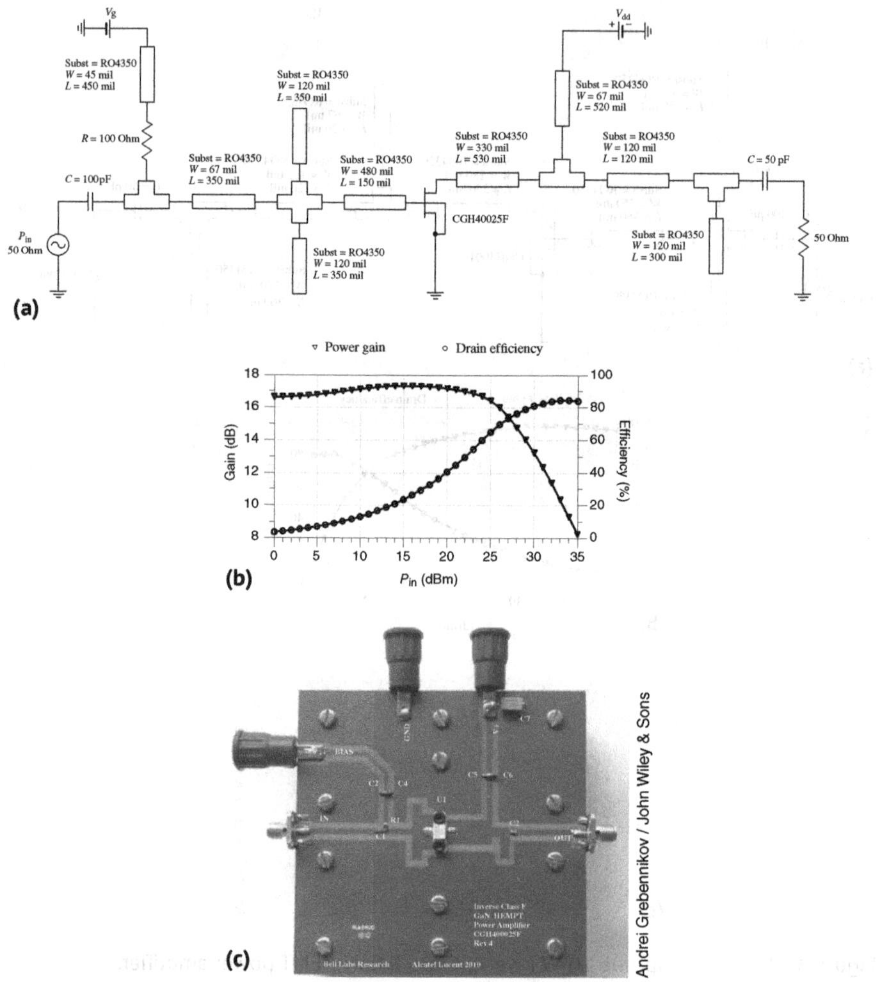

Figure 3.42 Microstrip-line 25-W inverse Class-F GaN HEMT power amplifier.

$$4pq + 2q \tan 2\theta_1 - 1 = 0 \tag{3.68}$$

$$3p + \tan 3\theta_1 = 0 \tag{3.69}$$

where $q = Z_1\omega_0 C_{out}$ and $p = \omega_0 L_{out}/Z_1$. Substituting Eq. (3.69) into Eq. (3.68) results in a single equation with one unknown parameter,

$$2q(2\tan 3\theta_1 - 3\tan 2\theta_1) + 3 = 0 \tag{3.70}$$

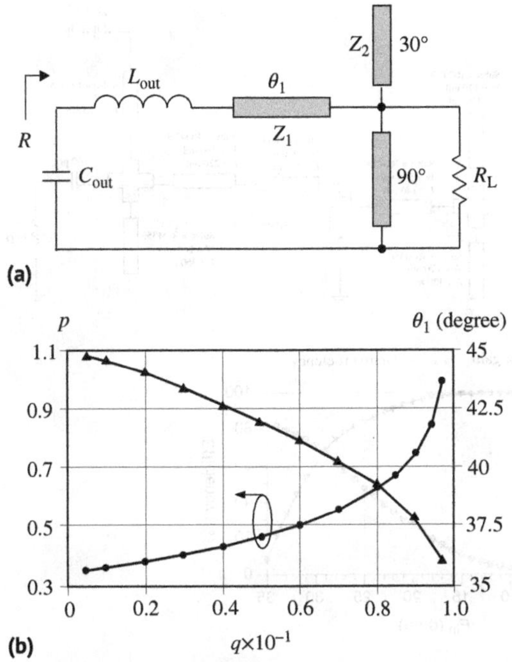

Figure 3.43 Schematic and performance of simplified inverse Class-F load network.

from which the required electrical length θ_1 can be implicitly obtained for proper q. Then, the characteristic impedance Z_1 can be explicitly derived from Eq. (3.69) by calculating the load-network parameter p for a given electrical length θ_1 and output inductance L_{out}.

The frequency dependences of the load-network parameters p and q characterizing the transmission-line inverse Class-F load network including a series transmission line with the characteristic impedance Z_1 and electrical length θ_1 is shown in Figure 3.43b, where the electrical length θ_1 varies within 35–45° when p decreases from 0.34 to 1.02 for $q < 0.1$. As an example, when $q = 0.09$, resulting in $p = 0.74$ and $\theta_1 = 38°$, as shown in Figure 3.42b, the characteristic impedance Z_1 and series inductance L_{out} are calculated for $C_{out} = 0.3$ pF as equal to $Z_1 = 22\ \Omega$ and $L_{out} = 1.2$ nH at 2.14 GHz.

Another example of the circuit schematic of an inverse Class-F GaN HEMT power amplifier based on a 28-V 60-W GaN HEMT CGH27060F transistor is shown in Figure 3.44a, with similar structure for the input matching circuit and load network but having a two-stepped series transmission line instead of a single transmission line with a shorter overall length [34]. In order to improve stability

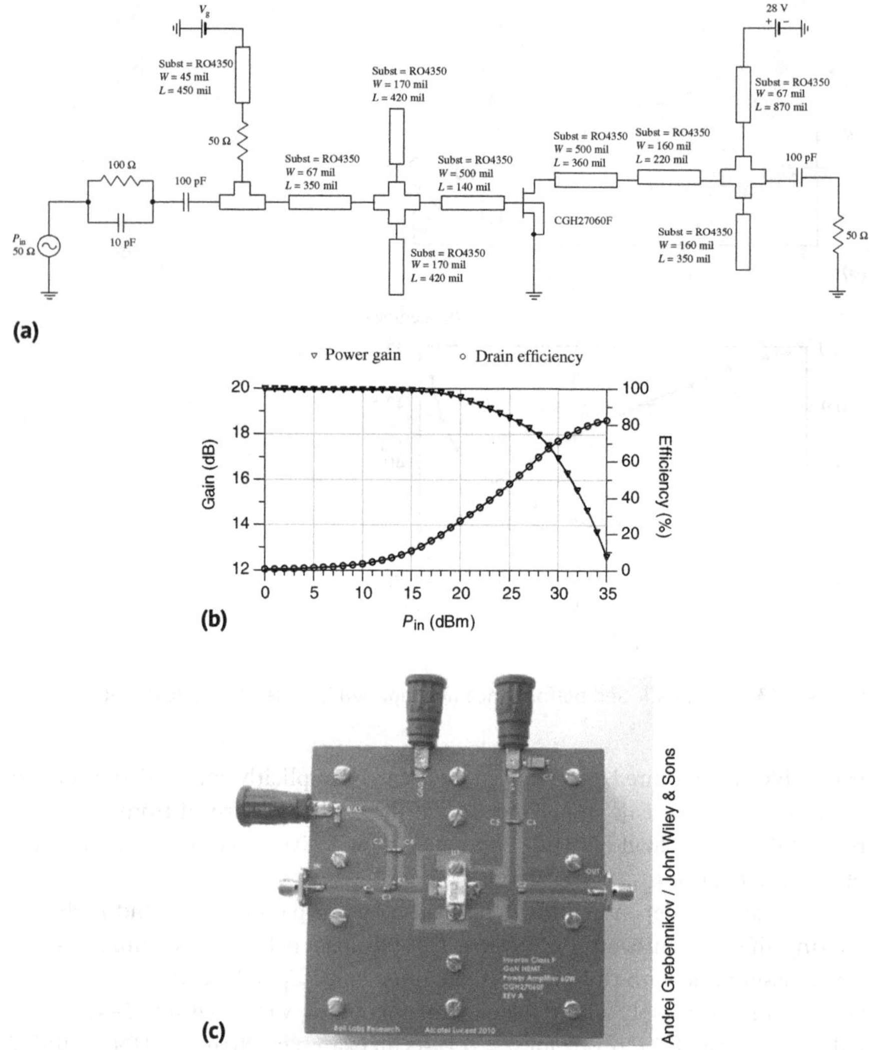

Figure 3.44 Microstrip-line 50-W inverse Class-F GaN HEMT power amplifier.

of the power amplifier operation, the parallel connected 10-pF capacitor and 100-Ω resistor are included in series into the input circuit. Figure 3.44b shows the simulation results of a 2.14-GHz transmission-line inverse Class-F GaN HEMT power amplifier assuming implementation on a 30-mil RO4350 substrate, where the maximum output power of 47.6 dBm (57.5 W) and drain efficiency of 82.4% were obtained with a power gain of 13.6 dB at a DC-supply voltage of 32 V. In this case, a high drain efficiency greater than 80% is achieved in a strong saturation mode when the power gain is compressed by more than 5 dB to minimize the transistor power losses due to finite switching time. Figure 3.44c shows the measured results of a high-power GaN HEMT power amplifier implemented on a 30-mil RO4350 substrate demonstrating a saturated output power of 47.3 dBm, a drain efficiency of 82.3%, and a power gain of 14.3 dB, which were also obtained without any tuning of the load network [34].

3.3 Class E with Shunt Capacitance

However, there is an alternative way to increase efficiency, which was first demonstrated experimentally in the late 1940s and early 1950s when, in the vacuum-tube amplifier operating in a saturation mode, the resonant circuit is detuned in the direction of frequencies higher than the operating frequency [35–37]. In this case, the second- and higher-order harmonics with proper phasing contribute to the anode voltage and current waveforms having asymmetric shapes and minimum overlapping, with the proper inductive impedance at the fundamental frequency and capacitive reactances at the second- and higher-order harmonics. The anode efficiencies of 93–94% were achieved with the phase angle of the load in the range of 30–40°. A decade later, it was analytically derived that, when the transistor is operated as a switch, the collector voltage and voltage derivative should be near zero at the time just prior to the conduction interval in the saturation mode, and there is no overlapping between the voltage and current waveforms at the collector, as shown in [38]. It was shown on the example of the 500-kHz bipolar power amplifier with shunt capacitance and series-tuned *LC* circuit, with the circuit schematic shown in Figure 3.45a, that a 20-W output power with a collector efficiency of 94% can be achieved.

The simplified collector circuit of a switchmode power amplifier is shown in Figure 3.45b, where C is the collector capacitance, C_0 is the DC-blocking capacitance, L is the series inductance, and R_L is the load resistance. The behavior of the power amplifier was analytically described by the linear first-order differential equations for the on-state and off-state conditions, and the collector efficiencies up to 85% were obtained in the frequency bandwidth of 48–70 MHz [39]. However, because of the absence of the series-tuned resonant circuit in the load network, not only fundamental current but also the currents corresponding to higher-order harmonics flow into the load.

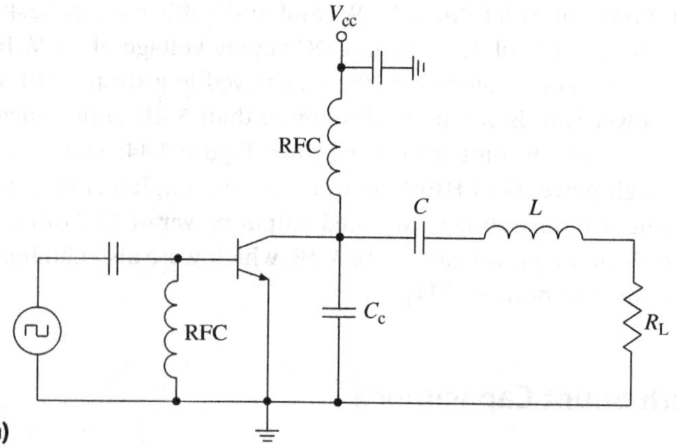

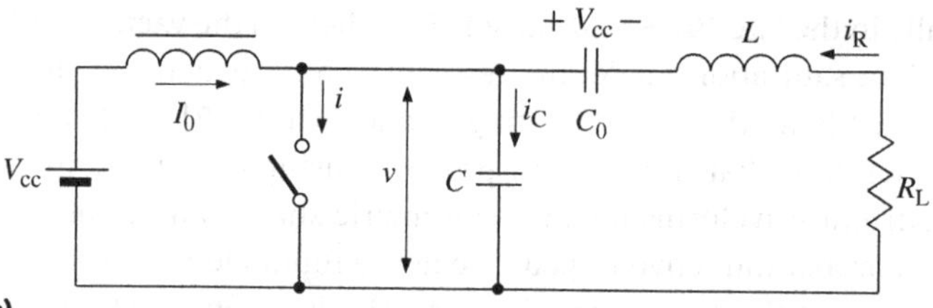

Figure 3.45 Basic circuits of switchmode transistor power amplifier with different load networks.

3.3.1 Optimum Load-network Parameters

In order to maximize efficiency of a Class-E power amplifier, it is very important to connect the load through the series resonant circuit tuned to the fundamental frequency to minimize power losses at the harmonics, for example, by using a series L_0C_0 circuit shown in Figure 3.46a that results in a sinusoidal current on the load with significant harmonic suppression [40, 41]. The switchmode power amplifier with shunt capacitance and series filter was first publicly introduced by the Sokals in 1975 as a Class-E power amplifier [42]. This type of high-efficiency power amplifier was then widely used in different frequency ranges and output power levels [43].

Figure 3.46b shows the simplified equivalent circuit of a Class-E power amplifier with shunt capacitance and series filter, with a shunt capacitance C including the collector capacitance, a series inductance L, a series fundamentally tuned L_0C_0 circuit, a DC-feed inductance with high reactance at the fundamental frequency, and a load resistance R. The transistor is considered an ideal switch to provide the

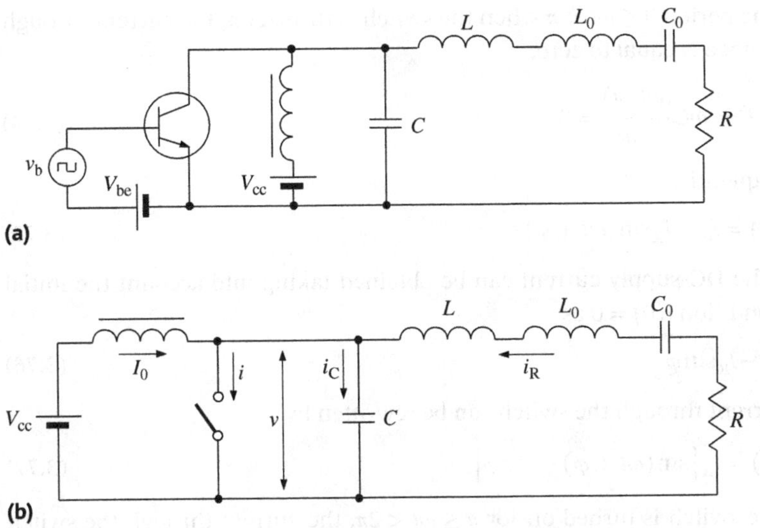

Figure 3.46 Basic circuits of a Class-E power amplifier with shunt capacitance.

instant device switching between its on-state and off-state operation conditions. As a result, the collector voltage waveform is determined by the switch when it is turned on and by the transient response of the load network when the switch is turned off.

To simplify the analysis of a Class-E power amplifier, the transistor should have zero saturation voltage with instantaneous switching action, DC-feed inductance is lossless and infinite impedance at the fundamental and high-order harmonics, the current flowing through the lossless series resonant L_0C_0 circuit tuned to the fundamental frequency is sinusoidal, and the duty ratio is 50%.

The optimum conditions for voltage $v(\omega t)$ on the switch at the end of the entire period when $\omega t = 2\pi$ should be satisfied for a lossless operation mode as

$$v(\omega t)\big|_{\omega t=2\pi} = 0 \tag{3.71}$$

$$\left.\frac{dv(\omega t)}{d\omega t}\right|_{\omega t=2\pi} = 0 \tag{3.72}$$

A detailed theoretical analysis of a Class-E power amplifier with shunt capacitance for any duty ratio is given in [44] under the assumption of the sinewave current on the load, written as

$$i_R(\omega t) = I_R \sin(\omega t + \varphi) \tag{3.73}$$

where φ is the initial phase shift.

For a time period $0 \le \omega t < \pi$ when the switch is turned on, the current through the capacitance is equal to zero,

$$i_C(\omega t) = \omega C \frac{dv(\omega t)}{d\omega t} = 0 \tag{3.74}$$

and, consequently,

$$i(\omega t) = I_0 + I_R \sin(\omega t + \varphi) \tag{3.75}$$

Hence, the DC-supply current can be obtained taking into account the initial on-state condition $i(0) = 0$ as

$$I_0 = -I_R \sin\varphi \tag{3.76}$$

and the current through the switch can be rewritten by

$$i(\omega t) = I_R \left[\sin(\omega t + \varphi) - \sin\varphi\right] \tag{3.77}$$

When the switch is turned off for $\pi \le \omega t < 2\pi$, the current through the switch $i(\omega t) = 0$, and the current through the capacitor C is written as

$$i_C(\omega t) = I_0 + I_R \sin(\omega t + \varphi) \tag{3.78}$$

producing the voltage across the switch by charging this capacitor according to

$$\begin{aligned} v(\omega t) &= \frac{1}{\omega C} \int_{\pi}^{\omega t} i_C(\omega t) d\omega t \\ &= -\frac{I_R}{\omega C} \left[\cos(\omega t + \varphi) + \cos\varphi + (\omega t - \pi)\sin\varphi\right] \end{aligned} \tag{3.79}$$

The phase angle φ is calculated from the optimum condition given by Eq. (3.71) as

$$\varphi = \tan^{-1}\left(-\frac{2}{\pi}\right) = -32.482° \tag{3.80}$$

with the trigonometric representation of

$$\sin\varphi = \frac{-2}{\sqrt{\pi^2 + 4}} \qquad \cos\varphi = \frac{\pi}{\sqrt{\pi^2 + 4}} \tag{3.81}$$

Then, by using Eqs. (3.76) and (3.81), the voltage on the switch given in Eq. (3.79) can be rewritten as

$$v(\omega t) = \frac{I_0}{\omega C}\left(\omega t - \frac{3\pi}{2} - \frac{\pi}{2}\cos\omega t - \sin\omega t\right) \tag{3.82}$$

From the Fourier series expansion, the DC-supply voltage V_{cc} is defined by

$$V_{cc} = \frac{1}{2\pi}\int_0^{2\pi} v(\omega t)d\omega t = \frac{I_0}{\pi\omega C} \tag{3.83}$$

Consequently, the normalized collector voltage waveform for the turned-off period of $\pi \le \omega t < 2\pi$ and the current waveform for the turn-on period of $0 \le \omega t < \pi$ are

$$\frac{v(\omega t)}{V_{cc}} = \pi\left(\omega t - \frac{3\pi}{2} - \frac{\pi}{2}\cos\omega t - \sin\omega t\right) \tag{3.84}$$

$$\frac{i(\omega t)}{I_0} = \frac{\pi}{2}\sin\omega t - \cos\omega t + 1 \tag{3.85}$$

Figure 3.47 shows the normalized (a) load current, (b) collector voltage, and (c) collector current waveforms for an idealized optimum (or nominal) Class-E operation mode with shunt capacitance and series filter. Since there is no voltage (zero voltage) across the switch when the transistor is turned on, the current $i(\omega t)$, represented by the load sinewave current and DC current, is flowing through the transistor. However, for the other half of the period when the transistor is turned off, this current flows through the shunt capacitance C. In this case, both derivatives dv/dt and di/dt must be nonzero during at least one of the switching transitions to achieve a nonzero output power at the fundamental frequency delivered to the load, with the jump in the idealized collector current waveform at the time moment of switching off [45].

Since there is no nonzero voltage and current simultaneously, resulting in a lack of power losses, an idealized collector efficiency is equal to 100% when the DC power and fundamental-frequency output power delivered to the load are equal,

$$I_0 V_{cc} = \frac{I_R^2}{2}R \tag{3.86}$$

Hence, the DC-supply current I_0 using Eqs. (3.76) and (3.81) is derived as

$$I_0 = \frac{V_{cc}}{R}\frac{8}{\pi^2+4} = 0.577\frac{V_{cc}}{R} \tag{3.87}$$

Then, the amplitude of the output voltage $V_R = I_R R$ is obtained by

$$V_R = \frac{4V_{cc}}{\sqrt{\pi^2+4}} = 1.074\,V_{cc} \tag{3.88}$$

By differentiating the appropriate waveforms from Eqs. (3.84) and (3.85) and setting the results to zero, the peak collector voltage V_{max} and current I_{max} can be defined as

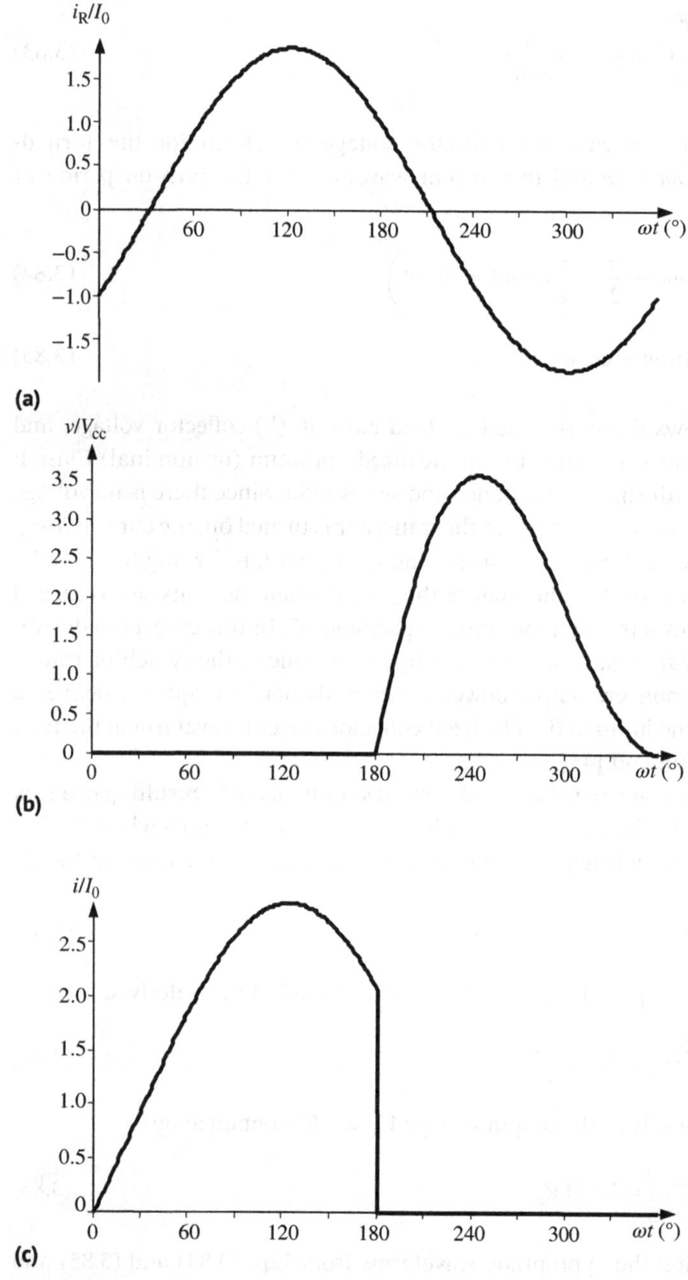

Figure 3.47 Load current, collector voltage, and collector current waveforms for ideal optimal Class-E with shunt capacitance.

$$V_{\max} = -2\pi\varphi V_{cc} = 3.562 V_{cc} \tag{3.89}$$

$$I_{\max} = \left(\frac{\sqrt{\pi^2+4}}{2} + 1\right) I_0 = 2.8621 I_0 \tag{3.90}$$

The amplitudes of the two quadrature components of the voltage on the switch at the fundamental frequency shown in Figure 3.48 are derived using the Fourier formulas and Eq. (3.84) as

$$V_R = -\frac{1}{\pi}\int_0^{2\pi} v(\omega t)\sin(\omega t + \varphi)d\omega t = \frac{I_R}{\pi\omega C}\left(\frac{\pi}{2}\sin 2\varphi + 2\cos 2\varphi\right) \tag{3.91}$$

$$\begin{aligned} V_L &= -\frac{1}{\pi}\int_0^{2\pi} v(\omega t)\cos(\omega t + \varphi)d\omega t \\ &= -\frac{I_R}{\pi\omega C}\left(\frac{\pi}{2} + \pi\sin^2\varphi + 2\sin 2\varphi\right) \end{aligned} \tag{3.92}$$

Finally, the optimum normalized series inductance L and shunt capacitance C are obtained respectively by

$$\frac{\omega L}{R} = \frac{V_L}{V_R} = 1.1525 \tag{3.93}$$

$$\omega CR = \frac{\omega C}{I_R} V_R = 0.1836 \tag{3.94}$$

whereas the optimum load resistance R is derived from Eqs. (3.86) and (3.88) as

$$R = \frac{8}{\pi^2 + 4}\frac{V_{cc}^2}{P_{out}} = 0.5768\frac{V_{cc}^2}{P_{out}} \tag{3.95}$$

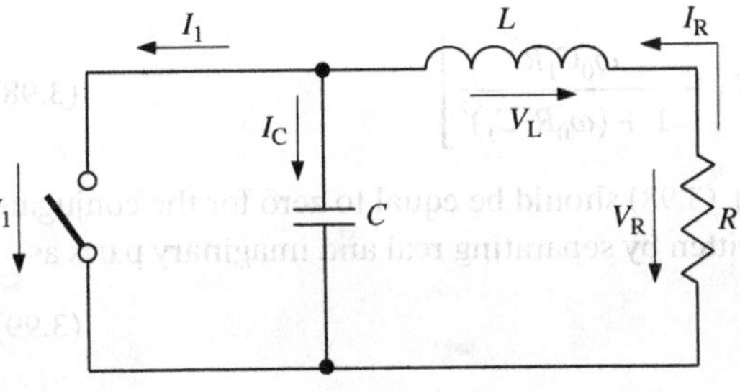

Figure 3.48 Schematic of Class-E load network at fundamental frequency.

where V_{cc} is the DC-supply voltage and P_{out} is the output power on the load at the fundamental frequency.

The phase angle of the load network seen by the switch, which is required to implement an idealized optimum (or nominal) Class-E mode with shunt capacitance, can be determined from Eqs. (3.93) and (3.94) through the load-network parameters as

$$\phi = \tan^{-1}\left(\frac{\omega L}{R}\right) - \tan^{-1}\left(\frac{\omega CR}{1 - \frac{\omega L}{R}\omega CR}\right) = 35.945^\circ \tag{3.96}$$

The maximum operation frequency f_{max} can be expressed through the shunt capacitance C, output power P_{out}, and DC-supply voltage V_{cc} using Eqs. (3.94) and (3.95) as

$$f_{max} = 0.0507\left(P_{out}/CV_{cc}^2\right) \tag{3.97}$$

where $C = C_{out}$ is the device output capacitance limiting the maximum operation frequency of an idealized optimum Class-E power amplifier with shunt capacitance and series filter.

The optimum parameters of a Class E with shunt capacitance and series filter are derived for an infinite loaded quality factor Q_L for the series resonant L_0C_0 circuit. However, if its value becomes too small in practical circuits, this can lead to considerable errors up to 10% for $Q_L = 7$ and less than 5% for $Q_L \geq 10$ [46].

3.3.2 Matching with Standard Load

Generally, it is necessary to match the optimum output resistance R of a Class-E power amplifier at the fundamental frequency with the load resistance R_L, usually equal to 50 Ω. The simplest and very convenient in practical implementation is the low-pass L-type matching circuit with a series inductance L_1 and a shunt capacitance C_1 shown in Figure 3.49a [47]. In this case, the optimum resistance R can be expressed through the parameters of the low-pass L-type matching circuit and the load resistance R_L as

$$R = \frac{R_L}{1 + (\omega_0 R_L C_1)^2} + j\left[\omega L_1 - \frac{\omega_0 C_1 R_L^2}{1 + (\omega_0 R_L C_1)^2}\right] \tag{3.98}$$

Since the imaginary part of Eq. (3.98) should be equal to zero for the conjugate matching, Eq. (3.98) can be rewritten by separating real and imaginary parts as

$$R_L = R\left(1 + Q_L^2\right) \tag{3.99}$$

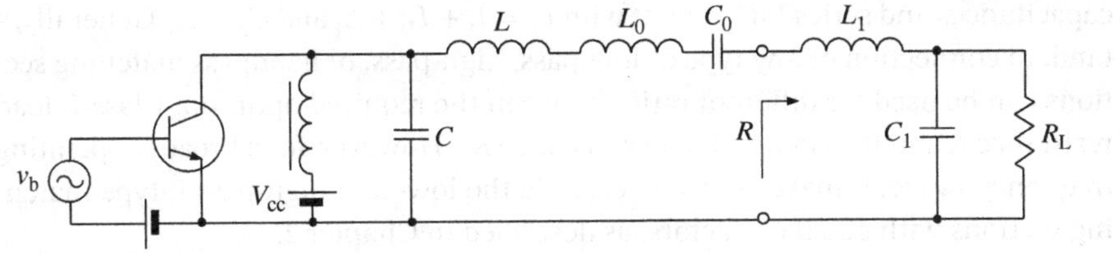

(a)

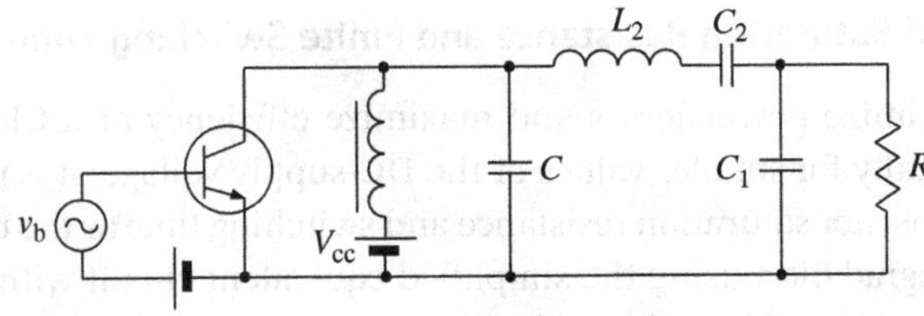

(b)

Figure 3.49 Schematics of Class-E load networks with matching circuits.

$$\frac{1}{\omega_0 C_1} = \omega_0 L_1 \left(1 + Q_L^{-2}\right) \tag{3.100}$$

where

$$Q_L = \omega_0 C_1 R_L = \frac{\omega_0 L_1}{R} \tag{3.101}$$

is the loaded quality factor equal being equal for both the series and parallel circuits at the fundamental frequency.

Thus, the loaded quality factor Q_L can be expressed through the ratio of the load resistance R_L and the optimum resistance R in a Class-E operation mode as

$$Q_L = \sqrt{\frac{R_L}{R} - 1} \tag{3.102}$$

and the parameters of the matching circuit can be obtained by

$$L_1 = \frac{R Q_L}{\omega_0} \tag{3.103}$$

$$C_1 = \frac{Q_L}{\omega_0 R_L} \tag{3.104}$$

Figure 3.49b shows the schematic of a Class-E power amplifier with shunt capacitance and series filter where the series high-Q_L filter and low-pass L-type matching circuit are combined into a low-pass Π-type circuit with two shunt capacitances and series L_2C_2 circuit with $L_2 = L + L_0 + L_1$ and $C_2 = C_0$. Generally, a tandem connection of any type of low-pass, high-pass, or bandpass matching sections can be used for different ratios between the required optimum Class-E load resistance R and the standard load resistance R_L. However, for a broader operating frequency range, it makes sense to cascade the low- and high-pass L-type matching sections with equal Q_L-factors, as described in Chapter 2.

3.3.3 Effect of Saturation Resistance and Finite Switching Time

In order to minimize power losses and maximize efficiency of a Class-E power amplifier, especially for smaller values of the DC-supply voltage, it is necessary to estimate the transistor saturation resistance and switching time as the main factors for efficiency degradation using the simplified equivalent circuit with the saturation resistance connected in series to the ideal switch, as shown in Figure 3.50.

In order to calculate the power losses due to the presence of the saturation resistance r_{sat}, assuming its constant value, the normalized saturated output power P_{sat} can be represented as

$$\frac{P_{sat}}{P_0} = \frac{r_{sat}}{2\pi I_0 V_{cc}} \int_0^{\pi} i^2(\omega t) d\omega t \tag{3.105}$$

where $P_0 = I_0 V_{cc}$ is the DC power and $i(\omega t)$ is the current through the saturation resistance r_{sat}.

Considering that

$$\int_0^{\pi} \left(\frac{\pi}{2}\sin\omega t - \cos\omega t + 1\right)^2 d\omega t = \frac{\pi}{8}\left(\pi^2 + 28\right) I_0^2 \tag{3.106}$$

Equation (3.105) can be rewritten using Eq. (3.87) by

$$\frac{P_{sat}}{P_0} = \frac{r_{sat}}{2\pi}\frac{I_0}{V_{cc}}\frac{\pi}{8}\left(\pi^2 + 28\right) = \frac{r_{sat}}{2R}\frac{\pi^2 + 28}{\pi^2 + 4} = 1.365\frac{r_{sat}}{R} \tag{3.107}$$

Generally, the collector (or drain) efficiency η can be defined as

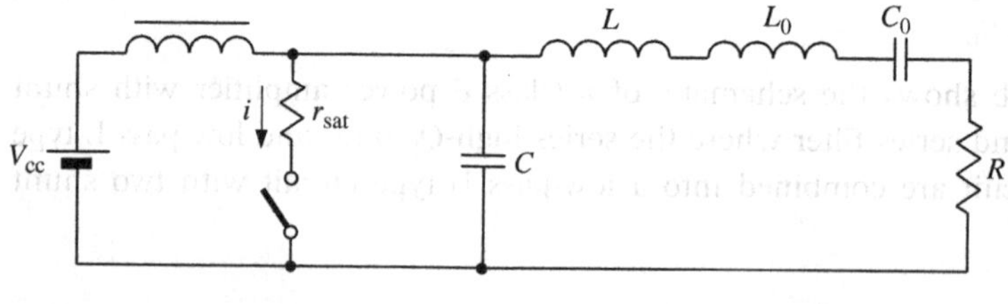

Figure 3.50 Schematic of Class-E load network with saturation resistance.

$$\eta = \frac{P_{out}}{P_0} = \frac{P_0 - P_{sat}}{P_0} = 1 - \frac{P_{sat}}{P_0} \tag{3.108}$$

Then, the finite value of the saturation voltage V_{sat} in view of the presence of the saturation resistance results can be calculated from

$$\frac{V_{sat}}{V_{cc}} = 1 - \frac{1}{1 + 1.365\frac{r_{sat}}{R}} \tag{3.109}$$

where V_{cc} is the DC-supply voltage [48].

At higher frequencies, it is impossible to realize the driving signal close to the rectangular form due to the low-pass *RC*-type of the transistor input circuit affecting the switching rise and fall times. Fortunately, to provide a high-efficiency operation mode, it is enough to apply a sinusoidal signal with slightly increased amplitude. In this case, the transition from the saturation mode to the pinch-off mode is happening at the point of the intersection of the curve corresponding to the base (or channel) charge process (curve 1) and the curve corresponding to the ideal Class-E current waveform provided by the load network (curve 2), resulting in an overall collector current waveform with finite switching time (curve 3), as shown in Figure 3.51a [49]. To minimize the switching time interval, it is sufficient to slightly overdrive the transistor with a signal amplitude by 20–30% higher than is required for a conventional Class-B power amplifier, as shown in Figure 3.51b. As an alternative, the second-harmonic component to approximate a half-sine waveform or the third-harmonic component to approximate a rectangular waveform can be added to the input driving signal with proper phase conditions to make the resulting driving waveform steeper compared to a sinewave driving signal.

The power dissipated due to finite switching time within this on-to-off transition can be derived for zero saturation resistance as

$$P_s = \frac{1}{2\pi} \int_{\theta_s}^{\pi} i(\omega t) v(\omega t) d\omega t \tag{3.110}$$

where, during the transition time $\tau_s = \pi - \theta_s$, the collector voltage is defined by

$$v(\theta_s) = \frac{1}{\omega C} \int_{\theta_s}^{\pi} i_C(\omega t) d\omega t \tag{3.111}$$

For a sufficiently short switching time, a linearly decreasing collector current during fall time $\tau_s = \pi - \theta_s$ starting at $i(\theta_s)$ at time θ_s and decaying to zero at time π can be assumed, resulting in

$$i(\omega t) = i(\theta_s)\left(1 - \frac{\omega t - \theta_s}{\tau_s}\right) \tag{3.112}$$

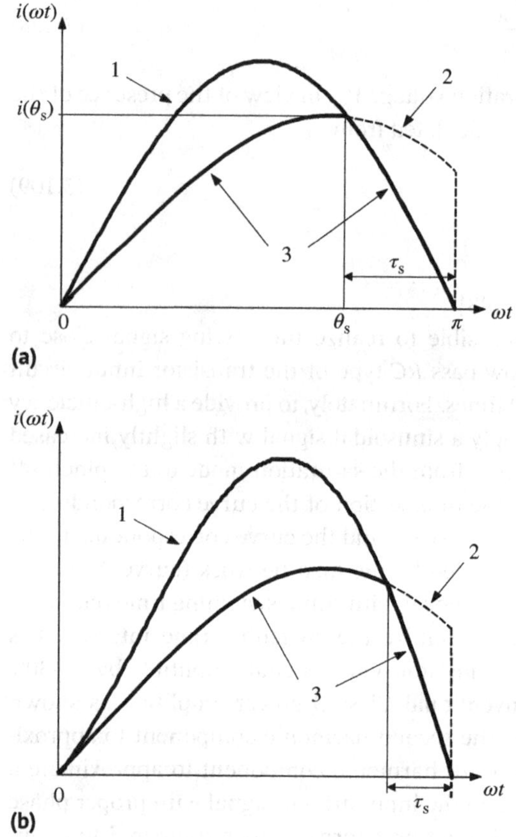

Figure 3.51 Effect of finite switching time in collector current waveform.

where $i(\theta_s)$ corresponds to peak collector current, as shown in Figure 3.51a [48, 49]. In this case, the current $i_C(\omega t) = i(\theta_s) - i(\omega t)$ charging the shunt capacitor varies linearly between zero and $i(\theta_s)$ during the on-to-off transition according to

$$i_C(\omega t) = i(\theta_s)\frac{\omega t - \theta_s}{\tau_s} \tag{3.113}$$

By substituting Eq. (3.113) into Eq. (3.111), a parabolic waveform of the collector voltage during the switching interval can be obtained as

$$v(\omega t) = \frac{i(\theta_s)}{2\omega C\tau_s}(\omega t - \theta_s)^2 \tag{3.114}$$

Thus, the power dissipated due to finite switching time from Eq. (3.110) is equal to

$$P_s = \frac{i^2(\theta_s)\,\tau_s^2}{48\pi\omega C} \tag{3.115}$$

In an optimum Class-E power amplifier with a short transition time when $i(\theta_s) = i(\pi)$, it follows from Eq. (3.85) that $i(\pi) = 2I_0$, hence

$$P_s = \frac{I_0^2\tau_s^2}{12\pi\omega C} \tag{3.116}$$

Consequently, the normalized switching loss power P_s when considering Eq. (3.87) can be obtained by

$$\frac{P_s}{P_0} = \frac{I_0\tau_s^2}{12\pi\omega CV_{cc}} = \frac{\tau_s^2}{12} \tag{3.117}$$

As a result, the collector efficiency η can be estimated as

$$\eta = 1 - \frac{P_s}{P_0} = 1 - \frac{\tau_s^2}{12} \tag{3.118}$$

As follows from Eq. (3.117), the power losses due to the nonzero switching time are sufficiently small, and they are only about 1% for $\tau_s = 0.35$ or 20°, increasing to approximately 10% for $\tau_s = 60°$. This result was obtained for an infinite Q_L factor of the series-tuned resonant circuit. However, for enough small values of Q_L, the collector efficiency increases by several percent, with the most desirable range of Q_L from 5 to 10 in practical implementation [50].

3.3.4 Optimum, Nominal, and Off-nominal Class-E Operation

An analysis of the collector (or drain) voltage and current waveforms for a nonzero saturation resistance r_{sat} shows that the idealized optimum (or nominal) conditions for a Class-E operation mode presented in Eqs. (3.71) and (3.72) do not correspond to the minimum dissipation power losses but the minimum overall power losses correspond to optimum nonzero values of the collector voltage and its derivative at switching time instant [51, 52]. For example, even for small losses with the normalized loss parameter $\omega Cr_{sat} = 0.1$ for a duty ratio of 50%, the optimum series inductance L is almost two times greater, and the optimum shunt capacitance C is about 20% greater than those obtained under idealized optimum conditions given by Eqs. (3.71) and (3.72) [49]. Nevertheless, the idealized optimum conditions can be considered as enough accurate initial guess for further optimization of the Class-E power amplifiers in simulations and practical implementation.

The term "optimum" for the idealized Class-E conditions implies zero voltage across the switch and a fully discharged shunt capacitor without any current jumps at the turn-on instants. Otherwise, the transistor is partly operated in an active region with nonzero current between its pinch-off and saturation modes and cannot be considered as an ideal switch, with overlapping of the collector current and voltage waveforms. Therefore, it is more appropriate to call the term "optimum" as nominal or idealized optimum for a lossless operation mode with an ideal switch.

Below the maximum operation frequency f_{max}, at which the shunt capacitance is fully represented by the transistor output capacitance, an external shunt capacitance can be added to the transistor output to achieve the optimum operation mode. However, above the maximum frequency f_{max}, an ideal 100% efficiency cannot be achieved by varying the load network parameters, and the maximum achievable collector (or drain) efficiency is reduced with an increased operating frequency above the maximum frequency. Nevertheless, efficiency above the maximum frequency can be maximized by proper selection of the series inductance and load resistance if it is possible [53]. For example, at an operating frequency of about $2.5f_{max}$, the collector efficiency of an ideal Class-E mode remains still high, being even higher than that for a Class-F mode with control of three collector voltage and three collector current harmonic components, but being lower than 80% above $3f_{max}$.

For off-nominal Class-E operation, when, for example, only the zero-voltage switching condition is satisfied, the shunt capacitance with a larger value than for nominal operation with the same load resistance can be used, resulting in higher peak collector (or drain) voltage and current [54]. However, for different turn-on switch voltage slope and duty ratio, the peak voltage and current values are different as well [55].

3.3.5 Load Network with Transmission Lines

In hybrid implementations of the power amplifiers at microwave frequencies or in the monolithic integrated circuits at millimeter waves, it is preferrable to design load networks using different types of microstrip lines including open-circuit and short-circuit stubs for harmonic control rather than lumped inductors and capacitors as having more predictable performance with less insertion loss and effect of the parasitic parameters, and being more convenient of their practical implementation. In this case, it is necessary to satisfy simultaneously the idealized optimum requirements at the fundamental-frequency impedance $Z_{net}(\omega_0)$ shown in

Figure 3.52a, and harmonic-component impedances $Z_{net}(n\omega_0)$, shown in Figure 3.52b, which can be written using Eq. (3.93) as

$$Z_{net}(\omega_0) = R + j\omega L = R\left(1 + j\frac{\omega L}{R}\right) = R(1 + j\,\tan 49.052°) \tag{3.119}$$

$$Z_{net}(n\omega_0) = \infty \tag{3.120}$$

where ω_0 is the radian fundamental frequency and $n = 2,3,\ldots,\infty$ is the harmonic number.

Since it is impossible to satisfy the optimum conditions for an infinite number of harmonics by using the transmission lines only, a close proximity to Class-E operation mode can be provided with the properly controlled fundamental-frequency and second-harmonic components of the voltage waveform on the switch, as follows from the Fourier-series analysis [56, 57]. In this case, the collector (or drain) efficiency will be similar to that in a maximum-efficiency Class-F mode with second-harmonic control, with the voltage waveform containing the fundamental-frequency and second-harmonic components (dashed line) plotted along with an ideal voltage waveform (solid line) shown in Figure 3.52c [21]. In a practical load-network implementation, as shown in Figure 3.53a, the series microstrip line l_1 and open-circuit microstrip stub l_2, each having an electrical length of about 45°

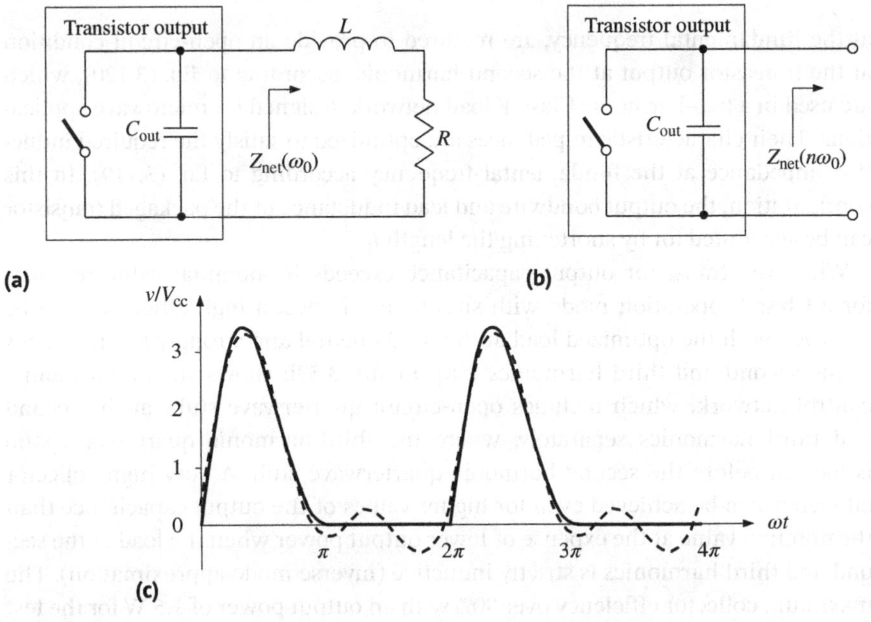

Figure 3.52 Impedances and waveforms of a two-harmonic Class-E load network.

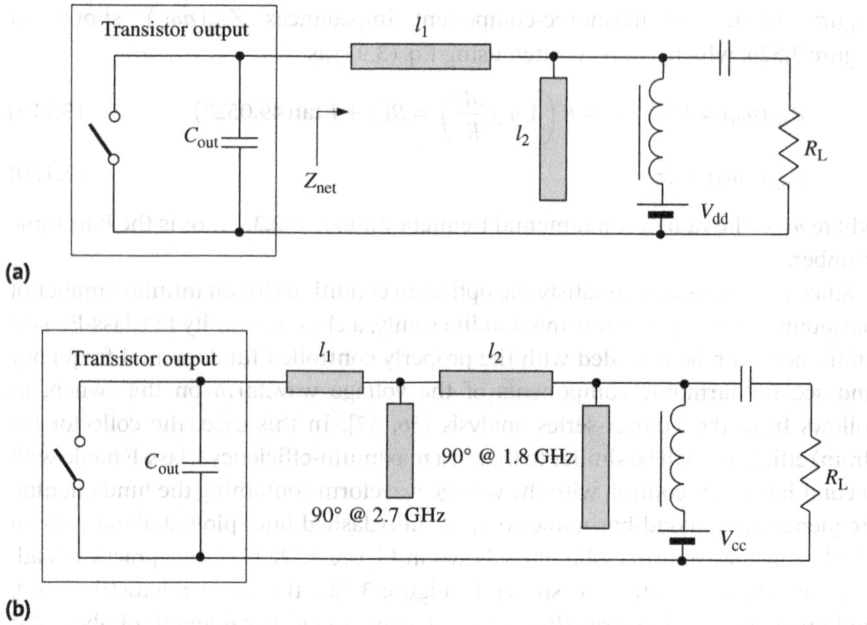

Figure 3.53 Schematics of Class-E load networks with transmission lines.

at the fundamental frequency, are required to provide an open-circuit condition at the transistor output at the second harmonic, according to Eq. (3.120), which are used in a two-harmonic Class-E load network designed for microwave applications. Their characteristic impedances are optimized to satisfy the required inductive impedance at the fundamental frequency according to Eq. (3.119). In this configuration, the output bondwire and lead inductance of the packaged transistor can be accounted for by shortening the length l_1.

When the transistor output capacitance exceeds its nominal value required for a Class-E operation mode with shunt capacitance, a high efficiency can be achieved with the optimized load at the fundamental and strongly reactive loads at the second and third harmonics [58]. Figure 3.53b shows such a harmonic-control network, which includes open-circuit quarterwave stubs at the second and third harmonics separately, where the third-harmonic quarterwave stub is located before the second-harmonic quarterwave stub. A very high collector efficiency can be achieved even for higher values of the output capacitance than the nominal value at the expense of lower output power when the load at the second and third harmonics is strictly inductive (inverse mode approximation). The maximum collector efficiency over 90% with an output power of 1.5 W for the test power amplifier using a bipolar transistor MRF557 was measured at 900 MHz.

In a Class-E operation mode analyzed in the frequency domain, the joint effect of the saturation resistance r_{sat} and second-harmonic termination had resulted in a shift of the nominal load impedance at the fundamental frequency [59]. Since the transmission-line open-circuit condition is difficult to realize in practical implementation due to package parasitic and on-board power losses, it is necessary to optimize the performance of the power amplifier by applying nonideal harmonic termination conditions. As an example, the voltage peak factor increases with the load impedance at the second harmonic, varying from capacitive to inductive, though with excessive current under capacitive load reactance. In this case, the maximum collector efficiency is obtained for a negative collector voltage-derivative condition at lower output power in a Class-E operation mode [60].

The transmission-line Class-E load network shown in Figure 3.53b can be modified in order to simplify the load-network implementation when the transmission-line parameters can be explicitly defined from analytical equations [61]. Figure 3.54 shows such a modified transmission-line Class-E load network, where the combined series quarterwave transmission line $TL_1 + TL_2$ provides an impedance transformation at the fundamental frequency, and open-circuit conditions

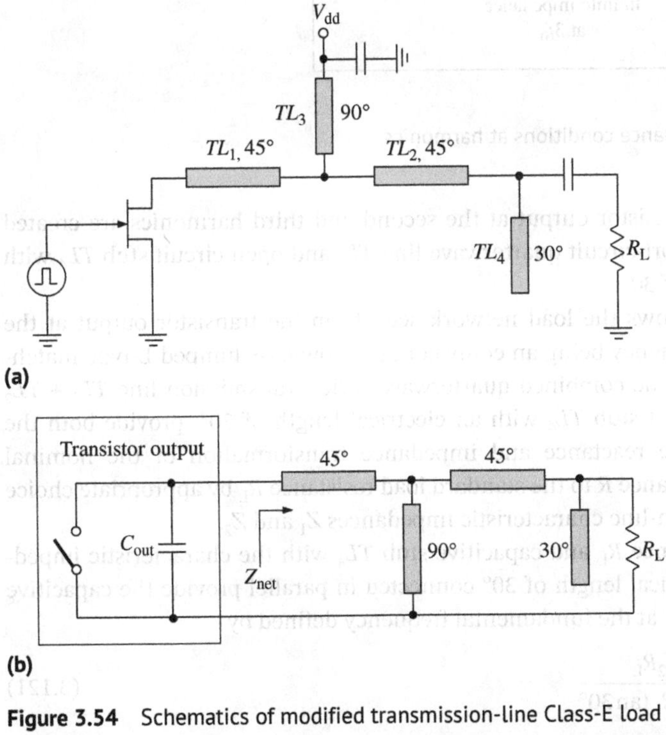

Figure 3.54 Schematics of modified transmission-line Class-E load network.

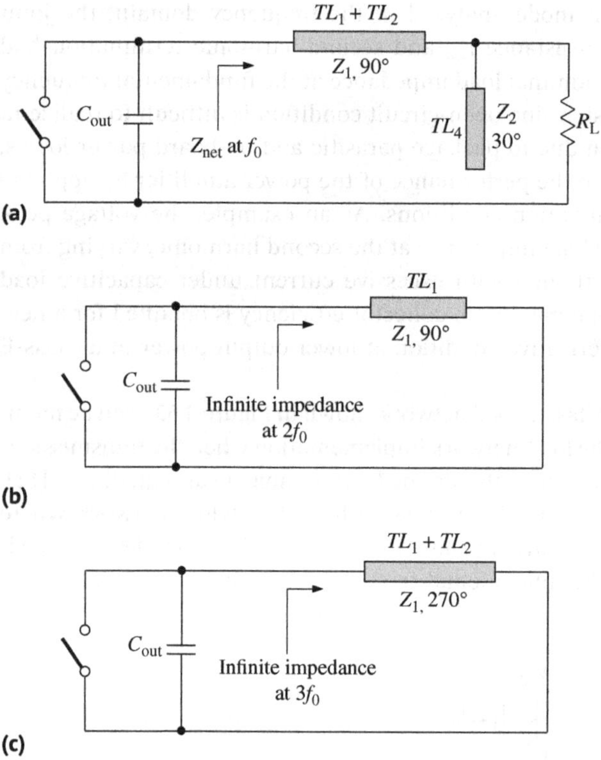

Figure 3.55 Impedance conditions at harmonics.

seen from the transistor output at the second and third harmonics are created by the parallel short-circuit quarterwave line TL_3 and open-circuit stub TL_4 with electrical length of 30°.

Figure 3.55a shows the load network seen from the transistor output at the fundamental frequency being an equivalent to a low-pass lumped L-type matching circuit where the combined quarterwave series transmission line $TL_1 + TL_2$ and an open-circuit stub TL_4 with an electrical length of 30°, provide both the required inductive reactance and impedance transformation of the nominal Class-E load resistance R to the standard load resistance R_L by appropriate choice of the transmission-line characteristic impedances Z_1 and Z_2.

The load resistance R_L and capacitive stub TL_4 with the characteristic impedance Z_2 and electrical length of 30° connected in parallel provide the capacitive load impedance Z_L at the fundamental frequency defined by

$$Z_L = \frac{Z_2 R_L}{Z_2 + jR_L \tan 30°} \tag{3.121}$$

and the input impedance Z_{net} of the loaded combined quarterwave transmission line is generally calculated from

$$Z_{net} = Z_1 \frac{Z_L + jZ_1 \tan\theta}{Z_1 + jZ_L \tan\theta} \tag{3.122}$$

where θ is the transmission-line electrical length. Then, substituting Eq. (3.121) into Eq. (3.122) for $\theta = 90°$ results in an inductive input impedance at the fundamental frequency as

$$Z_{net} = \frac{Z_1^2}{Z_L} = \frac{Z_1^2}{Z_2 R_L}(Z_2 + jR_L \tan 30°) \tag{3.123}$$

from which it follows that the required nominal or optimized Class-E load resistance can be defined by proper choice of the characteristic impedance Z_1, and the required optimum Class-E inductive reactance can be achieved with the corresponding value of the characteristic impedance Z_2.

The real and imaginary parts of Eq. (3.123) can be written as

$$\mathrm{Re}Z_{net} = \frac{Z_1^2}{R_L} \tag{3.124}$$

$$\mathrm{Im}Z_{net} = \frac{Z_1^2}{\sqrt{3}Z_2} \tag{3.125}$$

from where the characteristic impedances Z_1 and Z_2 can be directly calculated using Eq. (3.119) as

$$Z_1 = \sqrt{R_L R} \tag{3.126}$$

$$Z_2 = 0.5R_L \tag{3.127}$$

where $R_L = 50\ \Omega$ and R is calculated from Eq. (3.95).

Figure 3.55b shows the transmission-line Class-E load network seen from the transistor output at the second harmonic, with an open-circuit condition provided by the transmission line TL_1 at the second harmonic. The transmission-line Class-E load network at the third harmonic looks similar, as shown in Figure 3.55c, when the short-circuited combined transmission line $TL_1 + TL_2$ provides an open-circuit condition at the transistor output at the third harmonic. To account for the parasitic series bondwire and lead inductance of the packaged device, the higher characteristic impedance Z_2 should be chosen.

Figure 3.56 shows the simulated circuit schematic of a transmission-line Class-E power amplifier based on a commercial 28-V 5-W GaN HEMT power transistor NPTB00004 in a plastic package [61]. The input matching circuit with an open-circuit stub and a series transmission line provides the conjugate matching with the standard 50-Ω source. The load network represents the optimized transmission-line Class-E load network shown in Figure 3.27a. By using a 30-mil

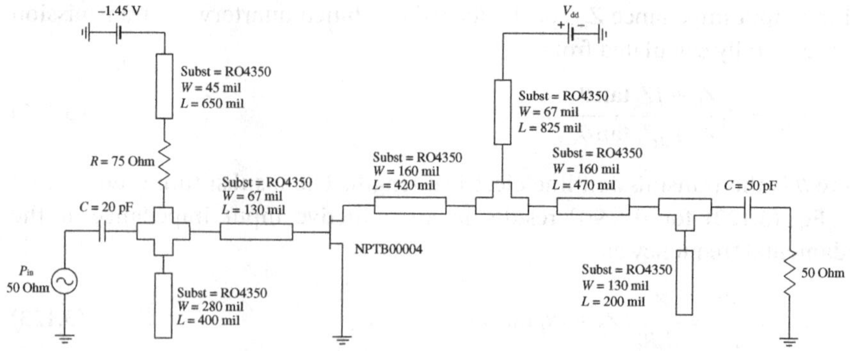

Figure 3.56 Simulation setup of transmission-line Class-E GaN HEMT power amplifier.

RO4350 substrate, an output power of 37 dBm, a drain efficiency of 73%, and a PAE of 71% at the operating frequency of 2.14 GHz were achieved with a power gain of 14 dB (linear gain of 19 dB) and a DC-supply voltage of 25 V.

3.4 Parallel-circuit Class E

In an idealized optimum Class-E with shunt capacitance, an RF choke with infinite impedance at the fundamental frequency and other harmonics is used to supply the DC-feed current to the transistor operating as a switch. However, in practice it is not possible to implement such an RF choke, therefore it should represent a finite DC-feed inductance, having an advantage of minimizing size, cost, and simplifying the overall circuit design. The detailed theoretical analysis of a switchmode parallel-circuit Class-E power amplifier using a parallel inductance and series filter was first published by Kozyrev with calculation of the voltage and current waveforms and some graphical results [35, 41]. This circuit configuration is very similar to a Class-E circuit with shunt capacitance and finite DC-feed inductance [62, 63]. However, in the parallel-circuit Class-E operation mode, the series phase-shifting inductance, which is an important element of the load network in the Class-E with shunt capacitance, should be set to zero. Hence, for the parallel-circuit Class-E power amplifiers, the circuit schematic, required waveforms, phase angles and load-network parameters differ from that for a Class-E power amplifier with shunt capacitance.

3.4.1 Optimum Load-network Parameters

The basic circuit of a switchmode parallel-circuit Class E power amplifier is shown in Figure 3.57a, with a parallel inductance L, a shunt capacitance C, and a

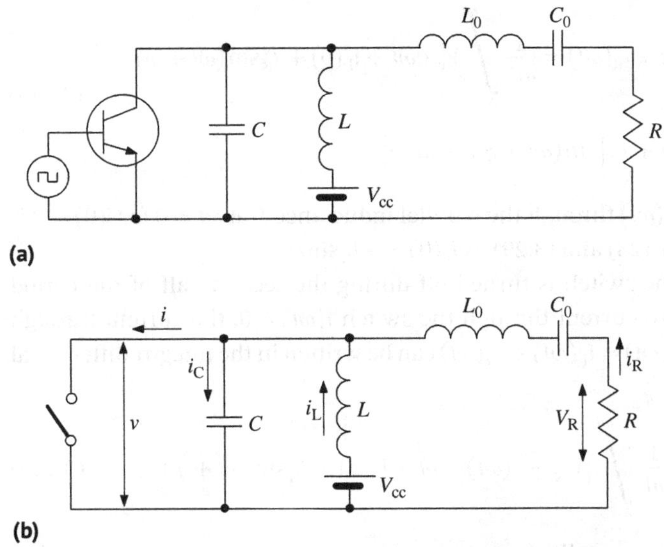

Figure 3.57 Circuit schematics of Class-E load network with parallel circuit.

series L_0C_0-resonant circuit tuned on the fundamental frequency composing the load network.

In order to analyze the parallel-circuit Class-E power amplifier, the simplified equivalent circuit of which is shown in Figure 3.57b, the preliminary assumptions similar to those for the Class-E power amplifier with shunt capacitance should be introduced, with negligible losses in reactive elements, duty ratio of 50%, and high loaded quality factor of the series L_0C_0 circuit. Similarly, it is required to provide the optimum zero-voltage and zero-voltage-derivative conditions for voltage $v(\omega t)$ on the switch at the switch-on moment $\omega t = 2\pi$ given by Eqs. (3.71) and (3.72). It should be noted that, for a nonideal switch with finite saturation resistance, the optimum conditions are satisfied for the parallel-circuit Class E in contrast to the Class E with shunt capacitance [41].

According to the initial assumption, the sinewave current should represent the output current delivered to the load, written as

$$i_R(\omega t) = I_R \sin(\omega t + \varphi) \tag{3.128}$$

where I_R is the load current amplitude and φ is the initial phase shift.

Since the switch is turned on during the first half of the period $0 \leq \omega t < \pi$, the voltage on the switch $v(\omega t)$ is equal to zero when $v_L(\omega t) = V_{cc}$ with zero the current through the capacitance $i_C(\omega t) = \omega C(di_L/d\omega t) = 0$, and the switch current can be written as

$$i(\omega t) = i_L(\omega t) + i_R(\omega t) = \frac{1}{\omega L}\int_0^{\omega t} V_{cc} d\omega t + i_L(0) + I_R \sin(\omega t + \varphi)$$
$$= \frac{V_{cc}}{\omega L}\omega t + I_R\left[\sin(\omega t + \varphi) - \sin\varphi\right] \tag{3.129}$$

where the current $i_L(\omega t)$ through the parallel inductance L at $\omega t = 0$ for $i(0) = 0$ is defined from Eqs. (3.128) and (3.29) as $i_L(0) = -I_R \sin\varphi$.

However, when the switch is turned off during the second half of the period $\pi \leq \omega t < 2\pi$ with zero current through the switch $i(\omega t) = 0$, the current through the capacitance C $i_C(\omega t) = i_L(\omega t) + i_R(\omega t)$ can be written in the integro-differential form as

$$\omega C \frac{dv(\omega t)}{d(\omega t)} = \frac{1}{\omega L}\int_\pi^{\omega t} \left[V_{cc} - v(\omega t)\right] d\omega t + i_L(\pi) + I_R \sin(\omega t + \varphi) \tag{3.130}$$

with the initial off-state conditions at $\omega t = \pi$ equal to $v(\pi) = 0$ and

$$i_L(\pi) = i(\pi) - i_R(\pi) = \frac{V_{cc}\pi}{\omega L} - \omega L I_R \sin\varphi$$

Equation (3.130) can be represented in the form of a linear nonhomogeneous second-order differential equation

$$\omega^2 LC \frac{d^2 v(\omega t)}{d(\omega t)^2} + v(\omega t) - V_{cc} - \omega L I_R \cos(\omega t + \varphi) = 0 \tag{3.131}$$

with the general solution in the normalized form as

$$\frac{v(\omega t)}{V_{cc}} = C_1 \cos(q\omega t) + C_2 \sin(q\omega t) + 1 - \frac{q^2 p}{1 - q^2}\cos(\omega t + \varphi) \tag{3.132}$$

where

$$q = \frac{1}{\omega\sqrt{LC}} \tag{3.133}$$

$$p = \frac{\omega L I_R}{V_{cc}} \tag{3.134}$$

and the coefficients C_1 and C_2 are determined from the initial off-state conditions by

$$C_1 = -(\cos q\pi + q\pi \sin q\pi)$$
$$-\frac{qp}{1 - q^2}\left[q \cos\varphi \cos q\pi - (1 - 2q^2)\sin\varphi \sin q\pi\right] \tag{3.135}$$

$$C_2 = (q\pi\cos q\pi - \sin q\pi) - \frac{qp}{1-q^2}\left[q\cos\varphi\sin q\pi + (1-2q^2)\sin\varphi\cos q\pi\right] \tag{3.136}$$

In the case of a parallel-circuit Class-E load network without series phase-shifting reactance, generally it is necessary to solve a system of three equations to define the three unknown parameters q, p, and φ. The first two equations are the result of applying the two optimum zero-voltage and zero-voltage-derivative Class-E conditions given by Eqs. (3.71) and (3.72) to Eq. (3.132). Because the fundamental-frequency collector voltage is fully applied to the load, this means that its reactive part must have zero value, resulting in an additional equation

$$V_X = -\frac{1}{\pi}\int_0^{2\pi} v(\omega t)\cos(\omega t + \varphi)d\omega t = 0 \tag{3.137}$$

This system of three equations with three unknown parameters can be solved numerically, resulting in the following values [64, 65]:

$$q = 1.412 \tag{3.138}$$

$$p = 1.210 \tag{3.139}$$

$$\varphi = 15.155° \tag{3.140}$$

The DC-supply current I_0 can be defined from Eq. (3.129) using the Fourier formula as

$$I_0 = \frac{1}{2\pi}\int_0^{2\pi} i(\omega t)d\omega t = \frac{I_R}{2\pi}\left(\frac{V_{cc}}{\omega L I_R}\frac{\pi^2}{2} + 2\cos\varphi - \pi\sin\varphi\right) = 0.826 I_R \tag{3.141}$$

Figure 3.58 shows the normalized (a) load current and collector, (b) voltage, and (c) current waveforms for an idealized optimum parallel-circuit Class-E operation. These waveforms are very similar to those for an idealized optimum Class E with shunt capacitance shown in Figure 3.47, with a small difference for the voltage waveform and initial phase shift for the sinusoidal current flowing to the load. Since there are no nonzero voltage and current simultaneously, no power loss occurs, and an idealized collector efficiency of 100% is achieved.

The fundamental-frequency current flowing through the switch consists of the two quadrature components, as shown in Figure 3.59, with amplitudes obtained from the Fourier formulas and Eq. (3.129) as

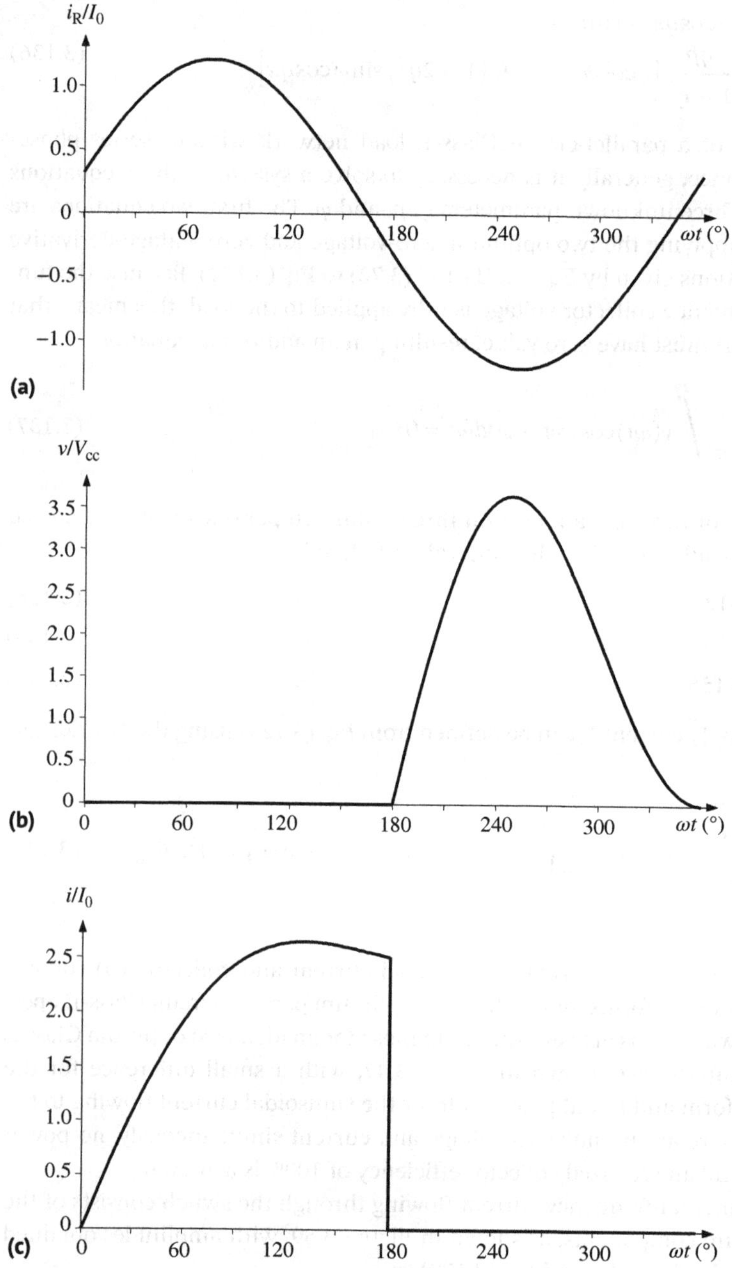

Figure 3.58 Ideal (a) load current and collector, (b) voltage, and (c) current waveforms for nominal parallel-circuit Class-E mode.

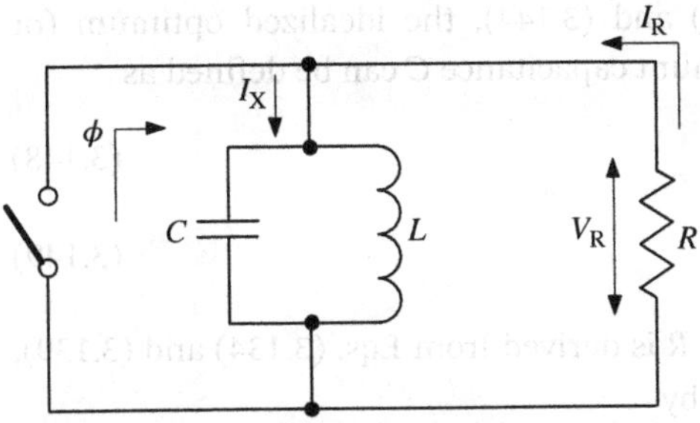

Figure 3.59 Equivalent circuit of parallel-circuit Class-E load network at fundamental.

$$I_R = \frac{1}{\pi}\int_0^{2\pi} i(\omega t)\sin(\omega t + \varphi)d\omega t = \frac{I_R}{\pi}\left[\frac{V_{cc}}{\omega L I_R}(\pi\cos\varphi - 2\sin\varphi) + \frac{\pi}{2} - \sin 2\varphi\right] \tag{3.142}$$

$$I_X = -\frac{1}{\pi}\int_0^{2\pi} i(\omega t)\cos(\omega t + \varphi)d\omega t = \frac{I_R}{\pi}\left[\frac{V_{cc}}{\omega L I_R}(\pi\sin\varphi + 2\cos\varphi) - 2\sin^2\varphi\right] \tag{3.143}$$

Consequently, the phase angle ϕ at the fundamental frequency seen from the transistor output is derived through the amplitudes of the quadrature components I_X and I_R as

$$\phi = \tan^{-1}\left(\frac{I_X}{I_R}\right) = 34.244^\circ \tag{3.144}$$

or, alternatively, the phase angle ϕ can be expressed through the load-network elements as

$$\tan\phi = \frac{R}{\omega L} - \omega RC \tag{3.145}$$

Hence, the normalized load-network parameters can be appropriately obtained using Eq. (3.133) by

$$\frac{\omega L}{R} = \frac{q^2 - 1}{q^2 \tan\phi} \tag{3.146}$$

$$\omega RC = \frac{\tan\phi}{q^2 - 1} \tag{3.147}$$

As a result, by using Eqs. (3.138) and (3.144), the idealized optimum (or nominal) parallel inductance L and shunt capacitance C can be defined as

$$L = 0.732\frac{R}{\omega} \tag{3.148}$$

$$C = \frac{0.685}{\omega R} \tag{3.149}$$

whereas the optimum load resistance R is derived from Eqs. (3.134) and (3.139), taking into account that $R = V_R^2/2P_{out}$, by

$$R = 1.365\frac{V_{cc}^2}{P_{out}} \tag{3.150}$$

where V_{cc} is the DC-supply voltage and P_{out} is the output power at the fundamental frequency.

The peak collector current I_{max} and peak collector voltage V_{max} can be determined from Eqs. (3.129), (3.132), and (3.141) as

$$I_{max} = 2.647\,I_0 \tag{3.151}$$

$$V_{max} = 3.647\,V_{cc} \tag{3.152}$$

The maximum operation frequency f_{max} can be expressed through the shunt capacitance $C = C_{out}$, output power P_{out}, and DC-supply voltage V_{cc} using Eqs. (3.149) and (3.150) as

$$f_{max} = 0.0798\frac{P_{out}}{C_{out}V_{cc}^2} \tag{3.153}$$

which is 1.57 times higher than the maximum operation frequency for an idealized optimum Class-E power amplifier with shunt capacitance [66].

3.4.2 Operation Beyond Maximum Class-E Frequency

When the value of a shunt capacitance exceeds its required value for a nominal (or idealized optimum) Class-E operation mode which usually happens at higher frequencies beyond maximum Class-E frequency due to a sufficiently large value of the transistor output (collector or drain–source) capacitance, the value of a parallel DC-feed inductance should be optimized to compensate, at least partly, for the negative effect of this excessive capacitance at the fundamental frequency, as well as, at the higher-order harmonics, maintaining efficiency at enough high level.

Figure 3.60a shows the circuit schematic of a parallel-circuit Class-E power amplifier, where C is the nominal parallel-circuit Class-E capacitance, C_{ex} is the

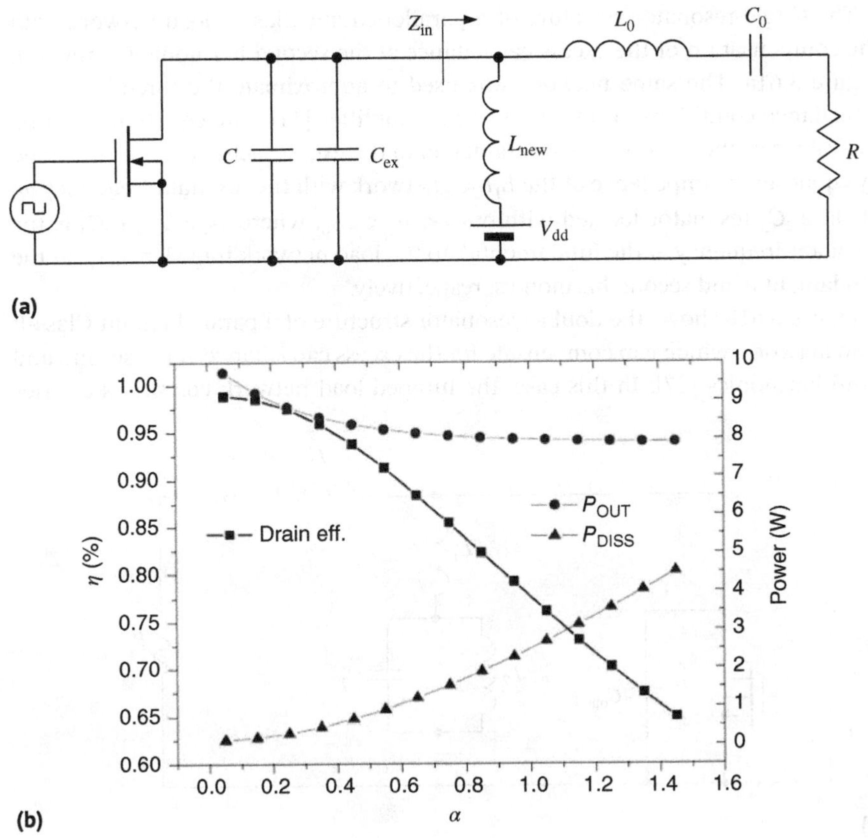

Figure 3.60 Schematic and performance of parallel-circuit Class-E amplifier with excess capacitance.

excess capacitance, and L_{new} is the required frequency-varying parallel DC-feed inductance [67]. Here, the susceptances of C_{ex} and L_{new} should result in the inductive reactance of an optimum Class-E parallel DC-feed inductance L. In this case, an inductance L_{new} becomes a frequency-dependent element, the value of which is different for the corresponding harmonic component and which can be implemented using an appropriate arrangement of lumped components, as well as the load-network impedance Z_{in} becomes a function of the excess factor $\alpha = C_{ex}/C$. Figure 3.60b shows the variations of the drain efficiency η, output power P_{out}, and power dissipation P_{diss} with the excess factor α, keeping constant the other circuit components.

The single-resonator structure of a parallel-circuit Class-E load network with the compensation of the excess capacitance at the second harmonic is shown in Figure 3.61a. The same network was used to approximate the three-harmonic impedance conditions of a Class-F power amplifier [11]. The equations relating the values of the load-network parameters to the excess factor α can be obtained by equating the impedance of the $L_1L_2C_2$ network with the resonant frequency ω_2 of the L_2C_2 resonator located within $\omega_0 < \omega_2 < 2\omega_0$, where $\omega_0 = 1/\sqrt{L_0C_0}$ is the resonant frequency at the fundamental, to the load network impedance Z_{in} at the fundamental and second harmonics, respectively.

Figure 3.61b shows the double-resonator structure of a parallel-circuit Class-E load network, which can compensate for the excess capacitance at the second and third harmonics [67]. In this case, the lumped load network comprises a series

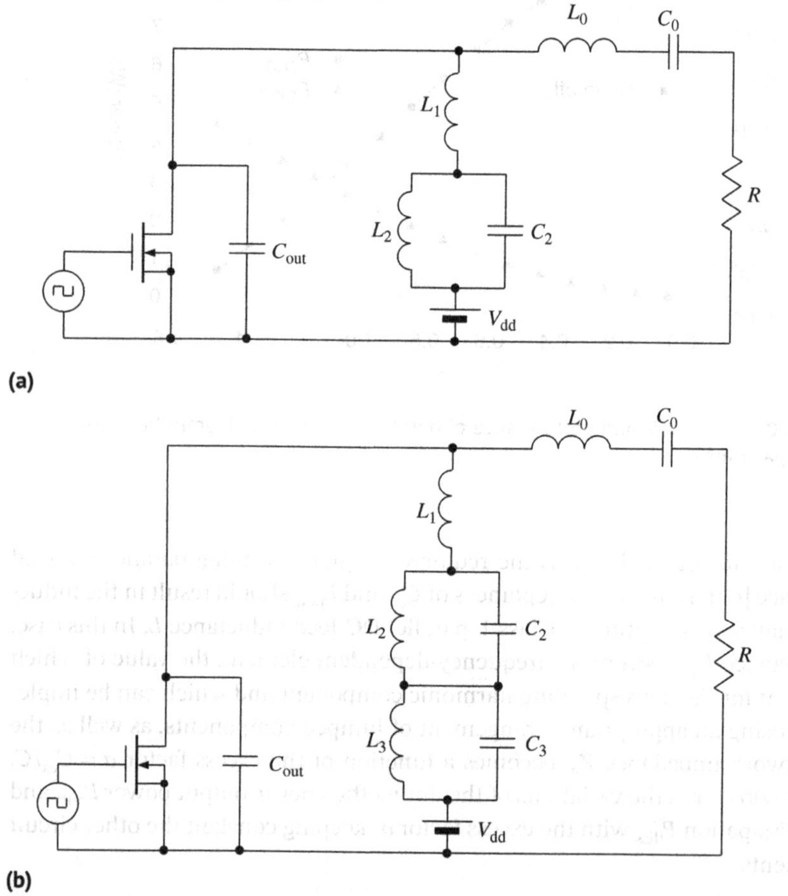

Figure 3.61 Single- and double-resonator circuits of parallel-circuit Class-E load network.

inductor and two parallel LC resonators. The resonant frequency ω_2 of the L_2C_2 resonator is located between ω_0 and $2\omega_0$, while the resonant frequency ω_3 of the L_3C_3 resonator is located between $2\omega_0$ and $3\omega_0$. Fixing the values of these resonances and selecting appropriate values for L_1, L_2, and L_3 allows the right value of harmonic reactances to be obtained.

The simulated voltage and current drain waveforms obtained for $\alpha = 1.174$ with an ideal switch and assuming the lossless load-network parameters demonstrate that, if the switch current exhibits a small negative peak during turn-on interval for a single-resonator topology, as shown in Figure 3.62a, this peak disappears for

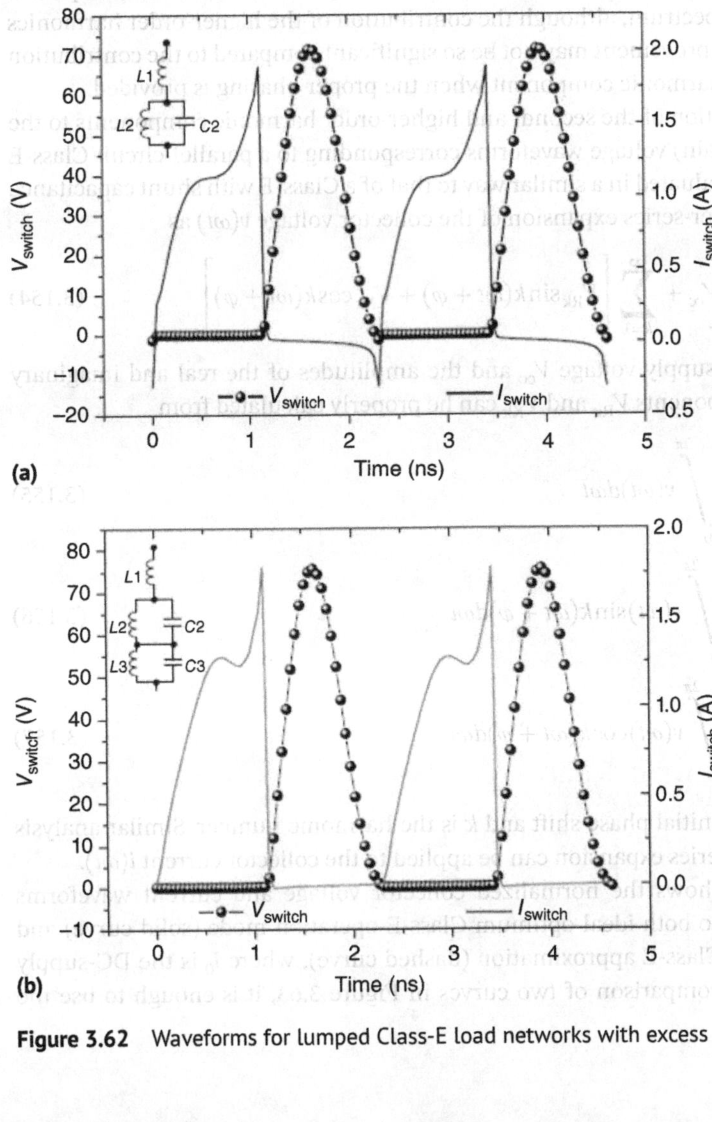

Figure 3.62 Waveforms for lumped Class-E load networks with excess capacitance.

a double-resonator topology, as shown in Figure 3.62b, thus clearly approximating the ideal optimum Class-E switching conditions. The component values of the series L_0C_0 resonant circuit were obtained by selecting $Q_L = 8$ and by setting the resonance frequency equal to 434 MHz.

3.4.3 Two-harmonic Approximation

In the frequency domain, an idealized calculation of the collector voltage and current waveforms required for an optimum Class-E mode implies the availability of an infinite number of the harmonics with their optimum amplitudes and phases in the output spectrum, although the contribution of the higher-order harmonics in efficiency improvement may not be so significant compared to the contribution of the second-harmonic component when the proper phasing is provided.

The contribution of the second- and higher-order harmonic components to the collector (or drain) voltage waveforms corresponding to a parallel-circuit Class-E mode can be evaluated in a similar way to that of a Class-E with shunt capacitance using the Fourier-series expansion of the collector voltage $v(\omega t)$ as

$$v(\omega t) = V_{cc} + \sum_{k=1}^{\infty} \left[V_{Rk} \sin k(\omega t + \varphi) + V_{Xk} \cos k(\omega t + \varphi) \right] \tag{3.154}$$

where the DC-supply voltage V_{cc} and the amplitudes of the real and imaginary harmonic components V_{Rk} and V_{Xk} can be properly calculated from

$$V_{cc} = \frac{1}{2\pi} \int_0^{2\pi} v(\omega t) d\omega t \tag{3.155}$$

$$V_{Rk} = \frac{1}{\pi} \int_0^{2\pi} v(\omega t) \sin k(\omega t + \varphi) d\omega t \tag{3.156}$$

$$V_{Xk} = \frac{1}{\pi} \int_0^{2\pi} v(\omega t) \cos k(\omega t + \varphi) d\omega t \tag{3.157}$$

where φ is the initial phase shift and k is the harmonic number. Similar analysis using Fourier-series expansion can be applied to the collector current $i(\omega t)$.

Figure 3.63 shows the normalized collector voltage and current waveforms corresponding to both ideal optimum Class-E operation mode (solid curve) and two-harmonic Class-E approximation (dashed curve), where I_0 is the DC-supply current. From comparison of two curves in Figure 3.63, it is enough to use the

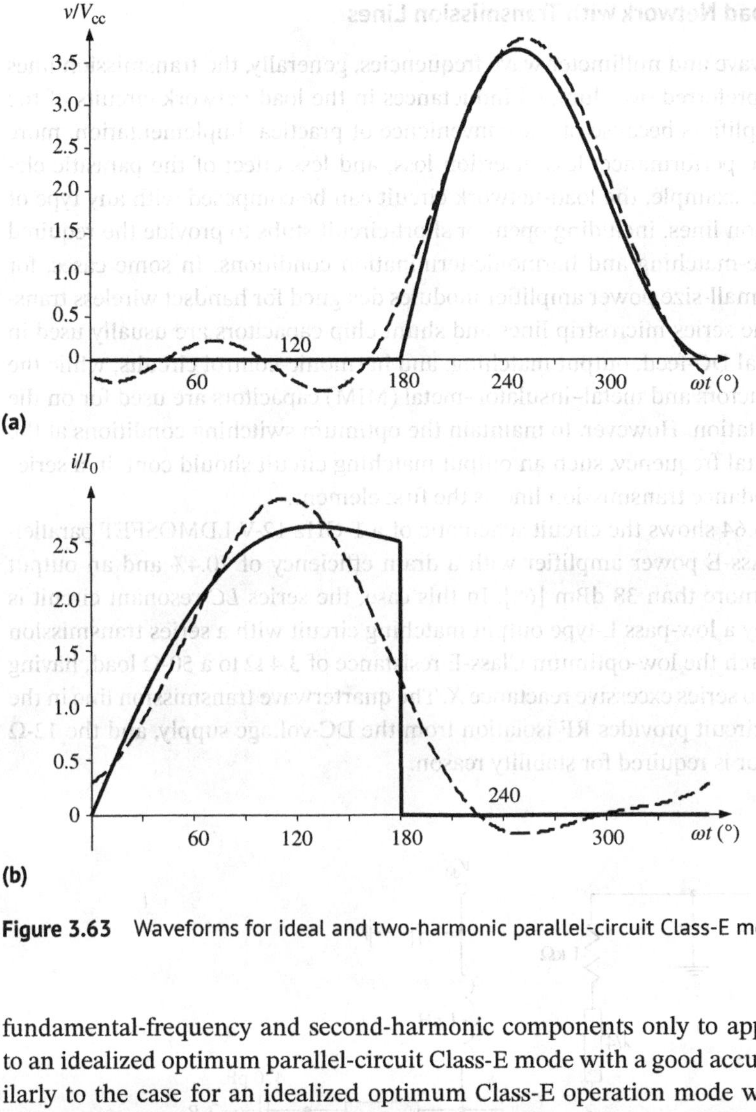

Figure 3.63 Waveforms for ideal and two-harmonic parallel-circuit Class-E mode.

fundamental-frequency and second-harmonic components only to approximate to an idealized optimum parallel-circuit Class-E mode with a good accuracy, similarly to the case for an idealized optimum Class-E operation mode with shunt capacitance shown in Figure 3.52c. Such an approximation of the high-efficiency parallel-circuit Class-E power amplifier circuit can be effectively used both in the hybrid integrated circuits intended to operate for base station applications and in the monolithic microwave integrated circuit (MMIC) design used for handset applications.

3.4.4 Load Network with Transmission Lines

At microwave and millimeter-wave frequencies, generally, the transmission lines are often preferred over lumped inductances in the load-network circuits of the power amplifiers because of the convenience of practical implementation, more predictable performance, less insertion loss, and less effect of the parasitic elements. For example, the load-network circuit can be composed with any type of transmission lines, including open- or short-circuit stubs to provide the required impedance-matching and harmonic-termination conditions. In some cases, for compact small-size power amplifier modules designed for handset wireless transmitters, the series microstrip lines and shunt chip capacitors are usually used in the external DC-feed, output matching, and harmonic-control circuits, while the spiral inductors and metal–insulator–metal (MIM) capacitors are used for on-die implementation. However, to maintain the optimum switching conditions at the fundamental frequency, such an output matching circuit should contain a series high-impedance transmission line as the first element.

Figure 3.64 shows the circuit schematic of a 1-GHz 12-V LDMOSFET parallel-circuit Class-E power amplifier with a drain efficiency of 70.4% and an output power of more than 38 dBm [68]. In this case, the series *LC* resonant circuit is replaced by a low-pass L-type output matching circuit with a series transmission line to match the low-optimum Class-E resistance of 3.4 Ω to a 50-Ω load, having almost zero series excessive reactance *X*. The quarterwave transmission line in the gate bias circuit provides RF isolation from the DC-voltage supply, and the 12-Ω gate resistor is required for stability reason.

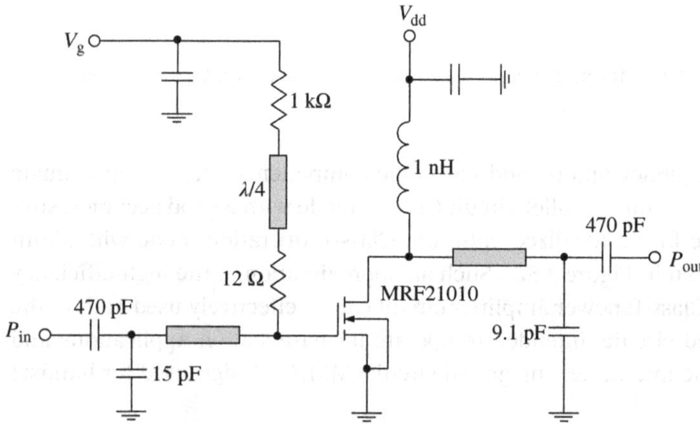

Figure 3.64 Circuit schematic of transmission-line MOSFET Class-E power amplifier.

The lumped inductor L in the drain DC-feed circuit can be replaced by a short-length short-circuited transmission line according to

$$Z_0 \tan\theta = \omega L \tag{3.158}$$

where Z_0 and θ are the characteristic impedance and electrical length of the transmission line TL, respectively [69]. In this case, the required idealized optimum input impedance of the load network at the fundamental frequency can be obtained by

$$Z_{net}(\omega_0) = \frac{R}{1 - j\tan 34.244^\circ} \tag{3.159}$$

which is derived from Eqs. (3.144), (3.148), and (3.149).

Substituting Eq. (3.158) into Eq. (3.148) results in

$$\tan\theta = 0.732\frac{R}{Z_0} \tag{3.160}$$

An example of the transmission-line Class-E load network of a two-stage 1.75-GHz GaAs HBT power amplifier with an output power of 33 dBm, designed for a handset transmitter including the series microstrip lines with two shunt chip capacitors, is shown in Figure 3.65a [64, 65]. Here, the series transmission

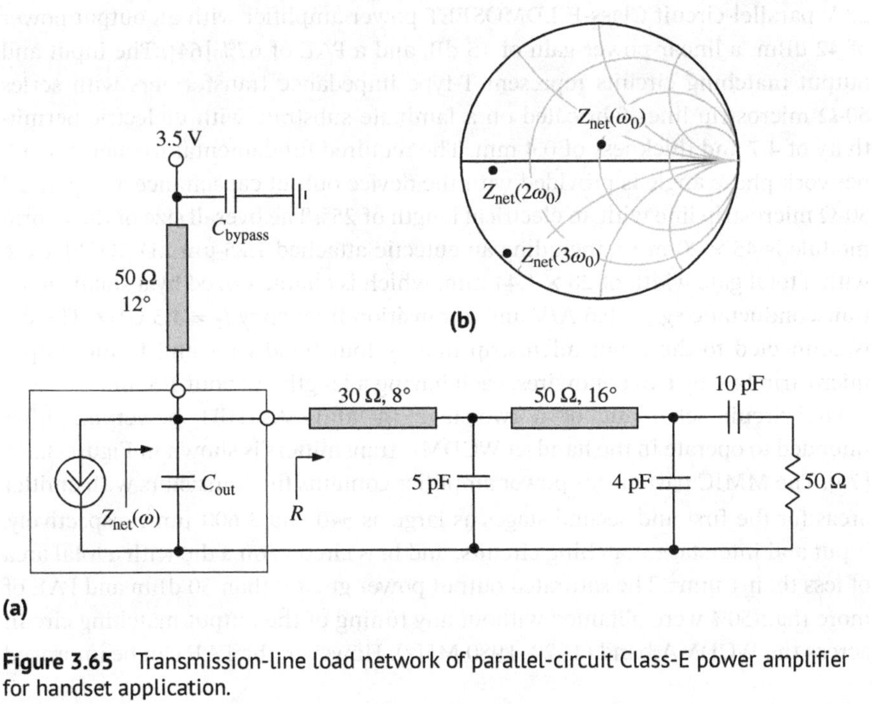

Figure 3.65 Transmission-line load network of parallel-circuit Class-E power amplifier for handset application.

line with electrical length less than 90° in the matching circuit cannot provide the required inductive impedance at the fundamental frequency with the purely capacitive reactances at higher-order harmonics. For example, the real part of the load-network impedance $Z_{net}(2\omega_0)$ at the second harmonic is high enough, as shown in Figure 3.65b. However, such a load network enables to provide a close proximity to the parallel-circuit Class-E parameters, resulting in a high operating efficiency. In this case, the short-length parallel microstrip connected to the DC supply is used instead of an RF choke, contributing also to an optimum inductive impedance at the fundamental frequency.

The main advantages of the parallel-circuit Class-E load network compared to other alternative structures are the low-value of the DC-feed inductance L, which can be provided by a short-length transmission line or bondwires, and a high-value of the optimum load resistance R, which is about 2.4 times greater in comparison with the Class E with shunt capacitance that is especially important for high-power and low DC-supply voltage applications with a high level of circuit integration. Besides, the parallel-circuit Class-E circuit can be easily implemented in a high-efficiency broadband design when it is necessary only to satisfy the required phase angle seen by the transistor collector (or drain) at the fundamental frequency to provide the corresponding inductive impedance [70, 71].

Figure 3.66a shows the circuit schematic and module structure of a 500-MHz 28-V parallel-circuit Class-E LDMOSFET power amplifier with an output power of 42 dBm, a linear power gain of 15 dB, and a PAE of 67% [64]. The input and output matching circuits represent T-type impedance transformers with series 50-Ω microstrip lines fabricated on a laminate substrate with dielectric permittivity of 4.7 and thickness of 0.4 mm. The required fundamental-frequency load-network phase angle is provided with the device output capacitance and parallel 50-Ω microstrip line with an electrical length of 25°. The overall size of the hybrid module is 45 × 20 mm^2, including an eutectic-attached 1.25-μm LDMOSFET die with a total gate width of 28 × 1.44 mm, which is characterized by a small-signal transconductance $g_m = 0.6$ A/V and a transition frequency $f_T = 4.5$ GHz. The die is connected to the input microstrip line by four bondwires and to the output microstrip line by five bondwires, each having a length of about 1.5 mm.

The circuit schematic of a two-stage InGaP/GaAs HBT power amplifier intended to operate in the handset WCDMA transmitters is shown in Figure 3.66b [70]. The MMIC part of this power amplifier contains the transistors with emitter areas for the first and second stages as large as 540 and 3 600 μm^2, respectively, input and interstage matching circuits, and bias circuits on a die with a total area of less than 1 mm^2. The saturated output power greater than 30 dBm and PAE of more than 50% were obtained without any tuning of the output matching circuit across the WCDMA band (1920–1980 MHz). However, the PAE can be improved

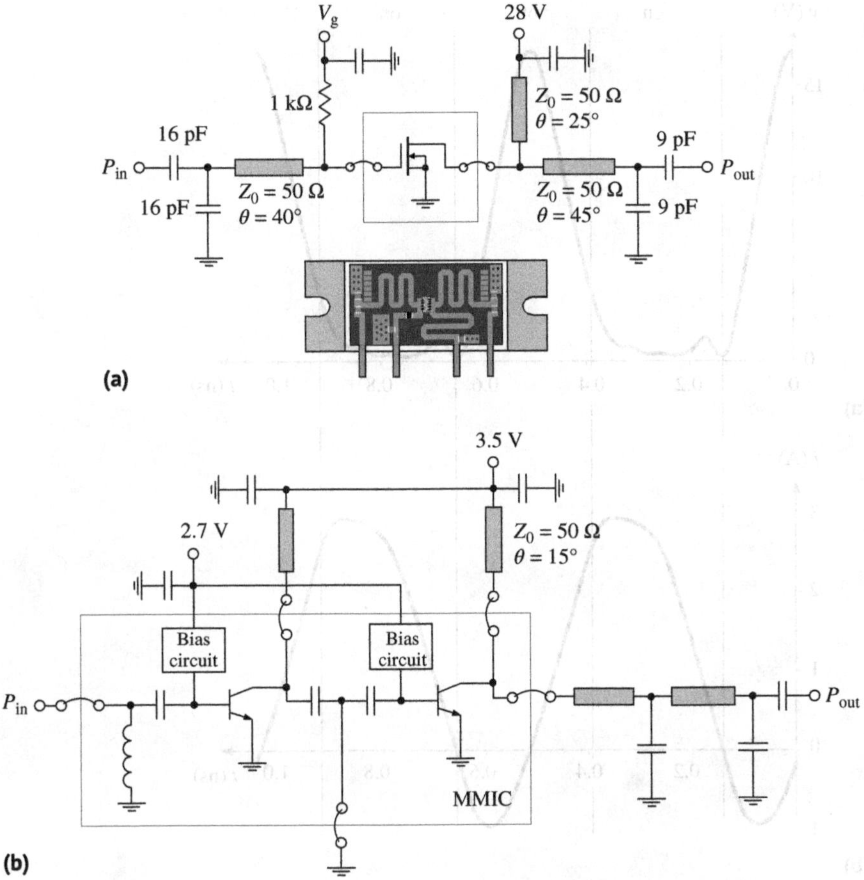

Figure 3.66 Schematics and module of parallel-circuit Class-E power amplifiers.

by about 8% using ceramic chip capacitors with high quality factors. At the same time, the high-linearity performance at a 3.5-dB backoff output power of 27 dBm with a power gain of 22.6 dB and a PAE of 38.3% at center bandwidth frequency of 1.95 GHz with an adjacent channel leakage power ratio (ACLR) of −37 dBc at a 5-MHz offset was measured.

As an example of a close proximity to the idealized optimum parallel-circuit Class-E mode, the simulated collector (a) voltage and (b) current waveforms shown in Figure 3.67 and collector efficiency of about 90% were achieved for a 1.75-GHz InGaP/GaAs HBT power amplifier at a supply voltage of 5 V [65]. During off-state operation mode, only the current through the device collector capacitance determines the total collector current. The collector voltage waveform is very similar to

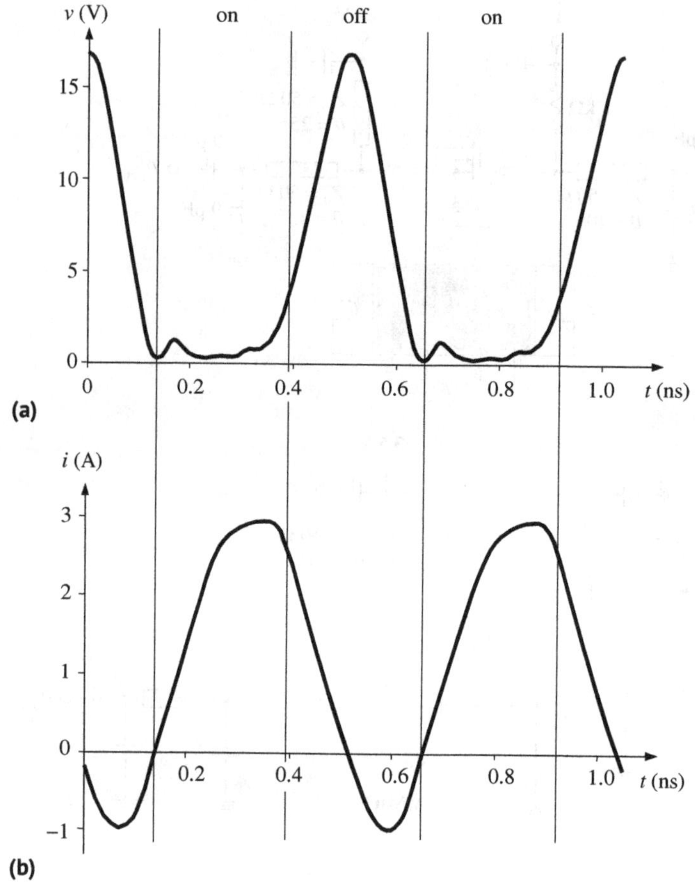

Figure 3.67 Simulated collector (a) voltage and (b) current waveforms of transmission-line parallel-circuit Class E HBT power amplifier.

the ideal one, with a peak factor of about 3. The main reason for efficiency degradation from an ideal 100% is the high value of the saturation voltage of about 0.5–0.8 V, which can result in 15–20% collector efficiency reduction for a DC-supply voltage of 3.5 V. Generally, the further decrease in collector efficiency in a practical circuit can be explained by a violation of the required idealized optimum impedance conditions because of the transmission-line effect at the second- and higher-order harmonics, device finite switching time and parasitics, and power losses in the load network.

The suboptimum Class-E design techniques can also be applied to microwave or millimeter-wave power amplifiers using a silicon-based technology when a high-power gain can be achieved by using a cascode configuration. Figure 3.68 shows the circuit schematic of a *W*-band cascode power amplifier using a parallel-circuit Class-E approximation, which is implemented in a 0.13-μm SiGe BiCMOS process using SiGe HBTs with f_T and f_{max} of 300 and 500 GHz, respectively [72]. The cascode configuration allows to protect devices from breakdown for the increased voltage swing corresponding to a Class-E mode. In this case, the optimum parallel-circuit Class-E capacitance is determined entirely by the parasitic output capacitance of the common-base stage representing a sum of the collector–substrate and collector–base capacitances, with RF shorted substrate and base nodes. The Class-E load network with a low-pass response consists of a short series microstrip line of 10° and an RF shorted parallel microstrip line of 60° to provide the required inductive impedance at the fundamental and the output L-type matching section consisting of a series microstrip line of 35° and an open-circuit microstrip stub of 35° to compensate for the parasitic capacitance of the output RF pad and to provide the output impedance matching with a standard 50-Ω load. As a result, a peak PAE of 40.4% and a saturated output power of 17.7 dBm were achieved at 93 GHz with a DC-supply voltage of 4 V. With the drain bias increased to 5.2 V, the peak PAE and saturated output power at 93 GHz were measured to be 37.6% and 19.3 dBm, respectively.

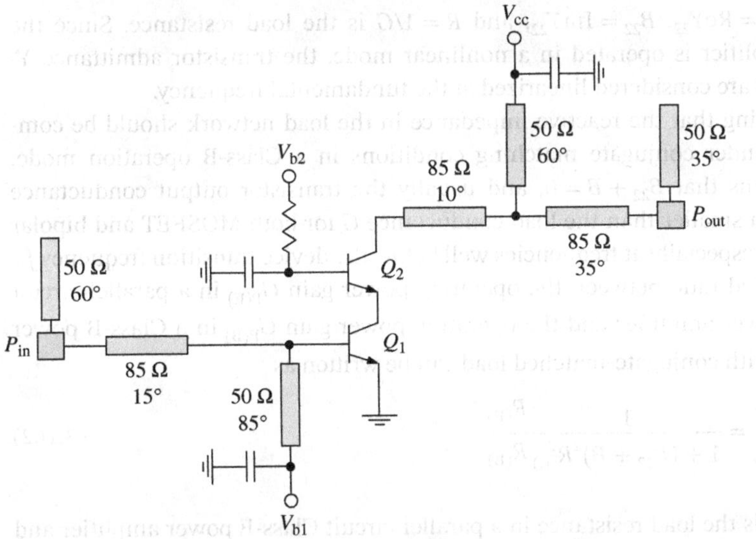

Figure 3.68 Circuit schematic of *W*-band cascode Class-E SiGe HBT power amplifier.

3.4.5 Power Gain

As opposed to a conventional Class-B operation mode where the conjugate matching condition is required to provide maximum power delivery to the load, in any type of Class-E operation mode, the load network should be tuned to provide an inductive or capacitive (inverse Class E) impedance at the fundamental frequency seen by the transistor collector (or drain). Therefore, the output voltage and current waves distributed in the load network generally represent both types of electromagnetic waves, incident and reflected, which means that, generally, the power gain in a Class-E operation mode is lower than in a Class-B operation mode because the higher driving voltage to operate under the device saturation conditions is required. However, the power gain is a function of not only the device transconductance as the main parameter but also the load resistance. Hence, it is very important to quantitatively compare the power gains in both operation classes. In this case, a parallel-circuit Class-E operation mode looks very attractive since it provides the highest value of the load resistance compared with other Class-E alternatives.

The operating power gain G_p, expressed through the transistor admittance Y-parameters and load admittance $Y_L = G + jB$, can be written as

$$G_P = \frac{|Y_{21}|^2}{\mathrm{Re}Y_{in}} \frac{R}{\left(1 + \frac{G_{22}}{G}\right)^2 + \left(\frac{B_{22} + B}{G}\right)^2} \tag{3.161}$$

where $G_{22} = \mathrm{Re}Y_{22}$, $B_{22} = \mathrm{Im}Y_{22}$, and $R = 1/G$ is the load resistance. Since the power amplifier is operated in a nonlinear mode, the transistor admittance Y-parameters are considered linearized at the fundamental frequency.

Considering that the reactive impedance in the load network should be compensated under conjugate matching conditions in a Class-B operation mode, which means that $B_{22} + B = 0$, and usually the transistor output conductance G_{22} is much smaller than the load conductance G for both MOSFET and bipolar transistors, especially at frequencies well below the device transition frequency f_T, the simplified ratio between the operating power gain $G_{P(E)}$ in a parallel-circuit Class-E power amplifier and the operating power gain $G_{P(B)}$ in a Class-B power amplifier with conjugate-matched load can be written as

$$\frac{G_{P(E)}}{G_{P(B)}} = \frac{1}{1 + (B_{22} + B)^2 R_{(E)}^2} \frac{R_{(E)}}{R_{(B)}} \tag{3.162}$$

where $R_{(E)}$ is the load resistance in a parallel-circuit Class-E power amplifier and $R_{(B)}$ is the load resistance in a Class-B power amplifier.

For an idealized optimum parallel-circuit Class-E operation mode with a 100% collector efficiency when $I_0V_{cc} = V_R^2/2R$ and $V_R = I_RR$, from Eq. (3.141), it follows that

$$V_{R(E)} = 1.652\,V_{cc} \tag{3.163}$$

Then, considering the fact that $V_{R(B)} = V_{cc}$ in a Class-B operation mode with zero saturation voltage for the same output power P_{out}, the ratio between load resistance in a parallel-circuit Class-E and Class-B operation modes can be written as

$$\frac{R_{(E)}}{R_{(B)}} = \frac{V_{R(E)}^2}{V_{R(B)}^2} = 2.729 \tag{3.164}$$

that demonstrates a much higher value for the load resistance in an idealized optimum (or nominal) parallel-circuit Class-E operation mode.

Thus, the power gain ratio given by Eq. (3.162) can be rewritten as

$$\frac{G_{P(E)}}{G_{P(B)}} = \frac{2.729}{1 + \tan^2\phi} = 1.865 \tag{3.165}$$

where $\phi = 34.244°$ is the phase angle seen by the device output required for a nominal parallel-circuit Class-E operation mode.

An analysis of Eq. (3.165) shows that the operating power gain in a nominal parallel-circuit Class-E operation mode in comparison with a Class-B operation mode is practically similar, although the load network is detuned. This happens due to the larger value of the load resistance required for an idealized optimum parallel-circuit Class-E load network. So, if the operating power gain in a Class-E mode with shunt capacitance is smaller in comparison with a Class-B power amplifier because its nominal load resistance is of about 2.4 times smaller than that in a parallel-circuit Class-E mode [73]. In view of sufficiently low switching power losses due to nonzero switching time according to Eq. (3.117), a slight overdrive of the transistor is required when the input power should be increased by 1–2 dB to minimize the switching time and maximize the collector efficiency of a parallel-circuit Class-E power amplifier. Thus, the resulting operating power gain in a nominal Class-E power amplifier becomes very close or even equal to the operating power gain in a Class-B power amplifier.

3.4.6 General Case of Class E with Finite DC-feed Inductance

Generally, a parallel-circuit Class-E load network represents a particular case of the Class-E load network with finite DC-feed inductance, which has multiple solutions because its operation is described by the second-order differential equation, unlike the Class-E with shunt capacitance having the only a single solution for its load-network parameters. Figure 3.69a shows the generalized second-order load network of a Class-E power amplifier with finite DC-feed inductance [74]. Here, in addition to the shunt capacitance C, series inductance L_b, parallel inductance L, and series resonant L_0C_0 circuit, the series reactance X is included. The series inductance L_b represents the bondwire and package lead inductance, the parallel inductance L is considered the finite DC-feed inductance, and the series reactance X can have positive (inductance), negative (capacitance), or zero value, depending on the particular Class-E operation mode. In an idealized optimum Class-E operation mode, the transistor is considered an ideal switch that is driven in such a way as to provide the device instant switching between its on-state and off-state operation modes.

An analysis of the Class-E power amplifier with the generalized load network is similar to that for the parallel-circuit Class-E power amplifier, based on the same assumptions and simplified equivalent circuit shown in Figure 3.69b. As a result, its electrical behavior can be described by the linear nonhomogeneous second-order differential equation

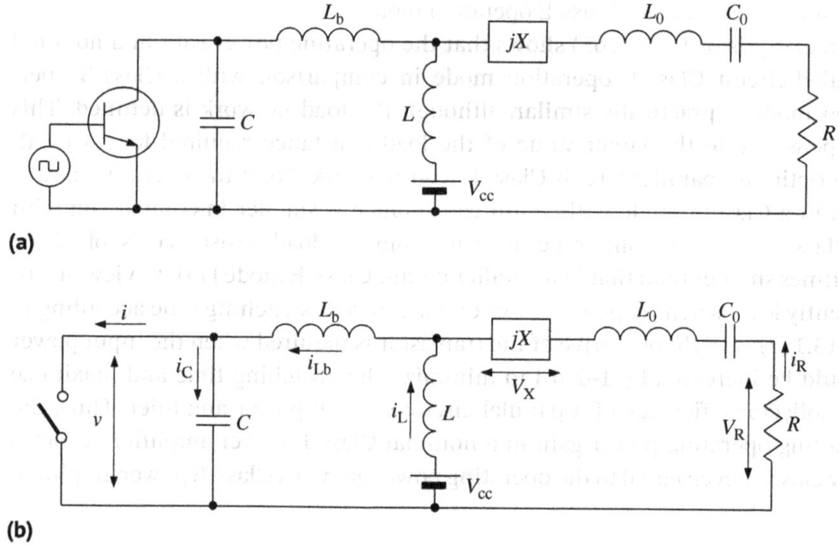

Figure 3.69 Circuit schematics of Class-E mode with generalized load network.

$$\omega^2(L+L_b)C\frac{d^2v(\omega t)}{d(\omega t)^2}+v(\omega t)-V_{cc}-\omega LI_R\cos(\omega t+\varphi)=0 \tag{3.166}$$

with the general solution in the normalized form as

$$\frac{v(\omega t)}{V_{cc}}=C_1\cos(q\omega t)+C_2\sin(q\omega t)+1-\frac{q^2p}{1-q^2}\cos(\omega t+\varphi) \tag{3.167}$$

where

$$q=\frac{1}{\omega\sqrt{C(L+L_b)}} \tag{3.168}$$

$$p=\frac{\omega LI_R}{V_{cc}} \tag{3.169}$$

and the coefficients C_1 and C_2 are determined from the initial off-state conditions.

As a result, the optimum load resistance R can be obtained by

$$R=\frac{1}{2}\left(\frac{V_R}{V_{cc}}\right)^2\frac{V_{cc}^2}{P_{out}} \tag{3.170}$$

where V_{cc} is a DC-supply voltage, P_{out} is a fundamental output power, and

$$\frac{V_R}{V_{cc}}=\frac{1}{\pi}\left(\frac{\pi^2}{2p}+2\cos\varphi-\pi\sin\varphi\right)\Big/\left(1+\frac{L_b}{L}\right) \tag{3.171}$$

By using Eqs. (6.46), (6.47), and (6.50), the normalized load-network inductance L and capacitance C as functions of the ratio L_b/L can be obtained as

$$\frac{\omega L}{R}=p\left(1+\frac{L_b}{L}\right)\Big/\left(\frac{\pi}{2p}+\frac{2}{\pi}\cos\varphi-\sin\varphi\right) \tag{3.172}$$

$$\omega CR=\frac{1}{q^2\left(1+\frac{L_b}{L}\right)\frac{\omega L}{R}} \tag{3.173}$$

Generally, Eq. (3.167) contains the three unknown parameters q, p, and φ, which must be calculated analytically or numerically to determine the collector voltage waveform. In a common case, the parameter q can be considered a variable, and the other two parameters p and φ are calculated from a system of two equations after applying two optimum zero-voltage and zero-voltage-derivative conditions given by Eq. (3.71) and (3.72) to Eq. (3.167). The dependencies of the parameters p and φ versus q for an idealized optimum Class E with finite DC-feed inductance when $L_b = 0$ are shown in Figure 3.70.

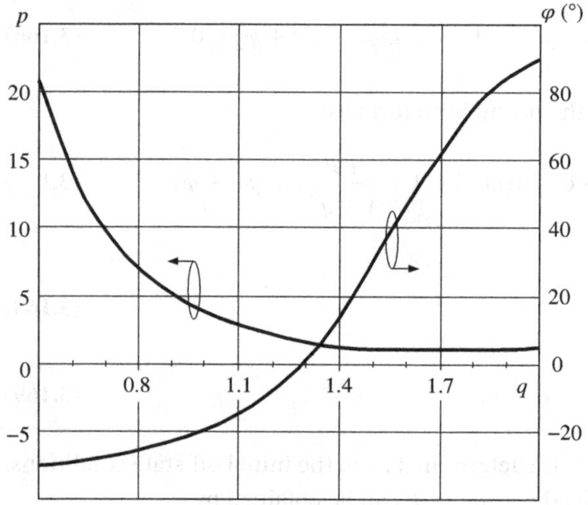

Figure 3.70 Optimum Class-E parameters p and φ versus q.

Based on the relationship of the two (real and imaginary) quadrature fundamental-frequency voltage Fourier components, the series reactance X can be calculated according to $X/R = V_X/V_R$. The dependences of the normalized optimum DC-feed inductance $\omega L/R$ and series reactance X/R are shown in Figure 3.71a, whereas the dependences of the normalized optimum shunt capacitance ωCR and load resistance RP_{out}/V_{cc}^2 are plotted in Figure 3.71b. Here, the value of the series reactance X changes its sign from positive to negative, which means that the inductive reactance is followed by the capacitive reactance. As a result, there is a special case of the load network with a parallel circuit and a load resistor only when $X = 0$ at $q = 1.412$. In this case, the maximum value of the optimum load resistance R can be provided for the same DC-supply voltage and output power, as shown in Figure 3.71b, thus simplifying the matching with the standard load of 50 Ω. In addition, the value of a DC-feed inductance L becomes enough small for convenient practical implementation and the maximum operation frequency f_{max} is realized at $q = 1.468$, where the normalized optimum shunt capacitance ωCR reaches its maximum. To speed up the design procedure, the analytical design equations represented by the second- and third-order polynomial functions can be used to replace the graphical solutions or numerical calculations for the optimum load-network parameters [75].

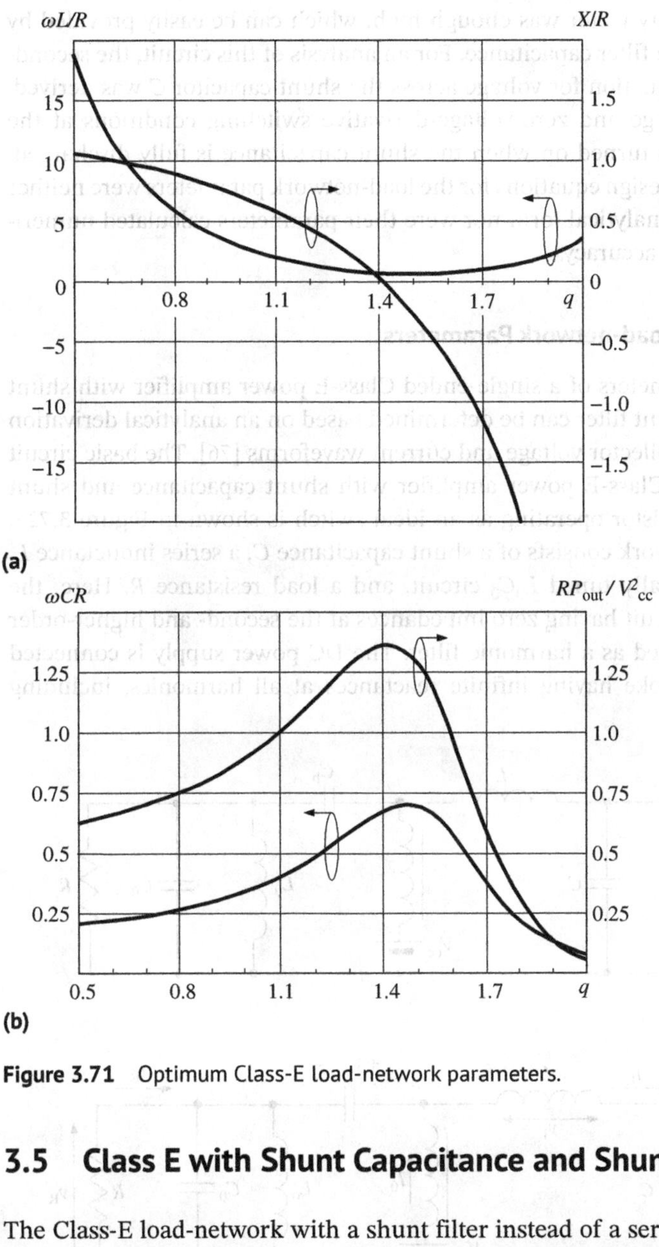

Figure 3.71 Optimum Class-E load-network parameters.

3.5 Class E with Shunt Capacitance and Shunt Filter

The Class-E load-network with a shunt filter instead of a series filter, which can improve the spectral performance, was first proposed by Artym at the end of 1960s [35, 40]. In this configuration, in order to provide close to sinewave load current, the shunt filter was tuned to the frequency close to the operating frequency and its loaded quality factor was enough high, which can be easily provided by increasing the shunt filter capacitance. For an analysis of this circuit, the second-order differential equation for voltage across the shunt capacitor C was derived, assuming zero-voltage and zero-voltage-derivative switching conditions at the moment the tube is turned on when the shunt capacitance is fully discharged. However, the final design equations for the load-network parameters were neither derived in explicit analytical form nor were their parameters calculated numerically with sufficient accuracy.

3.5.1 Optimum Load-network Parameters

The optimum parameters of a single-ended Class-E power amplifier with shunt capacitance and shunt filter can be determined based on an analytical derivation of its steady-state collector voltage and current waveforms [76]. The basic circuit configuration of a Class-E power amplifier with shunt capacitance and shunt filter with the transistor operating as an ideal switch is shown in Figure 3.72a, where the load network consists of a shunt capacitance C, a series inductance L, a shunt fundamentally tuned L_0C_0 circuit, and a load resistance R. Here, the ideal shunt L_0C_0 circuit having zero impedances at the second- and higher-order harmonics is operated as a harmonic filter. The DC power supply is connected through the RF choke having infinite reactances at all harmonics, including

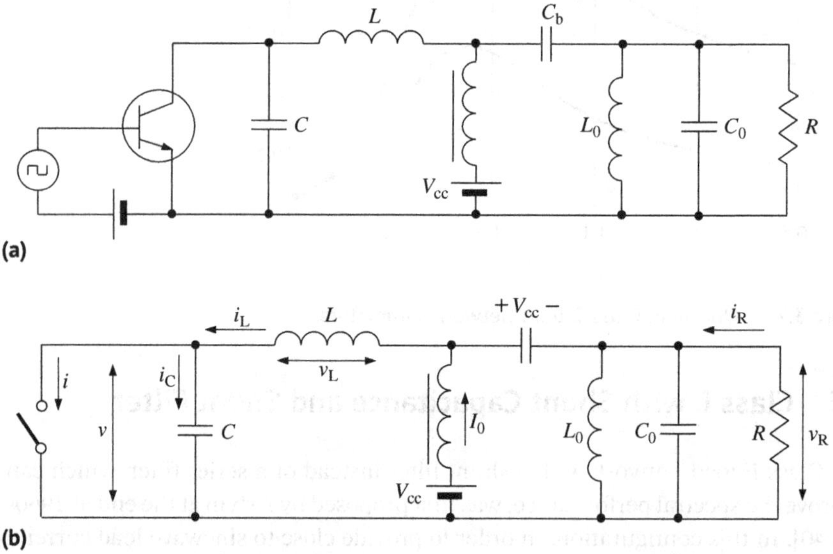

Figure 3.72 Circuit schematics of Class-E load network with shunt filter.

fundamental. However, in practical applications, the DC power supply can be connected through the filter inductance L_0 using the bypass capacitance. In this circuit configuration, the blocking capacitance should be located between the filter capacitance C_0 and load resistance R.

To simplify the analysis of a single-ended Class-E power amplifier with a shunt filter, whose equivalent circuit is shown in Figure 3.72b, the same assumptions, which were introduced for an analysis of a Class-E power amplifier with shunt capacitance and series filter, are used. For a lossless operation mode, it is necessary to provide the idealized optimum (or nominal) zero-voltage and zero-voltage-derivative conditions for voltage across the switch (just prior to the start of the switch on) at the moment $\omega t = 2\pi$, when the transistor is saturated, given by Eqs. (3.71) and (3.72). The output current i_R flowing through the load is assumed to be sinusoidal written as: $i_R(\omega t) = I_R \sin(\omega t + \varphi)$, where j is the initial phase shift.

When the switch is turned on for of $0 \leq \omega t \leq \pi$, the voltage on the switch is

$$v(\omega t) = V_{cc} - v_L(\omega t) - v_R(\omega t) = 0 \tag{3.174}$$

where

$$v_L(\omega t) = \omega L \frac{di_L(\omega t)}{d\omega t} \tag{3.175}$$

$$v_R(\omega t) = V_R \sin(\omega t + \varphi) \tag{3.176}$$

where $V_R = I_R R$ is the voltage amplitude across the load.

As the current flows through the shunt capacitance C is

$$i_C(\omega t) = \omega C \frac{dv(\omega t)}{d\omega t} = 0 \tag{3.177}$$

consequently,

$$i(\omega t) = i_L(\omega t) \tag{3.178}$$

$$i_L(\omega t) = \frac{1}{\omega L} \int_0^{\omega t} V_{cc} d\omega t + \frac{V_R}{\omega L} \cos(\omega t + \varphi) + i_L(0) \tag{3.179}$$

where V_{cc} is the DC-supply voltage.

Because $i(0) = 0$, the initial value for the current $i_L(\omega t)$ flowing through the series inductor L at time instant $\omega t = 0$ can be defined using Eqs. (3.178) and (3.179) as

$$i_L(0) = -\frac{V_R}{\omega L} \cos\varphi \tag{3.180}$$

As a result, from Eqs. (3.178)–(3.180), it follows that

$$i(\omega t) = \frac{V_{cc}}{\omega L}\omega t + \frac{V_R}{\omega L}\left[\cos(\omega t + \varphi) - \cos\varphi\right] \tag{3.181}$$

When the switch is turned off for $\pi \le \omega t < 2\pi$, there is no current flowing through the switch, that is, $i(\omega t) = 0$, and the current flowing through the capacitor C can be written as $i_C(\omega t) = i_L(\omega t)$, or

$$i_C(\omega t) = \omega C\frac{dv}{d\omega t} = \frac{1}{\omega L}\int_{\pi}^{\omega t}\left[V_{cc} - v(\omega t) - v_R(\omega t)\right]d\omega t + i_L(\pi) \tag{3.182}$$

under the initial off-state conditions $v(\pi) = 0$ and

$$i_C(\pi) = \frac{\pi V_{cc} - 2V_R\cos\varphi}{\omega L} \tag{3.183}$$

Equation (3.182) can be rewritten in the form of the linear nonhomogeneous second-order differential equation as

$$\omega^2 LC\frac{d^2v(\omega t)}{d(\omega t)^2} + v(\omega t) - V_{cc} + V_R\sin(\omega t + \varphi) = 0 \tag{3.184}$$

whose general solution can be given in the normalized form by

$$\frac{v(\omega t)}{V_{cc}} = C_1\cos q\omega t + C_2\sin q\omega t + 1 + \frac{q^2}{1-q^2}\frac{V_R}{V_{cc}}\sin(\omega t + \varphi) \tag{3.185}$$

where $q = 1/\omega\sqrt{LC}$, and the coefficients C_1 and C_2 are determined from the initial off-state conditions as

$$C_1 = -\cos q\pi - q\pi\sin q\pi + \frac{q}{1-q^2}\frac{V_R}{V_{cc}} \times \left[q\cos q\pi\sin\varphi + (1-2q^2)\sin q\pi\cos\varphi\right] \tag{3.186}$$

$$C_2 = -\sin q\pi + q\pi\cos q\pi + \frac{q}{1-q^2}\frac{V_R}{V_{cc}} \times \left[q\sin q\pi\sin\varphi - (1-2q^2)\cos q\pi\cos\varphi\right] \tag{3.187}$$

The fundamental-frequency voltage $v_1(\omega t)$ on the switch can be represented by the two quadrature (real and imaginary) components with the amplitudes V_R and V_L, as shown in Figure 3.73, and the amplitude of the real component is defined by the Fourier formula as

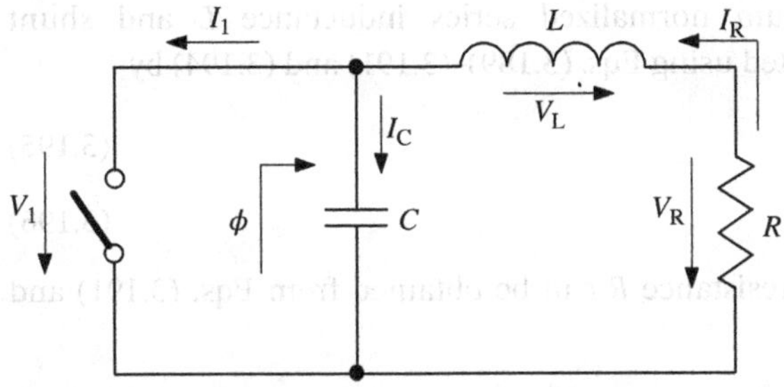

Figure 3.73 Class-E load network at fundamental frequency.

$$V_R = -\frac{1}{\pi}\int_0^{2\pi} v(\omega t)\sin(\omega t + \varphi)d\omega t \tag{3.188}$$

Hence, by solving a system of three equations, the first two of them defined by the Class-E switching conditions given by Eqs. (3.71) and (3.72) applying to Eq. (3.185) and the third equation for V_R given by Eq. (3.188), the three unknown parameters can be calculated as

$$\varphi = -41.614^\circ \tag{3.189}$$

$$q = 1.607 \tag{3.190}$$

$$\frac{V_R}{V_{cc}} = 0.9253 \tag{3.191}$$

For an idealized collector efficiency of 100%, the DC power $P_0 = I_0V_{cc}$ and the fundamental-frequency output power $P_{out} = V_R^2/(2R)$ delivered to the load are equal, written as

$$I_0V_{cc} = \frac{V_R^2}{2R} \tag{3.192}$$

The DC-supply current I_0 is defined from Eq. (3.181) using a Fourier-series expansion as

$$I_0 = \frac{1}{2\pi}\int_0^{2\pi} i(\omega t)d\omega t = \frac{V_{cc}}{2\pi\omega L}\left[\frac{\pi^2}{2} - \frac{V_R}{V_{cc}}(2\sin\varphi + \pi\cos\varphi)\right] \tag{3.193}$$

By using Eqs. (3.192) and (3.193), the optimum normalized series inductance $\omega L/R$ can be derived as

$$\frac{\omega L}{R} = \frac{1}{\pi}\left[\frac{\pi^2}{2} - \frac{V_R}{V_{cc}}(2\sin\varphi + \pi\cos\varphi)\right]\bigg/\left(\frac{V_R}{V_{cc}}\right)^2 \tag{3.194}$$

Consequently, the optimum normalized series inductance L and shunt capacitance C can be calculated using Eqs. (3.189)–(3.191) and (3.194) by

$$\frac{\omega L}{R} = 1.4836 \tag{3.195}$$

$$\omega CR = 0.261 \tag{3.196}$$

whereas the optimum load resistance R can be obtained from Eqs. (3.191) and (3.192) as

$$R = \frac{1}{2}\frac{V_{\rm R}^2}{P_{\rm out}} = \frac{1}{2}\left(\frac{V_{\rm R}}{V_{\rm cc}}\right)^2 \frac{V_{\rm cc}^2}{P_{\rm out}} = 0.4281\frac{V_{\rm cc}^2}{P_{\rm out}} \tag{3.197}$$

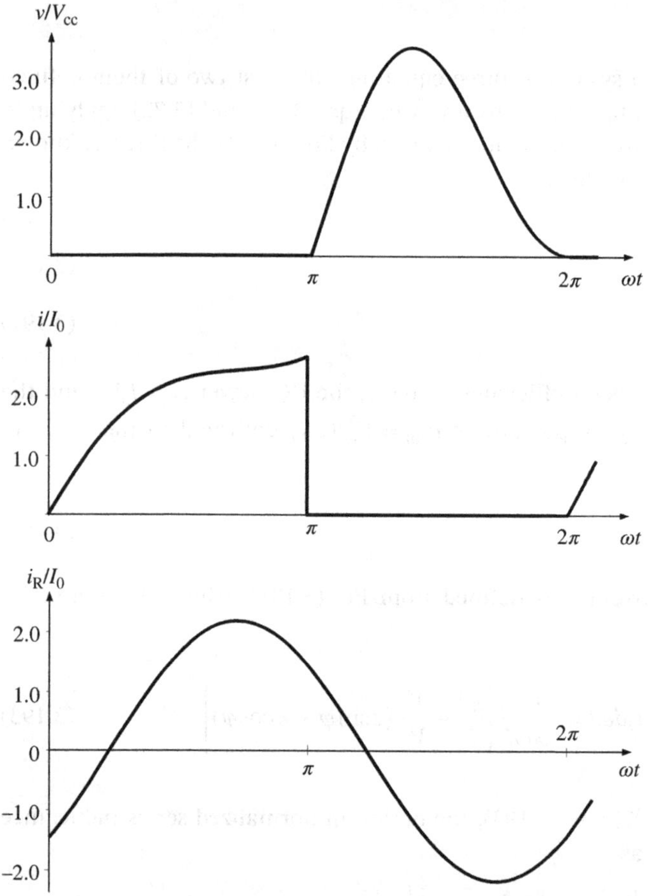

Figure 3.74 Ideal waveforms for nominal Class-E mode with shunt filter.

Figure 3.74 shows the normalized collector voltage and current waveforms and sinusoidal current waveform for an idealized optimum Class-E mode with shunt capacitance and shunt filter during the entire interval $0 \leq \omega t \leq 2\pi$ [76]. Since there is no nonzero collector voltage and current simultaneously, which means zero power losses, the resulting idealized collector efficiency is equal to 100%.

The phase angle ϕ of the load network at the fundamental frequency seen by the transistor operating as a switch, as shown in Figure 3.73, can be derived from the load-network parameters using Eqs. (3.195) and (3.196) as

$$\phi = \tan^{-1}\left(\frac{\omega L}{R}\right) - \tan^{-1}\left(\frac{\omega CR}{1 - \frac{\omega L}{R}\omega CR}\right) = 32.945° \tag{3.198}$$

The peak collector voltage V_{max} and current I_{max} can be calculated from Eqs. (3.181) and (3.185) using Eq. (3.193) as

$$\frac{V_{max}}{V_{cc}} = 3.677 \tag{3.199}$$

$$\frac{I_{max}}{I_0} = 2.768 \tag{3.200}$$

showing that the voltage peak factor is as high as in a Class E with shunt capacitance and series filter.

In view of the transistor output (collector or drain–source) capacitance C_{out} which is a limiting factor, the maximum operating frequency f_{max} in an idealized optimum Class-E mode with shunt capacitance and shunt filter can be obtained from Eqs. (3.196) and (3.197) as

$$f_{max} = 0.097\frac{P_{out}}{C_{out}V_{cc}^2} \tag{3.201}$$

which is 1.91 times higher than that for a Class E with shunt capacitance and series filter.

3.5.2 ADS Simulation Setup

Figure 3.75 shows the Advanced Design System (ADS) simulation setup used to obtain an idealized optimum Class-E operation mode with shunt capacitance and shunt filter in the frequency domain, where L_{ch} is the RF choke with a high inductance value and the input source Vf_Pulse represents a voltage source with a Fourier series expansion of the periodic rectangular wave used in a harmonic balance simulator [76]. The optimization procedure is applied with respect to efficiency as an optimization parameter, with an increased number of iterations for more accuracy. The normalized switch voltage and current waveforms

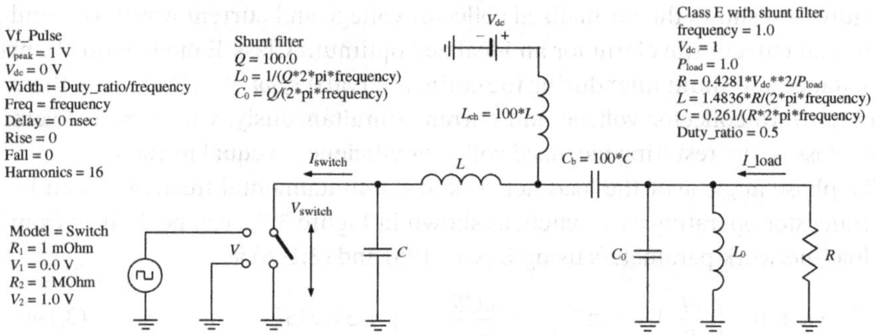

Figure 3.75 Simulation setup for idealized Class-E mode with series capacitance and shunt filter in frequency domain.

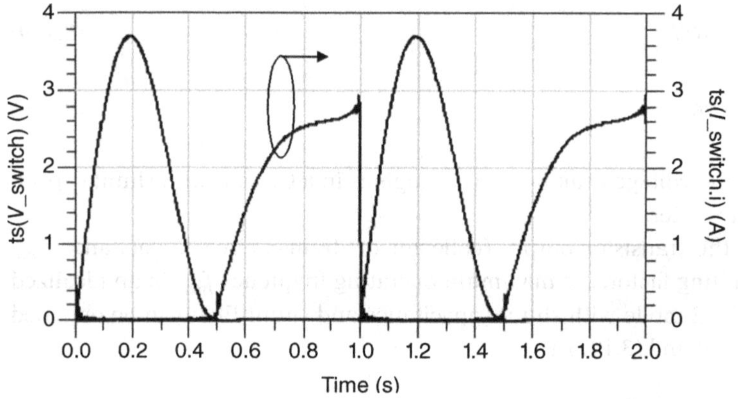

Figure 3.76 Simulated Class-E voltage and current waveforms in frequency domain.

are defined for the idealized optimum parameters given by Eqs. (3.195)–(3.197) with a 50% duty ratio are shown in Figure 3.76. Despite the smoother switching transitions for $r_{sat} = 1\ \text{m}\Omega$ and lossless circuit elements simulated in the frequency domain, the efficiency is equal to 98.8%, with the simulated switch voltage and current waveforms similar to that of the theoretical waveforms shown in Figure 3.74.

3.5.3 Load Network with Transmission Lines

Figure 3.77 shows the circuit schematics of a high-efficiency transmission-line Class-E power amplifier with shunt capacitance and shunt filter, where the

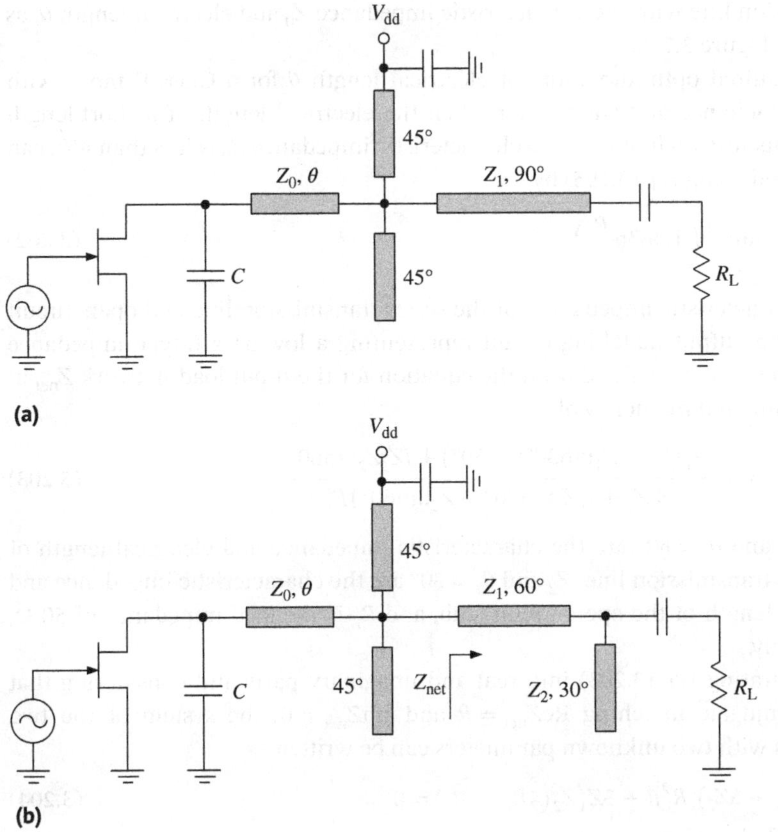

Figure 3.77 Schematics of transmission-line Class-E power amplifier with shunt filter.

quarterwave transmission line is used to match the optimum load resistance R to the standard load impedance $R_L = 50\ \Omega$ at the fundamental frequency, as shown in Figure 3.77a. As an alternative, the transmission-line low-pass L-type impedance transformer with a series transmission line and an open-circuit stub can be used, as shown in Figure 3.77b [76]. The shunt harmonic filter includes the short- and open-circuit stubs with electrical length of 45° each, having inductive and capacitive reactances at the fundamental, respectively, to create short-circuit conditions at even harmonics. The series transmission line with electrical length of 60° and an open-circuit stub with electrical length of 30° are used apart from the impedance matching to create a short-circuit condition at the third harmonic at the right-hand side of the series inductance represented by a short-length series transmission line with the characteristic impedance Z_0 and electrical length θ, as shown in Figure 3.77b.

The required optimum value of electrical length θ for a Class-E mode with shunt capacitance and shunt filter when the electrical length of a short-length series transmission line with the characteristic impedance Z_0 is less than 45°, can be obtained using Eq. (3.195) by

$$\theta = \tan^{-1}\left(1.4836\frac{R}{Z_0}\right) \tag{3.202}$$

The characteristic impedances of the series transmission line and open-circuit stub in the output matching circuit representing a low-pass L-type impedance transformer can be defined from the equation for the input load-network $Z_{\rm net}$ at the fundamental frequency of

$$Z_{\rm net} = Z_1\frac{R_{\rm L}(Z_2 - Z_1\tan 30°\tan 60°) + jZ_1Z_2\tan 60°}{Z_1Z_2 + j(Z_1\tan 30° + Z_2\tan 60°)R_{\rm L}} \tag{3.203}$$

where Z_1 and $\theta_1 = 60°$ are the characteristic impedance and electrical length of the series transmission line, Z_2 and $\theta_2 = 30°$ are the characteristic impedance and electrical length of the open-circuit stub, and $R_{\rm L}$ is the load impedance of 50 Ω, respectively.

By separating Eq. (3.203) into real and imaginary parts and considering that under conjugate matching $\text{Re}Z_{\rm net} = R$ and $\text{Im}Z_{\rm net} = 0$, the system of the two equations with two unknown parameters can be written as

$$(Z_1 + 3Z_2)^2R_{\rm L}^2R - 3Z_1^2Z_2^2(4R_L - 3R) = 0 \tag{3.204}$$

$$3Z_1^2Z_2^2 + R_{\rm L}^2(Z_1 + 3Z_2)(Z_2 - Z_1) = 0 \tag{3.205}$$

from which the characteristic impedances Z_1 and Z_2 can be appropriately calculated.

As a result, by introducing the parameter $r = R_{\rm L}/R$,

$$\frac{Z_1}{R_{\rm L}} = \frac{\sqrt{4r-1}}{\sqrt{3}r} \tag{3.206}$$

$$\frac{Z_1}{Z_2} = \frac{r-1}{r} \tag{3.207}$$

Consequently, first the characteristic impedance Z_1 is calculated from Eq. (3.206) for the specified value of the parameter r, then the characteristic impedance Z_2 is obtained from Eq. (3.207). As an example, if the required optimum load resistance R is equal to 20 Ω with $r = 2.5$, the characteristic impedance of the series transmission line Z_1 is equal to 35 Ω, and the characteristic impedance of the open-circuit stub Z_2 is equal to 58 Ω.

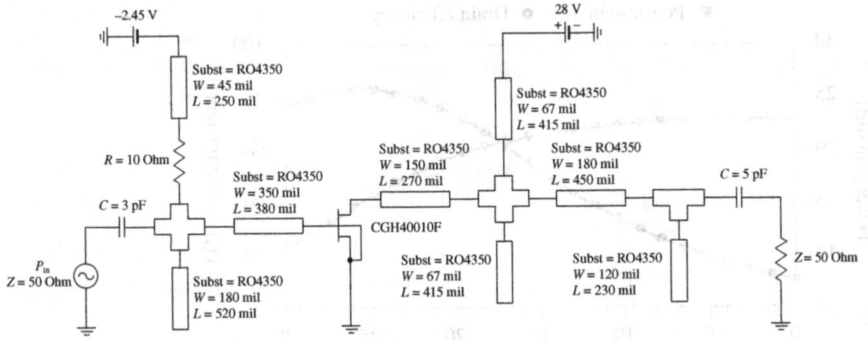

Figure 3.78 Simulation setup of transmission-line Class-E power amplifier with shunt filter.

Figure 3.78 shows the simulated circuit schematic of a transmission-line Class-E power amplifier with a shunt filter based on a 28-V 10-W GaN HEMT CGH40010F transistor and transmission-line load network according to Figure 3.77b. Here, the input transmission-line matching circuit in the form of low-pass L-type transformer provides conjugate matching with the standard 50-Ω source. During the simulation process, the load network was slightly modified by optimizing the parameters of the series and shunt transmission lines because the device output capacitance C_{out} and series inductance L_{out} formed by the drain bondwires and package (not shown in Figure 3.78) lead do not match the required exact values of the shunt capacitance C and series inductance L for an idealized optimum Class-E operation mode with shunt filter.

For the transmission-line GaN HEMT Class-E power amplifier with shunt capacitance and shunt filter of Figure 3.78 using a 30-mil RO4350 substrate, the simulated drain efficiency of 83%, a PAE of 80.4%, and a power gain of 15 dB at the output power of 40.3 dBm with a quiescent current of 30 mA are achieved at the operating frequency of 2.14 GHz with the DC-supply voltage of 28 V, as shown in Figure 3.79. Due to the presence of the shunt filter, the second and third harmonics were suppressed by more than 50 dB.

By considering the normalized parameter q as a variable, multiplicity of solutions for the load-network parameters can be achieved for a Class E with shunt capacitance and shunt filter, similarly to the generalized Class E with finite DC-feed inductance and series filter. In this case, an additional series reactance X should be inserted between the shunt filter and the load, which can generally have inductive, zero, or capacitive reactance [77]. The series reactance X changes from negative (capacitive) to positive (inductive) values as q increases.

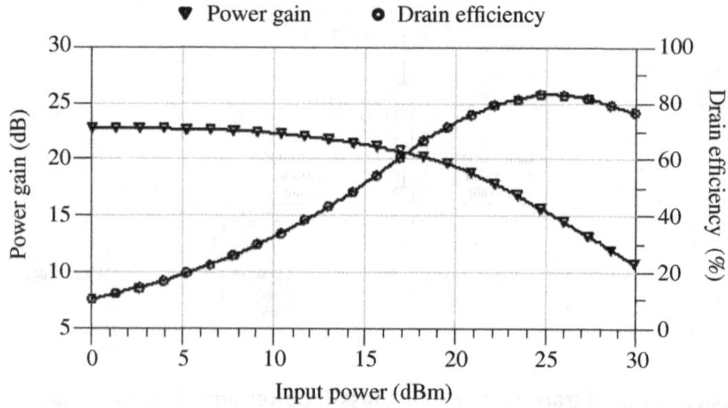

Figure 3.79 Simulated performance of transmission-line Class-E power amplifier with shunt filter.

3.6 Summary: Magic of Parallel Circuits

Modern commercial and military communication systems require the high efficiency, long-term operating conditions. For this purpose, the power amplifiers as the final and highly current-consuming transmitter elements are developed using the high-efficiency Class-B, Class-F, or Class-E operation modes depending on the technical requirements. In a Class-B operation mode, which is a special case of a Class C with a conduction angle of half a period, the collector voltage waveform is purely sine wave, and harmonics are present only in the collector current. In Class-F or inverse Class-F power amplifiers, the load impedances at the fundamental frequency and higher-order harmonic components are optimized by short-circuit and open-circuit conditions in order to control the voltage and current collector waveforms to obtain maximum efficiency. In Class-E power amplifiers, an efficiency improvement is achieved by realizing the on/off switching operation with special current and voltage waveforms so that high voltage and high current do not exist at the same time. And if all or most of the harmonics are present in both the voltage and current collector waveforms in a Class E, then only certain harmonic components are present either in the voltage or current collector waveform in a Class F or an inverse Class F. As a result, the impedance conditions are different when, for a Class-F or an inverse Class-F operation, it needs to create zero or infinite impedance at any harmonic component with an optimum impedance matching at the fundamental, whereas the reactances at the second- and higher-order harmonics should be capacitive with the specified particular values for a Class-E operation mode.

At first glance, when looking at the collector voltage and current waveforms, it looks like these high-efficiency operation modes do not have anything common between each other as it is necessary to provide different impedance conditions as well as to apply different circuit design approaches. However, to approximate the Class-F, inverse Class-F, or Class-E operation modes, it is enough to realize the required impedance conditions only for first several harmonics and, which is the most important, to apply the same circuit design technique – one or several harmonically tuned parallel circuits or the only parallel circuit mistuned at the fundamental frequency [78]. That is, all what basically needs to do when designing the high-efficiency RF power amplifier is a proper choice of the parameters of the parallel circuit or circuits in the load network. If the collector voltage represents a pure sinusoidal waveform in a Class-B operation mode due to the ideal parallel *LC* circuit tuned to the fundamental frequency, as shown in Figure 3.80a, that provides the suppression of the second- and higher-order harmonics of the collector (or drain) current, then the flattening of the collector voltage occurs in a Class-F operation mode with the third-harmonic voltage peaking due to the third-harmonic resonant circuit connected in series, as shown in Figure 3.80b, resulting in less overlapping with the current waveform, thus reducing the power loss and increasing efficiency. When the impedance conditions for higher-order current harmonics are satisfied, the efficiency can be ideally increased up to 90%, compared to the maximum collector (or drain) efficiency of 78.5% in a Class-B operation mode [79].

Higher efficiency can be obtained in an inverse Class-F mode with the second-harmonic voltage peaking because of the higher voltage peak factor, ideally by $\pi/2$ times more than in a Class-F mode with the second-harmonic short-circuit termination due to the second-harmonic resonant circuit connected in series, as shown in Figure 3.80c, thereby providing less value of the flattened current waveform when the transistor is turned on and, as a result, less power losses for the same output power, which is especially important for MMIC design when low-*Q* spiral inductors are used. To further maximize efficiency, it needs to increase the number of resonant circuits, each tuned to the corresponding harmonic component, which makes the design much more complicated. As an alternative, a Class-E operation mode can be used when only one resonant circuit is necessary, which is detuned between the fundamental-frequency and second-harmonic components, thus providing the inductive impedance at the fundamental frequency and capacitive reactances at the second- and higher-order harmonics and resulting in

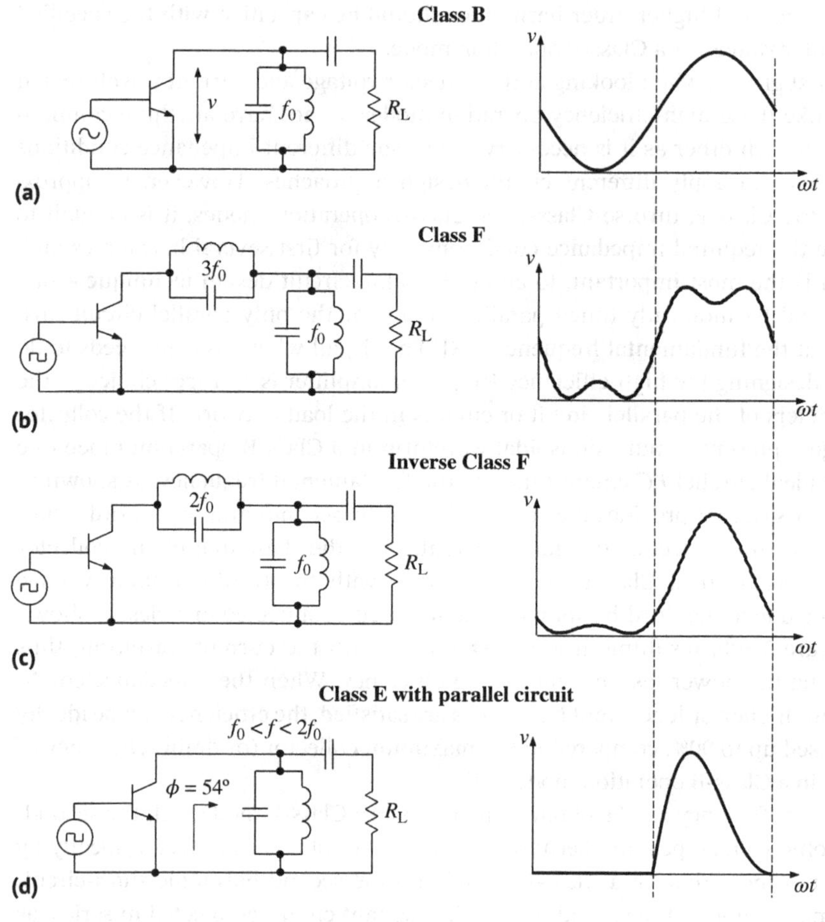

Figure 3.80 Comparison of different classes of high operation efficiency.

a collector voltage waveform shown in Figure 3.80d with zero voltage derivative at the end of the period. In this case, the voltage peak factor is slightly higher than that in an inverse Class-F operation mode and ideally zero voltage conditions are fulfilled when the switch, representing the ideal transistor, is turned on during an entire half a period [41, 80]. However, because the short-circuit termination of the second- and higher-order current harmonics are not provided in this case and these harmonics partly flow to the load, therefore a series resonant circuit tuned at the fundamental should follow the parallel resonant circuit, composing the parallel-circuit Class E, to provide the only fundamental to flow to the load and resulting ideally in 100% efficiency.

3.7 Summary: Comparison of Output Impedances and Load-network Parameters

Generally, the operation conditions and load-network implementations are different for different power amplifier operation modes, depending on the device DC biasing, operating frequency, and output power. The comparison of the output impedances for different operation modes is given in Table 3.1, where $R^{(B)}$, $R^{(F)}$, $R^{(invF)}$ are the equivalent output resistances or load resistances under conjugate matching conditions at the fundamental frequency in Class-B, Class-F, and inverse Class-F operation modes, respectively, derived in a simple idealized analytical form with zero saturation voltage V_{sat} where V_{cc} is the DC-supply voltage and P_{out} is the output power at the fundamental frequency, and $R^{(E)}$ is the load resistance at the fundamental frequency in an ideal parallel-circuit Class-E operation mode. As seen from Table 3.1, the load resistance in parallel-circuit Class-E mode is the highest, being 2.73 times higher than in a Class-B mode, 1.68 times greater than in a Class-F mode, and 1.11 higher than in an inverse Class-F mode. It should be noted that, because the parallel-circuit

Table 3.1 Load resistance for different Class-E modes.

Operation class	Load resistance
Class B	$R^{(B)} = \dfrac{V_{cc}^2}{2P_{out}}$
Class F	$R^{(F)} = \dfrac{8}{\pi^2}\dfrac{V_{cc}^2}{P_{out}} = \left(\dfrac{4}{\pi}\right)^2 R^{(B)}$
Inverse Class F	$R^{(invF)} = \dfrac{\pi^2}{8}\dfrac{V_{cc}^2}{P_{out}} = \dfrac{\pi^4}{64}R^{(F)} = \left(\dfrac{\pi}{2}\right)^2 R^{(B)}$
Parallel-circuit Class E	$R^{(E)} = 1.365\dfrac{V_{cc}^2}{P_{out}} = 2.73R^{(B)}$

Table 3.2 Load-network schematics at fundamental and harmonics for different Class-E modes.

Class-E load network	f_0 (Fundamental)	$2nf_0$ (Even harmonics)	$(2n+1)f_0$ (Odd harmonics)
Class-E with shunt capacitance and series filter	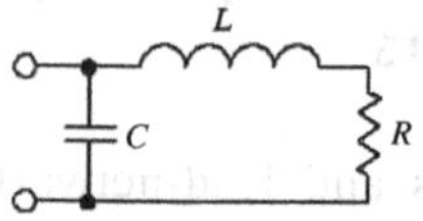	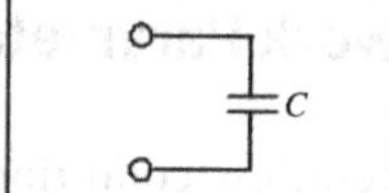	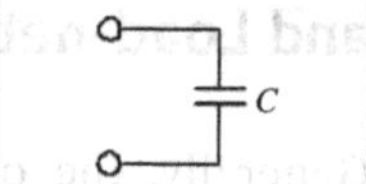
Parallel-circuit Class E	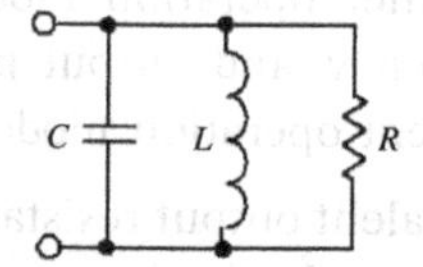	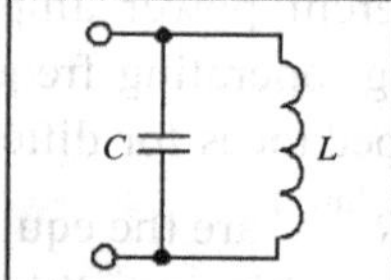	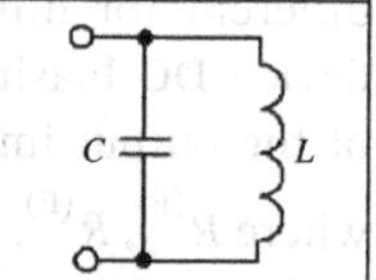
Class-E with shunt capacitance and shunt filter	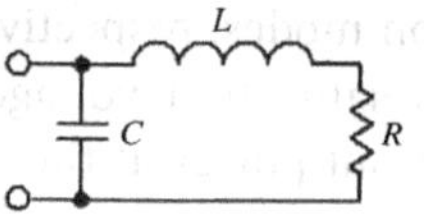	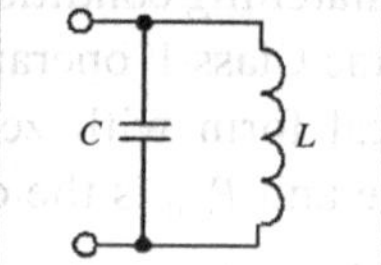	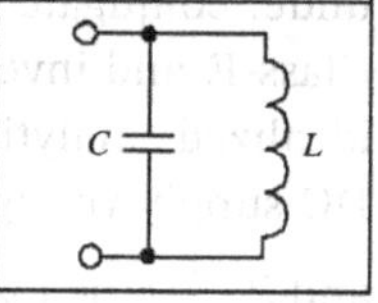

Class-E load network offers the largest value of R, this simplifies the matching design procedure when less matching sections are required and a wider frequency bandwidth is achieved.

Table 3.2 shows the load-network schematics seen by the device collector (or drain) at the fundamental-frequency and higher-order odd- and even-harmonics for different Class-E operation modes. For example, a Class-E operation mode with shunt capacitance and series filter and a Class-E operation mode with shunt capacitance and shunt filter are characterized by the same schematics at the fundamental frequency but different at the harmonics. At even and high-order odd harmonics, the optimum impedance of a Class-E operation mode with shunt capacitance and series filter can be established by the shunt capacitance, whereas the optimum impedance of a Class-E operation mode with shunt capacitance and shunt filter is defined by the parallel LC circuits, similar to a parallel-circuit Class-E operation mode.

The optimized load-network parameters corresponding to the different Class-E operation modes, such as a Class E with shunt capacitance and series filter, a parallel-circuit Class E, and a Class E with shunt capacitance and shunt filter, are shown in a normalized form in Table 3.3. As can be seen, a parallel-circuit Class-E operation mode offers the highest value of the load resistance R for the same DC-supply voltage V_{cc} and output power P_{out}, larger value of the shunt capacitance C, and lower value of the parallel inductance L for the same load R, that makes

Table 3.3 Load-network parameters for different Class-E modes.

Normalized load-network parameter	Class E with shunt capacitance and series filter	Parallel-circuit Class E	Class E with shunt capacitance and shunt filter
$\frac{\omega L}{R}$	1.1525	0.732	1.4836
ωCR	0.1836	0.685	0.261
$\frac{P_{out} R}{V_{cc}^2}$	0.5768	1.365	0.4281
$\frac{f_{max} C_{out} V_{cc}^2}{P_{out}}$	0.0507	0.0798	0.097

such power amplifier configuration very compact and less lossy, which is very attractive for MMIC implementation and better suited to the envelope tracking (ET) or envelope elimination and restoration (EER) transmitters, providing much wider modulation bandwidth, better modulation linearity, and ease in practical implementation using the same transistors [81]. At the same time, a Class E with shunt capacitance and shunt filter provides a much higher value of the maximum operating frequency f_{max} for the same DC-supply voltage V_{cc}, device output capacitance C_{out}, and output power P_{out}, compared to other Class-E alternatives.

3.8 Design Example: Two-stage High-efficiency 1.75-GHz MMIC HBT Power Amplifier Using Parallel-circuit Class-E Mode

Generally, the design procedure of the high-efficiency power amplifier for handset applications include several basic steps such as to understand the transistor nonlinear model in order to evaluate its input and output impedance, as well as stability of operation, to analytically calculate the optimum parameters of the proper load network to simplify the design procedure, to choose the proper bias circuit to optimize the quiescent current over temperature and minimize the reference current, the selection and design of the input and interstage matching circuits to provide the maximum input return loss and power gain, and the load-network parameters to maximize collector efficiency of the final stage, and to provide the stable operating conditions, and to finally to optimize the circuit parameters to overall PAE of the power amplifier.

As an example, consider the design procedure of a high-efficiency monolithic Class-E bipolar power amplifier operating at 1.75 GHz with an output fundamental-frequency power of 33 dBm, a power gain of 27 dB, and a DC-supply voltage $V_{cc} = 5$ V. To satisfy these requirements, first it is necessary to choose a two-stage structure and proper HBT devices with minimum saturation voltage and high transition frequency $f_T > 25$ GHz. The nonlinear equivalent circuit of the chosen InGaP/GaAs HBT cell with an emitter area of $3 \times 30\ \mu m^2$ (A = AREA/90 = 1), where input impedance Z_{in} is determined mainly by the transistor nonlinear elements R_{bd} and C_{bd}, representing the input diode forward-biased junction, is shown in Figure 3.81. Due to its high nonlinearity and very high voltage swing at the collector, the junction base–collector (or collector) capacitance C_{bc} is defined at a lower collector DC bias point, which is $C_{bc} = 4.25e - 14*A$ in farads in this case.

Then, assuming the collector efficiency of the second-stage transistor of 80%, the DC power is calculated to be equal to $P_0 = 2\ W/0.8 = 2.5$ W with the DC-supply current $I_0 = P_0/V_{cc} = 2.5\ W/5\ V = 500$ mA. The recommended maximum DC current density for operation reliability over wide temperature range should not exceed 15 mA per $90\ \mu m^2$ for the selected HBT cell, therefore, the overall emitter area was chosen with some margin to be equal to $5400\ \mu m^2$. Figure 3.82 shows the volt–ampere characteristics $I_{ce}(V_{ce})$ of such an InGaP/GaAs HBT cell for different base bias voltages V_{be}, from 1.2 to 1.5 V with a voltage step of 30 mV (thermal effect is not considered). Consequently, the peak collector current for a parallel-circuit Class-E operation mode according to Eq. (3.151) is equal to $I_{max} = 2.647 I_0 = 1.32$ A. The idealized Class-E load line is shown as a heavy line

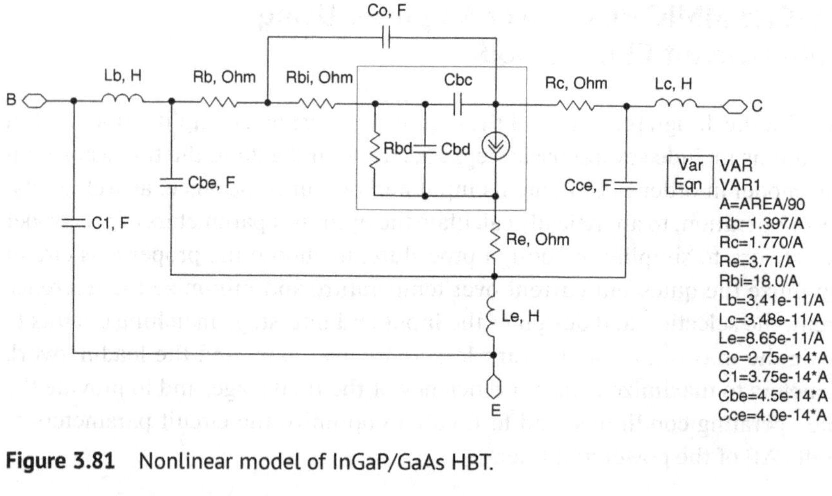

Figure 3.81 Nonlinear model of InGaP/GaAs HBT.

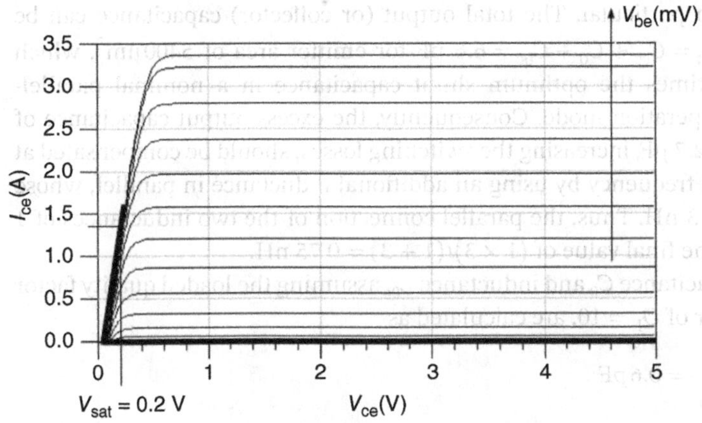

Figure 3.82 Volt-ampere characteristics and load line for HBT device.

in two sections: horizontal at zero current (transistor is pinched off) and slanted upward to the right at the left end of the figure (transistor conducting current). It can be seen that the operating point moves along the horizontal V_{ce} axis (pinch-off region) and then along the collector current saturated line (voltage-saturation region) until $I_{ce} = 1.32$ A. At this final point, the saturation voltage can be found as equal to $V_{sat} = 0.2$ V. This means that the power loss due to the finite device saturation resistance can be calculated as $P_{sat} = I_0 V_{sat} = 100$ mW, with degradation in the collector efficiency to $(1 - P_{sat}/P_0) = 0.96$ or by 4%.

The next step is to design the second stage with a maximum possible collector efficiency with an initially low value of the source resistance for simplicity to match it to the input impedance of the transistor. The nominal parameters of a parallel-circuit Class-E load network can be obtained from Eqs. (3.148)–(3.150) including the saturation voltage V_{sat}, as

$$R = 1.365\,\frac{(V_{cc} - V_{sat})^2}{P_{out}} = 15\,\Omega$$

$$L = 0.732\frac{R}{\omega} = 1.0\,\text{nH}$$

$$C = \frac{0.685}{\omega R} = 3.9\,\text{pF}$$

In a non-nominal Class-E operation mode with finite saturation resistance, the calculated load-network parameters generally cannot be considered optimum. However, they can be considered as an accurate enough initial guess for further design and optimization to maximize power amplifier performance in general and efficiency in particular. The total output (or collector) capacitance can be evaluated as $C_{out} = C_{ce} + C_0 + C_{bc} = 6.6$ pF for emitter area of 5400 μm^2, which exceeds by 1.5 times the optimum shunt capacitance in a nominal parallel-circuit Class-E operation mode. Consequently, the excess output capacitance of (6.6 − 3.9) pF = 2.7 pF, increasing the switching losses, should be compensated at the fundamental frequency by using an additional inductance in parallel, whose value is equal to 3 nH. Thus, the parallel connection of the two inductances of 1 and 3 nH gives the final value of (1 × 3)/(1 + 3) = 0.75 nH.

The series capacitance C_0 and inductance L_0, assuming the loaded quality factor of the series filter of $Q_L = 10$, are calculated as

$$C_0 = \frac{1}{\omega R Q_L} = 0.6\,\text{pF}$$

$$L_0 = \frac{1}{\omega^2 C_0} = 13.8\,\text{nH}$$

Because the input device impedance is sufficiently low, the intermediate source resistance is chosen to be equal to 5 Ω when it is enough to use a single input matching section. In terms of the circuit operation stability and convenience of its implementation, it is best to represent the input matching circuit in the form of a high-pass L-type section with a shunt inductance and a series capacitance, the parameters of which can be optimized in a large-signal simulation mode to maximize input return loss using the Smith chart tuning procedure.

The simulation setup of the second amplifying stage operating in a parallel-circuit Class-E mode with the current probes and voltage nodes is shown in Figure 3.83. Here, in order to match the low value of the transistor input impedance with the source impedance having an intermediate value of 5 Ω, the high-pass LC matching section is included. Since the bipolar transistor as a current-controlled device requires the DC base driving current, the emitter-follower bias circuit is used, providing some temperature compensation and minimizing the reference current.

In order to simulate the small- and large-signal characteristics of the second amplifying stage, the measurement equations are necessary, which are shown in Figure 3.84. Figure 3.85a shows the small-signal frequency response of the stability factor K, demonstrating the unconditional stability conditions across the entire frequency range when $K > 1$. As an alternative, it is necessary to analyze the dependence of the real part of the transistor base impedance ReZ_b, whose value should be positive for stable operation. Since the real part of the base impedance is slightly negative, as shown in Figure 3.85b, in order to compensate for this negative resistance, the resistance of 0.35 Ω is connected in series to the base

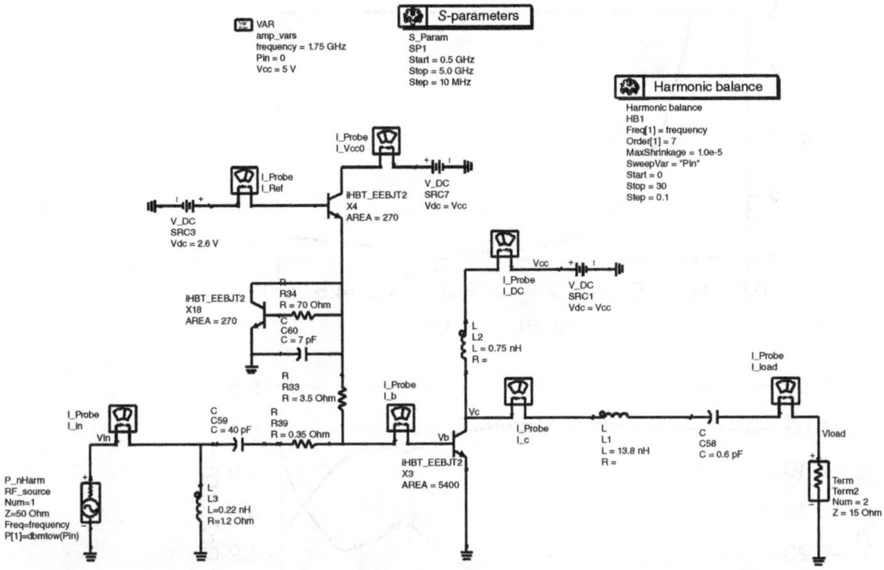

Figure 3.83 Simulation setup for Class-E second amplifying stage.

```
Eqn P_DC=Vcc[0]*(I_Ref.i[::,0]+I_Vcc0.i[::,0]+I_DC.i[::,0])
Eqn Pload_W=mag(Vload[::,1])*mag(I_load.i[::,1])/2
Eqn Pload_dBm=10*log(Pload_W)+30
Eqn Efficiency=100*Pload_W/Vcc[0]*I_DC.i[::,0])
Eqn Gain=Pload_dBm-Pin
Eqn K=stab_fact(S)
Eqn Pin_W=mag(Vin[::,1])*mag(I_in.i[::,1])/2
Eqn PAE=100*(Pload_W-Pin_W)/P_DC)
```

Figure 3.84 Measurement equations to simulate single amplifying stages in Figures 3.83 and 3.88.

assuming that the power gain will be high enough. In this case, the real part of the input impedance ReZ_{in} due to the presence of a high-pass *LC* matching section becomes close to the required 5 Ω, as shown in Figure 3.85b.

As a result of the simulations of the second parallel-circuit Class-E stage of a two-stage power amplifier, a saturated output power of 33.1 dBm and a linear

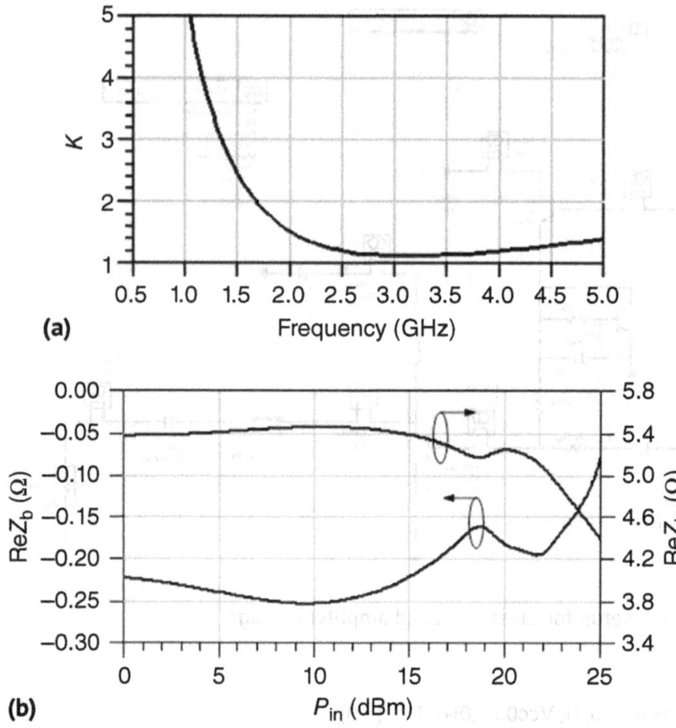

Figure 3.85 Simulated (a) stability factor and (b) input impedance.

power gain of about 13 dB, shown in Figure 3.86a, as well as a collector efficiency of 85.3% and a PAE of 60.5%, shown in Figure 3.86b, were achieved. It should be noted that the maximum collector efficiency is obtained in a deep saturation when the power gain decreases significantly. However, the PAE reaches its maximum value when the power gain is reduced by about 3 dB, because the contribution of the driving power with higher level becomes significant.

Figure 3.87 shows the collector and base voltage waveforms, with the collector voltage peak factor of about 3.6 for a DC-supply voltage $V_{cc} = 5$ V. In this case, the base voltage waveform with a flattened top part demonstrates the switching operation conditions with minimum switching losses to approximate the parallel-circuit Class-E mode.

In order to match the output resistance of 15 Ω, which is equal to a Class-E load resistance, with the required load resistance of 50 Ω, a low-pass L-type matching circuit with the series inductance as the first element to create the high impedance at the second- and higher-order harmonic components can be used instead

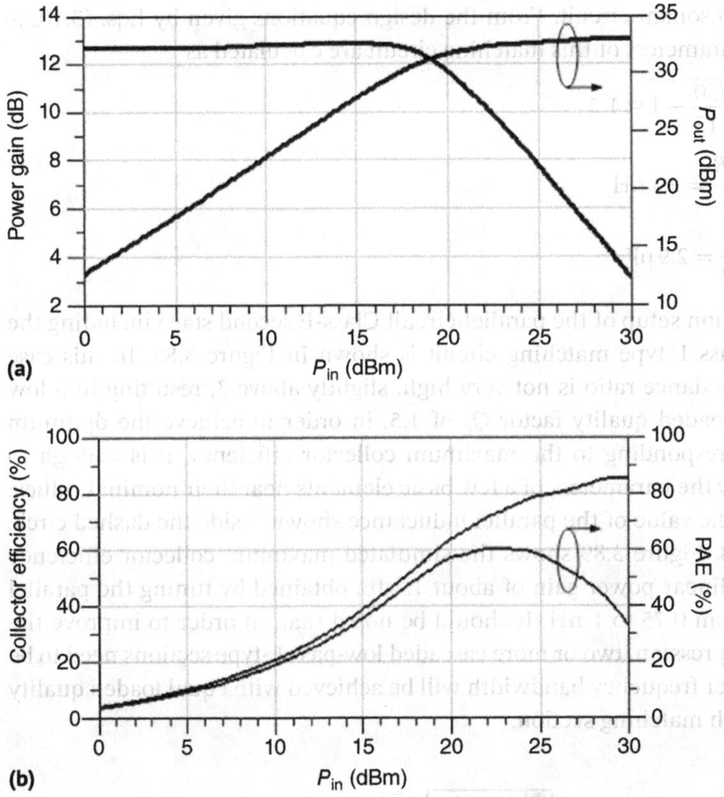

Figure 3.86 Simulated (a) power gain and output power, and (b) collector and power-added efficiencies.

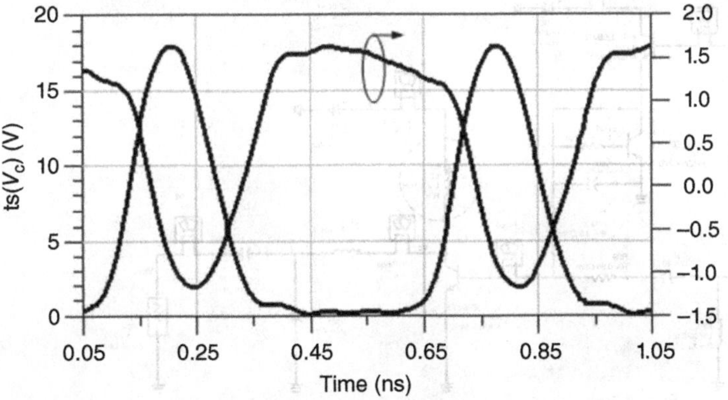

Figure 3.87 Collector and base voltage waveforms.

of the series resonant circuit. From the design equations given by Eqs. (3.102)–(3.104), the parameters of this matching circuit are calculated as

$$Q_L = \sqrt{\frac{50}{15} - 1} \cong 1.5$$

$$L = \frac{Q_L R}{\omega} = 2.1\,\text{nH}$$

$$C = \frac{Q_L}{\omega R} = 2.9\,\text{pF}$$

The simulation setup of the parallel-circuit Class-E second stage including the output low-pass L-type matching circuit is shown in Figure 3.88. In this case when the impedance ratio is not very high, slightly above 3, resulting in a low value of the loaded quality factor Q_L of 1.5, in order to achieve the optimum condition corresponding to the maximum collector efficiency, it is enough to tune manually the parameters of a few basic elements near their nominal values, for example, the value of the parallel inductance shown inside the dashed circle in Figure 3.88. Figure 3.89 shows the simulated maximum collector efficiency of 84.8% and linear power gain of about 13 dB, obtained by tuning the parallel inductance from 0.75 to 1 nH. It should be noted that, in order to improve the harmonic suppression, two or more cascaded low-pass L-type sections need to be used, and wider frequency bandwidth will be achieved with equal loaded quality factors for each matching section.

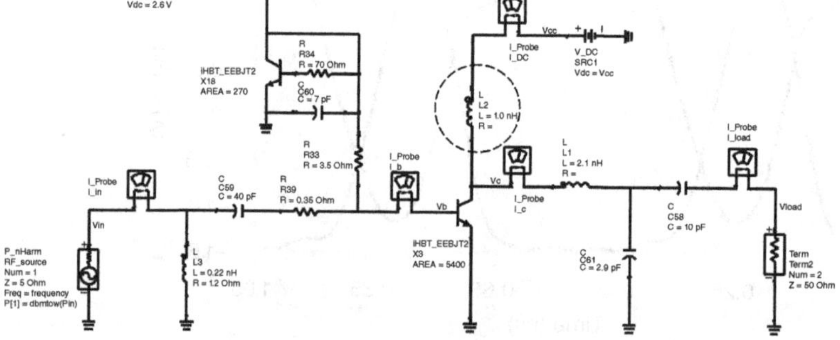

Figure 3.88 Simulation setup for Class-E second amplifying stage with output matching circuit.

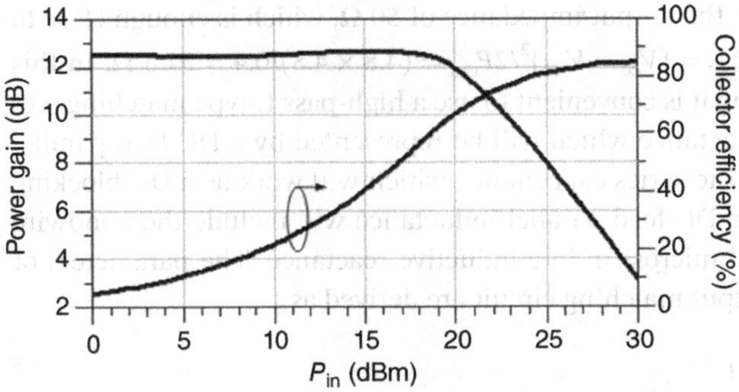

Figure 3.89 Power gain and collector efficiency.

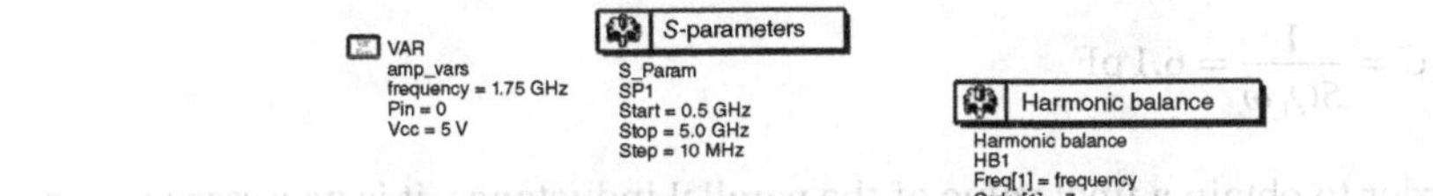

Figure 3.90 Simulation setup for Class-AB first amplifying stage.

In order to provide the maximum power gain with high efficiency level, the first amplifying stage should be designed to operate in a Class-AB mode with a small quiescent current, simulation setup of which is shown in Figure 3.90, with the load resistance of 5 Ω to directly match with the input impedance of the second amplifying stage. Considering the power gain of the second stage at the maximum efficiency level, it is sufficient to achieve the maximum output power from the first stage of about 200 mW (23 dBm), the device emitter area size was chosen to be 720 μm^2, resulting in the output impedance of 50 Ω, which is enough close to the idealized value of $R_{out} = (V_{cc} - V_{sat})^2/2P_{out} = (4.8 \times 4.8)/0.4 = 57.6\ \Omega$. In this case, for output matching it is convenient to use a high-pass L-type matching section with the shunt inductance which will be represented by a DC-feed parallel inductance followed by the series capacitance which will work as a DC-blocking capacitance as well. The DC-feed parallel inductance will include the bondwire inductance and external microstrip line inductive reactance. The parameters of the high-pass L-type output matching circuit are derived as

$$Q_L = \sqrt{\frac{50}{5} - 1} = 3$$

$$L = \frac{R_{out}}{Q_L \omega} = 1.5\,\text{nH}$$

$$C = \frac{1}{5 Q_L \omega} = 6.1\,\text{pF}$$

In order to obtain a final value of the parallel inductance, it is necessary to consider the device collector capacitance, whose total value (including its internal and external values) can be evaluated as $C_{out} = 6.6\,\text{pF} \times \left(720\ \mu\text{m}^2/5400\ \mu\text{m}^2\right) = 0.88\,\text{pF}$ for an emitter area of 720 μm^2. At the fundamental frequency, it is necessary to connect an additional parallel inductance of 9.4 nH to compensate for this output capacitance, resulting in a final value of 1.29 nH for the parallel connection of two inductances of 9.4 and 1.5 nH. The input matching circuit is also convenient to represent as a high-pass L-type section. In order to improve stability factor and provide a wider operating frequency bandwidth, the 8 Ω resistor is connected in series to the transistor base.

Figure 3.91a shows the small-signal frequency response of the stability factor K and magnitude S_{11}, demonstrating the unconditional stable conditions across the entire frequency range when $K > 1$ and the input return loss better than 30 dB at the operating frequency. In a large-signal operation mode, a maximum collector efficiency of about 72% and a linear power gain of about 15 dB were achieved for the first amplifying stage, as shown in Figure 3.91b. As expected, the high efficiency level of about 65% was obtained at 3-dB gain compression point.

Figure 3.92 shows the simulation schematic of a two-stage lumped parallel-circuit Class-E power amplifier, where the first Class-AB amplifying stage loaded by a 5-Ω impedance is combined together with the second parallel-circuit Class-E amplifying stage having a 5-Ω source impedance in a two-stage power amplifier. In order to maximize the overall performance of the power amplifier, a small tuning of the interstage matching circuit elements, composing of the two high-pass L-type matching sections, around their initial values shown in a dashed circle can be provided.

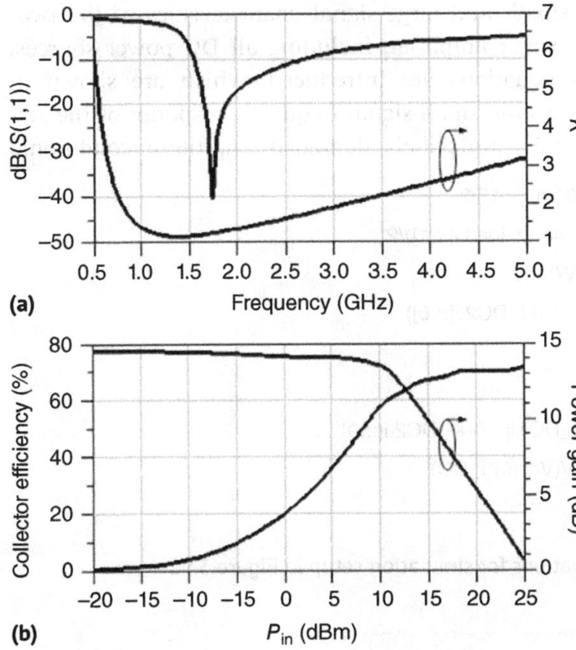

Figure 3.91 Simulated (a) input return loss and (b) collector efficiency of first amplifying stage.

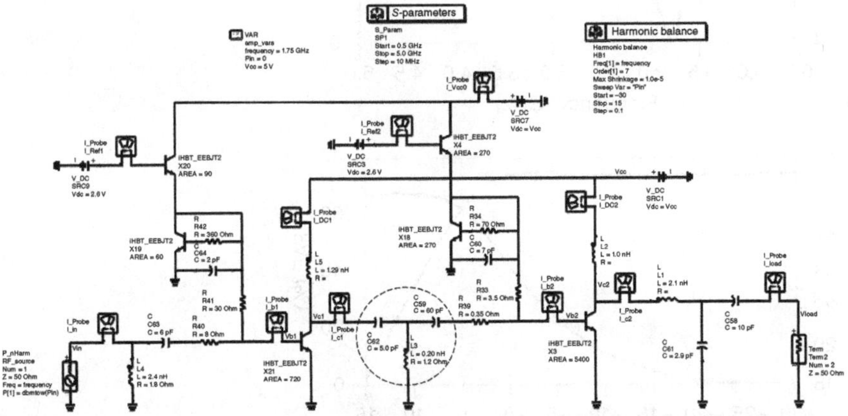

Figure 3.92 Simulation schematic of two-stage lumped parallel-circuit Class-E power amplifier.

In order to simulate the small- and large-signal characteristics of the two-stage lumped Class-E HBT power amplifier, including all DC power sources, the extended measurement equations are introduced, which are shown in Figure 3.93. Figure 3.94a shows the small-signal frequency response of the stability factor K and magnitude S_{11} in decibels, demonstrating the unconditional

```
Eqn Pin_W=mag(Vin[::,1])*mag(I_in.i[::,1])/2
Eqn Pload_W=mag(Vload[::,1])*mag(I_load.i[::,1])/2
Eqn Pload_dBm=10*log(Pload_W)+30
Eqn Efficiency=100*Pload_W/Vcc[0]*I_DC2.i[::,0])
Eqn Gain=Pload_dBm-Pin
Eqn K=stab_fact(S)
Eqn I_DC=I_Ref+I_Vcc0.i[::,0]+I_DC1.i[::,0]+I_DC2.i[::,0]
Eqn PAE=100*(Pload_W-Pin_W)/(Vcc[0]*I_DC)
Eqn I_Ref=I_Ref1.i[::,0]+I_Ref2.i[::,0]
```

Figure 3.93 Measurement equations for simulation setup in Figure 3.92.

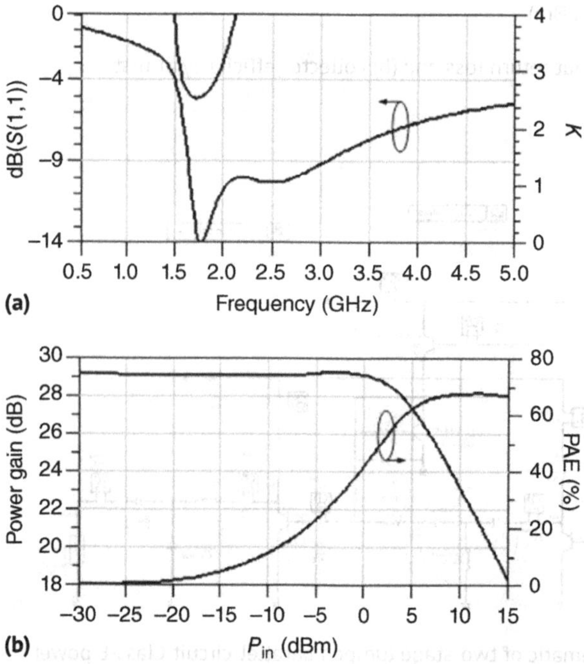

Figure 3.94 Simulated (a) input return loss and K-factor and (b) power gain and PAE of two-stage lumped Class-E power amplifier.

stability conditions across the entire frequency range with $K > 2$ and input return loss better than 13 dB at the operating frequency. In a large-signal operation mode, a maximum PAE of about 68% and a linear power gain over 29 dB were obtained, as shown in Figure 3.94b, with a high efficiency of about 57.5% at a 1-dB gain compression point.

As a final design step, it is necessary to replace the lumped inductors in the DC-supply and output matching circuits by the microstrip lines using standard FR4 or any other substrate, while the lumped inductors in the input and interstage circuits can be replaced by spiral inductors for MMIC implementation. The lumped capacitors in the input and interstage matching circuits should be implemented as the MIM capacitors, whereas the lumped capacitors in the load network can be represented by high-Q ceramic chip capacitors. The simplified equation to calculate the parameters of the short-length transmission line with the characteristic impedance much higher than the load impedance can be written as

$$Z_{\text{in}} = jZ_0 \tan\theta \tag{3.208}$$

Because the input impedance of a lumped inductance L is equal to $Z_{\text{in}} = j\omega L$, the electrical length θ is calculated using Eq. (3.208) as

$$\theta = \tan^{-1}\frac{\omega L}{Z_0} \tag{3.209}$$

Consequently, the electrical lengths of the parallel microstrip line in the first amplifying stage, parallel microstrip line in the second amplifying stage, and series microstrip line in the output matching circuit can be respectively obtained from Eq. (3.209) as $\theta_1 = \tan^{-1}(0.284) = 15.85°$, $\theta_2 = \tan^{-1}(0.220) = 12.41°$, and $\theta_3 = \tan^{-1}(0.462) = 24.79°$.

Finally, the corresponding lengths of the microstrip lines with a 29-mil width (characteristic impedance of approximately 50 Ω) implemented on a 16-mil FR4 substrate with their parameters calculated by the LineCalc program, available in ADS Tools, which are shown in Figure 3.95, are as follows: $l_1 = 4$ mm, $l_2 = 3.1$ mm, and $l_3 = 6.3$ mm.

Figure 3.95 shows the resulting simulation schematic of a two-stage parallel-circuit Class-E HBT power amplifier with the internal input and interstage matching and bias circuits for monolithic implementation on a single chip and external laminate based DC-feed microstrip lines and load network with a series microstrip line in a 50-Ω environment. In this design, an output power of 33.6 dBm and a maximum PAE of 67.1% with a power gain of 22.5 dB are simulated, as shown in Figure 3.96. For an increased power gain up to 27 dB, the output power of 33 dBm and a PAE of 62.7% can be achieved. Thus, such a simple and straightforward design procedure described above shows how effective can be a simplified analytical approach that can provide both the high performance and can significantly

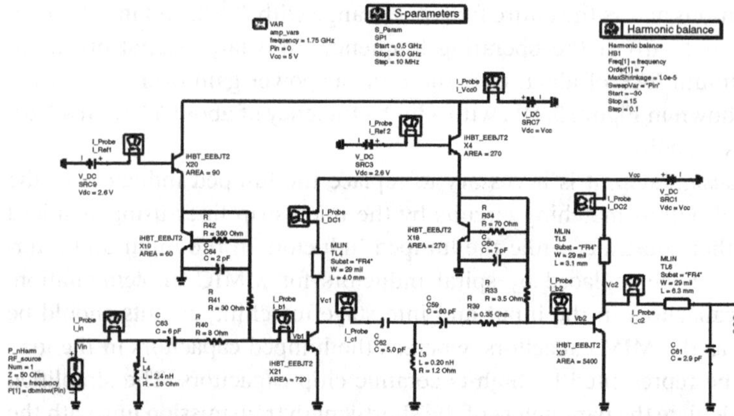

Figure 3.95 Simulation schematic of two-stage microstrip Class-E power amplifier.

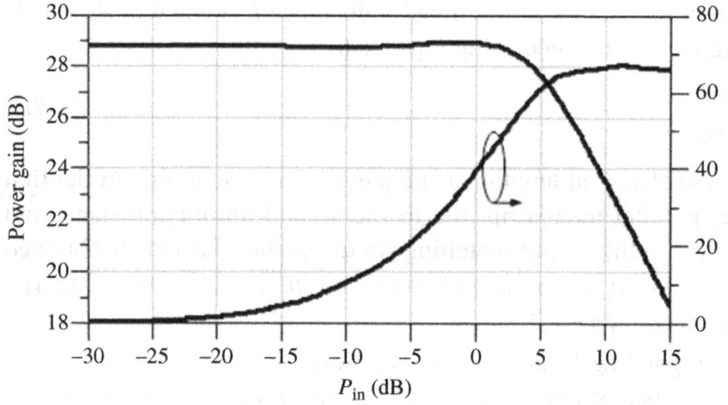

Figure 3.96 Simulated power gain and PAE of two-stage microstrip Class-E power amplifier.

speed up the entire design procedure, although the final circuit optimization can be helpful.

3.9 Broadband Class-E Power Amplifiers

The bandwidth performance of an idealized optimum Class-E power amplifier with shunt capacitance and series filter requiring an infinite value of the loaded quality factor of the series resonant circuit to provide open-circuit impedance conditions at the second- and higher-order harmonics at the transistor collector (or drain) output is limited to a single frequency. However, by reducing the loaded quality factor of the series resonant circuit, a high-efficiency broadband operation of the Class-E power amplifier becomes possible if the inductive impedance at the fundamental frequency seen by the transistor collector (or drain) maintains constant across some frequency range. For example, close approximation to a Class-E operation mode with a constant load phase angle of 50° in a frequency bandwidth of about 50% was achieved using a simple load network composing of a parallel inductance and a series resonant circuit with optimized loaded quality factor [82]. As it was found yet in the mid-1960s on the example of the parametric amplifier, it is possible to improve the bandwidth response using multiple-resonant bandpass filters for the signal and idling circuits rather than using simple resonant circuits [83, 84]. Moreover, it was analytically derived at the same time that, in order to optimally reduce the rate of change of reactance of both the signal and idling circuits, the added resonant circuits should have an appropriate loaded quality factor [85]. By adding additional resonators, the frequency bandwidth can be even further increased, but extension in the frequency bandwidth by using an additional resonator significantly decreases with the increased number of resonators [86, 87].

3.9.1 Reactance Compensation Technique

The reactance (or susceptance) compensation circuit technique can be clearly described using the simplified load network with a shunt resonant L_pC_p circuit and a series resonant L_sC_s circuit, as shown in Figure 3.69, where both resonant circuits are tuned to the fundamental frequency and R is the load resistance. Since the reactance of the series resonant circuit increases with positive slope and the reactance of the loaded shunt resonant circuit decreases with negative slope near the resonant frequency, the overall reactance slope of the load network can be controlled. As a result, the rate of reactance variation of the series resonant circuit with frequency can be adjusted to be exactly opposite to that of the shunt resonant circuit by correct choice of the components in the shunt circuit, thereby producing a zero total variation across a wide frequency range.

The load-network admittance $Y_{\text{net}} = 1/Z_{\text{net}}$ at the input of the reactance compensation circuit, as shown in Figure 3.97, can be expressed as

$$Y_{\text{net}}(\omega) = \left(j\omega C_p + \frac{1}{j\omega L_p} + \frac{1}{R + j\omega' L_s} \right) \tag{3.210}$$

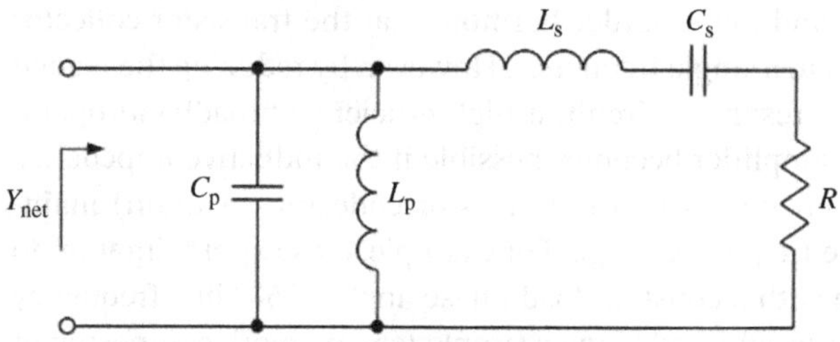

Figure 3.97 Reactance compensation circuit based on lumped elements.

where

$$\omega' = \omega\left(1 - \frac{\omega_0^2}{\omega^2}\right) \tag{3.211}$$

and $\omega_0 = 1/\sqrt{L_s C_s} = 1/\sqrt{L_p C_p}$ is the resonant frequency.

The frequency bandwidth can be maximized if at a resonant frequency ω_0,

$$\left.\frac{dB_{net}(\omega)}{d\omega}\right|_{\omega=\omega_0} = 0 \tag{3.212}$$

where

$$B_{net}(\omega) = \mathrm{Im}Y_{net}(\omega) = \omega C_p - \frac{1}{\omega L_p} - \frac{\omega' L_s}{R^2 + (\omega' L_s)^2} \tag{3.213}$$

is the load-network reactive admittance (or susceptance).

For zero susceptance when $B_{net} = 0$, an additional equation derived from Eq. (3.213) is written as

$$C_p + \frac{1}{\omega_0^2 L_p} - \frac{2L_s}{R^2} = 0 \tag{3.214}$$

from which the series inductance L_s and series capacitance C_s are obtained through the values of the shunt inductance L_p and shunt capacitance C_p by

$$L_s = C_p R^2 \tag{3.215}$$

$$C_s = \frac{L_p}{R^2} \tag{3.216}$$

As an example, for the reactance compensation load network shown in Figure 3.98a, almost constant conductance $\mathrm{Re}Y_{net}$ over the frequency bandwidth of 40% is provided from 4 to 6 MHz, according to Figure 3.98b. In this case, the susceptance

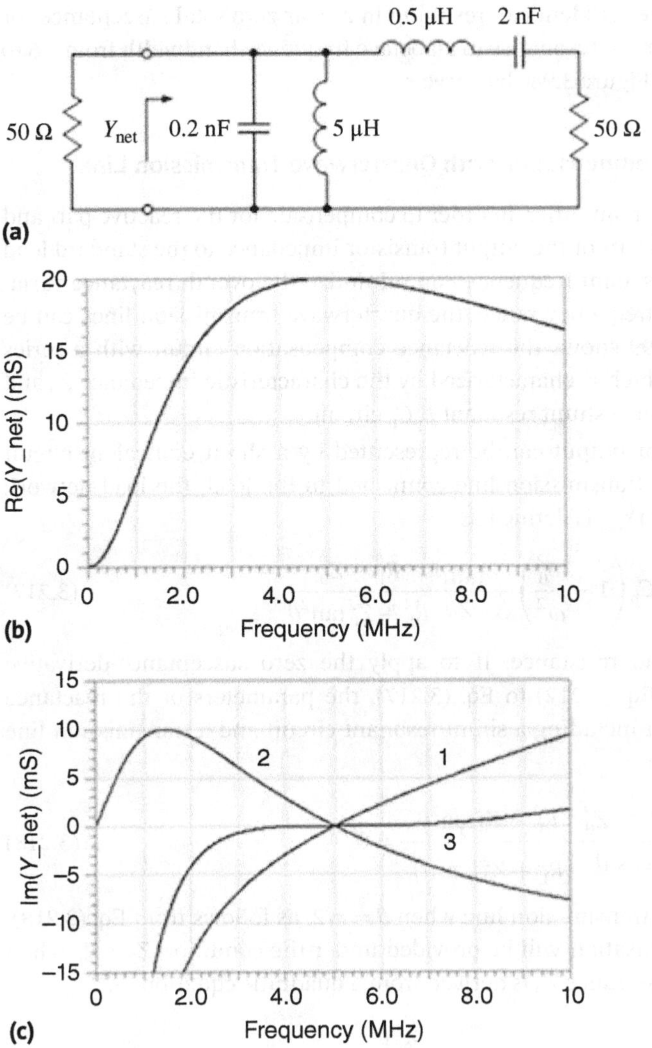

Figure 3.98 Low-frequency example of reactance compensation circuit and its performance.

$\mathrm{Im}Y_{net}$ of the shunt resonant circuit changes with frequency with the gradient of $2C_p$ at the resonant frequency ω_0, as shown in Figure 3.98c by curve 1, whereas the series circuit with the same resonant frequency of 5 MHz provides an additional susceptance with a negative slope, as shown in Figure 3.98c by curve 2. By properly selecting the parameters of the series resonant circuit, the magnitude of the two slopes can be set identical, resulting in a near zero total susceptance (or reactance) slope near the resonance in an octave frequency bandwidth from 3.5 to 7 MHz, as shown in Figure 3.98c by curve 3.

3.9.2 Reactance Compensation with Quarterwave Transmission Line

In a microwave power amplifier, in order to compensate for the reactive part and to transform the real part of the output transistor impedance to the standard load impedance at the resonant frequency and minimize the overall reactance variations across certain frequency range, the quarterwave transmission lines can be used [88]. Figure 3.99 shows the reactance compensation circuit with a series transmission line, which is characterized by the characteristic impedance Z_0 and electrical length θ, and a shunt resonant L_pC_p circuit.

Since the transistor output can be represented by a shunt equivalent circuit loaded by the series transmission line connected to the load, the load-network susceptance $B_{net} = \mathrm{Im}Y_{net}$ is defined as

$$B_{net}(\omega) = \omega L_p C_p \left(1 - \frac{\omega_0^2}{\omega^2}\right) + \frac{\tan\theta}{Z_0} \frac{R_L^2 - Z_0^2}{R_L^2 + Z_0^2 \tan^2\theta} \tag{3.217}$$

where R_L is the load resistance. If to apply the zero susceptance-derivative condition given by Eq. (3.212) to Eq. (3.217), the parameters of the reactance compensation circuit including a shunt resonant circuit and a transmission line can be obtained from

$$2C_p + \frac{\pi}{2Z_0\omega_0} \frac{R_L^2 - Z_0^2}{\cos^2\theta} \frac{R_L^2 - Z_0^2\tan^2\theta}{\left(R_L^2 + Z_0^2\tan^2\theta\right)^2} = 0 \tag{3.218}$$

For a quarterwave transmission line when $\theta = \pi/2$, as follows from Eq. (3.218), the reactance compensation will be provided under the condition $Z_0 < R_L$ when the characteristic impedance Z_0 is derived from a quadratic equation

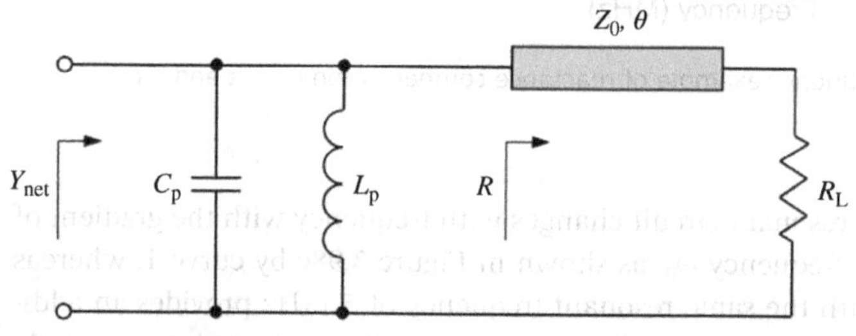

Figure 3.99 Transmission-line reactance compensation circuit.

$$Z_0^2 + 4\frac{QR_L}{\pi}Z_0 - R_L^2 = 0 \tag{3.219}$$

where $Q = \omega_0 C_p R$ and $R = Z_0^2/R_L$.

Figure 3.100a shows an example of the reactance compensation load network using a series quarterwave transmission line with the characteristic impedance of 61.2 Ω, which is required to match a 50-Ω real part of the transistor output

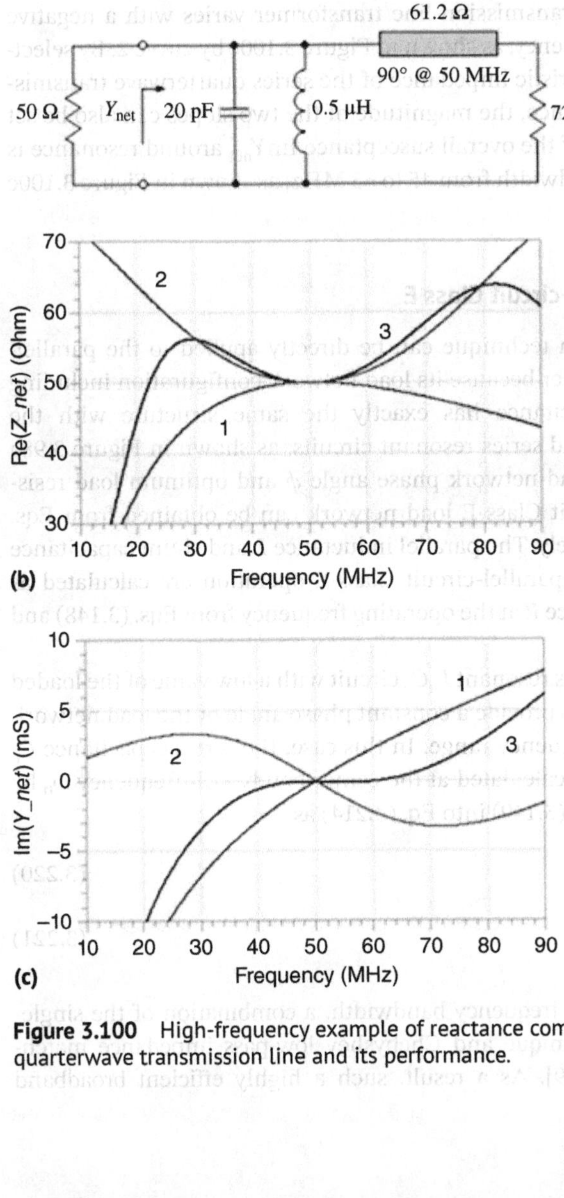

Figure 3.100 High-frequency example of reactance compensation circuit with quarterwave transmission line and its performance.

impedance to a 75-Ω load at the center bandwidth frequency of 50 MHz. For the corresponding variations of the real parts of a shunt *LC* circuit (curve 1) and a series quarterwave transmission line (curve 2), the shunt resonant circuit loaded by the input impedance of the quarterwave line minimizes the variations of the total resistance $\mathrm{Re}Z_{\mathrm{net}}$ around 50 Ω in a very wide frequency bandwidth, as shown in Figure 3.100b by curve 3. At the same time, the susceptance of a shunt circuit having a resonant frequency of 50 MHz varies with frequency with a positive slope, as shown in Figure 3.100c by curve 1, whereas the input susceptance a loaded series quarterwave transmission-line transformer varies with a negative slope near the resonant frequency, as shown in Figure 3.100c by curve 2. By selecting the appropriate characteristic impedance of the series quarterwave transmission line and the load resistance, the magnitude of the two slopes can also be set identical, so that the slope of the overall susceptance $\mathrm{Im}Y_{\mathrm{net}}$ around resonance is near zero in a frequency bandwidth from 45 to 65 MHz, as shown in Figure 3.100c by curve 3.

3.9.3 Broadband Parallel-circuit Class E

The reactance compensation technique can be directly applied to the parallel-circuit Class E power amplifier because its load network configuration including the transistor output capacitance has exactly the same structure with the combination of the shunt and series resonant circuits, as shown in Figure 3.98a [70, 71]. In this case, the load-network phase angle ϕ and optimum load resistance R of the parallel-circuit Class-E load network can be obtained from Eqs. (3.144) and (3.150), respectively. The parallel inductance L and shunt capacitance C required for an optimum parallel-circuit Class-E operation are calculated as functions of the load resistance R at the operating frequency from Eqs. (3.148) and (3.149), respectively.

The parameters of the series resonant L_0C_0 circuit with a low value of the loaded quality factor should be set to provide a constant phase angle of the load network across the required wide frequency range. In this case, the series capacitance C_0 and series inductance L_0 are calculated at the center bandwidth frequency ω_0 by substituting Eqs. (3.148) and (3.149) into Eq. (3.214) as

$$L_0 = 1.026\frac{R}{\omega_0} \tag{3.220}$$

$$C_0 = \frac{1}{\omega_0^2 L_0} \tag{3.221}$$

To maximize the operating frequency bandwidth, a combination of the single-reactance compensation technique and Chebyshev low-pass impedance matching network can be used [89]. As a result, such a highly efficient broadband parallel-circuit Class-E power amplifier operating at 12-V drain DC-supply voltage achieved the drain efficiencies of 82.15–86.3% at output power levels of 36.92–38.57 dBm with a power gain of more than 19 dB in the frequency bandwidth from 7 to 9.4 MHz.

Similarly, the reactance (or susceptance) compensation technique can also be applied to a parallel-circuit Class-E power amplifier, where the series quarterwave transmission line is used instead of a series L_0C_0 resonant circuit, as shown in Figure 3.101a. In some practical implementations when it is necessary to reduce an overall size of a broadband parallel-circuit Class-E power amplifier, the series quarterwave line can be replaced by an equivalent low-pass T-type circuit composing of a series transmission line with higher characteristic impedance and electrical length of less than 90° and a shunt capacitance connected to the middle of the transmission line. The series quarterwave line can also be replaced by an equivalent low-pass Π-type circuit composing of a series transmission line with higher characteristic impedance and electrical length much less than 90° and two shunt capacitors, and the capacitance adjacent to the transistor output can be counted within the total shunt capacitance including the collector capacitance required for an optimum

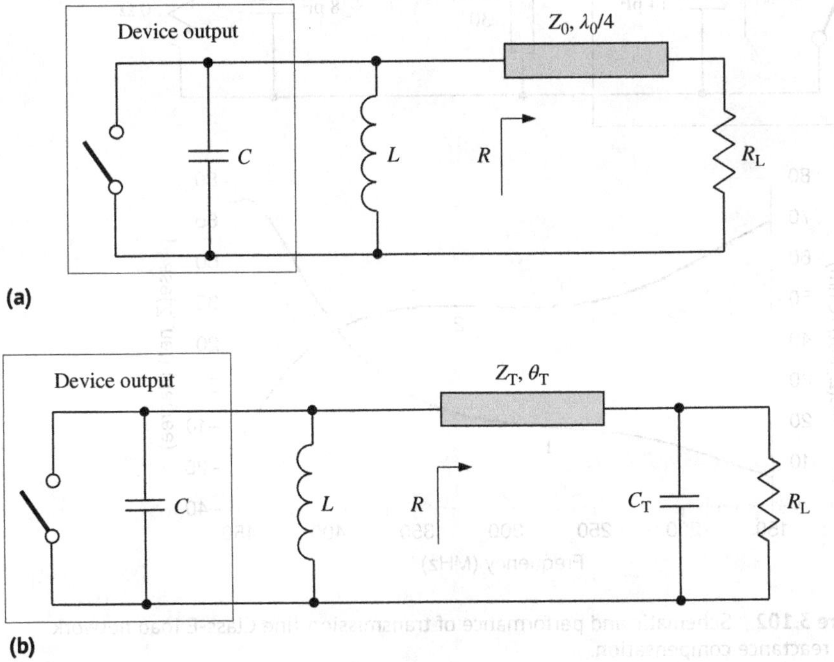

Figure 3.101 Transmission-line reactance compensation circuits.

(or nominal) parallel-circuit Class-E operation mode. As an alternative, a series quarterwave line can be replaced by a low-pass L-type matching circuit with a series transmission line with higher characteristic impedance and lower electrical length and a shunt capacitance, as shown in Figure 3.101b.

Besides, a short-length short-circuited transmission line can replace the parallel inductance, as shown in Figure 3.102a for the transmission-line broadband Class-E load network, for convenience of practical implementation on a printed-circuit board as a microstrip line with minimum insertion loss. Here, the electrical lengths of the parallel and series transmission lines are given at the center bandwidth frequency of 300 MHz, and the real part of the input load-network impedance $\mathrm{Re}Z_{net}$ varies from 17 Ω at 225 MHz to 47.5 Ω at 400 MHz, as shown in Figure 3.102b by curve 1, with much less variation from 18.5 to 27 Ω in a frequency bandwidth from 250 to 350 MHz. In this case, the phase angle remains almost constant

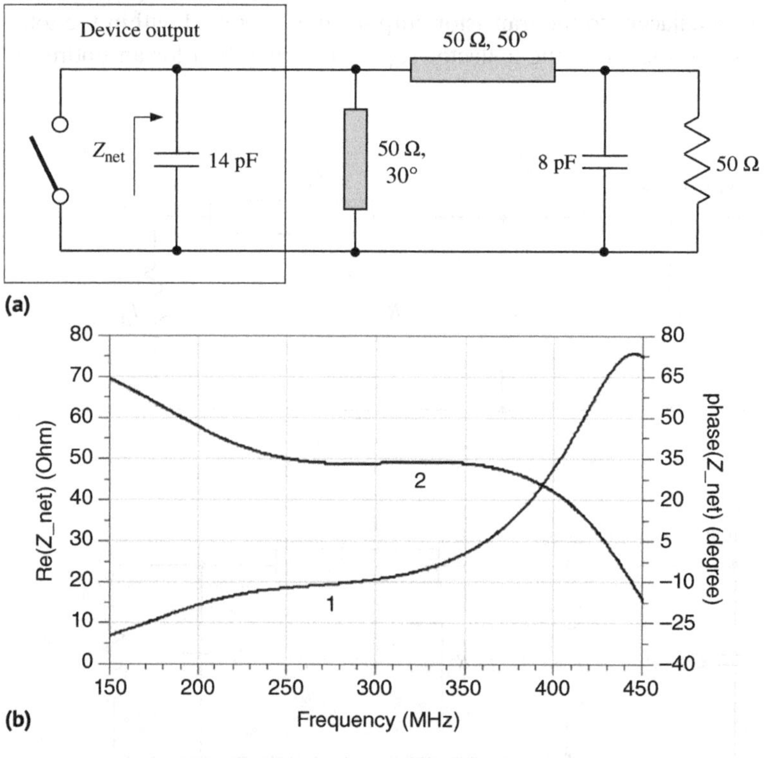

Figure 3.102 Schematic and performance of transmission-line Class-E load network with reactance compensation.

around 33° in a frequency bandwidth from 250 to 350 MHz and varies from 22.5° at 225 MHz to 39.5° at 400 MHz (curve 2).

The circuit schematic of a broadband high-efficiency transmission-line LDMOSFET power amplifier is shown in Figure 3.103a, providing an output power of around 20 W and a power gain greater than 12 dB in a frequency range from 225 to 400 MHz. Here, in order to apply a reactance compensation technique to the parallel-circuit Class-E power amplifier, the load network includes a parallel short-length transmission line to supply DC power to the transistor and a single-section L-type transmission-line transformer in view of a sufficiently small difference between the transistor output impedance required for an optimum Class-E mode and the standard load of 50 Ω. The input matching circuit includes the two low-pass transmission-line L-type matching sections for conjugate matching with the 50-Ω source, and the parallel resistor of 75 Ω in the gate bias circuit is necessary to simplify the matching procedure and to improve the input return loss. The simulated broadband high-efficiency power amplifier performance was

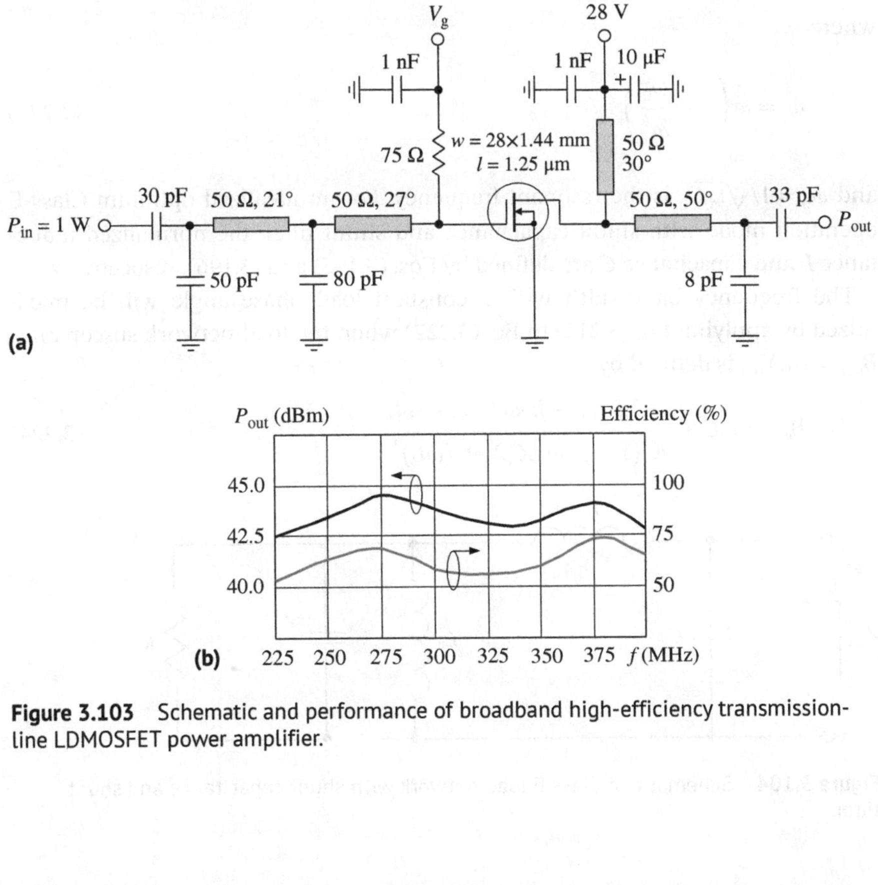

Figure 3.103 Schematic and performance of broadband high-efficiency transmission-line LDMOSFET power amplifier.

obtained by optimizing separately the input and output circuits to speed up the optimization process, achieving an output power of 42.5–44.5 dBm, a power gain of 13.5 ± 1 dB, and a drain efficiency of 64 ± 10% in a frequency bandwidth from 225 to 400 MHz is shown in Figure 3.103b.

3.9.4 Broadband Class E with Shunt Capacitance and Shunt Filter

The reactance compensation technique can be extended to a Class E with shunt capacitance and shunt filter, whose load network with a shunt capacitance C, a series inductance L, and a shunt L_0C_0 circuit tuned to the fundamental is shown in Figure 3.104 [76]. Here, the positive slope of the reactance provided by the low-pass L-type circuit cancel the negative slope of the reactance provided by the shunt L_0C_0 circuit, thus maintaining the load phase angle constant, and the load-network admittance $Y_{net} = 1/Z_{net}$ can be written as

$$Y_{net} = j\omega C + \frac{1 + j\omega' C_0 R}{R(1 - j\omega\omega' LC_0) + j\omega L} \tag{3.222}$$

where

$$\omega' = \omega\left(1 - \frac{\omega_0^2}{\omega^2}\right) \tag{3.223}$$

and $\omega_0 = 1/\sqrt{L_0C_0}$ is the resonant frequency. For an idealized optimum Class-E operation mode with shunt capacitance and shunt filter, the normalized inductance L and capacitance C are defined by Eqs. (3.195) and (3.196), respectively.

The frequency bandwidth with a constant load phase angle will be maximized by applying Eq. (3.212) to Eq. (3.222) when the load-network susceptance $B_{net} = \text{Im}Y_{net}$ is defined by

$$B_{net} = \omega C + \frac{\omega' C_0 R^2(1 - j\omega\omega' LC_0) - \omega L}{R^2(1 - j\omega\omega' LC_0)^2 + (\omega L)^2} \tag{3.224}$$

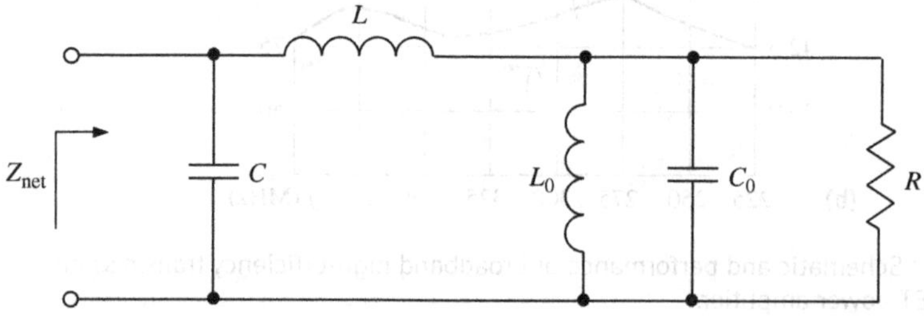

Figure 3.104 Schematic of Class E load network with shunt capacitance and shunt filter.

As a result, the required additional equation can be derived as

$$2\omega_0 C_0 R - \frac{\omega_0 L}{R} - \omega_0 C R \frac{\left(\frac{\omega_0 L}{R}\right)^2 + 1}{\left(\frac{\omega_0 L}{R}\right)^2 - 1} = 0 \tag{3.225}$$

and the shunt capacitance C_0 and inductance L_0 can be calculated from

$$C_0 = \frac{1.0896}{\omega_0 R} \tag{3.226}$$

$$L_0 = \frac{1}{\omega_0^2 C_0} \tag{3.227}$$

The reactance compensation load network for a Class-E power amplifier with shunt capacitance and shunt filter using lumped elements is shown in Figure 3.105a. Here, by selecting the proper circuit parameters according to

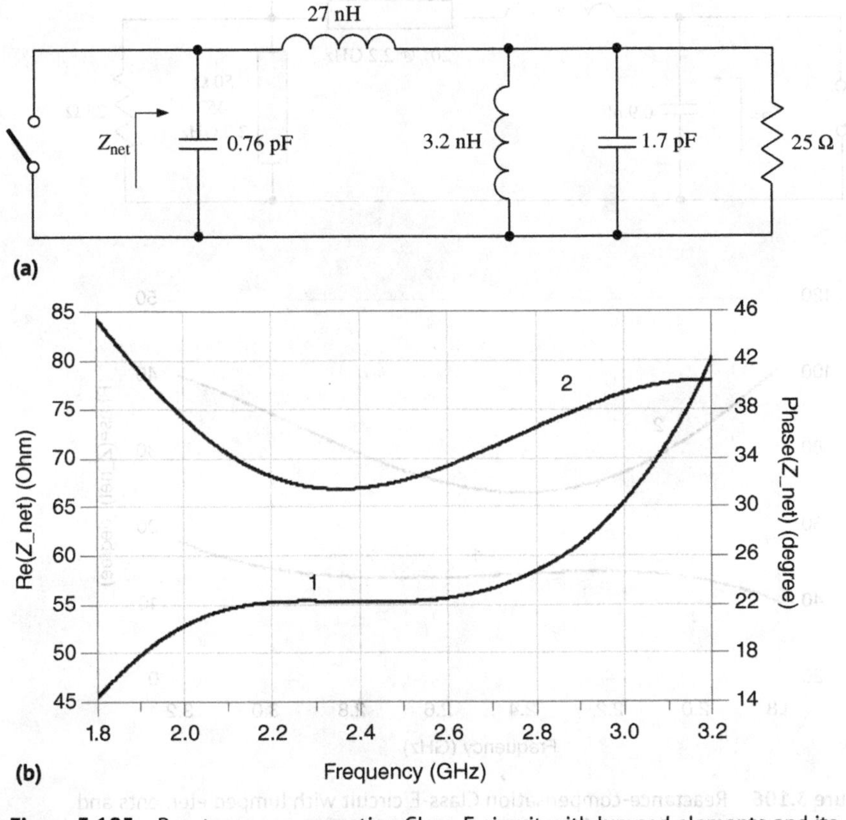

Figure 3.105 Reactance compensation Class-E circuit with lumped elements and its performance.

Eqs. (3.195)–(3.197) for a Class-E mode with shunt capacitance and shunt filter together with Eqs. (3.226) and (3.227) for shunt filter elements, a constant total reactance and phase of the load-network impedance Z_{net} over a wide frequency range is achieved as well. The simulation results at the fundamental frequency show that the resistance ReZ_{net} varies from 50 Ω at 1.9 GHz to 65 Ω at 3 GHz, as shown in Figure 3.105b by curve 1, whereas the load-network phase varies between 31.5 and 41° from 1.9 to 3.2 GHz (curve 2).

The reactance compensation load network for a Class-E power amplifier with shunt capacitance and shunt filter using both lumped elements and transmission lines is shown in Figure 3.106a, where the shunt filter is represented by open- and short-circuit stubs replacing lumped capacitor and inductor, respectively,

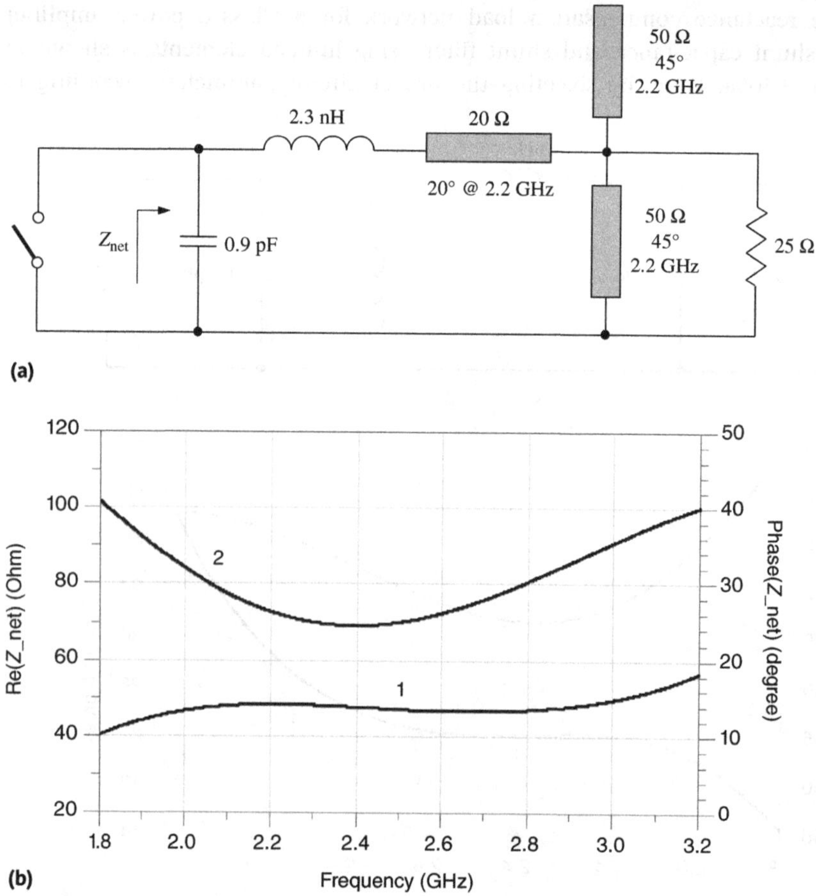

Figure 3.106 Reactance-compensation Class-E circuit with lumped elements and transmission lines and its performance.

each having the characteristic impedance of 50 Ω and electrical length of 45° at 2.2 GHz. The series inductive reactance is provided by lumped inductor that can be represented by the transistor bondwire and package lead and short transmission line. The simulation results at the fundamental frequency show that the load-network resistance ReZ_{net} and phase vary between 40 and 57 Ω (curve 1) and between 25 and 41° (curve 2) from 1.8 to 3.2 GHz, as shown in Figure 3.106b.

Generally, using an input lossy *LCR* matching circuit can provide a very broadband power amplifier design by minimizing the input return loss and output power variations with an output low-pass *LC* circuit to compensate for the transistor output reactance [90, 91]. However, at microwave frequencies, it is very convenient to provide an input broadband matching over an octave bandwidth by using a multi-section matching transformer composed of the stepped transmission-line sections with different characteristic impedances and electrical lengths [92]. In this case, there is no need to use any tuning capacitors.

The idealized simulation setup of a 10-W 28-V broadband Class-E power amplifier circuit using a GaN HEMT CGH40010 transistor is shown in Figure 3.107a,

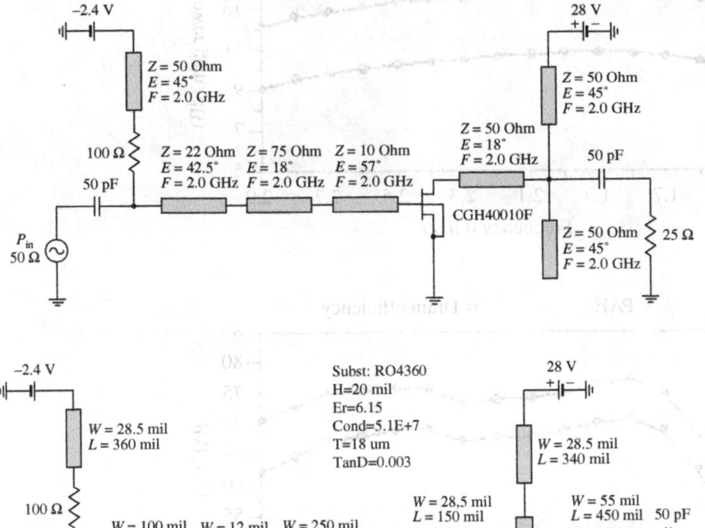

(a)

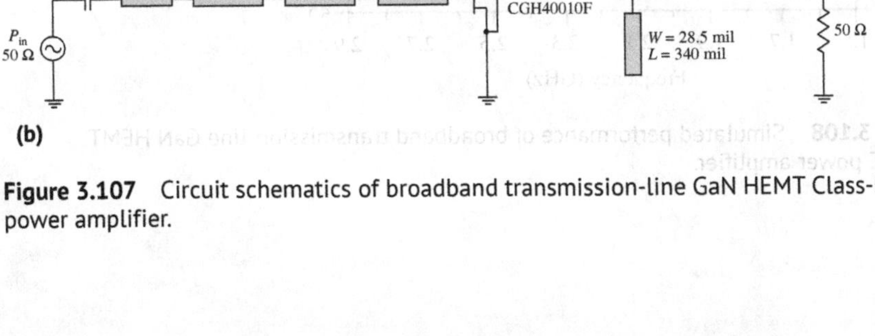

(b)

Figure 3.107 Circuit schematics of broadband transmission-line GaN HEMT Class-E power amplifier.

where both the input matching circuit representing an ideal three stepped transmission-line matching transformer and consisting of the transmission-line sections with different characteristic impedances and electrical lengths and an ideal Class-E transmission-line load network are designed to operate over a frequency bandwidth from 1.7 to 2.7 GHz [93]. The optimum Class-E load resistance is calculated for an output power $P_{out} = 12$ W, DC-supply voltage $V_{dd} = 28$ V, and saturation voltage $V_{sat} = 1.7$ V according to Eq. (3.197) as

$$R = 0.4281\frac{(V_{dd} - V_{sat})^2}{P_{out}} = 25\,\Omega \tag{3.228}$$

where the saturation voltage V_{sat} is defined from the device output current–voltage characteristics. The parallel resonant circuit in a transmission-line

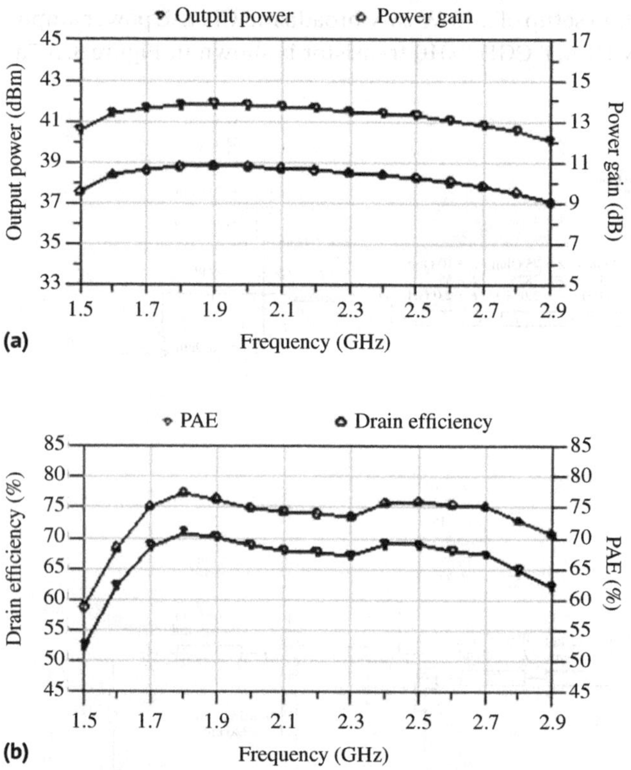

Figure 3.108 Simulated performance of broadband transmission-line GaN HEMT Class-E power amplifier.

broadband Class-E load network is connected in parallel to a 25-Ω load, as shown in Figure 3.107a, composing of both open- and short-circuit stubs, each having a characteristic impedance of 50 Ω and electrical length of 45° at 2 GHz.

In order to match an optimum Class-E impedance $R = 25\ \Omega$ with a standard 50-Ω load, an additional series quarterwave transmission line with the characteristic impedance of around 35 Ω and electrical length of 90° at high bandwidth frequency can be used. The simulation setup with an implementation of an idealized circuit of the broadband GaN HEMT Class-E power amplifier on a 20-mil RO4360 substrate is shown in Figure 3.107b, where an additional series quarterwave microstrip line is included to match an idealized 25-Ω load with a standard 50-Ω load. Figure 3.108a,b show the simulated results with an output power of

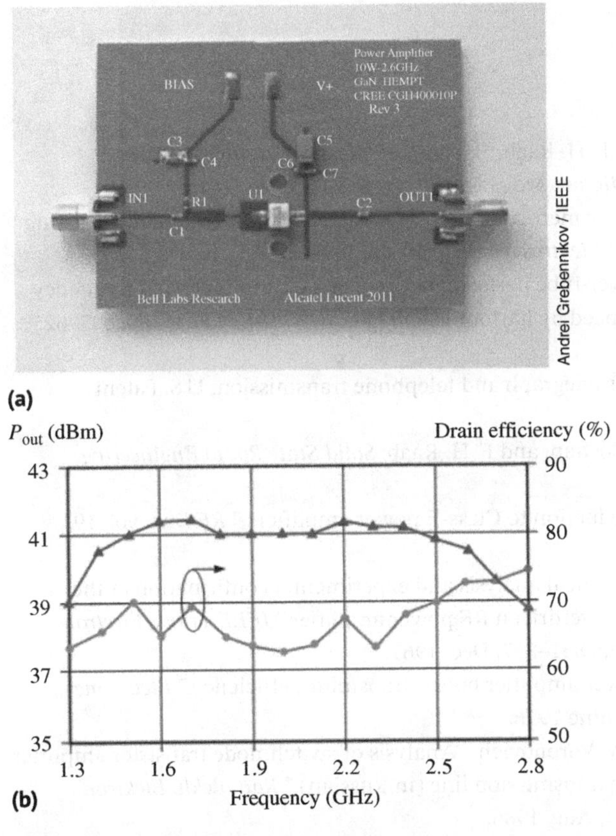

Figure 3.109 Test board and performance of broadband Class-E GaN HEMT amplifier.

more than 41 dBm with a power gain of about 10 dB, and a drain efficiency over 73% with a PAE of more than 65% within the required frequency range from 1.7 to 2.7 GHz, respectively.

The test PCB board of a broadband Class-E GaN HEMT power amplifier using a GaN HEMT CGH40010P transistor in a pill-type package to minimize the overall size of a 20-mil RO4360 substrate is shown in Figure 3.109a [93]. In this case, due to the accurate transistor model and minimum parasitics in the transistor and microstrip-line implementations, there was no need for any additional tuning of the input matching circuit and load network. Figure 3.109b shows the measured results of a transmission-line broadband Class-E GaN HEMT power amplifier with an output power of 40.6 ± 0.9 dBm, a drain efficiency varying from 63 to 73%, and a power gain around 9.5 dB were achieved at a drain supply voltage of 24 V in a frequency range from 1.4 to 2.7 GHz, as shown in Figure 3.109b.

References

1 A. Grebennikov and F. H. Raab, "History of Class-F and inverse Class-F techniques," *IEEE Microw. Mag.*, vol. 19, pp. 99–115, Nov. 2018.

2 I. N. Fomichev, "A new method to increase efficiency of the radio broadcasting station (in Russian)," *Elektrosvyaz*, pp. 58–66, June 1938.

3 R. I. Sarbacher, "Power-tube performance in Class C amplifiers and frequency multipliers as influenced by harmonic voltage," *Proc. IRE*, vol. 31, pp. 607–625, Nov. 1943.

4 H. J. Round, Wireless telegraph and telephone transmission, U.S. Patent 1,564,627, Dec. 1925.

5 H. L. Krauss, C. W. Bostian, and F. H. Raab, *Solid State Radio Engineering*, Wiley, 1980.

6 F. H. Raab, "An introduction to Class-F power amplifiers," *RF Des.*, vol. 19, pp. 79–84, May 1996.

7 D. M. Snider, "A theoretical analysis and experimental confirmation of the optimally loaded and overdriven RF power amplifier," *IEEE Trans. Electron Devices*, vol. ED-14, pp. 851–857, Dec. 1967.

8 F. H. Raab, "FET power amplifier boosts transmitter efficiency," *Electronics*, vol. 49, pp. 122–126, June 1976.

9 V. A. Borisov and V. V. Voronovich, "Analysis of switchmode transistor amplifier with parallel forming transmission line (in Russian)," *Radiotekh. Elektron.*, vol. 31, pp. 1590–1597, Aug. 1986.

10 A. Grebennikov, "High-efficiency Doherty amplifiers using Class-F and inverse Class-F load networks with a quarterwave transmission line," *High Freq. Electron.*, vol. 14, pp. 22–32, Sep. 2015.

11 A. V. Grebennikov, "Circuit design technique for high efficiency Class F amplifiers," *2000 IEEE MTT-S Int. Microwave Symp. Dig.*, vol. 2, pp. 771–774, June 2000.

12 C. Trask, "Class-F amplifier loading networks: a unified design approach," *1999 IEEE MTT-S Int. Microwave Symp. Dig.*, vol. 1, pp. 351–354, June 1999.

13 F. H. Raab, "Class-F power amplifiers with maximally flat waveforms," *IEEE Trans. Microw. Theory Tech.*, vol. MTT-45, pp. 2007–2012, Nov. 1997.

14 D. Schmelzer and S. I. Long, "A GaN HEMT Class F amplifier at 2 GHz with 80% PAE," *IEEE J. Solid-State Circuits*, vol. SC-42, pp. 2130–2136, Oct. 2007.

15 P. Butterworth, S. Gao, S. F. Ooi, and A. Sambell, "High-efficiency Class-F power amplifier with broadband performance," *Microw. Opt. Technol. Lett.*, vol. 44, pp. 243–247, Feb. 2005.

16 A. Ramadan, T. Reveyrand, A. Martin, J. M. Nebus, P. Bouysee, and L. Lapierre, "Two-stage GaN HEMT amplifier with gate-source voltage shaping for efficiency versus bandwidth enhancements," *IEEE Trans. Microw. Theory Tech.*, vol. MTT-59, pp. 699–706, Mar. 2011.

17 N. Tuffy, A. Zhu, T. J. Brazil, "Novel realization of a broadband high-efficiency continuous Class-F power amplifier," *Proc. 6th Europ. Microwave Integrated Circuits Conf.*, pp. 1220–1223, Oct. 2011.

18 Z. Yang, Y. Yao, L. Liu, M. Li, T. Li, and Z. Dai, "Design of high efficiency broadband continuous Class-F power amplifier using real frequency technique with finite transmission zero," *IEEE Access*, vol. 6, pp. 61983–61993, Nov. 2018.

19 M. Yang, J. Xia, Y. Guo, and A. Zhu, "Highly efficient broadband continuous inverse Class-F power amplifier design using modified elliptic low-pass filtering matching network," *IEEE Trans. Microw. Theory Tech.*, vol. MTT-64, pp. 1515–1525, May 2016.

20 K. Chen and D. Peroulis, "Design of broadband highly efficient harmonic-tuned power amplifier using in-band continuous Class-F^{-1}/F mode transferring," *IEEE Trans. Microw. Theory Tech.*, vol. MTT-60, pp. 4107–4116, Dec. 2012.

21 F. H. Raab, "Class-E, Class-C, and Class-F power amplifiers based upon a finite number of harmonics," *IEEE Trans. Microw. Theory Tech.*, vol. MTT-49, pp. 1462–1468, Aug. 2001.

22 A. I. Kolesnikov, "A new method to improve efficiency and to increase power of the transmitter (in Russian)," *Master Svyazi*, vol. 6, pp. 27–41, June 1940.

23 E. S. Glazman, L. B. Kalinin, and Y. I. Mikhailov, "Improving VHF transmitter efficiency by using the biharmonic mode," *Telecommun. Radio Eng., Part 1*,vol. 30, pp. 46–51, July 1975.

24 V. J. Tyler, "A new high-efficiency high-power amplifier," *Marconi Rev.*, vol. 21, pp. 96–109, Fall 1958.

25 M. K. Kazimierczuk, "A new concept of Class F tuned power amplifier," *Proc. 27th Midwest Circuits and Systems Symp*, pp. 425–428, 1984.

26 P. Heymann, R. Doerner, and M. Rudolph, "Harmonic tuning of power transistors by active load-pull measurement," *Microw. J.*, vol. 43, pp. 22–37, June 2000.

27 C. J. Wei, P. DiCarlo, Y.A. Tkachenko, R. McMorrow, and D. Bartle, "Analysis and experimental waveform study on inverse Class-F mode of microwave power FETs," *2000 IEEE MTT-S Int. Microwave Symp. Dig.*, vol. 1, pp. 525–528, June 2000.

28 M. J. Franco, "An efficient, 35 dBm, inverse Class-F, UHF RF power amplifier module on a 12 mm^2 footprint designed in first pass through accurate modeling and simulation," *2010 IEEE MTT-S Int. Microwave Symp. Dig.*, pp. 928–931, May 2010.

29 A. Grebennikov, "High-efficiency transmission-line GaN HEMT inverse Class F power amplifier for active antenna arrays," *Proc. 2009 Asia-Pacific Microwave Conf.*, pp. 317–320, Dec. 2009.

30 A. Grebennikov, "Load network design technique for Class F and inverse Class F PAs," *High Freq. Electron.*, vol. 10, pp. 58–76, May 2011.

31 A. Jarndal, P. Aflaki, R. Negra, A. K. Kouki, and F. Ghannouchi, "Large-signal modeling methodology for GaN HEMTs for RF switching-mode power amplifiers design," *Int. J. RF Microw. Comput.-Aided Eng.*, vol. 21, pp. 45–51, Jan. 2011.

32 A. Grebennikov, "A high-efficiency 100-W four-stage Doherty GaN HEMT power amplifier module for WCDMA systems," *2011 IEEE MTT-S Int. Microwave Symp. Dig.*, pp. 1–4, June 2011.

33 A. Grebennikov, "High-efficiency advanced multistage Doherty GaN HEMT power amplifiers," *RF Technol. Int.*, vol. 1, pp. 10–18, Feb. 2012.

34 A. Grebennikov, "High-efficiency transmission-line inverse Class F power amplifiers for 2-GHz WCDMA systems," *Int. J. RF Microw. Comput.-Aided Eng.*, vol. 21, pp. 446–456, July 2011.

35 A. Grebennikov and F. H. Raab, "A history of switching-mode Class-E techniques," *IEEE Microw. Mag.*, vol. 19, pp. 26–41, July 2018.

36 A. G. Khvilivitsky, "Designing the vacuum-tube generator, operating on detuned load (in Russian)," *Radiotekhnika*, vol. 8, pp. 21–26, Apr. 1953.

37 E. P. Khmelnitsky, "A new concept of significant increasing of fundamental power and efficiency of generator, operating in saturation mode (in Russian)," *Radiotekhnika*, vol. 10, pp. 59–63, Aug. 1955.

38 G. D. Ewing, High-efficiency radio-frequency power amplifiers, PhD dissertation, Oregon State University, June 1964.

39 D. R. Lohrmann, "Amplifiers has 85% efficiency while providing up to 10 watts power over a wide frequency band," *Electron. Des.*, vol. 14, pp. 38–43, Mar. 1966.
40 A. D. Artym, "Switching mode of high frequency power amplifiers (in Russian)," *Radiotekhnika*, vol. 24, pp. 58–64, June 1969.
41 V. B. Kozyrev, "Single-ended switching-mode power amplifier with filtering circuit (in Russian)," = *Poluprovodnikovye Pribory v Tekhnike Svyazi*, vol. 8, pp. 152–166, 1971.
42 N. O. Sokal and A.D. Sokal, "Class E – A new class of high-efficiency tuned single-ended switching power amplifiers," *IEEE J. Solid-State Circuits*, vol. SC-10, pp. 168–176, June 1975.
43 N. O. Sokal, "Class E high-efficiency power amplifiers, from HF to microwave," *1998 IEEE MTT-S Int. Microwave Sym. Dig.*, vol. 2, pp. 1109–1112, June 1998.
44 F. H. Raab, "Idealized operation of the Class E tuned power amplifier," *IEEE Trans. Circuits Syst.*, vol. CAS-24, pp. 725–735, Dec. 1977.
45 B. Molnar, "Basic limitations on waveforms achievable in single-ended switching-mode tuned (Class E) power amplifiers," *IEEE J. Solid-State Circuits*, vol. SC-19, pp. 144–146, Feb. 1984.
46 M. Kazimierczuk and K. Puczko, "Exact analysis of Class E tuned power amplifier at any Q and switch duty cycle," *IEEE Trans. Circuits Syst.*, vol. CAS-34, pp. 149–158, Feb. 1987.
47 I. A. Popov, "Switching mode of single-ended transistor power amplifier (in Russian)," *Poluprovodnikovye Pribory v Tekhnike Svyazi*, vol. 5, pp. 15–35, 1970.
48 F. H. Raab and N. O. Sokal, "Transistor power losses in the Class E tuned power amplifier," *IEEE J. Solid-State Circuits*, vol. SC-13, pp. 912–914, Dec. 1978.
49 I. A. Popov (ed.), *Transistor Generators of Harmonic Oscillations in Switching Mode* (in Russian), Radio i Svyaz, 1985.
50 S. H. Tu and C. Toumazou, "Effect of the loaded quality factor on power efficiency for CMOS Class-E RF tuned power amplifiers," *IEEE Trans. Circuits Syst.-I: Fundam. Theory Appl.*, vol. CAS-I-46, pp. 628–634, May 1999.
51 A. N. Bruevich, "About optimum parameters of switching-mode tuned power amplifier with filtering resonant circuit (in Russian)," *Poluprovodnikovaya Elektronika v Tekhnike Svyazi*, vol. 18, pp. 43–48, 1977.
52 N. O. Sokal and A. Mediano, "Redefining the optimum RF Class-E switch-voltage waveform, to correct a long-used incorrect waveform," *2013 IEEE MTT-S Int. Microwave Sym. Dig.*, pp. 1–3, June 2013.
53 F. H. Raab, "Suboptimum operation of Class-E power amplifiers," *Proc. RF Technology Expo'89*, Santa Clara, CA, pp. 85–98, Feb. 1989.
54 F. H. Raab, "Effects of circuit variations on the Class E tuned power amplifier," *IEEE J. Solid-State Circuits*, vol. SC-13, pp. 239–246, Apr. 1978.

55 T. Suetsugu and M. K. Kazimierczuk, "Design procedure of Class-E amplifier for off-nominal operation at 50% duty cycle," *IEEE Trans. Circuits Syst.-I: Regul. Pap.*, vol. CAS-I-53, pp. 1468–1476, July 2006.

56 T. B. Mader, E. W. Bryerton, M. Marcovic, M. Forman, and Z. Popovic, "Switched-mode high-efficiency microwave power amplifiers in a free-space power-combiner array," *IEEE Trans. Microw. Theory Tech.*, vol. MTT-46, pp. 1391–1398, Oct. 1998.

57 T. B. Mader and Z. B. Popovic, "The transmission-line high-efficiency Class-E amplifier," *IEEE Microw. Guided Wave Lett.*, vol. 5, pp. 290–292, Sep. 1995.

58 F. J. Ortega-Gonzalez, J. L. Jimenez-Martin, A. Asensio-Lopez, and G. Torregrosa-Penalva, "High-efficiency load-pull harmonic controlled Class-E power amplifier," *IEEE Microw. Guided Wave Lett.*, vol. 8, pp. 348–350, Oct. 1998.

59 Y. B. Choi and K. K. Cheng, "Generalised frequency-domain analysis of microwave Class-E power amplifiers," *IEE Proc. Microw. Antennas Propag.*, vol. 148, pp. 403–409, Dec. 2001.

60 P. Watson, R. Neidhard, L. Kehias, R. Welch, T. Quach, R. Worley, M. Pacer, R. Pappaterra, R. Schweller, and T. Jenkins, "Ultra-high efficiency operation based on an alternative Class-E mode," *2000 IEEE GaAs IC Symp. Dig.*, pp. 53–56, Nov. 2000.

61 A. Grebennikov, "A high-efficiency transmission-line GaN HEMT Class E power amplifier," *High Freq. Electron.*, vol. 8, pp. 16–24, Dec. 2009.

62 R. E. Zulinski and J. W. Steadman, "Class E power amplifiers and frequency multipliers with finite DC-feed inductance," *IEEE Trans. Circuits Syst.*, vol. 34, pp. 1074–1087, Sep. 1987.

63 C. P. Avratoglou, N. C. Voulgaris, and F. I. Ioannidou, "Analysis and design of a generalized Class E tuned power amplifier," *IEEE Trans. Circuits Syst.*, vol. 36, pp. 1067–1079, Aug. 1989.

64 A. V. Grebennikov and H. Jaeger, "Class E with parallel circuit – A new challenge for high efficiency RF and microwave power amplifiers," *2002 IEEE MTT-S Int. Microwave Sym. Dig.*, vol. 3, pp. 1627–1630, June 2002.

65 A. Grebennikov, "Switched-mode RF and microwave parallel-circuit Class E power amplifiers," *Int. J. RF Microw. Comput.-Aided Eng.*, vol. 14, pp. 21–35, Jan. 2004.

66 M. K. Kazimierczuk and W. A. Tabisz, "Class C-E high-efficiency tuned power amplifier," *IEEE Trans. Circuits Syst.*, vol. CAS-36, pp. 421–428, Mar. 1989.

67 J. Cumana, A. Grebennikov, G. Sun, N. Kumar, and R. H. Jansen, "An extended topology of parallel-circuit Class-E power amplifier to account for larger output capacitances," *IEEE Trans. Microw. Theory Tech.*, vol. MTT-59, pp. 3174–3183, Dec. 2011.

68 Y. Xu, X. Zhu, and C. You, "Analysis and design of Class-E power amplifier using equivalent LDMOS model with drift region effect," *Microw. Opt. Technol. Lett.*, vol. 52, pp. 1836–1842, Aug. 2010.

69 A. V. Grebennikov and H. Jaeger, High efficiency transmission line tuned power amplifier, U.S. Patent 6,552,610, April 2003.

70 H. Jaeger, A. V. Grebennikov, E. P. Heaney, and R. Weigel, "Broadband high-efficiency monolithic InGaP/GaAs HBT power amplifiers for 3G handset applications," *2002 IEEE MTT-S Int. Microwave Symp. Dig.*, vol. 2, pp. 1035–1038, June 2002.

71 H. Jaeger, A. Grebennikov, E. Heaney, and R. Weigel, "Broadband high-efficiency monolithic InGaP/GaAs HBT power amplifiers for wireless applications," *Int. J. RF Microw. Comput.-Aided Eng.*, vol. 13, pp. 496–510, Nov. 2003.

72 P. Song, M. A. Oakley, A. C. Ulusoy, M. Kaynak, B. Tillack, G. A. Sadowy, and J. D. Cressler, "A Class-E tuned W-band SiGe power amplifier with 40.4% power-added efficiency at 93 GHz," *IEEE Microw. Wirel. Compon. Lett.*, vol. 25, pp. 663–665, Oct. 2015.

73 J. Modzelewski, "Power gain of Class E and Class B VHF tuned power amplifiers," *Proc. 15th Int. Conf. Microwaves Radar Wireless Commun*, vol. 1, pp. 41–44, 2004.

74 A. Grebennikov, "Load network design techniques for Class E RF and microwave amplifiers," *High Freq. Electron.*, vol. 3, pp. 18–32, July 2004.

75 M. Acar, A. J. Annema, and B. Nauta, "Analytical design equations for Class-E power amplifiers," *IEEE Trans. Circuits Syst.-I: Regul. Pap.*, vol. CAS-I-54, pp. 2706–2717, Dec. 2007.

76 A. Grebennikov, "High-efficiency Class-E power amplifier with shunt capacitance and shunt filter," *IEEE Trans. Circuits Syst.-I: Regul. Pap.*, vol. CAS-I-63, pp. 12–22, Jan. 2016.

77 M. S. Mogisho, D. G. Makarov, Y. V. Rassokhina, V. G. Krizhanovski, A. Grebennikov, and M. Thian, "Generalized Class-E power amplifier with shunt capacitance and shunt filter," *IEEE Trans. Microw. Theory Tech.*, vol. MTT-67, pp. 3464–3474, Aug. 2019.

78 A. Grebennikov, "High-efficiency Class-B, -E, -F power amplifiers: Magic of parallel circuits," *Microw. J.*, vol. 45, pp. 20–34, Dec. 2002.

79 F. H. Raab, "Maximum efficiency and output of Class-F power amplifiers," *IEEE Trans. Microw. Theory Tech.*, vol. MTT-49, pp. 1162–1166, June 2001.

80 M. Kazimierczuk, "Exact analysis of Class E tuned power amplifier with only one inductor and one capacitor in load network," *IEEE Trans. Solid-State Circuits*, vol. SC-18, pp. 214–221, Apr. 1983.

81 A. Heiskanen and T. Rahkonen, "Comparison of two Class-E amplifiers for EER transmitter," *2005 IEEE Int. Circuits and Systems Symp. Dig.*, pp. 704–707, May 2005.

82 J. K. A. Everard and A. J. King, "Broadband power efficient Class E amplifiers with a nonlinear CAD model of the active MOS device," *J. IERE*, vol. 57, pp. 52–58, Mar. 1987.

83 G. L. Matthaei, "A study of the optimum design of wide-band parametric amplifiers and up-converters," *IRE Trans. Microw. Theory Tech.*, vol. MTT-9, pp. 23–28, Jan. 1961.

84 J. T. DeJaeger, "Maximum bandwidth performance of a nondegenerate parametric amplifier with single-tuned idler circuit," *IEEE Trans. Microw. Theory Tech.*, vol. MTT-12, pp. 459–467, July 1964.

85 B. L. Humphreys, "Characteristics of broadband parametric amplifiers using filter networks," *Proc. IEE*, vol. 111, pp. 264–274, Feb. 1964.

86 C. S. Aitchison and R. V. Gelsthorpe, "A circuit technique for broadbanding the electronic tuning range of Gunn oscillators," *IEEE J. Solid-State Circuits*, vol. SC-12, pp. 21–28, Feb. 1977.

87 R. V. Gelsthorpe and C. S. Aitchison, "Analytical evaluation of the components necessary for double reactance compensation of an oscillator," *Electron. Lett.*, vol. 12, pp. 485–486, Sep. 1976.

88 A. G. Chapman and C. S. Aitchison, "Circuit technique for broadband impedance matching of passive loads," *IEE J. Microw. Opt. Acoust.*, vol. 3, pp. 43–50, Mar. 1979.

89 H. Lu, J. Jiang, and H. Jang, "Optimizing power losses and efficiency of broadband parallel-circuit Class-E power amplifier," *IEEE Access*, vol. 12, pp. 21936–21946, Dec. 2024.

90 Y. F. Wu, R. A. York, S. Keller, B. P. Keller, and U. K. Mishra, "3-9-GHz GaN-based microwave power amplifiers with L-C-R broad-band matching," *IEEE Microw. Guided Wave Lett.*, vol. 9, pp. 314–316, Aug. 1999.

91 K. Krishnamurthy, D. Green, R. Vetury, M. Poulton, and J. Martin, "0.5-2.5, 10 W MMIC power amplifier in GaN HEMT technology," *2009 IEEE Compound Semiconductor Integrated Circuits Symp. Dig.*, pp. 1–4, Oct. 2009.

92 V. P. Meschanov, I. A. Rasukova, and V. D. Tupikin, "Stepped transformers on TEM-transmission lines," *IEEE Trans. Microw. Theory Tech.*, vol. MTT-44, pp. 793–798, June 1996.

93 A. Grebennikov, "Multiband Doherty amplifiers for wireless applications," *High Freq. Electron.*, vol. 13, pp. 30–46, May 2014.

4
High-efficiency Oscillator Design

Generally, RF and microwave transistor oscillator design is not an easy task and represents a complex problem. Depending on the technical requirements, it is necessary to define the configuration of the oscillator circuit, choose a proper transistor type, evaluate and measure the parameters of the transistor nonlinear model under small- and large-signal conditions. To provide high oscillator efficiency, the corresponding circuit techniques, in particular Class-F or Class-E mode, which can simplify the circuit implementation, can be used. Finally, an appropriate nonlinear simulator must be used to simulate the oscillator performance in the time and frequency domains. An oscillator analysis can be based on using the two-port network approach to describe the active device and the feedback circuit. In this case, the basic parameters of the transistor equivalent circuit can be directly measured, or approximated based on experimental data, with sufficient accuracy across a wide frequency range. However, the values of the external feedback circuit elements are initially unknown. The process of determining the optimum values of the feedback and load parameters can be time-consuming, requiring significant simulation time. Consequently, it is convenient to use an analytic method of optimizing the oscillator design based on the matrix technique.

Use of an analytic method involves a two-step procedure. First, the optimum combination of feedback elements required to achieve a small-signal maximum negative resistance to permit oscillations at the largest amplitude is defined. Second, considering the nonlinearities of the transistor equivalent circuit elements, the maximum negative resistance condition under large-signal operation is characterized to determine the optimum load impedance when maximum output power is generated for a given oscillator circuit configuration.

At lower RF frequencies when the transformer-coupled oscillator structure can be easily implemented, the oscillator design can be based on a Class-E power amplifier to maximize its efficiency by providing just a small transformer coupling from the oscillator output to its input when the power gain of the Class-E power amplifier is sufficiently high. In this case, the load network and feedback circuit are separate, thus making the design quite straightforward.

4.1 Oscillator Basic Structures

The first oscillator configurations using vacuum tubes were based on the electromagnetic coupling between the output and input circuits. By providing the proper coupling with the output, sufficient energy is supplied to the input circuit to keep the continuous self-sustaining oscillations. In this case, the oscillation frequency is approximately equal to the resonant frequency of the parallel LC circuit if the coupling and active device parasitics are small enough. Generally, there are two basic configurations of a transformer-coupled oscillator. The oscillator configuration with a parallel resonant circuit at the input electromagnetically coupled with the output, as shown in Figure 4.1a, is called the *Armstrong oscillator* in honor of E.H. Armstrong who first developed and described the condition of obtaining the self-sustaining oscillations using such a configuration [1]. Approximately at the same time, A. Meissner in Germany described the transformer-coupled oscillator with a parallel resonant circuit at the output electromagnetically coupled with the input, as shown in Figure 4.1b, which is called the *Meissner oscillator* [2]. High-purity stable oscillations can be obtained by a fully balanced version of the Meissner oscillator with a three-turn link providing an excellent isolation of the resonant circuit when very little energy is taken from the resonator to provide the voltages to drive the input and sustain the oscillations [3].

To realize a phase balance condition in the transformer-coupled oscillator, the transformer should provide a phase shift of 180°. If the primary and secondary windings of the transformer have the same direction of winding, it is necessary to connect the secondary winding in the opposite direction relative to the primary one, that is, to connect the end of the secondary winding where the

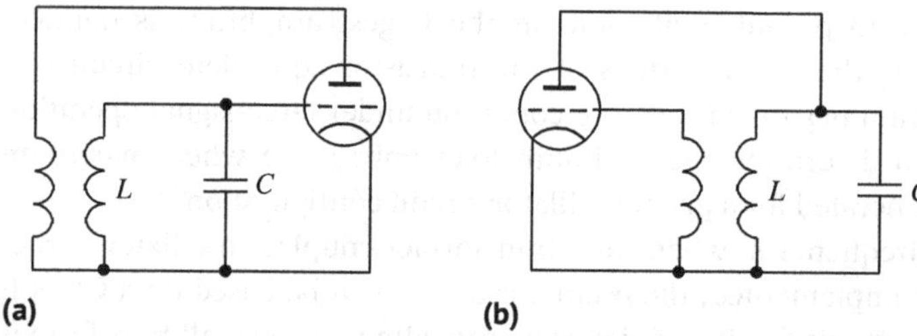

Figure 4.1 Schematics of (a) Armstrong and (b) Meissner oscillators.

voltage is in-phase with the anode (drain or collector) voltage to the ground. The coupling coefficient of the transformer is chosen to provide a soft start-up oscillation mode.

The schematic of the *Hartley oscillator* is very close to the schematic of the Meissner oscillator [4]. The difference is that the inductor L having an additional output replaces the transformer, as shown in Figure 4.2a, and the inductance ratio determines the feedback coefficient. The Hartley oscillator can be represented by a well-known inductive three-point configuration shown in Figure 4.2b, where the feedback element from the output to the input is the capacitor C. The inductors L_1 and L_2 represent the output and input circuits, respectively.

Unlike the Hartley oscillator with electromagnetic coupling, the schematic of the *Colpitts oscillator* is based on electrostatic coupling using the capacitive divider, the capacitance ratio of which determines the feedback coefficient, as shown in Figure 4.3a [5]. The Colpitts oscillator can be represented by a well-known capacitive three-point configuration shown in Figure 4.3b, where the feedback element from the output to the input is the inductor L. At the same time, the capacitors C_1 and C_2 represent the output and input circuits, respectively.

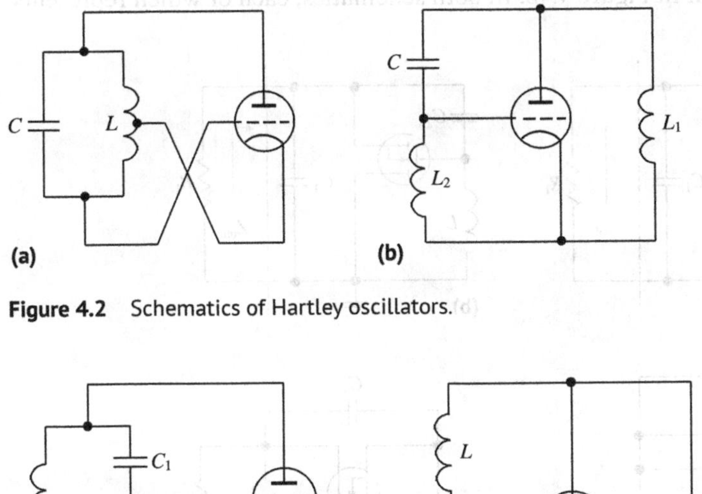

Figure 4.2 Schematics of Hartley oscillators.

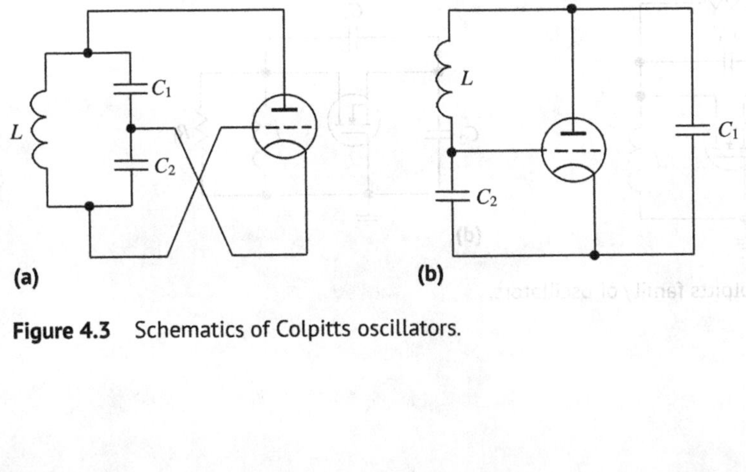

Figure 4.3 Schematics of Colpitts oscillators.

Depending on which electrode of the active device is grounded, the three basic configurations representing a Colpitts family of oscillators can be distinguished, as shown in Figure 4.4 with the MOSFET devices as an example. The Colpitts oscillator with grounded source is shown in Figure 4.4a, the Colpitts oscillator with grounded drain is shown in Figure 4.4b, and the Colpitts oscillators with grounded gates are shown in Figure 4.4c,d. The connection of the appropriate electrode of the tube (or transistor) to the ground has important effects upon the following:

- the manner in which the DC power is fed to the oscillator and the corresponding loading effects on the resonant circuit;
- the manner in which the output power is fed to the external load;
- the distribution of the stray elements to the ground plane, which has important effects, particularly for microwaves.

To understand whether the electrical performance of the same oscillator can change depending on the grounded electrode or terminal of the active device, the two basic schematics of the Colpitts oscillator can be considered, with a grounded source (emitter) shown in Figure 4.4a and a grounded drain (collector) shown in Figure 4.4b. In both schematics, each of which represents

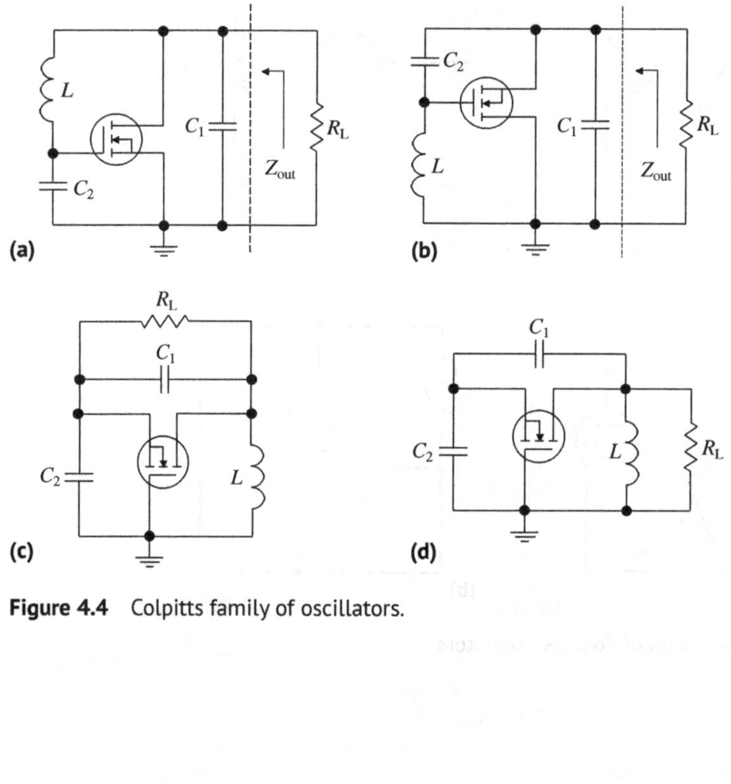

Figure 4.4 Colpitts family of oscillators.

the three-port network in a common case, the load resistance R_L is connected in parallel to the capacitance R_L, that is, between the drain and source terminals, respectively.

The indefinite admittance Y-matrix of a three-port network can be written as

$$[Y] = \begin{bmatrix} Y_{11} & Y_{12} & -(Y_{11}+Y_{12}) \\ Y_{21} & Y_{22} & -(Y_{21}+Y_{22}) \\ -(Y_{11}+Y_{21}) & -(Y_{12}+Y_{22}) & Y_{11}+Y_{12}+Y_{21}+Y_{22} \end{bmatrix} \tag{4.1}$$

By choosing successively terminal 1, 2, and 3 as the datum terminal of the active device shown in Figure 4.5, the appropriate three definite two-port admittance matrices of the initial three-port network can be obtained. Consider finding these two-port matrices in the form of an admittance matrix of (a) the common source, (b) the common gate, and (c) the common drain circuit connection of the MOSFET transistor, as shown in Figure 4.5, where the term "common terminal" for the grounded terminal is used similarly to that of using at the power amplifier design. If the common source device is treated as a two-port circuit and characterized by four Y-parameters (Y_{11}, Y_{12}, Y_{21}, and Y_{22}), the two-port matrix of the common drain circuit with grounded-drain terminal is simply obtained by deleting the second row and the second column in the matrix given by Eq. (4.1). For the common gate circuit with grounded-gate terminal, the first row and the first column should be deleted because the source terminal is considered the input terminal.

A similar approach can be applied to the indefinite three-port impedance network. This allows the impedance Z-parameters of the common gate (base) and common drain (collector) circuits to be determined through the initially known impedance Z-parameters of the common source (emitter) circuit. Parameters of the three-port network with different common terminals for bipolar and FET devices are given in Table 4.1.

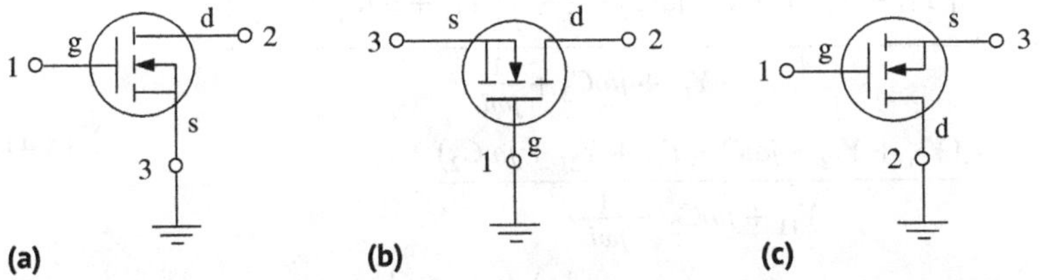

Figure 4.5 MOSFET devices with different common terminals.

Table 4.1 Active device *Y*- and *Z*-parameters with different common terminals.

	***Y*-parameters**		***Z*-parameters**	
Common source (emitter)	Y_{11}	Y_{12}	Z_{11}	Z_{12}
	Y_{21}	Y_{22}	Z_{21}	Z_{22}
Common gate (base)	$Y_{11}+Y_{12}+Y_{21}+Y_{22}$	$-(Y_{12}+Y_{22})$	$Z_{11}+Z_{12}+Z_{21}+Z_{22}$	$-(Z_{12}+Z_{22})$
	$-(Y_{21}+Y_{22})$	Y_{22}	$-(Z_{21}+Z_{22})$	Z_{22}
Common drain (collector)	Y_{11}	$-(Y_{11}+Y_{12})$	Z_{11}	$-(Z_{11}+Z_{12})$
	$-(Y_{11}+Y_{21})$	$Y_{11}+Y_{12}+Y_{21}+Y_{22}$	$-(Z_{11}+Z_{21})$	$Z_{11}+Z_{12}+Z_{21}+Z_{22}$

For the schematic with a common source shown in Figure 4.4a,

$$Y_{out} = Y_{22} + j\omega C_1 + \frac{1}{j\omega L} - \frac{\left(Y_{12} - \frac{1}{j\omega L}\right)\left(Y_{21} - \frac{1}{j\omega L}\right)}{Y_{11} + j\omega C_2 + \frac{1}{j\omega L}} \tag{4.2}$$

where $Y_{out} = 1/Z_{out}$ is the equivalent output admittance.

For the oscillator schematic with common drain shown in Figure 4.4b, by using the relationships from Table 4.1,

$$Y_{out} = Y_{11} + Y_{12} + Y_{21} + Y_{22} + j\omega C_1 + j\omega C_2 - \frac{(-Y_{11} - Y_{12} - j\omega C_2)(-Y_{11} - Y_{21} - j\omega C_2)}{Y_{11} + j\omega C_2 + \frac{1}{j\omega L}} \tag{4.3}$$

Equation (4.3) can be rewritten as

$$\begin{aligned} Y_{out} &= Y_{22} + j\omega C_1 + \frac{1_1}{j\omega L} \\ &\quad + \frac{\left(Y_{11} + Y_{12} + Y_{21} + j\omega C_2 - \frac{1}{j\omega L}\right)\left(Y_{11} + j\omega C_2 + \frac{1}{j\omega L}\right)}{Y_{11} + j\omega C_2 + \frac{1}{j\omega L}} \\ &\quad - \frac{(Y_{11} + Y_{12} + j\omega C_2)(Y_{11} + Y_{21} + j\omega C_2)}{Y_{11} + j\omega C_2 + \frac{1}{j\omega L}} \\ &= Y_{22} + j\omega C_1 + \frac{1}{j\omega L} - \frac{\left(Y_{12} - \frac{1}{j\omega L}\right)\left(Y_{21} - \frac{1}{j\omega L}\right)}{Y_{11} + j\omega C_2 + \frac{1}{j\omega L}} \end{aligned} \tag{4.4}$$

which is identical to Eq. (4.2), corresponding to the output admittance Y_{out} of the oscillator schematic with a common source. Consequently, grounding of any terminal of the oscillator circuit does not change its electrical performance, provided there are no changes in the connection of the feedback elements and load to the active device.

The oscillator schematic with a common gate shown in Figure 4.4c is characterized by the same electrical behavior as the oscillator circuits with common source and common drain. However, in such a configuration, there is no connection of any load terminal to the ground, which makes its practical implementation more complicated. Figure 4.4d shows the oscillator schematic with a common gate with the load connection in parallel to the inductance L, that is, between the drain and gate terminals. However, another load connection with one grounded port results in different electrical properties of the oscillator compared with common source or common drain configurations.

The Gouriet–Clapp oscillator shown in Figure 4.6a is a variation of the Colpitts oscillator with a tank inductor replaced by the series combination of the inductor L and capacitor C. In this case, frequency stability is improved because the reactance of such a series circuit varies more rapidly with frequency than that of a single inductor. However, the possibility of improving the oscillator stability by connecting a capacitor in series to one or each circuit inductor was first found based on a Hartley type of the oscillator [6]. In modern practical oscillator design, the Gouriet–Clapp oscillator configuration is called the *Clapp oscillator*, although

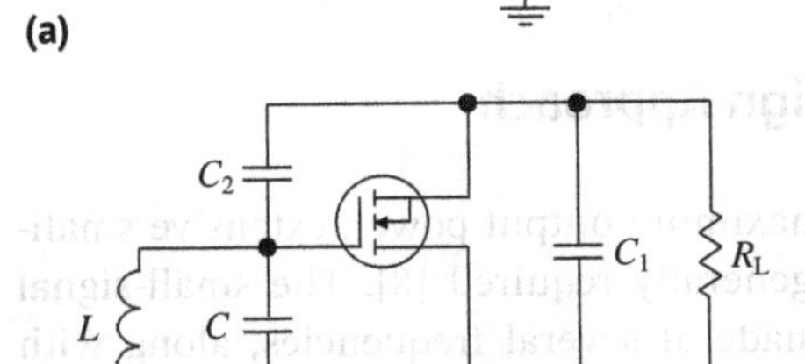

(a)

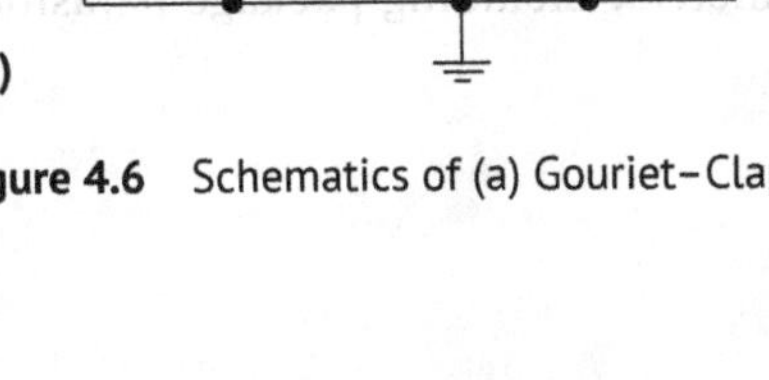

(b)

Figure 4.6 Schematics of (a) Gouriet–Clapp and (b) Seiler oscillators.

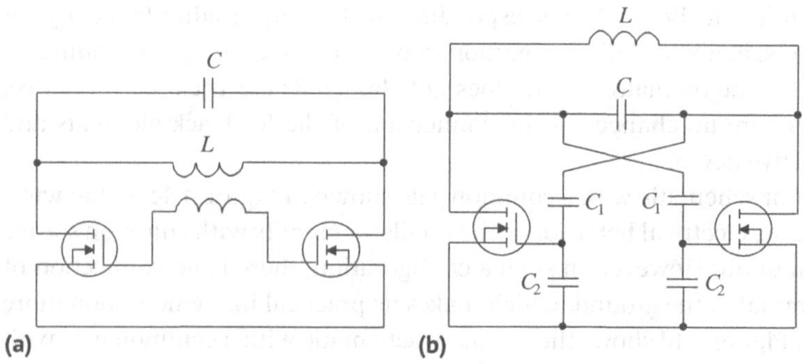

Figure 4.7 Schematics of push–pull oscillators.

G. Gouriet and J. Clapp had developed it independently [7]. A parallel counterpart of the Clapp oscillator shown in Figure 4.6b was described by E. Seiler and, therefore, is called the *Seiler oscillator* [7]. Such an oscillator configuration is useful for wideband frequency tuning when the capacitance C in the parallel circuit is variable.

The push–pull connection of the transistors in power amplifiers is usually used to increase the resulting output power, simplify the output matching with the load, and improve spectral performance by suppressing even-order harmonics. The same concept can also be applied to the oscillator design. The push–pull oscillator circuit shown in Figure 4.7a is based on the two single-ended Meissner oscillators where the transistors are turned on and off alternately. Since the voltage at the gate of the one transistor is in-phase with the voltage at the drain of the other transistor, there is no need to invert phase using the secondary winding. Another push–pull oscillator configuration based on the two single-ended Seiler oscillators is shown in Figure 4.7b, where the positive feedback is formed by the capacitive divider based on the capacitors C_1 and C_2.

4.2 Empirical Optimum Design Approach

For optimum oscillator design with maximum output power, extensive small- and large-signal measurements are generally required [8]. The small-signal S-parameter measurements can be made at several frequencies, along with estimated device equivalent circuit parameters, including package parasitics.

Then, a computer optimization is used to match the measured *S*-parameters to the *S*-parameters, computed for the active device from its equivalent circuit. The next step is to vary those elements of the equivalent circuit, which can vary under large signals. By varying the active device nonlinear equivalent circuit parameters, the set of large-signal *S*-parameters can be obtained corresponding to the saturation condition where maximum oscillator output power can be achieved.

The simplified design approach assumes that all of the *S*-parameters except the magnitude of S_{21} are constant under large signals. For example, for a MESFET device up to saturation with small limitations in accuracy, it is possible to derive the large-signal behavior of the main active device nonlinear elements as functions of the device transconductance g_m [8]. Consequently, the *S*-parameters become functions of g_m only and, at each incremental reduction of g_m, are recomputed and optimized along with power gain. If one were interested in the 1-dB compression point, the *S*-parameters used would be those at that point.

For a common source (emitter) power amplifier, it is helpful to characterize the maximum power through the saturated output power P_{sat} and small-signal transducer power gain G_T of the corresponding power amplifier. In this case, an empirical expression for the output power of the common source MESFET power amplifier can be written as

$$P_{out} = P_{sat}\left[1 - \exp\left(-\frac{G_T P_{in}}{P_{sat}}\right)\right] \tag{4.5}$$

where P_{in} is the input power [8].

If the required objective is to maximize the oscillator output power $P_{osc} = P_{out} - P_{in}$, the following condition should be provided:

$$\frac{dP_{osc}}{dP_{in}} = \frac{d(P_{out} - P_{in})}{dP_{in}} = 0 \tag{4.6}$$

Then, substituting Eq. (4.5) into Eq. (4.6) yields

$$\frac{dP_{out}}{dP_{in}} = G_T \exp\left(-\frac{G_T P_{in}}{P_{sat}}\right) = 1 \tag{4.7}$$

From Eq. (4.7), it follows that

$$\exp\left(\frac{G_T P_{in}}{P_{sat}}\right) = G_T \tag{4.8}$$

or

$$\frac{P_{in}}{P_{sat}} = \frac{\ln G_T}{G_T} \tag{4.9}$$

As a result, from Eqs. (4.5) and (4.8),

$$P_{out} = P_{sat}\left(1 - \frac{1}{G_T}\right) \tag{4.10}$$

and the maximum output power of the oscillator can be approximated using Eq. (4.9) by

$$P_{osc} = P_{sat}\left(1 - \frac{1}{G_T} - \frac{\ln G_T}{G_T}\right) \tag{4.11}$$

Consequently, the oscillator's maximum power can be predicted from the saturated power and small-signal transducer power gain of a common-source power amplifier. From Eq. (4.11), it follows that the oscillator output power P_{osc} will approach P_{sat} at low frequencies where the small-signal transducer power gain G_T is large. As the transducer power gain approaches unity, the oscillator output power approaches zero.

In this case, the maximum efficiency power gain G_{ME} can be calculated from

$$G_{ME} = \frac{G_T \;\; - 1}{\ln G_T} \tag{4.12}$$

For example, a MESFET device, which is characterized by the small-signal transducer power gain $G_T = 7.5$ dB and saturated amplifier output power of 1 W, would be capable of a maximum oscillator power of 515 mW at 10 GHz. The maximum efficient gain at this point calculated from Eq. (4.12) is equal to $G_{ME} = 4.3$ dB. The expression for maximum efficient gain given by Eq. (4.12) can be used to determine at what gain level the large-signal *S*-parameters are to be used for optimum oscillator design [8].

Generally, there are a multitude of possible oscillator configurations with feedback inductive or capacitive elements depending on the load connection points. Having computed the large-signal *S*-parameters following the procedure outlined above, the embedding elements for six oscillator topologies with three parallel feedback configurations are shown in Figure 4.8a–c and three series feedback configurations shown in Figure 4.8d–f may be computed [9]. To represent the design equations for the feedback elements and load in terms of admittance

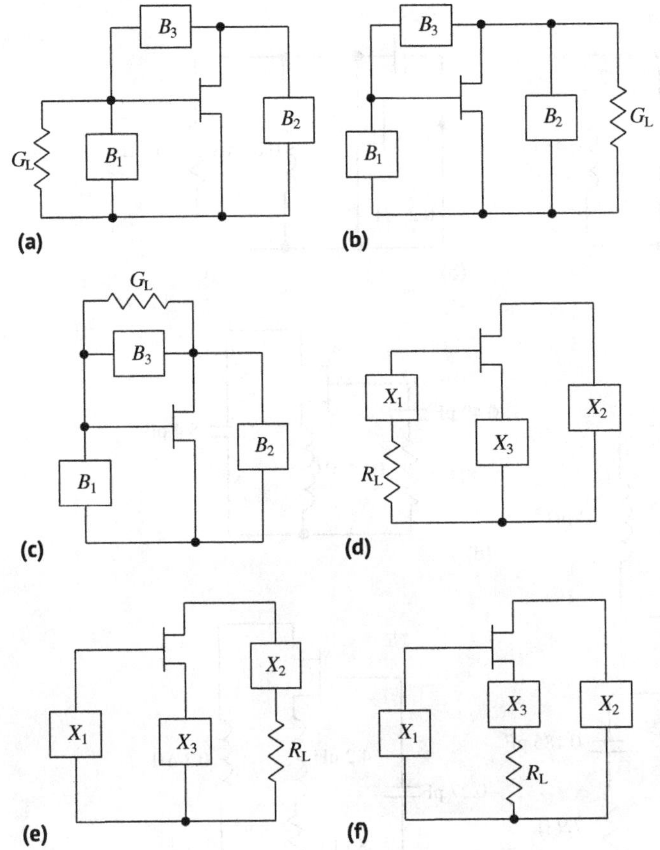

Figure 4.8 Oscillator topologies with parallel and series feedback.

Y-parameters or impedance *Z*-parameters, it is necessary to transform the *S*-parameters by the usual *S*- to *Y*- and *Z*-matrix conversion formulas.

Based on an MSX801G device, the six MESFET oscillator circuits embedding element values are shown in Figure 4.9 [8]. It should be noted that not all of these circuit configurations can be readily realized in practice. In particular, all configurations with parallel feedback when it is not so easy to physically allocate a capacitor between the gate and drain terminals. The circuit configurations with series feedback are more convenient for practical implementation.

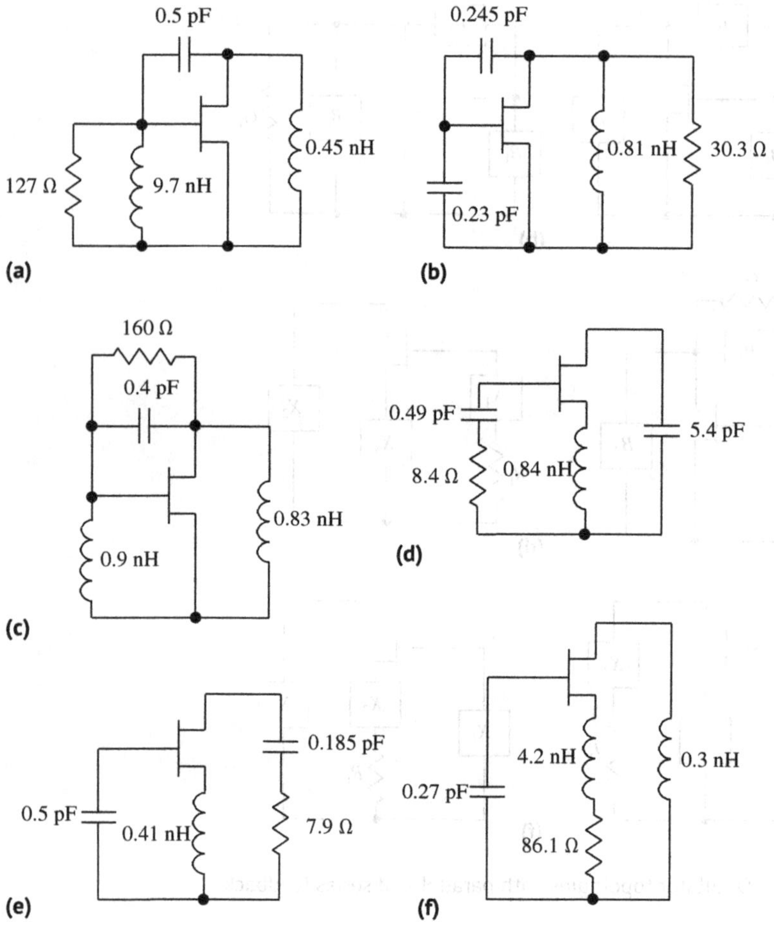

Figure 4.9 MESFET oscillator circuit schematics with parallel and series feedback.

4.3 Analytic Optimum Design Approach

The oscillator analysis using matrix techniques brings out the similarity between several types of oscillators and results in one group of equations, which can be used to analyze different oscillator configurations [10]. In this case, a two-port network can represent both the active device and the feedback element. Depending on the oscillator circuit configuration with parallel or series feedback, the corresponding *Y*- or *Z*-parameters can be used for analysis and optimization.

4.3.1 Parallel and Series Feedback Oscillators

Depending on the type of the transistor used, frequency band, required output power level, and spectral characteristics, most of the oscillator schematics can be reduced to two basic arrangements with external positive (a) parallel or (b) series feedback as shown in Figure 4.10. One model may be preferred over another, depending on the oscillator configuration and characteristics.

A generalized equivalent circuit of the parallel feedback oscillator is shown in Figure 4.10a, where Y_1, Y_2, and Y_3 are the feedback elements. To calculate the steady-state stationary operation mode, it is convenient to add the matrix of the parallel feedback element Y_2 and the matrix of the active device Y-parameters according to

$$[Y] + [Y_2] = \begin{bmatrix} Y_{11} + Y_2 & Y_{12} - Y_2 \\ Y_{21} - Y_2 & Y_{22} + Y_2 \end{bmatrix} \tag{4.13}$$

In this case, a system of two equations for the input current I_{in}, output current I_{out}, input voltage V_{in}, and output voltage V_{out}, characterizing the two-port network, can be written as

$$I_{in} = (Y_{11} + Y_2)V_{in} + (Y_{12} - Y_2)V_{out} \tag{4.14}$$

$$I_{out} = (Y_{21} + Y_2)V_{in} + (Y_{22} - Y_2)V_{out} \tag{4.15}$$

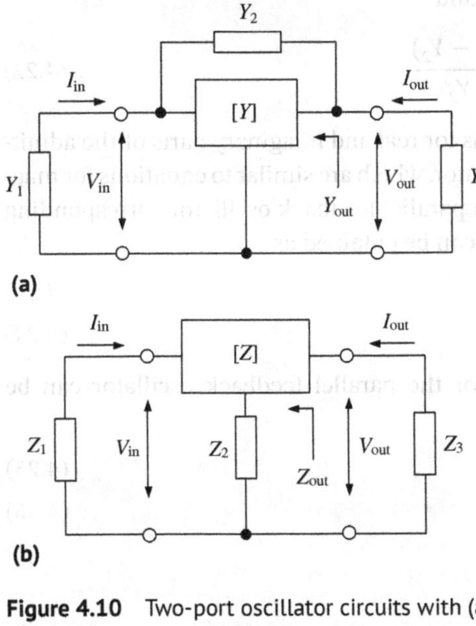

Figure 4.10 Two-port oscillator circuits with (a) parallel and (b) series feedback.

Since, for the parallel feedback oscillator shown in Figure 4.10a, the boundary conditions are obtained according to

$$I_{in} = -Y_1 V_{in} \tag{4.16}$$

$$I_{out} = -Y_3 V_{out} \tag{4.17}$$

the corresponding matrix equation can be written as

$$\begin{bmatrix} Y_{11} + Y_1 + Y_2 & Y_{12} - Y_2 \\ Y_{21} - Y_2 & Y_{22} + Y_2 + Y_3 \end{bmatrix} \begin{bmatrix} V_{in} \\ V_{out} \end{bmatrix} = 0 \tag{4.18}$$

Thus, the oscillator steady-state stationary condition can be derived by equating the determinant to zero,

$$\begin{vmatrix} Y_{11} + Y_1 + Y_2 & Y_{12} - Y_2 \\ Y_{21} - Y_2 & Y_{22} + Y_2 + Y_3 \end{vmatrix} = 0 \tag{4.19}$$

After some simplification, Eq. (4.19) can be rewritten as

$$Y_{22} + Y_2 + Y_3 - \frac{(Y_{12} - Y_2)(Y_{21} - Y_2)}{Y_{11} + Y_1 + Y_2} = 0 \tag{4.20}$$

As a result, the steady-state oscillation condition for the parallel feedback oscillator represented as a one-port negative conductance oscillator can be written as

$$Y_{out} + Y_L = 0 \tag{4.21}$$

where $Y_3 = Y_L$ is the load admittance, and

$$Y_{out} = Y_{22} + Y_2 - \frac{(Y_{12} - Y_2)(Y_{21} - Y_2)}{Y_{11} + Y_1 + Y_2} \tag{4.22}$$

Consequently, the separate equations for real and imaginary parts of the admittances of a negative conductance oscillator, which are similar to equations for magnitude and phase of the loop gain of a parallel feedback oscillator corresponding to the steady-state oscillation process, can be obtained as

$$\mathrm{Re}Y_{out} + \mathrm{Re}Y_L = 0 \tag{4.23}$$

$$\mathrm{Im}Y_{out} + \mathrm{Im}Y_L = 0 \tag{4.24}$$

Similarly, the start-up conditions for the parallel feedback oscillator can be written as

$$\mathrm{Re}Y_{out} + \mathrm{Re}Y_L < 0 \tag{4.25}$$

$$\mathrm{Im}Y_{out} + \mathrm{Im}Y_L = 0 \tag{4.26}$$

Similarly, the steady-state oscillation condition for the series feedback oscillator of Figure 4.10b represented as a one-port negative resistance oscillator can be written by

$$Z_{\text{out}} + Z_{\text{L}} = 0 \tag{4.27}$$

where $Z_3 = Z_L$ and

$$Z_{\text{out}} = Z_{22} + Z_2 - \frac{(Z_{12} + Z_2)(Z_{21} + Z_2)}{Z_{11} + Z_1 + Z_2} \tag{4.28}$$

4.3.2 Optimum Oscillation Conditions

As a next step in the optimum design approach, it is necessary to obtain optimum feedback and load impedance parameters through the device admittance Y-parameters in the case of a parallel feedback oscillator and impedance Z-parameters in the case of a series feedback oscillator. For the parallel feedback oscillator shown in Figure 4.10a, by analyzing Eq. (4.22) at the extremum, it is possible to obtain the optimum values of feedback susceptances $B_1 = \text{Im}(Y_1)$ and $B_2 = \text{Im}(Y_2)$, at which the negative value $\text{Re}Y_{\text{out}}$ is maximal, by solving the following two equations [11, 12]:

$$\frac{\partial \text{Re}Y_{\text{out}}}{\partial B_1} = 0 \qquad \frac{\partial \text{Re}Y_{\text{out}}}{\partial B_2} = 0 \tag{4.29}$$

As a result, for $Y_1 = jB_1$ and $Y_2 = jB_2$, the optimum values of feedback susceptances B_1^{o} and B_2^{o} can be expressed through the two-port active device Y-parameters as

$$B_1^{\text{o}} = \frac{G_{21} - G_{12}}{B_{21} - B_{12}}\left(\frac{G_{12} + G_{21}}{2} + G_{11}\right) - B_{11} - \frac{B_{12} + B_{21}}{2} \tag{4.30}$$

$$B_2^{\text{o}} = \frac{(G_{21} + G_{12})(G_{21} - G_{12})}{2(B_{21} - B_{12})} + \frac{B_{12} + B_{21}}{2} \tag{4.31}$$

In this case, the optimum values of the real and imaginary parts of the output admittance $Y_{\text{out}}^{\text{o}} = G_{\text{out}}^{\text{o}} + jB_{\text{out}}^{\text{o}}$ are written as

$$G_{\text{out}}^{\text{o}} = G_{22} - \frac{(G_{21} + G_{12})^2 + (B_{21} - B_{12})^2}{4G_{11}} \tag{4.32}$$

$$B_{\text{out}}^{\text{o}} = B_{22} + \frac{G_{21} - G_{12}}{B_{21} - B_{12}}\left(\frac{G_{21} + G_{12}}{2} + G_{22} - G_{\text{out}}^{\text{o}}\right) + \frac{B_{21} + B_{12}}{2} \tag{4.33}$$

Thus, in the steady-state stationary oscillation mode with admittance $Y_3 = G_L + jB_3$, where $G_L = 1/R_L$ and R_L is the load resistance, Eq. (4.21) can be rewritten in the form of the following amplitude and phase balance conditions:

$$G^{o}_{out} + G_L = 0 \tag{4.34}$$

$$B^{o}_{out} + B^{o}_3 = 0 \tag{4.35}$$

The output power can be calculated using Eq. (4.34) according to

$$P_{out} = \frac{1}{2} V_L^2 G_L \tag{4.36}$$

where $V_L = V_{out}$ is the voltage amplitude across the load, which should be considered as the voltage amplitude across the output and common terminals of the active device.

4.3.3 Optimum MOSFET Oscillator

The electrical behavior of the MOSFET device in the high frequency ranges can be described with sufficient accuracy by the intrinsic device equivalent circuit shown in Figure 4.11. According to this equivalent circuit, the transistor Y-parameters in a common source small-signal operation are written as

$$\begin{aligned} Y_{11} &= \frac{j\omega C_{gs}}{1 + j\omega C_{gs} R_{gs}} + j\omega C_{gd} \qquad & Y_{12} &= -j\omega C_{gd} \\ Y_{21} &= \frac{g_m}{1 + j\omega C_{gs} R_{gs}} - j\omega C_{gd} \qquad & Y_{22} &= \frac{1}{R_{ds}} + j\omega \left(C_{ds} + C_{gd} \right) \end{aligned} \tag{4.37}$$

By substituting the expressions for real and imaginary parts of Y-parameters from Eq. (4.37) into Eqs. (4.30) and (4.31), the optimum values of the feedback susceptances B^{o}_1 and B^{o}_2 expressed through the parameters of the MOSFET equivalent circuit can be calculated as

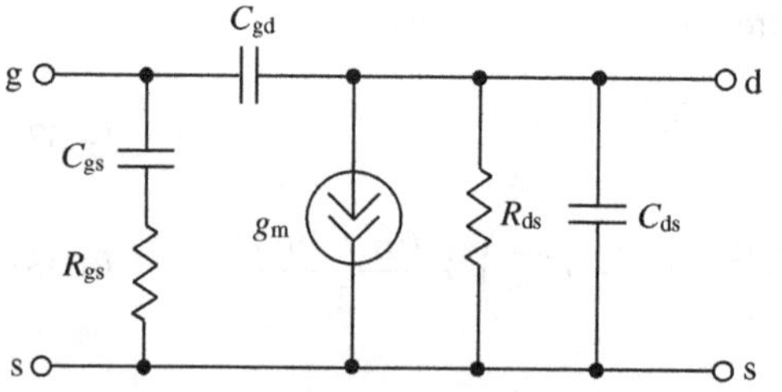

Figure 4.11 Small-signal intrinsic MOSFET model.

$$B_1^o = \frac{g_m}{2\omega C_{gs} R_{gs}} \tag{4.38}$$

$$B_2^o = -\omega C_{gd} - \frac{g_m}{2\omega C_{gs} R_{gs}} \tag{4.39}$$

According to Eqs. (4.38) and (4.39), the optimum values of the susceptances B_1^o and B_2^o should be capacitive and inductive, respectively. Taking into account that the value of the feedback capacitance C_{gd} is sufficiently small and variations of the radian transition frequency $\omega_T = g_m/C_{gs}$ and gate–source resistance R_{gs} in a large-signal operation are not significant, Eqs. (4.38) and (4.39) give a close approximation for the feedback elements in a steady-state oscillation mode based on the small-signal parameters of the MOSFET device at the operating bias point.

The optimum values of the real and imaginary parts of the output admittance Y_{out}^o can be obtained as

$$G_{out}^o = \frac{1}{R_{ds}} - \frac{1}{R_{gs}} \left(\frac{g_m}{2\omega\, C_{gs}} \right)^2 \tag{4.40}$$

$$B_{out}^o = \omega C_{gd} - \frac{g_m}{2\omega\, C_{gs} R_{gs}} \left(1 - \frac{1}{\omega\, C_{gs} R_{gs}} \frac{g_m}{2\omega\, C_{gs}} \right) \tag{4.41}$$

From Eq. (4.40), it follows that the absolute value of the negative conductance G_{out}^o reduces as the frequency increases, and the negative conductance becomes zero at the maximum oscillation frequency f_{max}, resulting from Eq. (4.40) in

$$f_{max} = \frac{g_m}{4\pi C_{gs}} \sqrt{\frac{R_{ds}}{R_{gs}}} \tag{4.42}$$

that coincides with a well-known expression for maximum frequency, at which the unilateral power gain of the FET power amplifier becomes equal to unity. At the same time, from Eq. (4.41), it follows that the value of the optimum output susceptance B_{out}^o, which may be either positive or negative, depends mostly on the values of the device transconductance g_m and gate time constant $\tau_g = R_{gs}C_{gs}$ in an optimal oscillator.

It should be noted that despite the fact that such an optimum analytic approach to obtain maximum negative conductance is based on a small-signal intrinsic MOSFET model, by considering R_{ds} as a fundamentally averaged drain–source resistance R_{ds1} under large-signal operation, the optimum load resistance can be calculated to correspond to a maximum available output power that can be achieved under final computer optimization, similarly to the same approach which was applied to the design of an optimum negative resistance series feedback oscillator [12, 13].

As an example, consider a 200-MHz optimum MOSFET power oscillator with the following electrical characteristics and small-signal parameters of the device equivalent circuit: transconductance $g_m = 30\,\text{mA/V}$, gate resistance $R_{gs} = 70\,\Omega$, gate–source capacitance $C_{gs} = 8\,\text{pF}$, gate–drain capacitance $C_{gd} = 0.4\,\text{pF}$, drain–source $C_{ds} = 2\,\text{pF}$, maximum drain current $I_{dmax} = 0.3\,\text{A}$, and transition frequency $f_T = 1\,\text{GHz}$. According to Eqs. (4.38) and (4.39), the feedback elements of the optimum oscillator shown in Figure 4.12a must have the approximate values of $C_1^o = 17\,\text{pF}$ and $L_2^o = 36\,\text{nH}$. In this case, the value of B_{out}^o is positive and can be represented by the output capacitance $C_{out}^o = 26\,\text{pF}$. Consequently, to satisfy the phase balance condition, an inductive character of the load impedance must be provided. In the case of the parallel inductor, its inductance is determined by

$$L_3^o = \frac{1}{\omega B_3^o} = \frac{1}{\omega B_{out}^o} = \frac{1}{\omega^2 C_{out}^o} \tag{4.43}$$

For a common drain MOSFET oscillator with varactor wideband tuning, it is easy to achieve the wideband tuning of the oscillation frequency using a variable capacitor connected in parallel to the feedback inductor. The equivalent circuit of such an oscillator is shown in Figure 4.12b. By tuning the parallel resonant circuit capacitor in the capacitance range from 2 to 40 pF, it is possible to easily tune the oscillator in the frequency range from 50 to 200 MHz with the output power of 100–200 mW delivered to the load $R_L = 200\,\Omega$. Maximum flatness of

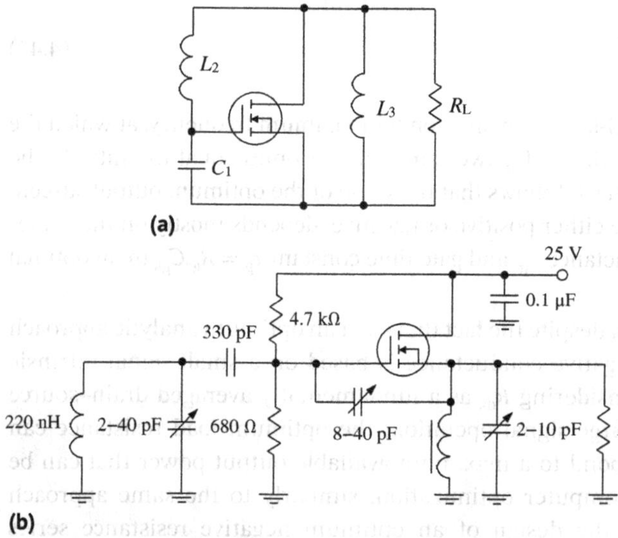

Figure 4.12 Schematics of MOSFET oscillators.

the amplitude characteristic can be achieved by choosing a proper value of the feedback capacitances. This can be accomplished by tuning the variable capacitors in the capacitance ranges of 2–10 and 8–40 pF, respectively.

The oscillator performance is invariant to the point of grounding, which is considered a common point, provided there are no changes in the connection of the feedback elements and the load. This occurs, for example, when the load for the oscillators with a common source configuration shown in Figure 4.13a or with a common drain configuration shown in Figure 4.13b is connected between the drain and the source, having in each case a direct ground connection. However, when the gate is grounded, the load has no connection with the ground. Therefore, for a common gate oscillator shown in Figure 4.13c, when the load is connected between the drain and the gate, the output power, frequency spectrum, or stability of the oscillator circuit can change because of the different effects of the intrinsic MOSFET device parameters on the amplitude and phase oscillation conditions. The attractiveness of using the common drain or common gate oscillator is attributable to the schematic simplicity, realizing wideband frequency tuning. The wide frequency tuning in a common source oscillator is limited by the negative influence of the source lead inductance and the inconvenience of the parallel feedback circuit implementation, especially at higher frequencies.

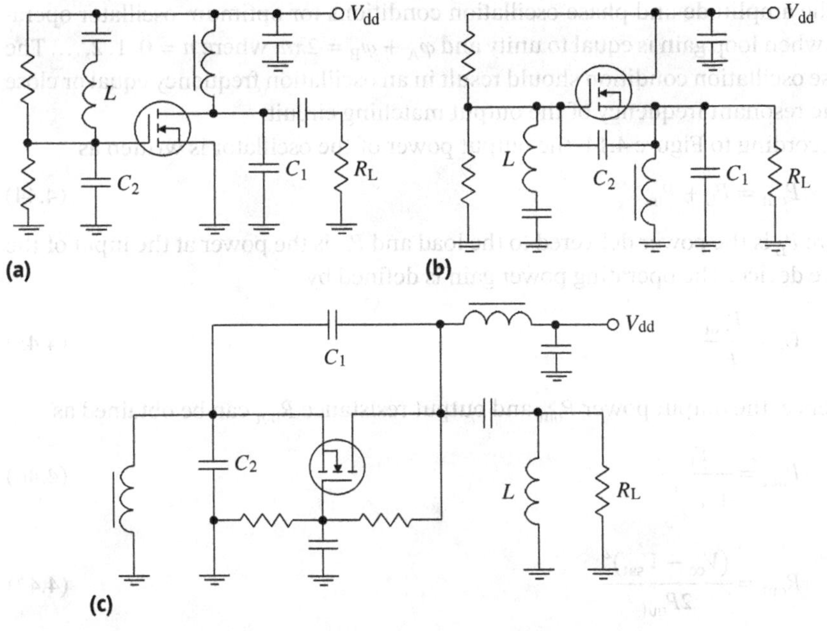

Figure 4.13 MOSFET oscillators with different grounded terminals.

4.4 High-efficiency Design Technique

4.4.1 Class-C Operation Mode

Achieving high efficiency in oscillators is very important in order to increase the battery lifetime, output power with minimum requirements for heat dissipation, and reliability of measurement equipment and radio transmitters. The popular design approach is based on the method of equivalent two-port networks, representing the active device and a parallel feedback circuit. The oscillator can be designed in a Class-C operation mode so that the phase oscillation condition is fulfilled at the resonant frequency of the output matching circuit [14]. Therefore, the output power and collector efficiency of the oscillator are close to their high values in the corresponding power amplifiers when high operating power gain is provided.

Figure 4.14a shows a block diagram of the power oscillator where the feedback network is connected between the output and input matching circuits providing the corresponding amplitude and phase conditions for steady-state oscillations. However, the feedback network elements can be simultaneously a part of the output and input matching networks, and both matching circuits can be combined. In this case, the oscillator block diagram can be simplified to the form shown in Figure 4.14b. The resulting feedback and matching three-port network should fulfill the amplitude and phase oscillation conditions for optimum oscillator operation when loop gain is equal to unity and $\varphi_A + \varphi_B = 2\pi n$, where $n = 0, 1, 2, \ldots$. The phase oscillation condition should result in an oscillation frequency equal or close to the resonant frequency of the output matching circuit.

According to Figure 4.14, the output power of the oscillator is written as

$$P_{out} = P_L + P_{in} \tag{4.44}$$

where P_L is the power delivered to the load and P_{in} is the power at the input of the active device. The operating power gain is defined by

$$G_p = \frac{P_{out}}{P_{in}} \tag{4.45}$$

Hence, the output power P_{out} and output resistance R_{out} can be obtained as

$$P_{out} = \frac{P_L}{1 + \frac{1}{G_p}} \tag{4.46}$$

$$R_{out} = \frac{(V_{cc} - V_{sat})^2}{2P_{out}} \tag{4.47}$$

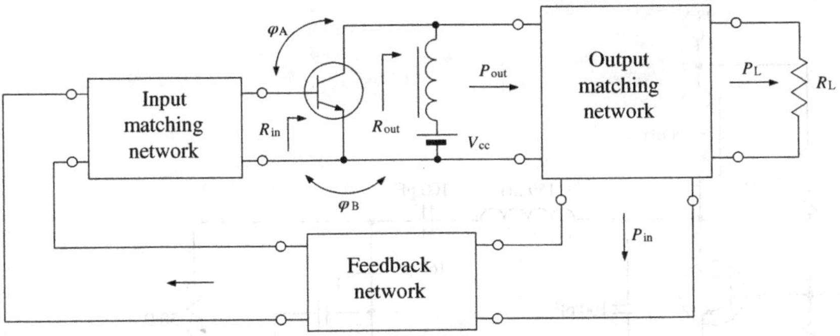

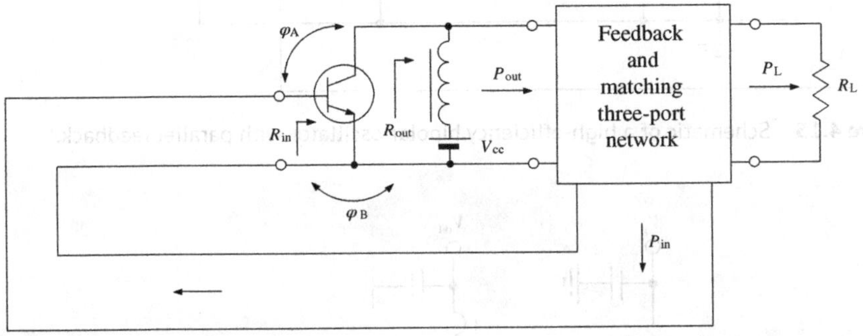

Figure 4.14 Block diagrams of bipolar oscillators with parallel feedback.

where V_{cc} is the DC-supply voltage and V_{sat} is the saturation voltage determined from the output voltage–current characteristics of the active device. It is assumed that the collector voltage is sinusoidal. To calculate the values of the oscillator circuit elements, the method of equivalent two-terminal networks connected in parallel and in series is used.

Figure 4.15 shows the circuit schematic of the Class-C bipolar power oscillator operating at a resonant frequency of 60 MHz and providing load power of 5 W [14]. The input of the bipolar transistor is connected to the output matching circuit representing a low-pass Π-type *LC* matching circuit through the series capacitance 1.5 nF and a capacitive divider formed by the capacitances of 160 pF and 1.3 nF to provide a small amount of the output power to flow to the input of the bipolar transistor.

At higher frequencies, it is convenient to use a directional coupler with proper coupling to provide the amplitude oscillation condition and a transmission line with the electrical length required to satisfy the phase balance condition.

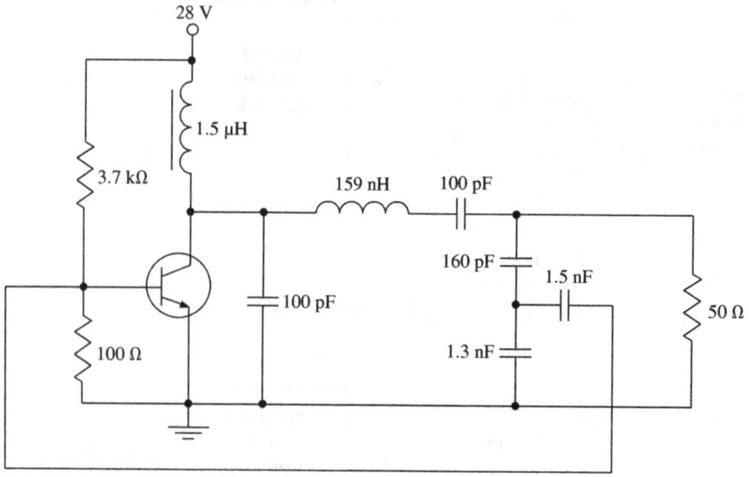

Figure 4.15 Schematic of a high-efficiency bipolar oscillator with parallel feedback.

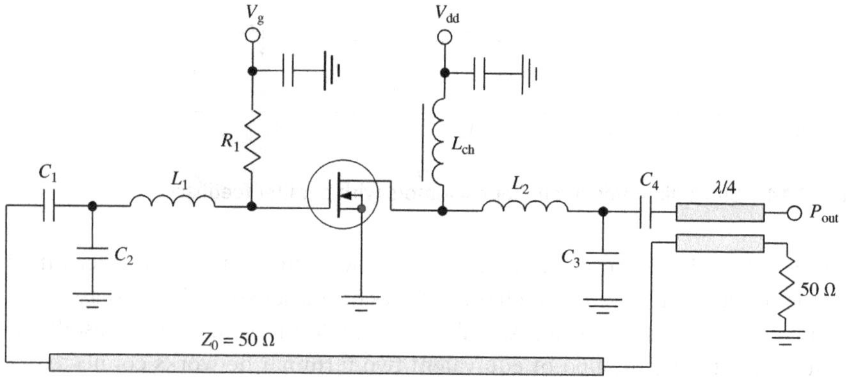

Figure 4.16 Schematic of a high-efficiency MOSFET oscillator with parallel feedback.

The circuit schematic of such a high-efficiency MOSFET power oscillator is shown in Figure 4.16, where the input and output matching circuits represent low-pass T-type lumped transformers, both matched to the source and load impedances of 50 Ω of the corresponding power amplifier. The directional coupler and feedback transmission line can easily be implemented using microstrip lines with a characteristic impedance of 50 Ω. Such a design method has an important advantage because the same transistor parameters and operating frequency, as well as the same parameters of the input and output matching circuits, are used to design both a high-efficiency power amplifier and an oscillator. An additional transistor parameter is its phase shift between the fundamental-frequency components of the drain–source and gate–source voltages, which should be compensated by a proper choice of the electrical length of the feedback microstrip line.

In order to enhance the negative resistance at higher frequencies, the oscillator circuit configuration can include both the parallel feedback circuit and the series feedback element. For example, each of two oscillators with a common source configuration shown in Figure 4.17 uses a dual feedback scheme consisting of a series feedback element from the source terminal to the ground and a parallel feedback circuit from the drain terminal to the gate terminal. The MESFET oscillator configuration shown in Figure 4.17a is composed of an input open-ended microstrip line connected to the gate terminal, a parallel feedback circuit with two series microstrip lines and a blocking capacitor, and a short-circuited microstrip line connected to the source terminal [15]. The output matching circuit consists of a series microstrip line connected between the drain terminal and the load.

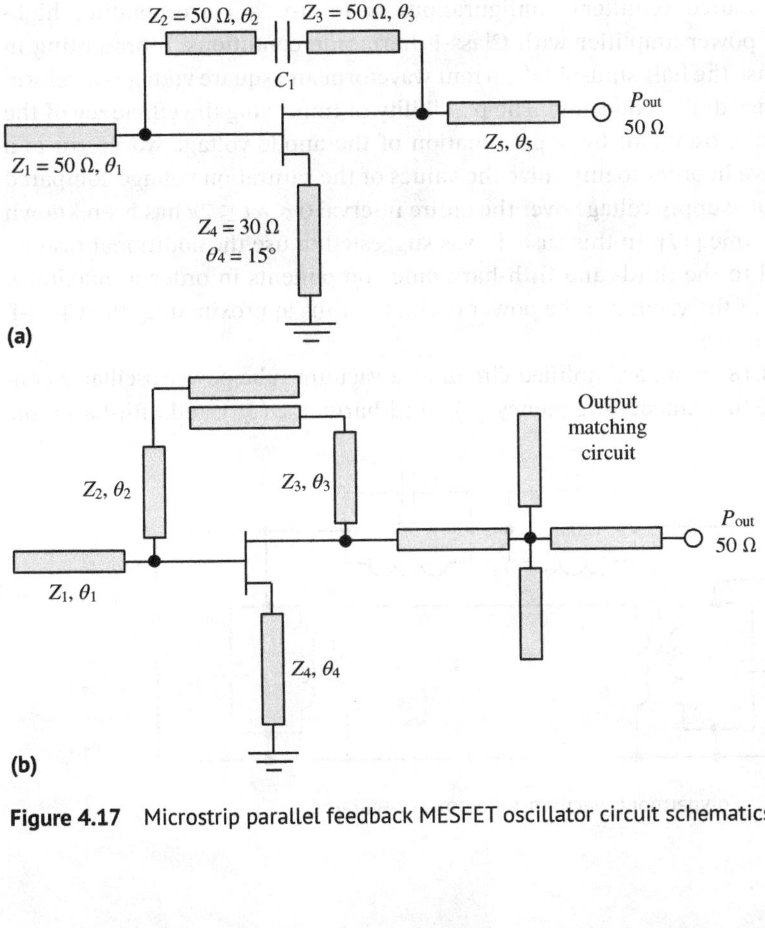

Figure 4.17 Microstrip parallel feedback MESFET oscillator circuit schematics.

With the device gate geometry of 1 μm × 1500 μm, the output power of 350 mW with 26% efficiency can be achieved at 6.5 GHz. The parallel feedback circuit can be designed as a combination of the phase-delay microstrip lines and edge-coupled microstrip lines shown in Figure 4.17b [16]. The edge-coupled lines are used for DC blocking instead of a metal–insulator–metal (MIM) capacitor having high losses at high frequencies above 100 GHz. The output matching circuit, consisting of the two series microstrip lines and two microstrip stubs, provides the optimum impedance conditions at the drain terminal. The open-ended microstrip line connected to the gate terminal determines the oscillation frequency as a main frequency-tuning element. This monolithic fundamental HEMT oscillator with the device geometry of 0.1 μm × 90 μm was designed to operate at 130.7 GHz with an output power of −7.9 dBm at the drain DC-supply voltage of 1.3 V.

4.4.2 Class-F and Inverse Class-F Oscillators

Another way to design a high-efficiency oscillator, based on the active device common source (emitter) configuration, is to use a corresponding high-efficiency power amplifier with Class-F harmonic conditions, representing in an ideal case the half-sinusoidal current waveform and square voltage waveform at the device drain (collector). The possibility of improving the efficiency of the vacuum-tube oscillator by approximation of the anode voltage waveform to a square wave in order to minimize the values of the saturation voltage compared with the DC-supply voltage over the entire interval $0 \leq \omega t \leq 2\pi$ has been known for a long time [17]. In this case, it was suggested to use the additional resonators tuned to the third- and fifth-harmonic components in order to maximize efficiency of the vacuum-tube power oscillator, thus approximating the Class-F operation mode.

Figure 4.18 shows a simplified circuit of a vacuum-tube power oscillator containing the fundamental-frequency (f_0), third-harmonic ($3f_0$), and fifth-harmonic

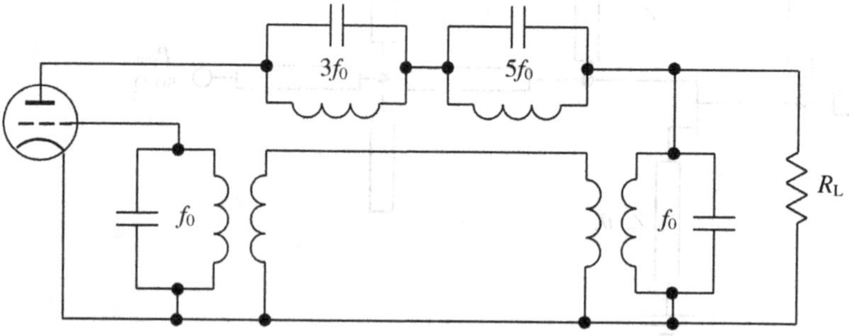

Figure 4.18 Polyharmonic vacuum-tube power oscillator.

($5f_0$) LC-resonators in the anode circuit to provide a close approximation to the rectangular anode voltage waveform. The anode fundamental-frequency resonator is inductively coupled with the tank grid circuit. As a result, the anode voltage represents a sum of the fundamental-frequency, third-harmonic, and fifth-harmonic voltages, while the only fundamental-frequency output signal flows to the load resistance R_L.

At the same time, it was experimentally found that the presence of the second-harmonic resonator in the anode circuit can significantly improve the oscillator efficiency by approximating inverse Class-F operation conditions with a half-sinusoidal voltage waveform and providing the biharmonic driving signal due to a much steeper driving waveform, resulting in a lower power loss in the active device [18]. In this case, the resultant driving waveform shown in Figure 4.19a consists of the fundamental-frequency and second-harmonic components being in phase at their maximum amplitudes, and the amplitude of the second harmonic

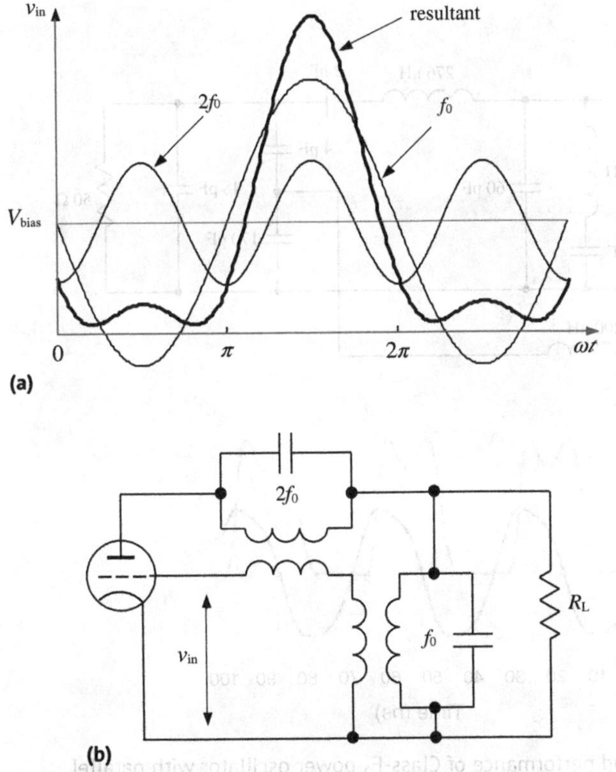

Figure 4.19 Oscillator with fundamental and second-harmonic resonators.

is preferably chosen to have approximately one-quarter the amplitude of the fundamental. Figure 4.19b shows a simplified circuit of a vacuum-tube power oscillator containing the fundamental-frequency (f_0) and second-harmonic ($2f_0$) *LC* resonators in the anode circuit inductively coupled with the grid circuit. As a result, the driving feedback voltage represents a sum of the fundamental-frequency and second-harmonic voltages, while the only fundamental-frequency output signal flows to the load resistance R_L.

In modern industrial applications, in particular using GaN-on-Silicon technology, a Class-F power oscillator with a second-harmonic series resonator (Class-F_2 oscillator) can be efficiently used as an inverter at very high frequencies with a high output power. Figure 4.20a shows the circuit schematic of a Class-F_2

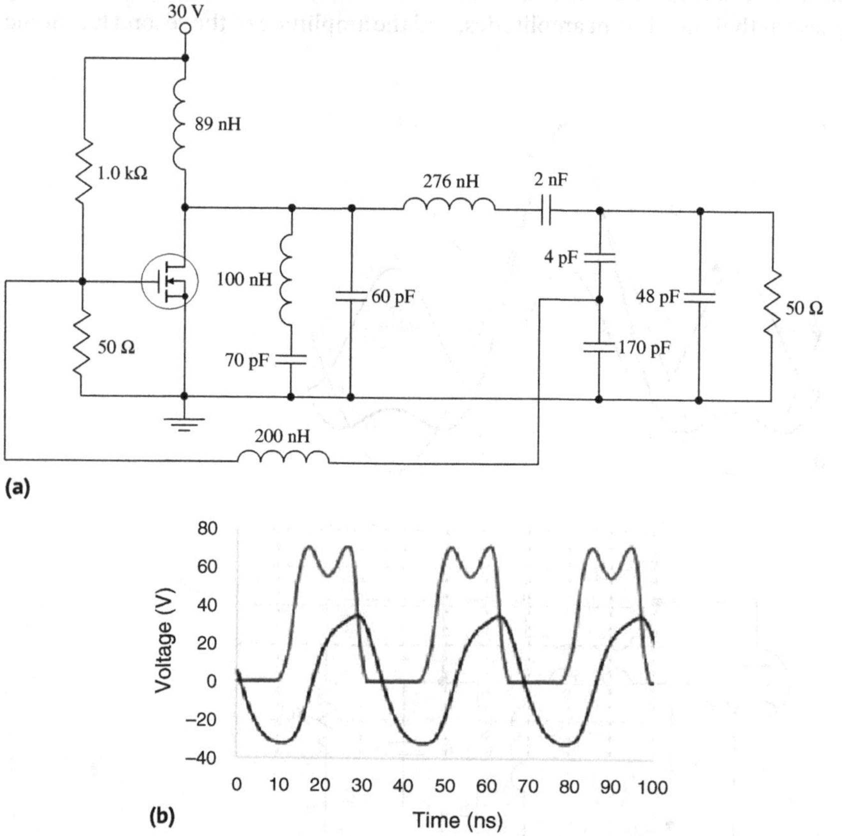

Figure 4.20 Schematic and performance of Class-F_2 power oscillator with parallel feedback.

power oscillator with parallel feedback based on the GS61004B transistor [19]. Here, a low-pass L-type matching section is used at the oscillator output to match to a 50-Ω load. The oscillator operates at 30 MHz and provides an output almost sinusoidal voltage waveform, oscillating between +30 and −30 V. The drain-to-source voltage has the expected waveform with depression at its center and maximum voltage amplitude of 70V, as shown in Figure 4.20b which is approximately twice the DC-supply voltage, providing zero-voltage switching behavior.

In practical implementation at microwave frequencies, the matching circuit with a series lumped inductance is replaced by the matching circuit with a series transmission line, as shown in Figure 4.21a. The directional coupler at the output of the oscillator with a coupling coefficient less than the operating power gain of

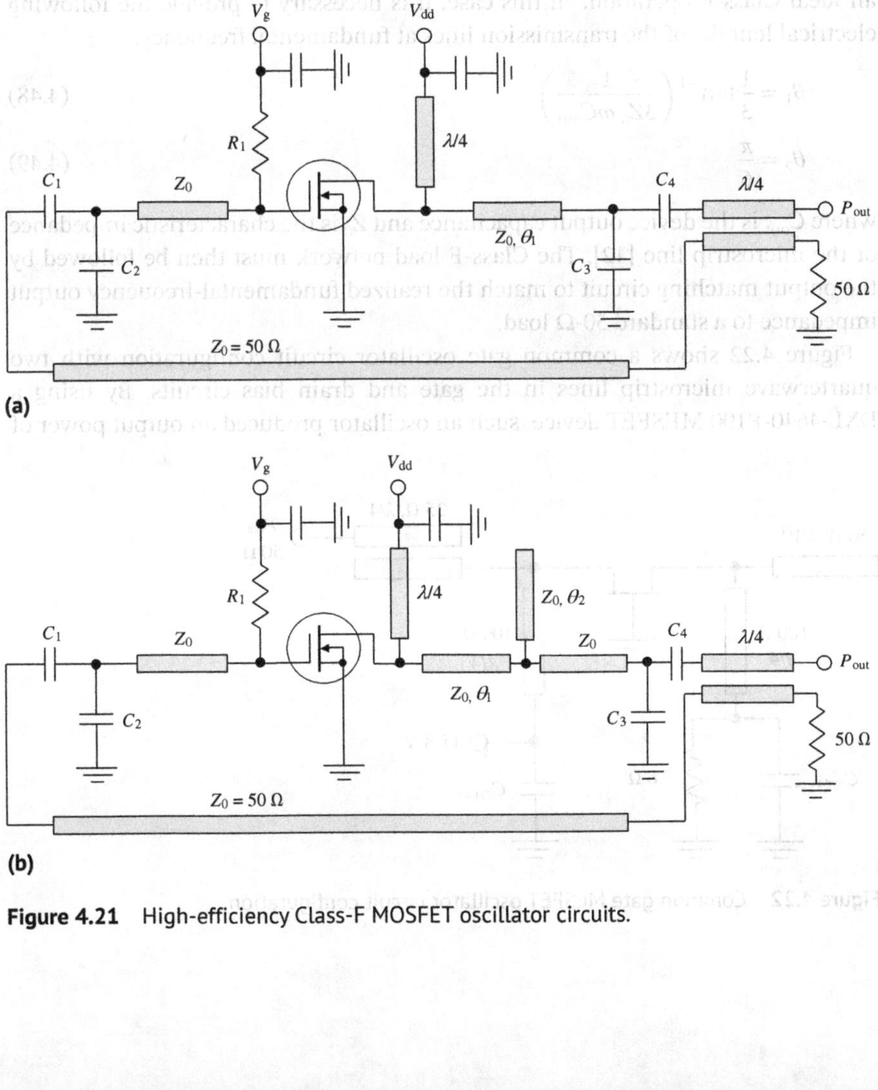

Figure 4.21 High-efficiency Class-F MOSFET oscillator circuits.

the equivalent power amplifier and the transmission line with proper electrical length in the feedback circuit provide the necessary phase shift. By using an additional third-harmonic impedance peaking circuit, the total oscillator efficiency can be increased. Note that the ideal Class-F power amplifier with all even-harmonic short-circuit termination due to the quarterwave transmission line and third-harmonic peaking having maximally flat waveforms achieves a maximum drain efficiency of 88.4% [20].

It is easy to realize such impedance conditions using the series transmission line with electrical length of θ_1 and an open-ended stub with electrical length of θ_2, as shown in Figure 4.21b. An appropriate control of short-circuit termination at the second harmonic and open-circuit termination at the third harmonic results in a drain (collector) voltage waveform close to a square wave, thus approximating an ideal Class-F operation. In this case, it is necessary to provide the following electrical lengths of the transmission lines at fundamental frequency:

$$\theta_1 = \frac{1}{3}\tan^{-1}\left(\frac{1}{3Z_0\,\omega C_{\text{out}}}\right) \tag{4.48}$$

$$\theta_2 = \frac{\pi}{6} \tag{4.49}$$

where C_{out} is the device output capacitance and Z_0 is the characteristic impedance of the microstrip line [12]. The Class-F load network must then be followed by the output matching circuit to match the realized fundamental-frequency output impedance to a standard 50-Ω load.

Figure 4.22 shows a common gate oscillator circuit configuration with two quarterwave microstrip lines in the gate and drain bias circuits. By using a DXL-4640-P100 MESFET device, such an oscillator produced an output power of

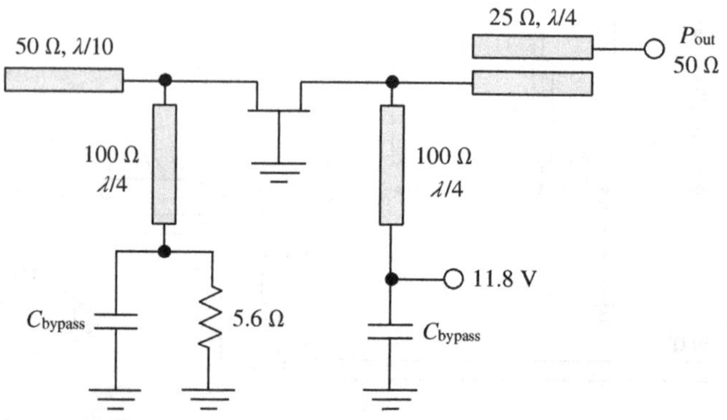

Figure 4.22 Common gate MESFET oscillator circuit configuration.

1.2 W at an operating frequency of 8 GHz [21]. The transition from self-oscillating small-signal conditions to large-signal steady-state operation mode is provided by a self-bias resistor of 5.6 Ω connected to the source circuit in parallel with a bypass capacitor. Both RF short-circuited quarterwave microstrip lines connected to the source and drain, respectively, eliminate the influence of device bias circuits on the electrical properties of the resonant circuit. An open-ended microstrip line provides the capacitive reactance seen by the source terminal. Since the output resonator is a quarterwave microstrip stub, a simple adjacent line coupling section was chosen and adjusted for the required coupling to achieve maximum output power. A quarterwave microstrip line connected to the drain improves the oscillator efficiency and spectral characteristics due to the short-circuit conditions for the even voltage harmonics.

In some cases, it is more convenient to use an inverse Class F rather than a conventional Class F, especially at higher frequencies when the effect of the parasitic device package parameters is significant, and it is easier to provide an open-circuit rather than short-circuit condition for the second harmonic at the device drain inside the package. The schematic of an inverse Class-F power oscillator with LDMOS MRF281 transistor is shown in Figure 4.23a [22]. Here, the load matching network provides the second- and third-harmonic control using the three open-circuit stubs (one with an electrical length of $\lambda/8$ and two with an electrical length of $\lambda/12$ connected in parallel for implementation convenience), resulting in a close approximation to the square current and half-sinusoidal voltage drain waveforms. The series microstrip lines TL_3 and TL_4 are necessary to compensate for the package parasitics and to provide high impedance at the second harmonic and low impedance at the third harmonic at the device drain, respectively. The oscillator circuit includes a feedback loop with a coupling capacitor, whose value was optimized for maximum efficiency, providing a simple and easy soft excitation of self-oscillations. As a result, the output power and conversion efficiency were measured to be 38.15 dBm and 73.2% at 981.5 MHz, respectively.

Figure 4.23b shows the other implementation of an inverse Class-F power oscillator with separate feedback and load-network matching circuits [23]. Here, a T-type feedback circuit with two series microstrip lines and a shunt capacitor is connected between the device gate and drain terminals to present an open circuit at harmonic frequencies due to L_0C_0 a resonant circuit tuned to the fundamental, thus minimizing its effect on the harmonic impedance conditions seen by the device output. Therefore, the load network can be designed separately from the feedback network to match the device output impedance with the load at the fundamental and to provide the required harmonic impedance conditions with high impedance at the second and low impedance at the third harmonics. This can be provided by using open-circuit stubs with electrical lengths

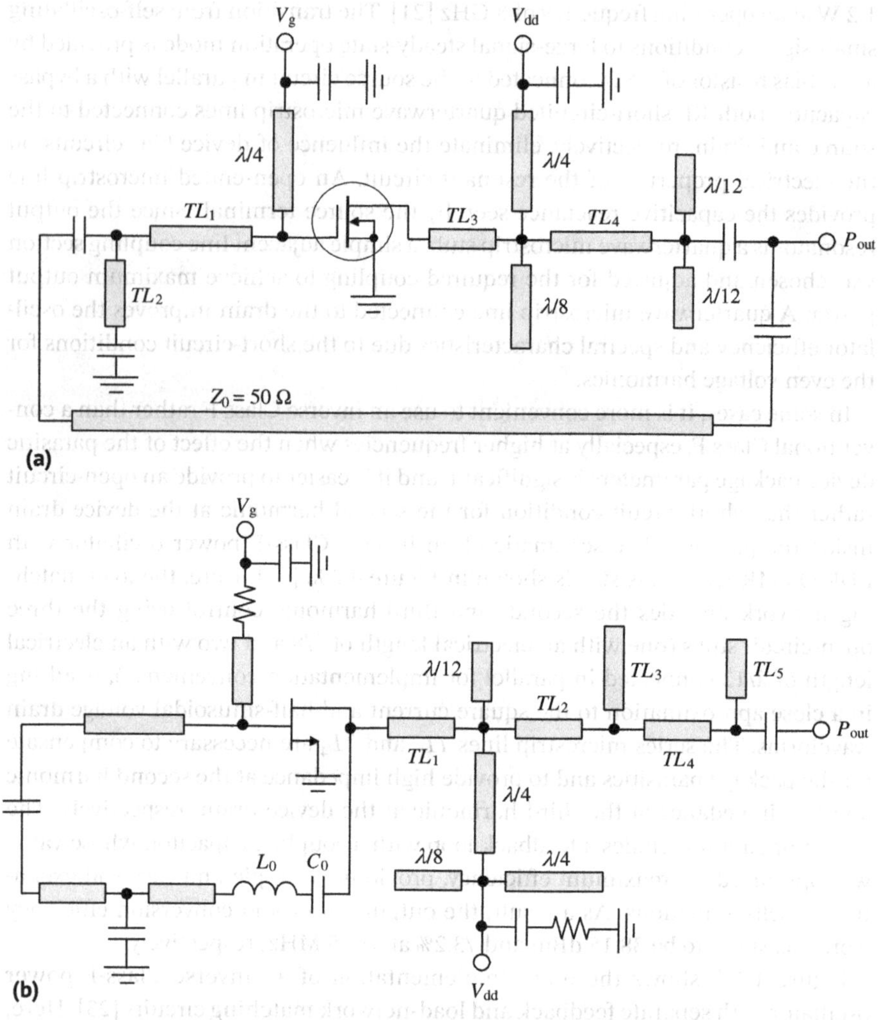

Figure 4.23 High-efficiency inverse Class-F oscillator circuits.

of λ/4, λ/8, and λ/12, respectively, followed by a two-section output matching circuit consisting of two series microstrip lines (TL_2 and TL_4) and two open-circuit stubs (TL_3 and TL_5). As a result, a very high efficiency of 80.2% with an output power of 35.1 dBm at 2.42 GHz was achieved for such an inverse Class-F power oscillator using a GaN HEMT CGH40006P device.

The high-power and high-efficiency GaN-HEMT power oscillator with a harmonic control and parasitic parameter compensation circuits based on a power amplifier operating in an inverse Class-F mode using a GaN HEMT CGH400010F device for efficiency enhancement was described in [24]. The feedback circuit of the described oscillator uses a frequency-tunable stepped-impedance resonator and a coupled-line coupler instead of the traditional coupling capacitor to control the feedback power precisely. As a result, such a power oscillator achieved a maximum conversion efficiency of 74.5% and an output power of 40.2 dBm at 2.44 GHz.

4.4.3 Class-E Power Oscillators: From RF to Millimeter Waves

It is very attractive to apply the Class-E operation principles to the tuned power oscillator considering the potential simplicity of the resonant circuit implementation. In this case, to achieve Class-E operation mode for a power oscillator based on the corresponding Class-E power amplifier, it is necessary to design a three-port matching and feedback reactive network, satisfying the following conditions that were first suggested in [25]:

- The collector-emitter voltage of the bipolar transistor should be the same as in the corresponding Class-E power amplifier.
- The energy transmission from the collector to the load should be provided in the same way as in the corresponding Class-E power amplifier, taking into account the driving power.
- The energy stored in the load network should be high enough to ensure the required frequency stability.
- The feedback voltage should have the same amplitude and phase (with respect to the phase of the collector voltage) as the input voltage of the corresponding Class-E power amplifier.

One of the important requirements for the output spectrum of the oscillator is a sufficiently small level of harmonics. The loaded quality factor of the resonant circuit must be not less than 5.1 to obtain the second harmonic level below −20 dBc without additional filtering [26]. Therefore, the close to sinusoidal output voltage will represent the input voltage of the feedback network or driving voltage of the oscillator. This means that the active device cannot operate either in pinch-off or saturation regions, and there is a time period when it is operated in the active region, which is characterized by power losses. However, the effect of finite switching time is not so significant compared with losses provided by the finite value of the saturation resistance, as shown in Section 3.3.3.

Generally, the design of a Class-E power oscillator is divided into two parts. The first step is to design the load network of the corresponding Class-E power amplifier. Then, the second step is to provide the proper phase shift between the input and output ports of the feedback network. Since some part of the power amplifier's output power is transmitted to its input, this means that the load resistance R of the power amplifier will be equal to a parallel combination of the load resistance R_L and the input resistance of the feedback network R_{in}. Figure 4.24a shows an example of the Class-E MOSFET power oscillator with a series inductance of 48 μH in the feedback network having an efficiency of 86% and an output power of 1 W at an oscillation frequency of 800 kHz with a DC-supply voltage of 4.5 V [27]. Similar results with the description of a complete design procedure of a Class-E MOSFET power oscillator have been reported in [28]. By using an IRF530 MOSFET device, the oscillator efficiency achieved 94.6% at an output power of

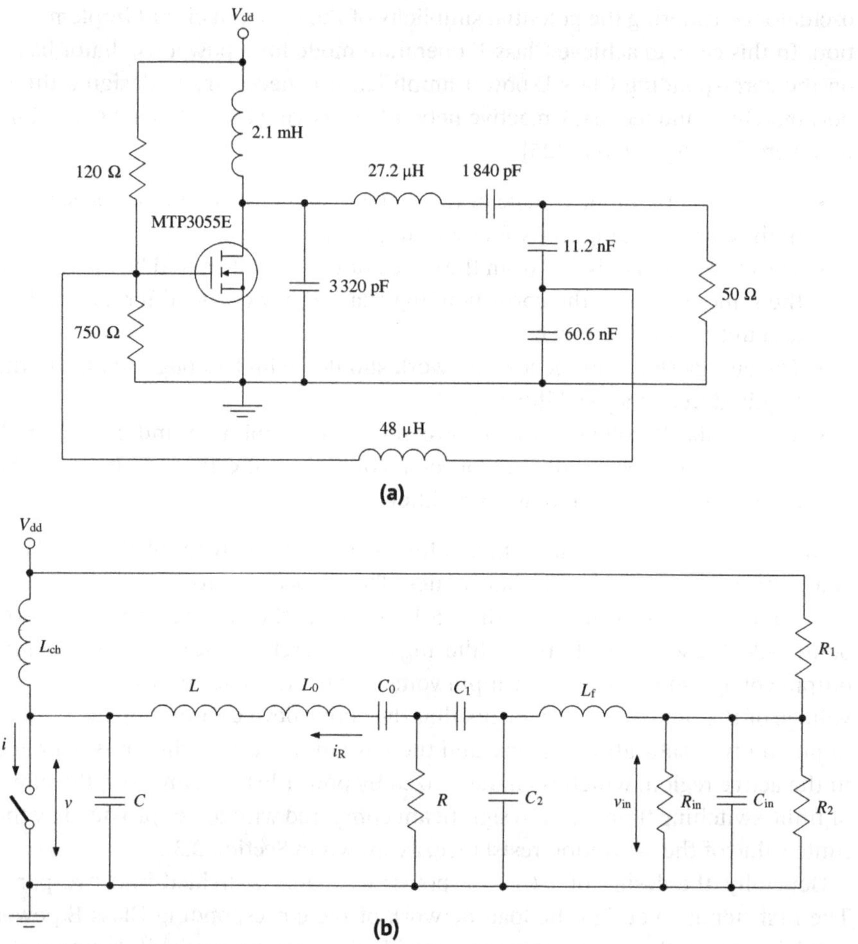

Figure 4.24 Basic circuits of Class-E power oscillators with a three-port network.

7.8 W operating at 1 MHz with a DC-supply voltage of 12 V for the same Class-E oscillator configuration [29].

To simplify the analysis of such a Class-E power oscillator with shunt capacitance, the active device should be replaced by an ideal switch, as shown in Figure 4.24b. In this case, the input resistance R_{in} and capacitance C_{in} represent a parallel equivalent of the input circuit of the MOSFET device equivalent circuit consisting of the gate resistance R_{gs} and gate–source capacitance C_{gs} connected in series. The optimum parameters of the Class-E load network with shunt capacitance C, series inductance L, and load resistance R are given by

$$L = \frac{1.1525R}{\omega} \tag{4.50}$$

$$C = \frac{0.1836}{\omega R} \tag{4.51}$$

$$R = 0.5768\frac{V_{dd}^2}{P_{out}} \tag{4.52}$$

where V_{dd} is the DC-supply voltage and P_{out} is the output power at the fundamental frequency (Section 3.3.1).

Thus, the phase angle of the load network (between the fundamental voltage and current phasors) seen from the device collector and required for an idealized optimum Class-E mode with shunt capacitance can be obtained through the load network parameters using Eqs. (4.50) and (4.51) as

$$\phi = \tan^{-1}\left(\frac{\omega L}{R}\right) - \tan^{-1}\left(\frac{\omega CR}{1 - \frac{\omega L}{R}\omega CR}\right) = 35.945^\circ \tag{4.53}$$

which means that the load network is slightly mistuned with respect to the fundamental frequency. The values of the load-network parameters should be chosen to create an inductive reactance at the fundamental frequency and the capacitive reactances at other harmonic components. The feedback inductance L_f is necessary to provide a greater than 180° phase shift between the fundamental components of the gate and drain voltages.

In practical design, it is impossible to realize choke inductance L_{ch} with infinite impedance at the fundamental and any harmonics. Moreover, using a finite DC-feed inductance has the advantage of minimizing size, cost, and complexity of the overall circuit. When the ratio L/L_{ch} is sufficiently small, corresponding to only DC current flowing through the choke inductance L_{ch}, the optimum load-network parameters defined by Eqs. (4.50)–(4.52) are almost constant. However, in the range of large L/L_{ch}, the optimum load-network parameters vary since the inductance L_{ch} no longer works as a finite DC-feed inductance [30]. The experimental results for the MOSFET Class-E power oscillator show the efficiency of

90.7% and output power of 6.8 W at the oscillation frequency of 2.02 MHz for a high quality factor $Q = \omega L/R$ and low ratio L/L_{ch}. For the case of low Q and high L/L_{ch}, the efficiency was 89.7% with an output power of 2.8 W at an oscillation frequency of 1.97 MHz. By using the same topology for the bipolar Class-E power oscillator, the collector efficiency over 95% with an output power of 3 W at 2 MHz was measured [25].

If the theoretical value of the resistance R for the optimum Class-E load network given by Eq. (4.52) is too small or differs significantly from the required load impedance (typically of 50 Ω), it is necessary to use a matching circuit instead of a series L_0C_0 resonant circuit to deliver maximum output power to the load, as shown in Figure 4.25. In a practical oscillator, the series inductors L and L_1 are combined and considered as a single inductor. To provide a sufficient level of harmonic suppression, the loaded quality factor of such a transformer must be high enough. In addition, the first series element of the output matching circuit which can represent the low-pass L-or T-type configuration should be the inductor with a sufficiently large inductance value to realize high impedance conditions for harmonics. However, the output matching circuit can also be designed using any combination of the lumped capacitors and inductors with a series inductor as the first element.

To determine the circuit parameters of the low-pass L-type matching circuit, it is necessary first to calculate the loaded quality factor from

$$Q_L = \sqrt{\left(\frac{R_L}{R}\right) - 1} \tag{4.54}$$

Then, the matching circuit parameters (series inductance L_1 and shunt capacitance C_1) can be obtained by

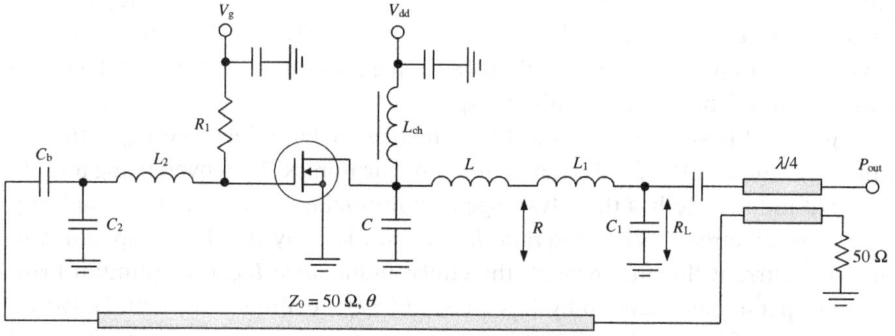

Figure 4.25 Schematic of Class-E power oscillator with impedance transformation.

$$L_1 = \frac{Q_L R}{\omega} \tag{4.55}$$

$$C_1 = \frac{Q_L}{\omega R_L} \tag{4.56}$$

At microwave frequencies, to minimize the insertion losses, it is preferable that any lumped inductors in the output matching circuit should be replaced by transmission lines. For example, any type of transmission lines, including open-circuit or short-circuit stubs to provide the required matching and harmonic suppression conditions, can compose the matching circuit. As a result, to approximate the idealized Class-E mode with shunt capacitance, it is necessary to design the transmission-line load network satisfying the required idealized optimum impedance Z_{net1} at the fundamental frequency defined by

$$Z_{net1} = R(1 + j\tan 49.052°) \tag{4.57}$$

which can be obtained from Eqs. (4.50) and (4.51). At the same time, the open-circuit conditions should be realized for all higher-order harmonics. However, as it follows from the Fourier analysis, a good approximation to a Class-E operation mode may be obtained with only two harmonics (fundamental and second) of the voltage waveform across the switch [31].

The parallel feedback can be used to design the Class-E power oscillators at microwave frequencies when the load network has impedance at the fundamental frequency of

$$Z_{net1} = \frac{0.28015}{\omega C_{out}} \exp(j49.052°) \tag{4.58}$$

where C_{out} is the transistor output capacitance [32]. Equation (4.58) establishes a relationship between the load-network impedance seen by the device output and transistor output capacitance, which is similar to Eq. (4.57). Here, the output device capacitance fully represents the shunt capacitance required for optimum Class-E operation mode. The circuit schematic of a parallel-feedback Class-E power oscillator with a variable-length microstrip line in the feedback path and an asymmetric microstrip branch-line coupler at the output, as shown in Figure 4.26, achieved a maximum efficiency of 59% with an output power of 300 mW at the operating frequency of 5 GHz and a DC-supply voltage of 6.5 V. The matching circuits and variable feedback microstrip lines of 50 Ω (1.6-mm wide) were fabricated on the 0.508-mm thick substrate with dielectric permittivity $\varepsilon_r = 2.2$, when $l_1 = 9$ mm, $l_2 = 1.8$ mm, $l_3 = 5.3$ mm, and $l_4 = 6.2$ mm. An asymmetric microstrip branch-line coupler is used to provide the feedback through a microstrip line of the appropriate length.

Depending on the implementation technology (hybrid or monolithic) at different frequencies and output power levels, the quarterwave transmission line in the feedback path of the oscillator can be replaced by its low-pass *LC* equivalent which can represent a low-pass Π-type structure with a series inductance and two shunt capacitors. In this case, by equating the *ABCD*-matrices of the transmission line with the characteristic impedance Z_0 and electrical length of θ, as shown in Figure 4.27a, and a low-pass Π-type lumped circuit consisting of a series inductance *L* and two shunt capacitors *C*, as shown in Figure 4.27b, according to

$$\begin{bmatrix} \cos\theta & jZ_0\omega L \\ j\sin\theta/Z_0 & \cos\theta \end{bmatrix} = \begin{bmatrix} 1-\omega^2 LC & j\omega L \\ j\omega C\left(2-\omega^2 LC\right) & 1-\omega^2 LC \end{bmatrix} \tag{4.59}$$

the ratio of the circuit parameters corresponding to the transmission line and its low-pass Π-type lumped equivalent can be obtained as a function of electrical length θ by

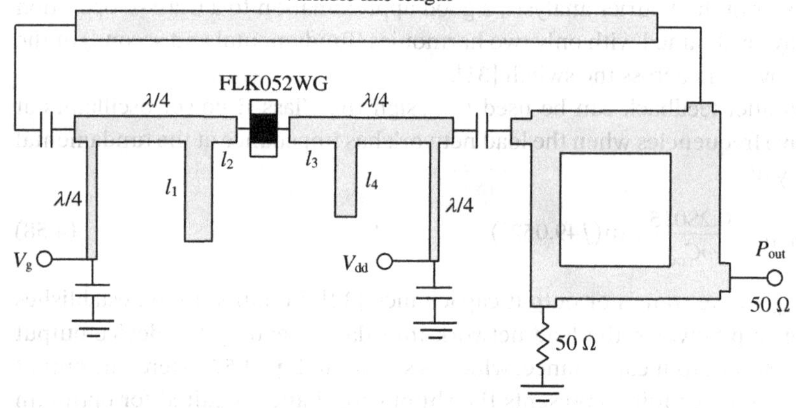

Figure 4.26 Microwave Class-E MESFET oscillator topology.

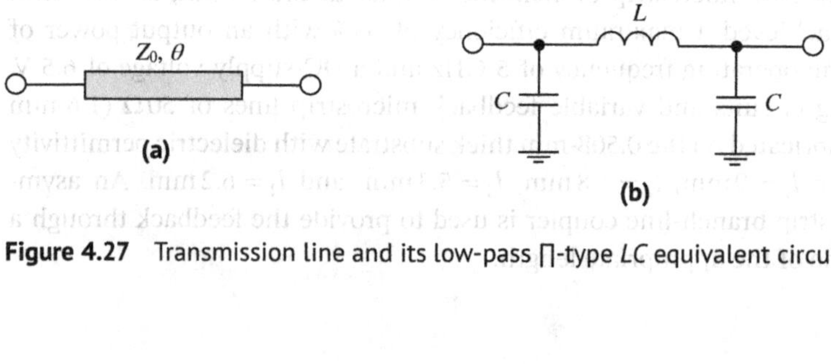

Figure 4.27 Transmission line and its low-pass Π-type *LC* equivalent circuit.

$$\omega L = Z_0 \sin\theta \tag{4.60}$$

$$\omega C = \frac{1 - \cos\theta}{Z_0 \sin\theta} \tag{4.61}$$

Figure 4.28a shows the circuit schematic of a Class-E power oscillator with a low-pass Π-type feedback network, which includes a Class-E load network with shunt capacitance, a high-pass L-type input matching circuit, and a low-pass L-type output matching network [33]. The Class-E load network consists of a shunt capacitor C_p (including the device output capacitance), a series L_sC_s resonant circuit

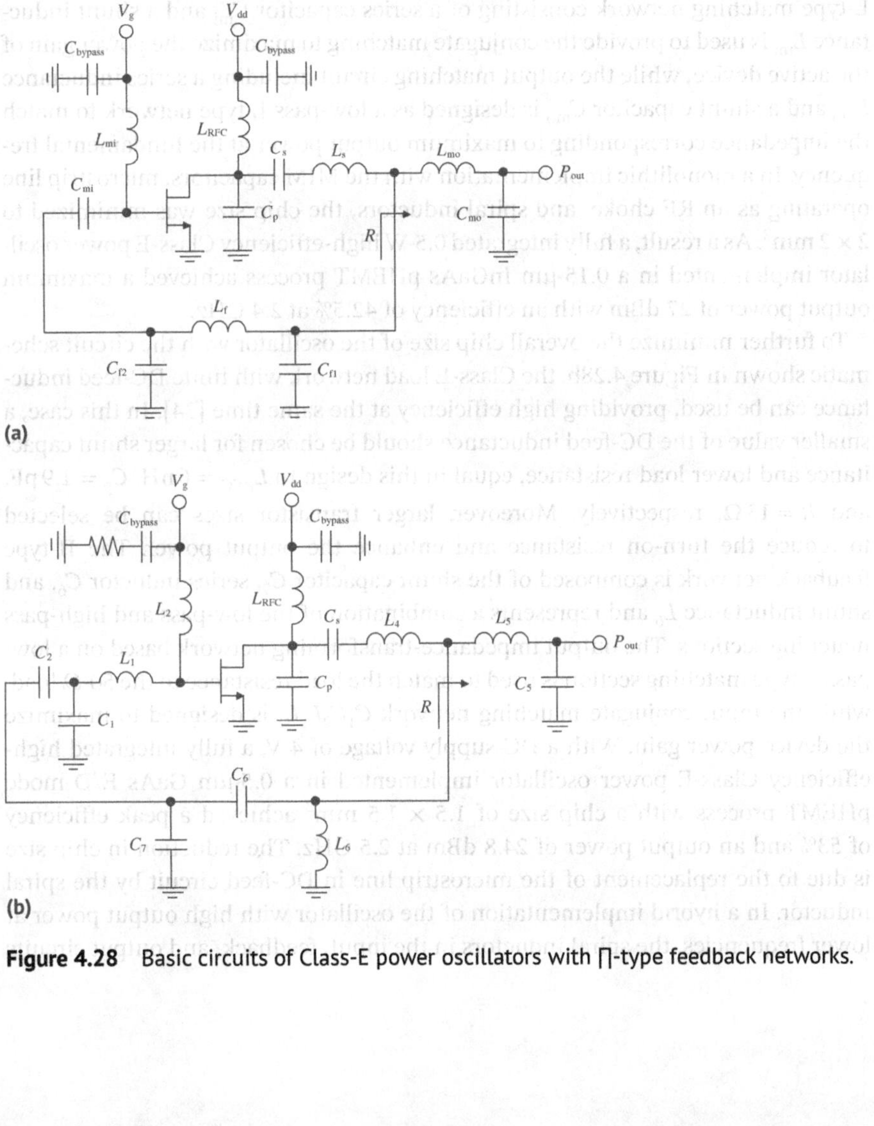

Figure 4.28 Basic circuits of Class-E power oscillators with Π-type feedback networks.

having the required inductive reactance at the fundamental frequency, and an RF choke. The low-pass Π-type feedback network is composed of a series inductor L_f and two shunt capacitors C_{f1} and C_{f2} and can also be used to suppress the higher-order harmonics. The two bypass capacitors C_{bypass} are included to enhance the isolation between the RF and DC paths, to reduce the influence of low-frequency noise sources, and to minimize the possibilities of the parasitic self-oscillations due to the bias circuits. Since the optimum load resistance $R = 17.7\,\Omega$ to deliver an output power of 0.8 W under a 5-V DC-supply voltage is required in a Class-E operation mode, an additional impedance matching section is necessary at the output to match the standard load $R_L = 50\,\Omega$. In this case, the input high-pass L-type matching network consisting of a series capacitor C_{mi} and a shunt inductance L_{mi} is used to provide the conjugate matching to maximize the power gain of the active device, while the output matching circuit including a series inductance L_{mo} and a shunt capacitor C_{mo} is designed as a low-pass L-type network to match the impedance corresponding to maximum output power at the fundamental frequency. In a monolithic implementation with the MIM capacitors, microstrip line operating as an RF choke, and spiral inductors, the chip size was minimized to $2 \times 2\,mm^2$. As a result, a fully integrated 0.5-W high-efficiency Class-E power oscillator implemented in a 0.15-μm InGaAs pHEMT process achieved a maximum output power of 27 dBm with an efficiency of 42.5% at 2.4 GHz.

To further minimize the overall chip size of the oscillator with the circuit schematic shown in Figure 4.28b, the Class-E load network with finite DC-feed inductance can be used, providing high efficiency at the same time [34]. In this case, a smaller value of the DC-feed inductance should be chosen for larger shunt capacitance and lower load resistance, equal in this design to $L_{RFC} = 6\,nH$, $C_p = 1.9\,pF$, and $R = 13\,\Omega$, respectively. Moreover, larger transistor sizes can be selected to reduce the turn-on resistance and enhance the output power. The Π-type feedback network is composed of the shunt capacitor C_7, series inductor C_6, and shunt inductance L_6 and represents a combination of the low-pass and high-pass matching sections. The output impedance-transforming network based on a low-pass L-type matching section is used to match the load resistance to the 50-Ω load, while the input conjugate matching network $C_1C_2L_1L_2$ is designed to maximize the device power gain. With a DC-supply voltage of 4 V, a fully integrated high-efficiency Class-E power oscillator implemented in a 0.5-μm GaAs E/D-mode pHEMT process with a chip size of $1.5 \times 1.5\,mm^2$ achieved a peak efficiency of 53% and an output power of 24.8 dBm at 2.5 GHz. The reduction in chip size is due to the replacement of the microstrip line in DC-feed circuit by the spiral inductor. In a hybrid implementation of the oscillator with high output power at lower frequencies, the spiral inductors in the input, feedback, and output circuits should be replaced by high-impedance short-length microstrip lines when an output power of 75 W with a 67% efficiency can be achieved at 410 MHz [35].

To provide a tuning capability of the power oscillator, an additional GaN HEMT device can be used instead of the varactor since the varactor is not available in a GaN process. Figure 4.29a shows an *X*-band tunable Class-E voltage-controlled oscillator (VCO) using a 0.25-μm GaN HEMT device [36]. Here, a low-pass T-type lumped-element feedback network with two series inductors L_1 and L_2 and shunt

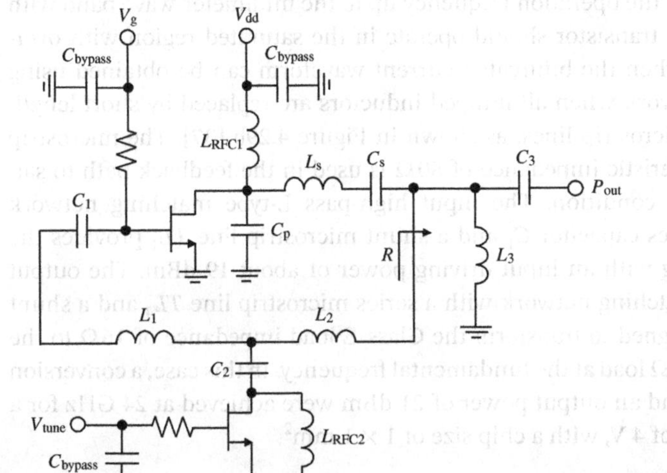

Figure 4.29 Basic circuits of tunable Class-E power oscillators.

capacitor C_2 is placed between the input and output ports to provide the required positive feedback phase shift to meet the oscillation conditions. An additional GaN HEMT device with grounded drain through the RF choke inductor is connected in series to the shunt capacitor C_2, thus providing a frequency tuning due to its nonlinear voltage–capacitance behavior by changing the gate bias voltage. As a result, an output power close to 28 dBm and a conversion efficiency of about 28% were achieved at 9.4 GHz, with a tuning frequency range from 9.35 to 9.46 GHz and a chip size of $2 \times 1.5\ \text{mm}^2$.

To further extend the operation frequency up to the millimeter-wave band with high efficiency, the transistor should operate in the saturated region with overdriven condition when the bifurcated current waveform can be obtained using a Class-E load network when all lumped inductors are replaced by short-length high-impedance microstrip lines, as shown in Figure 4.29b [37]. The microstrip line with a characteristic impedance of 80 Ω is used in the feedback path to satisfy the oscillation condition. The input high-pass L-type matching network composed of a series capacitor C_1 and a shunt microstrip line TL_1 provides the conjugate matching with an input driving power of about 19 dBm. The output low-pass L-type matching network with a series microstrip line TL_4 and a shunt capacitor C_4 is designed to transform the Class-E load impedance of 16 Ω to the standard output 50-Ω load at the fundamental frequency. In this case, a conversion efficiency of 19% and an output power of 21 dBm were achieved at 24 GHz for a DC-supply voltage of 4 V, with a chip size of $1 \times 1\ \text{mm}^2$.

4.4.4 Load-independent Class-E Power Oscillator

A conventional Class-E power oscillator is highly sensitive to load variations due to the load-dependent phase shift which may affect significantly the self-sustained oscillations under varying load conditions. The load variations cause efficiency deterioration due to the nonzero voltage switching and varying oscillator amplitude and phase balance conditions when, in the worst case, the self-oscillations may stop. By incorporating a two-section low-pass *LC* filter with a resonant-capacitor feedback network which can satisfy a phase-shift requirement for sustained oscillations regardless of the load resistance, the load-independent self-oscillations can be realized when load independence in zero-voltage switching, output current, and gate-drive voltage can be achieved.

Figure 4.30a shows the circuit schematic of a load-independent Class-E power oscillator which includes a finite inductor L_1, a shunt capacitor C_1, a two-section low-pass filter $L_2C_2L_3C_3$, a feedback capacitor C_4, and a load resistance R_L [38]. The capacitor C_3 is divided into two series capacitors C_{31} and C_{32} (capacitive divider) to adjust the amplitude of the gate-drive voltage, and the voltage across the capacitor C_{32}

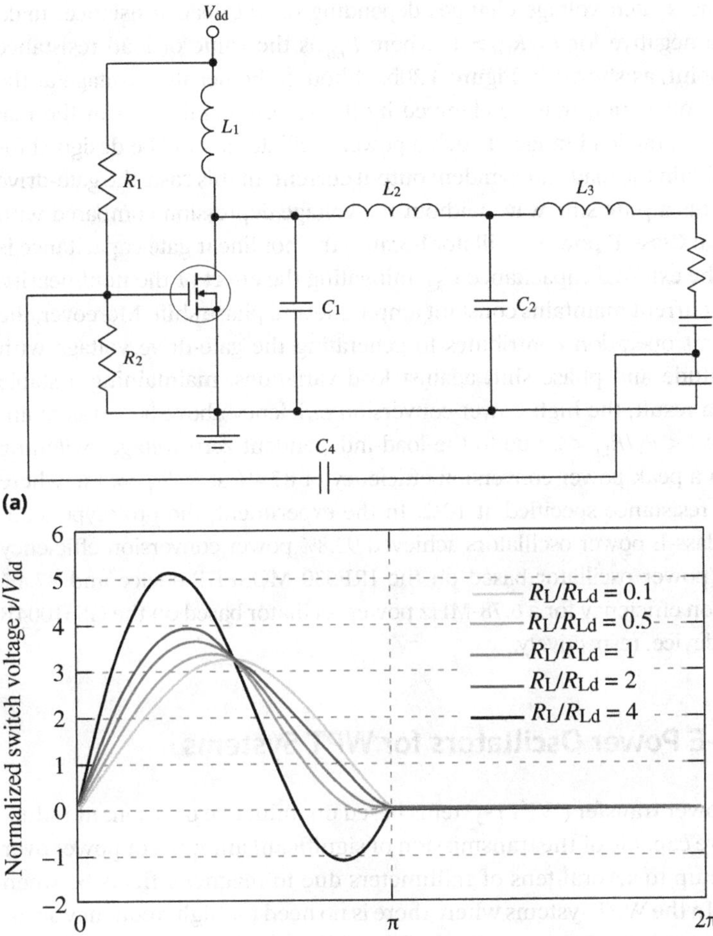

Figure 4.30 Schematic and voltage waveforms of a load-independent Class-E power oscillator.

becomes in phase with the gate-drive voltage to satisfy the phase-shift requirement when it should be fixed and independent of the load resistance. Besides, the input impedance of the two-section low-pass filter becomes purely resistive which does not interfere with the load-independent operation of the power oscillator.

Figure 4.30b shows the waveforms of the normalized idealized switch voltage of the load-independent Class-E power oscillator [38]. The load-independent zero-voltage-switching condition is theoretically guaranteed for any load resistance but the shape of the switch voltage changes depending on the load resistance, making the voltage negative for $R_L/R_{Ld} > 1$, where R_{Ld} is the value of load resistance at the design point, as shown in Figure 4.30b, although the negative voltage at the end of the turn-off period may be clamped by the reverse conduction in the real devices. Therefore, the load range of such a power oscillator should be designed for $R_L/R_{Ld} > 1$ to obtain the load-independent output current. In this case, the gate-drive voltage represents a pure sine wave without the voltage depression compared with the conventional Class-E power oscillator because the nonlinear gate capacitance is included into the external capacitance C_{32}, mitigating the effect of the nonlinearity, and the output current maintains constant amplitude and phase shift. Moreover, the load-independent operation contributes to generating the gate-drive voltage with constant amplitude and phase shift against load variations, maintaining a stable oscillation. As a result, the high power-conversion efficiency above 80% was maintained within $0.4 < R_L/R_{Lr} < 2$ due to the load-independent zero-voltage-switching operation, with a peak power conversion efficiency of 87.4% at $R_L/R_{Lr} = 1.6$, where R_{Lr} is the load resistance specified at 10Ω. In the experiment, the prototype load-independent Class-E power oscillators achieved 93.8% power conversion efficiency for a 0.8-MHz power oscillator based on the IRF530 MOSFET device and 87.4% power conversion efficiency for a 6.78-MHz power oscillator based on the GS61004B e-GaN HEMT device, respectively.

4.5 Class-E Power Oscillators for WPT Systems

The wireless power transfer (WPT) systems based on inductive or resonant inductive coupling are capable of the transmission of significant amounts of power over short distances up to several tens of millimeters due to magnetic fields between coupled wires. In the WPT systems where there is no need for high frequency accuracy, the power oscillator became very useful as it lowers the system complexity, form factor, and increases the transmitter efficiency using the high-efficiency load-network technique. Figure 4.31a shows the circuit schematic of a parallel-feedback power oscillator using an LDMOSFET device where a series inductor L_f and a capacitive divider with two capacitors C_{f1} and C_{f2} represent the feedback path, and the Class-E load network with a shunt capacitance C, a series inductance L, and a fundamentally tuned series resonant circuit L_0C_0 is used to maximize efficiency. Here, both inductances L and L_0 are combined, and high-Q inductors are used to minimize the power losses. At 20-V DC-supply voltage, the conversion efficiency of 86.5% and output power of 9.2 W were achieved at an operating frequency of 6.96 MHz [39]. To implement the oscillator circuit into the WPT system satisfying the load requirements, it is necessary to insert the two-coupled

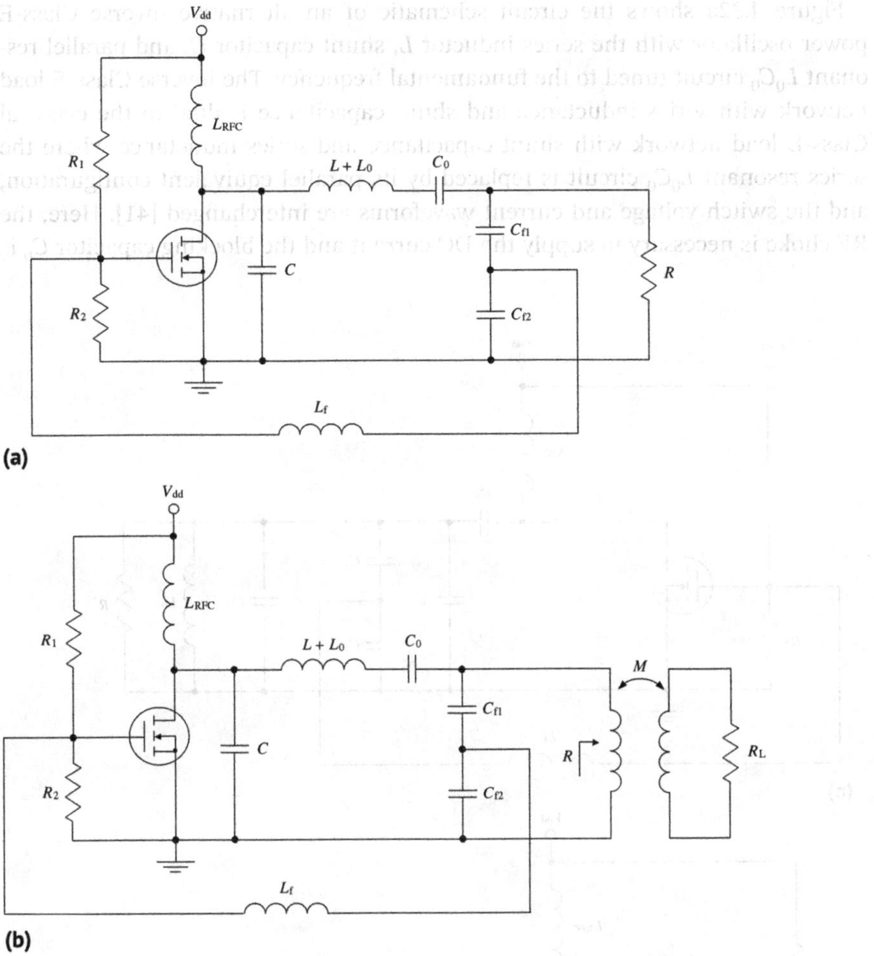

Figure 4.31 Basic circuits of Class-E power oscillators with parallel feedback for WPT systems.

inductors at the output where the proper coupling coefficient M could provide the required impedance transformation and the corresponding amount of output power flowing to the load R_L, as shown in Figure 4.31b. Such a parallel-feedback Class-E power oscillator can potentially be used at higher operating frequencies for far-field wireless charging as it achieves an output power of 24.9 W and a conversion efficiency of 64% at 433 MHz using EPC2037 e-GaN HEMT transistor with a drain bias voltage of 20 V [40].

Figure 4.32a shows the circuit schematic of an alternative inverse Class-E power oscillator with the series inductor L, shunt capacitor C, and parallel resonant L_0C_0 circuit tuned to the fundamental frequency. The inverse Class-E load network with series inductance and shunt capacitance is dual to the classical Class-E load network with shunt capacitance and series inductance where the series resonant L_0C_0 circuit is replaced by its parallel equivalent configuration, and the switch voltage and current waveforms are interchanged [41]. Here, the RF choke is necessary to supply the DC current and the blocking capacitor C_b is

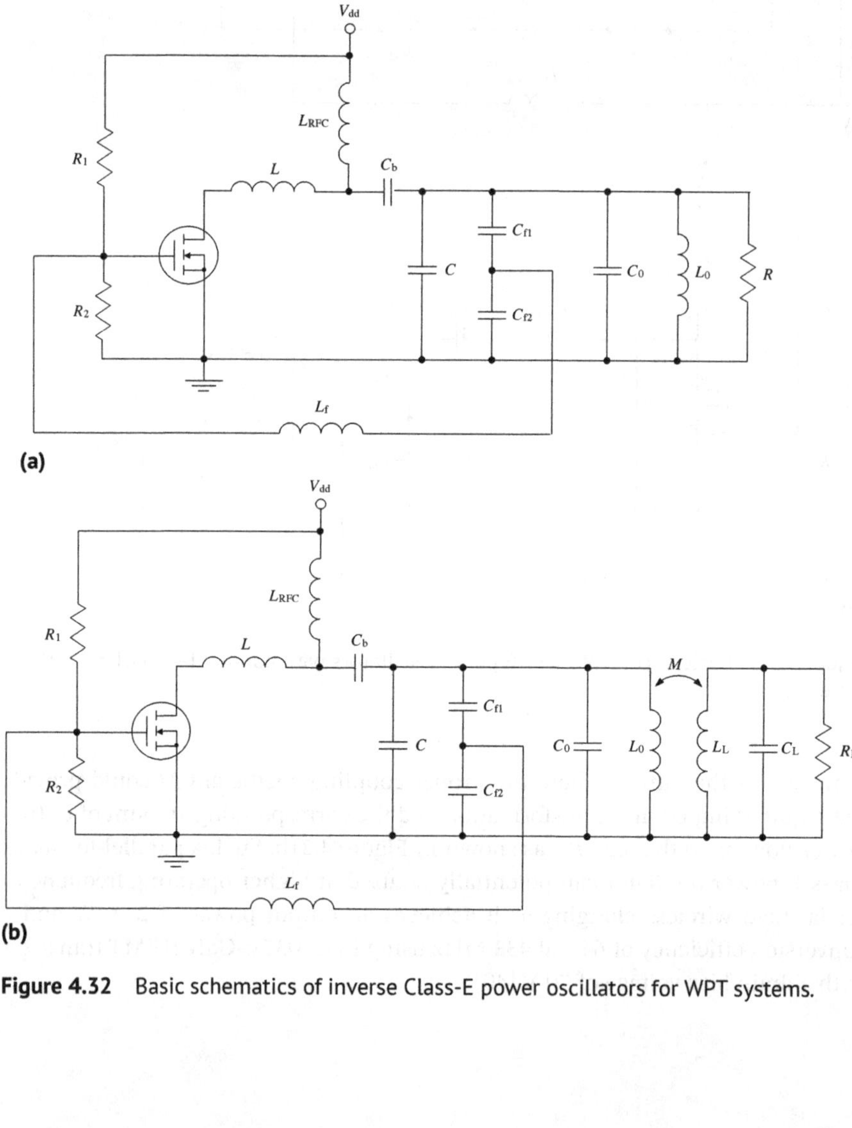

Figure 4.32 Basic schematics of inverse Class-E power oscillators for WPT systems.

required to decouple the parallel resonant circuit from the DC path. However, unlike the classical Class-E load network where the parasitic device output capacitance can be considered a part of the required shunt capacitance, it represents the parasitic element contributing to efficiency degradation in an inverse Class E and its negative effect should be minimized, choosing the transistor with lower output capacitance such as GaN HEMT device, applying special compensation schemes, or using at lower operating frequencies.

The optimum series inductance L, shunt capacitance C, and load resistance R of the Class-E load network with series inductance and shunt capacitance can be calculated as

$$\frac{\omega L}{R} = \frac{8}{\pi\left(\pi^2 + 4\right)} = 0.1836 \tag{4.62}$$

$$\omega CR = \frac{\pi\left(\pi^2 - 4\right)}{16} = 1.1525 \tag{4.63}$$

$$R = \frac{\pi^2 + 4}{8}\frac{V_{\mathrm{dd}}^2}{P_{\mathrm{out}}} = 1.7337\frac{V_{\mathrm{dd}}^2}{P_{\mathrm{out}}} \tag{4.64}$$

where V_{dd} is the drain supply voltage, and P_{out} is the fundamental-frequency output power delivered to the load.

The circuit schematic of an inverse Class-E power oscillator shown in Figure 4.32a can be employed as the basis of the power oscillator for WPT applications with resonant inductive coupling, as shown in Figure 4.32b. Here, the power transmission from the oscillator to the load represented by a coupled parallel $L_{\mathrm{L}}C_{\mathrm{L}}$ resonator connected to the load resistance R_{L} in parallel was provided through the transformer mutual inductance $M = k\sqrt{L_0 L_{\mathrm{L}}}$, where k is a coupling coefficient, reaching 1.2 W at 27 MHz [42].

Since the load for the power oscillator in the WPT system may vary significantly depending on the coupling factor of the inductances in the load network, it is important to minimize the variation of the output power from the power oscillator delivered to the varying load. Figure 4.33 shows the circuit schematic of the so-called load-independent inverse Class-E oscillator [43]. The load-independent design ensures the zero-current switching and constant-output voltage against load variation. The load-independent inverse Class-E oscillator generates the driving voltage by the feedback network from the voltage across the inductor L_2. Because the voltage across the inductor L_2 is almost constant despite the load variations due to the presence of a series-resonant filter composed of the inductor L_3 and capacitor C_4, this power oscillator can generate a stable driving voltage against load variations. The required phase shift between the output and driving voltage is provided by the proper coupling of the inductance L_4 with an inverting transformer formed by the coupled inductors L_2 and L_3. Since the impedance of

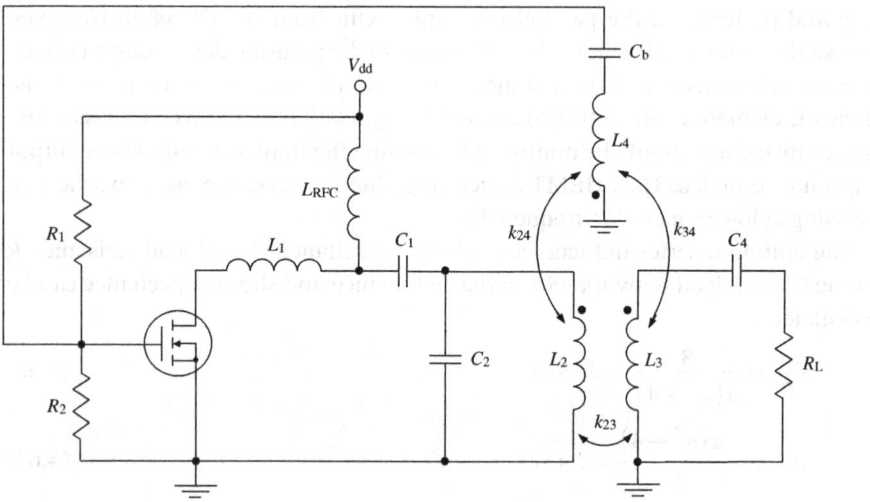

Figure 4.33 Circuit schematic of an inverse Class-E power oscillator with three-coupled inductors.

the feedback network is sufficiently high, the effect of the feedback network on the oscillator performance is negligible. In the experiment, such a WPT system based on the load-independent inverse Class-E power oscillator achieved a conversion efficiency of 76.5% with an output power of 20 W delivered to the 25-Ω load at the operating frequency of 1 MHz using the IRF510 MOSFET device with a DC-supply voltage of 24 V. At the same time, a four-coil WPT system based on a parallel-circuit Class-E inverter, which is very convenient in implementation due to a sufficiently small value of the parallel inductance used instead of RF choke, could achieve at 1 MHz the inverter efficiency of 96% and the system efficiency of 86.5% at the rated output power of 60 W with a coupling coefficient of 0.078 (transfer distance of 12 cm) between the receiving and transmitting coils [44].

Another possibility to design the Class-E power oscillator is to use an alternative series feedback common gate structure that can reduce the overall size of the oscillator because there is no need for an external feedback loop [12]. In this case, to maximize the negative resistance at the device output, it is necessary to provide an inductive reactance in the gate circuit using the series inductance L_g and the capacitive reactance in the source circuit using the shunt capacitance C_s, as shown in Figure 4.34. Here, a Class-E power oscillator is terminated with a coupled resonator, considering an intermediate coupling factor $k = 0.2$ and setting the load resistance to 50 Ω. The load-network shunt capacitance C, series capacitance C_1, and series inductance L_1 are calculated to operate in a Class-E mode with shunt

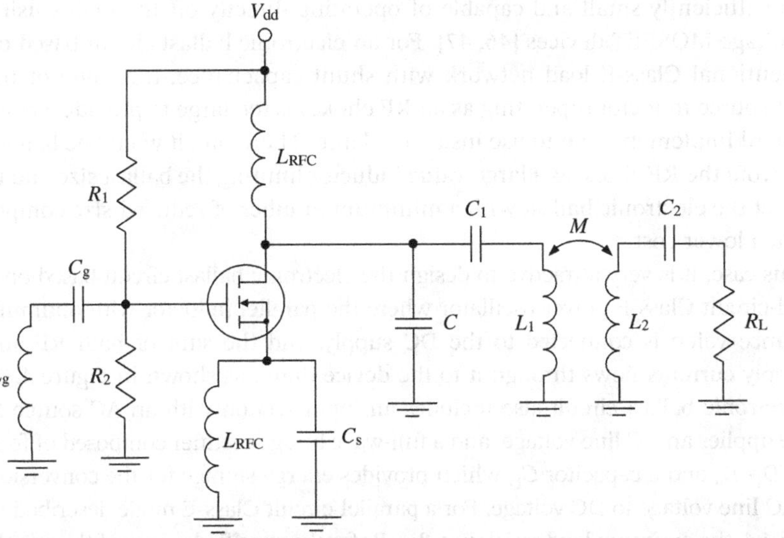

Figure 4.34 Circuit schematic of Class-E power oscillator with series feedback for WPT systems.

capacitance. The oscillation frequency varies with the coupling factor k and thus with the distance between the coupled coils. However, if the external-resonator frequency corresponds to the oscillation frequency in a standalone operation, the oscillation frequency will be equal to the resonant frequency $\omega_0 = 1/\sqrt{L_2C_2}$. By using the MOSFET device, the resulting oscillator output power of 0.5 W and conversion efficiency of 83% were achieved at an oscillation frequency of 12 MHz with a DC-supply voltage of 6 V [45].

4.6 Design Example: Parallel-circuit Class-E Oscillator for Compact Fluorescent Lamps

The high-efficiency Class-E power oscillators can be used inside compact fluorescent lamps (or compact RF lamps) to generate a high-frequency electromagnetic field which interacts with the gas inside the lamps. To ignite plasma in a compact fluorescent lamp using an RF power oscillator, a radio-frequency electromagnetic field excites the gas inside, causing ionization and emission of ultraviolet light, which then excites the fluorescent coating to produce visible light. At the same time, the Class-E power oscillator provides a very high conversion efficiency to minimize heating effect since the size of the lamp base is sufficiently small and capable of operating directly off the mains using high-voltage MOSFET devices [46, 47]. For an electronic ballast circuit based on a conventional Class-E load network with shunt capacitance, the value of the current source inductor (operating as an RF choke) is too large to provide a compact board implementation to use inside the lamp. Therefore, it would be beneficial to avoid the RF choke as a large-value inductor limiting the ballast size and to construct the electronic ballast with a minimum number of reduced-size components at a lower cost.

In this case, it is very attractive to design the electronic ballast circuit based on a parallel-circuit Class-E power oscillator where the parallel inductor with optimum inductance value is connected to the DC supply, and the sum of both RF and DC-supply currents flows through it to the device drain, as shown in Figure 4.35. The electronic ballast circuit also includes an input section with an AC source *S*, which supplies an AC line voltage, and a full-wave bridge rectifier composed of four diodes D_1–D_4 and a capacitor C_1, which provides energy storage for the conversion from AC line voltage to DC voltage. For a parallel-circuit Class-E mode described in Section 4.4, the optimum load resistance $R = R_L$ for the specified values of the supply voltage V_{dd} and fundamental output power P_{out} can be obtained by

$$R = 1.365V_{dd}^2/P_{out} \tag{4.65}$$

whereas the optimum parallel inductance *L* and shunt capacitance *C* can be determined as

$$L = 0.732R/\omega \tag{4.66}$$

$$C = 0.685/\omega R \tag{4.67}$$

The parameters of the series resonant L_0C_0 circuit depending on the chosen loaded quality factor Q_L are calculated by

$$C_0 = 1/Q_L\omega R \tag{4.68}$$

$$L_0 = 1/\omega^2C_0 \tag{4.69}$$

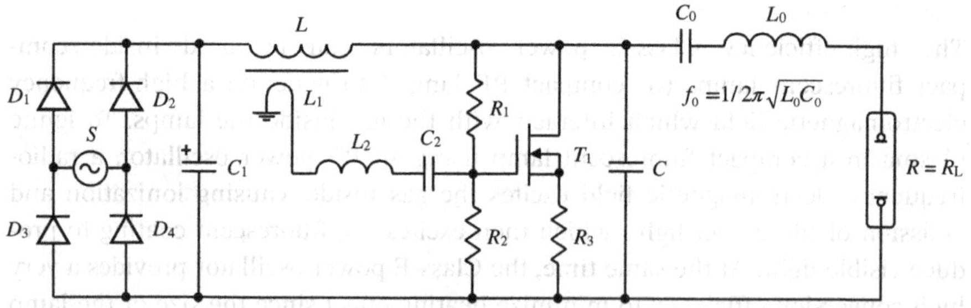

Figure 4.35 Electronic ballast circuit based on a parallel-circuit Class-E power oscillator.

Initially, due to the soft start-up oscillation conditions, the MOSFET device T_1, biased in a Class-AB mode with small quiescent current, starts oscillating at a frequency determined by the feedback inductor L_2 and parasitic input capacitance of T_1. Since the feedback current is regenerative, the amplitude of oscillations grows, causing the MOSFET device T_1 to operate in a switching mode. When it turns on and off, a periodic voltage is applied to the series resonant circuit consisting of the capacitor C_0 and inductor L_0. The current-limiting resistor R_3 limits the DC current when the MOSFET device T_1 first conducts and the capacitor C discharges, and the voltage will increase across the resistor R_3 providing a soft transition to the steady-state oscillator operation condition. The gain of the feedback loop L_1, L_2, C_2 determines whether oscillations will increase exponentially until the system reaches the switching mode. Self-oscillations are obtainable by the proper small coupling of the primary transformer coil L to its secondary coil L_1 consuming only a very small fraction of the oscillator output power, typically less than a few hundred mW. The value of the series inductance L_0 may vary to compensate for the parasitic reactance of the lamp. The important advantage of a parallel-circuit Class-E mode is the significantly increased value of the load resistance R (by more than 2.3 times) compared to the conventional Class-E mode with shunt capacitance that minimizes the number of matching sections in the load network.

As an example, for an electronic ballast based on a parallel-circuit Class-E power oscillator shown in Figure 4.36 operating at 2.65 MHz with a DC voltage of 100 V to supply a 15 W into a fluorescent lamp, the load-network parameters are calculated as follows:

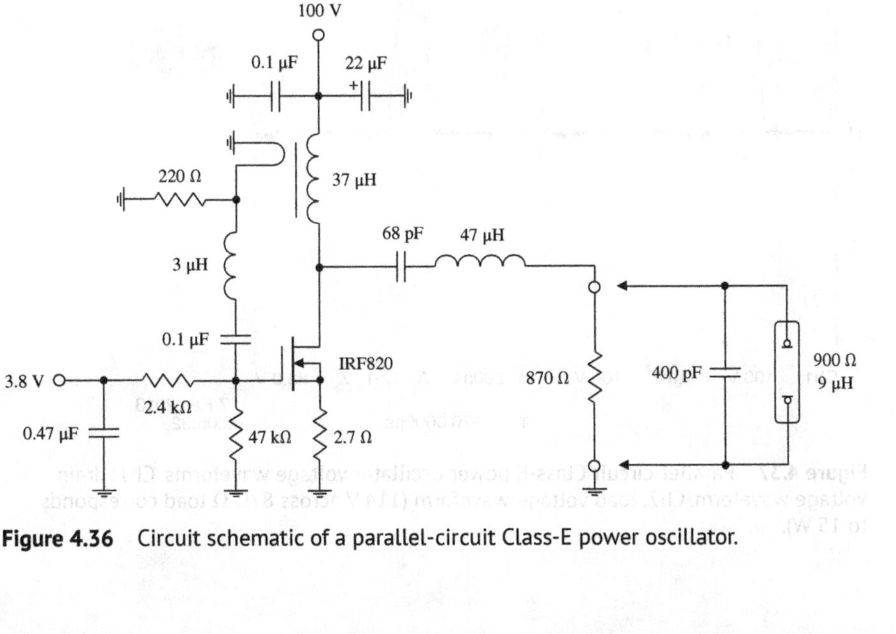

Figure 4.36 Circuit schematic of a parallel-circuit Class-E power oscillator.

2.5-A, 500-V *N*-channel power MOSFET IRF820:

- input capacitance $C_{iss} = 360\,\text{pF}$ ($V_{dd} = 25\,\text{V}$, $V_g = 0\,\text{V}$, $f = 1\,\text{MHz}$);
- output capacitance $C_{oss} = 60\,\text{pF}$ ($V_{dd} = 25\,\text{V}$, $V_g = 0\,\text{V}$, $f = 1\,\text{MHz}$);
- on-resistance $r_{sat} = 3\,\Omega$ ($V_g = 10\,\text{V}$, $I_d = 1.4\,\text{A}$);
- gate–source resistance $R_{gs} = 18\,\Omega$.

Load-network parameters:

- load resistance $R = 1.365\,V_{dd}^2/P_{out} = 910\,\Omega$;
- shunt DC-feed inductance $L = 0.732\,R/\omega = 40\,\mu\text{H}$;
- shunt capacitance $C = 0.685/\omega R = 56\,\text{pF}$;
- series resonant circuit ($Q_L = 1$): $L_0 = 55\,\mu\text{H}$, $C_0 = 66\,\text{pF}$.

Feedback circuit (tuning and optimization):

- half-turn coupling with output parallel inductor;
- shunt resistor of $220\,\Omega$ to improve stability;
- series inductor of $3\,\mu\text{H}$ to compensate for the device input capacitance $C_{iss} = 360\,\text{pF}$.

Figure 4.37 shows the measured drain and load voltage waveforms of the parallel-circuit Class-E power oscillator with transformer feedback. Since the

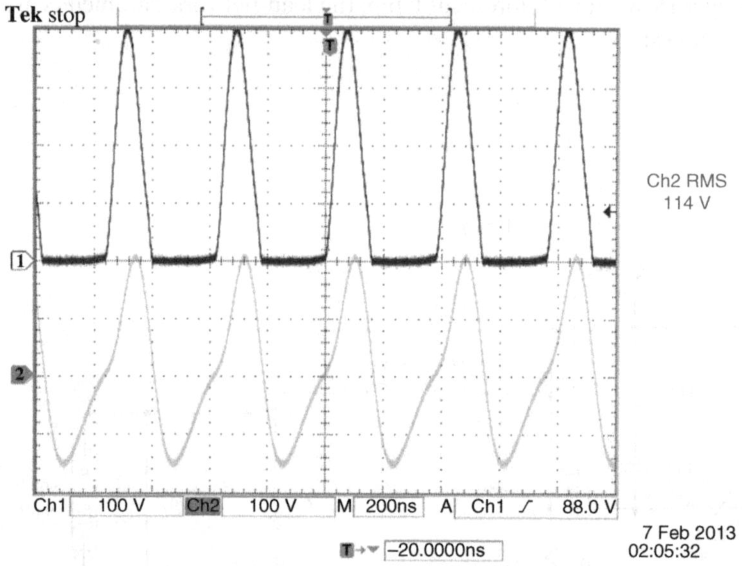

Figure 4.37 Parallel-circuit Class-E power oscillator voltage waveforms. Ch1: drain voltage waveform. Ch2: load voltage waveform (114 V across 870-Ω load corresponds to 15 W).

loaded quality factor of the series resonant circuit was chosen to be quite low to minimize the size of the load-network series coil, the output signal is not purely sinusoidal and contains a significant level of the second harmonic. However, generally, to further minimize the size of the load-network coils, the operating frequency of the parallel-circuit Class-E power oscillator can be increased to up to 30 MHz, still maintaining high efficiency, and choosing the MOSFET device with higher maximum operating frequency. The measured oscillator conversion efficiency η is calculated for the DC-supply current of 160 mA and $V_{dd} = 102\,\text{V}$ at the nominal output power of 15 W as

$$\eta = 15\,\text{W}/(102\,\text{V} \times 160\,\text{mA}) = 92\%$$

The practical verification of the parallel-circuit Class-E power oscillator for implementation into the electronic ballast circuit is shown in Figure 4.38a. To ignite plasma in a compact fluorescent lamp having high impedance at the

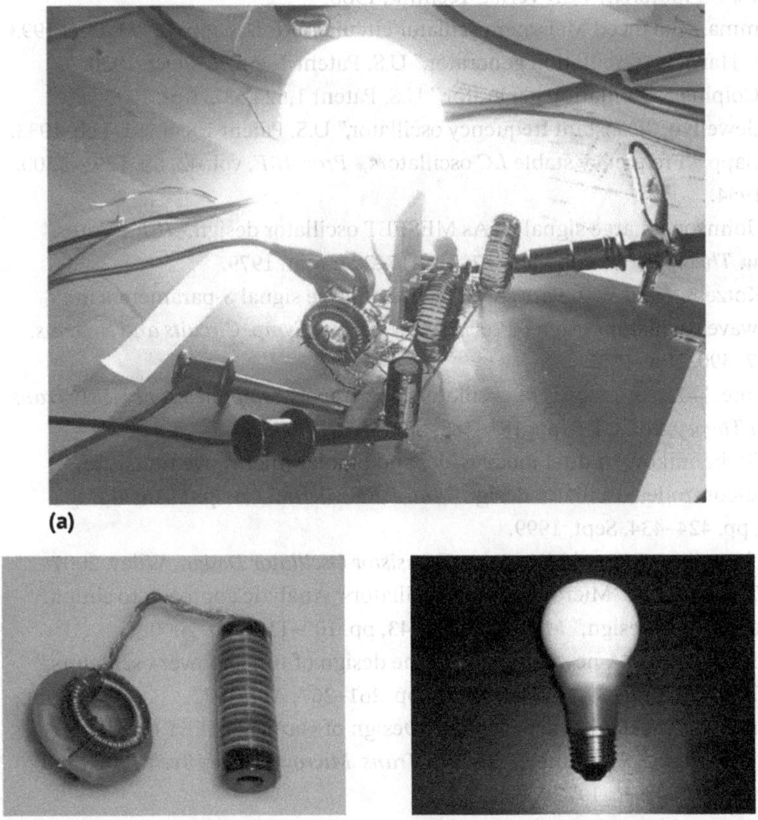

(a) (b) (c)

Figure 4.38 Practical implementation of Class-E power oscillator.

operating frequency, an inductive coupler shown in Figure 4.38b with an inductance of 9 μH is inserted into the lamp. To compensate for this inductance at the operating frequency, an external shunt capacitor of 400 pF was connected in parallel, as shown in Figure 4.36. However, in practical implementation, this coupler will represent a part of the inductor in a series resonant circuit, and an electronic ballast circuit will be placed inside the base of the compact fluorescent lamp, shown in Figure 4.38c.

References

1 E. H. Armstrong, "Some recent developments in the audion receiver," *Proc. IRE*, vol. 3, pp. 215–247, Sep. 1915; reprinted in *Proc. IEEE*, vol. 85, pp. 685–697, Apr. 1997.

2 G. Kurz, *Oszillatoren*, VEB Verlag Technik, 1988.

3 N. Demma, "Balanced Meissner oscillator circuits," *RF Des.*, pp. 72–74, Dec. 1993.

4 R. V. L. Hartley, "Oscillation generator," U.S. Patent 1,356,763, Oct. 1920.

5 E. H. Colpitts, "Oscillation generator," U.S. Patent 1,624,537, Apr. 1927.

6 F. B. Llewellyn, "Constant frequency oscillator," U.S. Patent 1,896,781, Feb. 1933.

7 J. K. Clapp, "Frequency stable *LC* oscillators," *Proc. IRE*, vol. 42, pp. 1295–1300, Aug. 1954.

8 K. M. Johnson, "Large signal GaAs MESFET oscillator design," *IEEE Trans. Microw. Theory Tech.*, vol. MTT-27, pp. 217–227, Mar. 1979.

9 K. L. Kotzebue and W. J. Parrish, "The use of large signal *S*-parameters in microwave oscillator design," *Proc. 1975 IEEE Int. Symp. Circuits and Systems*, pp. 487–490, Nov. 1975.

10 A. J. Cote, "Matrix analysis of oscillators and transistor applications," *IRE Trans. Circuit Theory*, vol. CT-5, pp. 181–188, Sep. 1958.

11 A. V. Grebennikov, "A dual mode wideband bipolar microwave transistor voltage-controlled oscillator design," *Int. J. RF Microw. Comput.-Aided Eng.*, Vol. 9, pp. 424–434, Sept. 1999.

12 A. Grebennikov, *RF and Microwave Transistor Oscillator Design*, Wiley, 2007.

13 A. V. Grebennikov, "Microwave FET oscillators: Analytic approach to simplify computer-aided design," *Microw. J.*, vol. 43, pp. 100–110, Jan. 2000.

14 M. Kazimierczuk, "A new approach to the design of tuned power oscillators," *IEEE Trans. Circuits Syst.*, vol. CAS-29, pp. 261–267, Apr. 1972.

15 Y. Mitsui, M. Nakatani, and S. Mitsui, "Design of GaAs MESFET oscillator using large-signal *S*-parameters," *IEEE Trans. Microw. Theory Tech.*, vol. MTT-25, pp. 981–984, Dec. 1977.

16 Y. Kwon, D. Pavlidis, T. L. Brock, and D. C. Streit, "A *D*-band monolithic fundamental oscillator using InP-Based HEMT's," *IEEE Trans. Microw. Theory Tech.*, vol. MTT-41, pp. 2336–2344, Dec. 1993.

17 H. J. Round, "Wireless telegraph and telephone transmission," U.S. Patent 1,564,627, Dec. 1925.

18 J. C. Shelleng, "Amplifying system," U.S. Patent 1,484,967, Feb. 1924.

19 R. Makhoul, J. Zhuang, X. Maynard, P. Perichon, D. Frey, P. O. Jeannin, and Y. Lembeye, "A very high frequency self-oscillating inverter based on a novel free-running oscillator," *IEEE Trans. Power Electron.*, vol. PE-34, pp. 8289–8292, Sep. 2019.

20 F. H. Raab, "Class-F power amplifiers with maximally flat waveforms," *IEEE Trans. Microwave Theory Tech.*, vol. MTT-45, pp. 2007–2012, Nov. 1997.

21 R. M. Rector and G. Vendelin, "A 1.0-Watt GaAs MESFET oscillator at *X*-Band," *1978 IEEE MTT-S Int. Microwave Symp. Dig.*, pp. 145–146, June 1978.

22 W. J. Hwang, S. W. Shin, G. W. Choi, H. J. Kim, and J. J. Choi, "High-efficiency power oscillator using harmonic-tuned matching network," *2009 IEEE MTT-S Int. Microwave Symp. Dig.*, pp. 1–4, June 2009.

23 S. Lee, S. Jeon, and J. Jeong, "Harmonic-tuned high efficiency RF oscillator using GaN HEMTs," *IEEE Microw. Wirel. Compon. Letts.*, vol. 22, pp. 318–320, June 2012.

24 J. Mi, R. Fan, L. Yan, Y. Feng, and C. Liu, "Design of a high-power and high-efficiency GaN-HEMT VCO based on an inverse Class-F amplifier," *IEEE Microw. Wirel. Technol. Letts.*, vol. 35, pp. 241–244, Feb. 2025.

25 J. Ebert and M. Kazimierczuk, "Class E high-efficiency tuned power oscillator," *IEEE J. Solid-State Circuits*, vol. SC-16, pp. 62–66, Apr. 1981.

26 N. O. Sokal and F. H. Raab, "Harmonic output of Class-E RF power amplifiers and load coupling network design," *IEEE J. Solid-State Circuits*, vol. SC-12, pp. 86–88, Feb. 1977.

27 D. V. Chernov, M. Kazimierczuk, and V. G. Krizhanovski, "Class-E MOSFET low-voltage power oscillator," *Proc. 2002 IEEE Int. Symp. Circuits and Systems*, vol. 5, pp. 509–512, 2002.

28 M. K. Kazimierczuk, V. G. Krizhanovski, J. V. Rassokhina, and D. V. Chernov, "Class-E MOSFET tuned power oscillator design procedure," *IEEE Trans. Circuits Syst.– I: Regul. Pap.*, vol. CAS-I-52, pp. 1138–1147, June 2005.

29 F. Ebihara, X. Wei, H. Sekiya, and T. Suetsugu, "A novel analysis procedure for Class E oscillator," *Proc. 2018 IEEE Int. Symp. Circuits and Systems*, pp. 1–4, May 2018.

30 H. Hase, H. Sekiya, J. Lu, and T. Yahagi, "Novel design procedure for MOSFET Class E oscillator," *IEICE Trans. Fundam.*, vol. E87-A, pp. 2241–2247, Sep. 2004.

31 T. B. Mader, E. W. Bryerton, M. Marcovic, M. Forman, and Z. Popovic, "Switched-mode high-efficiency microwave power amplifiers in a free-space power-combiner array," *IEEE Trans. Microw. Theory Tech.*, vol. MTT-46, pp. 1391–1398, Oct. 1998.

32 E. W. Bryerton, W. A. Shiroma, and Z. B. Popovic, "A 5-GHz high-efficiency Class-E oscillator," *IEEE Microw. Guided Wave Lett.*, vol. 6, pp. 441–443, Dec. 1996.

33 C. H. Lin, W. P. Li, and H. Y. Chang, "A fully integrated 2.4-GHz 0.5-W high efficiency class-E voltage-controlled oscillator in 0.15-μm PHEMT process," *Proc.* 2011 *Asia-Pacific Microwave Conf.*, pp. 864–867, Dec. 2011.

34 H. Y. Chang, C. H. Lin, Y. C. Liu, W. P. Li, and Y. C. Wang, "A 2.5 GHz high efficiency high power low phase noise monolithic microwave power oscillator," *IEEE Microw. Wirel. Compon. Lett.*, vol. 25, pp. 730–732, Nov. 2015.

35 S. Jeon, A. Suarez, and D. B. Rutledge, "Nonlinear design technique for high-power switching-mode oscillators," *IEEE Trans. Microw. Theory Tech.*, vol. MTT-54, pp. 3630–3640, Oct. 2006.

36 C. C. Chuang and H. K. Chiou, "A low phase noise X-band Class E power VCO in 0.25 μm GaN/SiC technology," *Proc. 2021 IEEE Int. Symp. Radio-Frequency Integration Technol.*, pp. 1–4, Aug. 2021.

37 H. Y. Chang, C. H. Lin, Y. C. Liu, W. P. Li, and Y. C. Wang, "A *K*-band high power monolithic GaAs power oscillator using Class-E network," *IEEE Microw. Wirel. Compon. Lett.*, vol. 25, pp. 730–732, Nov. 2015.

38 Y. Komiyama, W. Zhu, A. Konishi, K. Nguyen, and H. Sekiya, "Analysis and design of load-independent Class-E zero-voltage-switching power oscillator," *IEEE Trans. Circuits Syst.– I: Regul. Pap.*, vol. CAS-I-73, pp. 721–734, Jan. 2026.

39 J. R. Yang, "A Class E power oscillator for 6.78-MHz wireless power transfer system," *J. Electr. Eng. Technol.*, vol. 13 pp. 220–225, Jan. 2018.

40 T. M. Rezk, G. A. Fahmy, S. A. Ibrahim, and H. F. Ragai, "A 433 MHz E-GaN HEMT based power oscillator for far field wireless power transfer," *Proc.* 2020 *8th Int. Japan-Africa Conf. Electronics Commun. Comp. (JAC-ECC)*, pp. 80–83, 2020.

41 M. Thian and V. F. Fusco, "Series-L/parallel-tuned comparison with shunt-C/series-tuned Class-E power amplifier," *IEE Proc. Circuits Devices Syst.*, vol. 152, pp. 709–717, Dec. 2005.

42 Y. Yamashita and K. Wada, "Wireless power transmitter using parallel-tuned class-E power oscillator," *Proc. 2017 IEEE Int. Symp. Electronics and Smart Devices*, pp. 287–290, 2017.

43 Y. Komiyama, S. Matsuhashi, W. Zhu, K. Nguyen, T. Uematsu, Y. Ito, T. Mishima, and H. Sekiya, "Wireless power transfer system with load-independent inverse Class-E oscillator," *Nonlinear Theory Its Appl., IECE*, vol. 13, pp. 465–470, Mar. 2012.

44 X. Huang, F. Liang, Y. Kong, and S. Lin, "Analysis and design of Class-E ZVS inverter with parallel circuit for MHz wireless power transfer," *Proc. 2018 IEEE Int. Power Electronics Appl. Conf.*, pp. 1–5, Nov. 2018.
45 V. Ardila, F. Ramirez, and A. Suarez, "Nonlinear analysis of high-power oscillator inductively coupled to external resonator," *IEEE Microw. Wirel. Compon. Lett.*, vol. 31, pp. 737–740, June 2021.
46 L. R. Nerone, "Analysis and design of self-oscillating Class E ballast for compact fluorescent lamps," *IEEE Trans. Ind. Electron.*, vol. IE-48, pp. 151–160, Feb. 2001.
47 L. R. Nerone, "Novel self-oscillating Class E ballast for compact fluorescent lamps," *IEEE Trans. Power Electron.*, vol. PE-16, pp. 175–183, Mar. 2001.

5
High-efficiency Doherty Amplifiers

A new efficient technique for amplitude-modulated (AM) radio-frequency signals to improve performance of vacuum-tube amplifiers in broadcasting compared to both conventional amplitude-modulation techniques and Chireix outphasing was introduced by William H. Doherty in the mid-1930s [1, 2]. As a result, by using such a technique, the plate efficiencies of up to 60–65% were achieved due to the load impedance variations over the modulation cycle. After joining the radio development department of the Bell Telephone Laboratories in June 1929 and being involved in the development of high-power radio transmitters for transoceanic radiotelephony and broadcasting, Doherty invented a technique to significantly improve the efficiency of the vacuum-tube amplifiers in 1936, which was then named the "Doherty amplifier." It was first used in a 50-kW transmitter with audio-frequency feedback providing a resulting distortion level from less than 1% at lower frequencies to a few percent at high audio frequencies, demonstrating an efficiency of 60% and representing a reduction of nearly one-half in the all-day power consumption as compared with the power required in the conventional type of a linear power amplifier operating at 33% efficiency [3]. For his improvement in the efficiency of radio-frequency vacuum-tube amplifiers, the IRE Morris Liebmann Memorial Award was voted to Doherty in May 1937 [4]. By 1940, the Doherty amplifiers had been incorporated in 35 commercial radio stations worldwide, at powers up to 50 kW.

Doherty amplifiers continued to be used in a number of medium- and high-power low-frequency (LF) and medium-frequency (MF) vacuum-tube AM transmitters [5, 6]. In August 1953, a 1-MW vacuum-tube transmitter where the outputs of the two 500-kW Doherty amplifiers were connected in a bridge-type combiner began regular operation in the long-wave band in Europe. The Doherty amplifiers had also been considered for use in solid-state MF and high-frequency (HF) systems, as well as in high-power ultra-high-frequency (UHF) transmitters [7–9]. However, the practical implementation of a classic triode-based Doherty scheme was restricted by its substantial nonlinearity that required complicated envelope correction and feedback linearization circuits. At the same time, the Doherty amplifiers using tetrode transmitting tubes could improve their overall performance when the modulation was applied to the grids of both the carrier and peaking tubes [10].

The Doherty amplifier continue to be in use in very high-power AM transmitters, but for lower-power AM transmitters, the vacuum-tube amplifiers were replaced in the 1980s by the transistor power amplifiers due to their advantages of smaller size and cost, lower operating voltages, higher reliability, insensitivity to mechanical shock and vibration, and high level of integration, although with a significant nonlinearity of their performance which was difficult to improve in analog domain at that time. But with significant progress in digital linearization techniques which can directly be applied to reduce nonlinear distortion, Doherty amplifiers of different architectures have found widespread application in handset and cellular base station transmitters operating at higher frequencies, including the microwave frequency band. Recent developments of the Doherty concept demonstrate that it can be successfully used in the base station and handset applications, as well as in other energy-efficient wireless systems [11, 12].

5.1 Conventional Doherty Architecture

5.1.1 Basic Structures

A Doherty amplifier architecture is based on the outputs of two (or more) power amplifiers combining through an impedance-transforming network, which can be represented by the *LC* elements, a quarterwave transmission line, or combination of the *LC* elements and transmission line. The basic forms of the Doherty amplifier with a shunt-connected load and with a series-connected load are shown in Figures 5.1a,b, respectively [2]. In the case of a shunt-connected load with equal-power tubes (or transistors), the load impedance is equal to $R_L = R/2$, similar to the conventional type of power amplifier where two tubes are connected in parallel, where R is the load impedance at maximum output power for each tube. For a series-connected load, the impedance-transforming network provides zero impedance when terminated as an open circuit, with a load impedance $R_L = 2R$ for the carrier tube. However, at high power levels, when the peaking tube is turned on, each tube operates into the impedance R, delivering twice the carrier power at the peak of modulation, so that the value of total output power is four times the carrier power. The series connection is less convenient for practical implementations because the load is not directly grounded compared to the shunt connection when the load circuit is grounded.

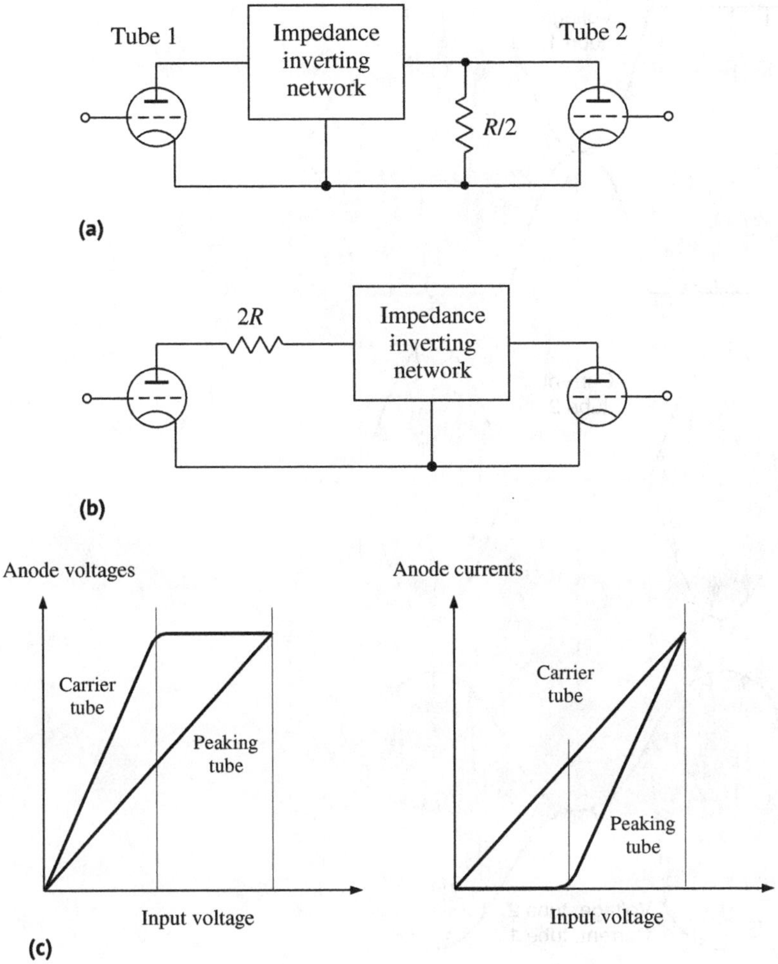

Figure 5.1 Load-network structure and ideal anode voltage and current behaviors of Doherty amplifier.

Figure 5.1c shows the ideal anode voltage and current behavior in the carrier and peaking tubes (or transistors) with varying amplitude of the grid (base or gate) excitation, with the transition voltage equal to half the peak-envelope point (PEP) for power tubes with equal size and the load-modulation region between the transition point and PEP values where the voltage on the carrier tube is constant at the PEP level. The carrier amplifier operates linearly at low output power levels, achieving saturation with a maximum efficiency at the transition voltage, while the carrier amplifier operates at saturation at higher output power levels when the peaking amplifier is in a linear region.

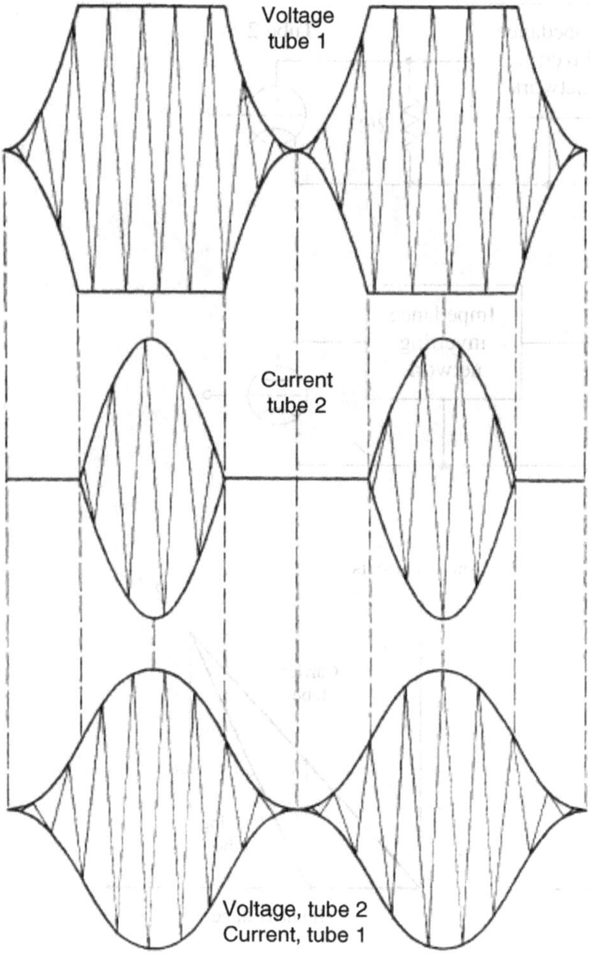

Figure 5.2 Anode envelopes of currents and voltages with varying load.

The waveforms of the anode (collector or drain) envelopes of currents and voltages with varying load are shown in Figure 5.2 [2]. For a varying voltage amplitude $V = V_0 + v_m(t)$, where V_0 is the carrier amplitude and $v_m(t) = V_m \cos \Omega t$ is the modulating signal, where V_m is the modulating amplitude and Ω is the modulating frequency, the average modulated output power P_{avr} can be obtained as

$$P_{avr} = \frac{V_0^2}{4\pi R_L} \int_0^{2\pi} (1 + m\cos \Omega t)^2 \, d\Omega t = P_0 \left(1 + \frac{m^2}{2}\right) \tag{5.1}$$

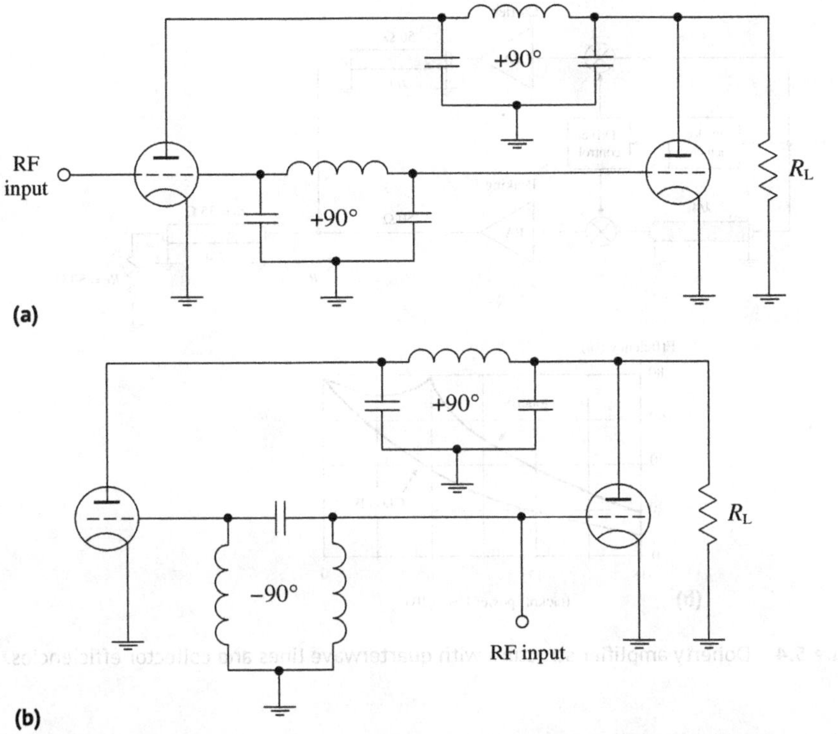

Figure 5.3 Circuit schematics of a Doherty amplifier with *LC* elements.

where $m = V_m/V_0$ is the modulation index and $P_0 = V_0^2/2R_L$ is the carrier power without modulation.

In the circuit schematic of a Doherty amplifier shown in Figure 5.3a, the anodes (collector or drains) of the carrier and peaking tubes (or transistors) are connected through a low-pass Π-type 90° circuit, where a positive phase shift of 90° is introduced between the anodes of the carrier and peaking tube [2, 10]. Such a low-pass Π-type 90° circuit is used due to its impedance-transforming properties, which means that, if the peaking tube is turned off, the impedance seen by the carrier tube will increase by four times, for example, from 35 to 140 Ω, where 35 Ω is the load-network impedance. In this case, the series inductance and two shunt capacitors are properly selected to provide a system characteristic impedance equal to $\sqrt{35 \times 140} = 70\ \Omega$. To compensate for the output phase shift of 90°, a high-pass Π-type lumped network composed of the series capacitance and two shunt inductances is included between the grids of the carrier and peaking tubes, as shown in Figure 5.3b.

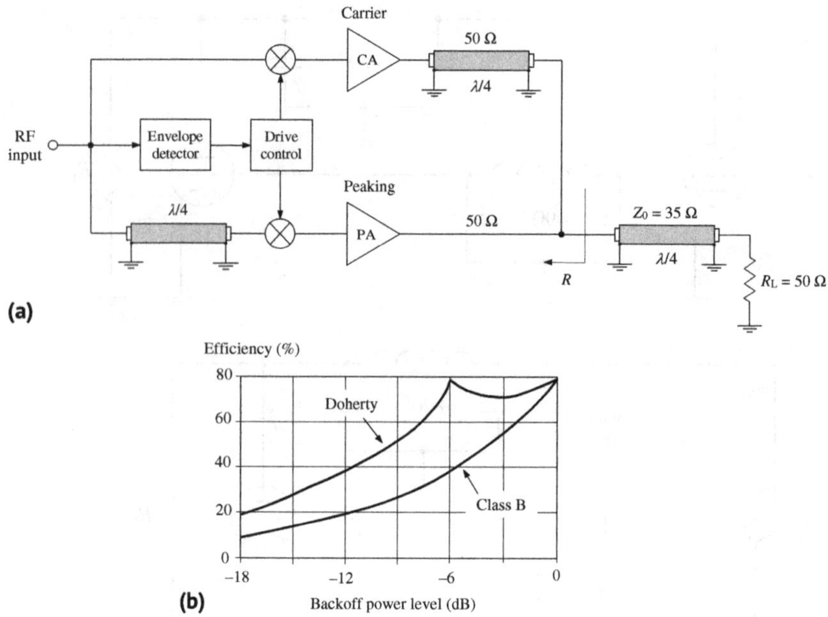

Figure 5.4 Doherty amplifier structure with quarterwave lines and collector efficiencies.

Figure 5.4a shows a two-stage transmission-line Doherty amplifier structure which incorporates the carrier (or main) and peaking (or auxiliary) power amplifiers, with a quarterwave transmission line in the carrier amplifying path for impedance transformation [13, 14]. The characteristic impedance Z_0 of an impedance-transforming quarterwave transmission line can be written as

$$Z_0 = \sqrt{Z_{in} Z_{out}} \tag{5.2}$$

where Z_{in} and Z_{out} are the input and output impedances. To clearly understand the impedance-transforming property of a quarterwave line, Eq. (5.2) can be rewritten as

$$Z_{out} = \frac{Z_0^2}{Z_{in}} \tag{5.3}$$

which shows that the output impedance Z_{out} is inversely proportional to the input impedance Z_{in} when the transmission-line characteristic impedance Z_0 is constant The output quarterwave line with $Z_0 = \sqrt{25 \times 50} = 35.4\ \Omega$ is required to match the system impedance R with the standard load impedance of 50 Ω, and the compensating quarterwave transmission line is necessary at the input of peaking amplifier.

In the particular case of an idealized system analysis or in a practical case with a power gain variation of the Doherty amplifier, an input drive control is required to turn on the peaking amplifier when the carrier amplifier is beginning to saturate and, in this case, both the carrier and peaking amplifiers will be biased in a Class-B operation mode. However, in most practical implementations, the peaking amplifier is operated in a Class-C mode, resulting in a maximum efficiency of 78.5% at a backoff power level of −6 dB from its peak value, which is twice that of a conventional Class-B power amplifier, as shown in Figure 5.4b.

5.1.2 Operation Principles

The operation principle of a conventional transmission-line Doherty amplifier with the input in-phase splitter and output impedance-transforming quarterwave transmission line shown in Figure 5.5a can be analyzed for low, medium, and peak output power regions separately [14]. Figure 5.5b shows the output current and voltage behaviors with the input voltage using idealized devices and a lossless impedance-transforming circuit, where V_L and I_L are the load voltage and current,

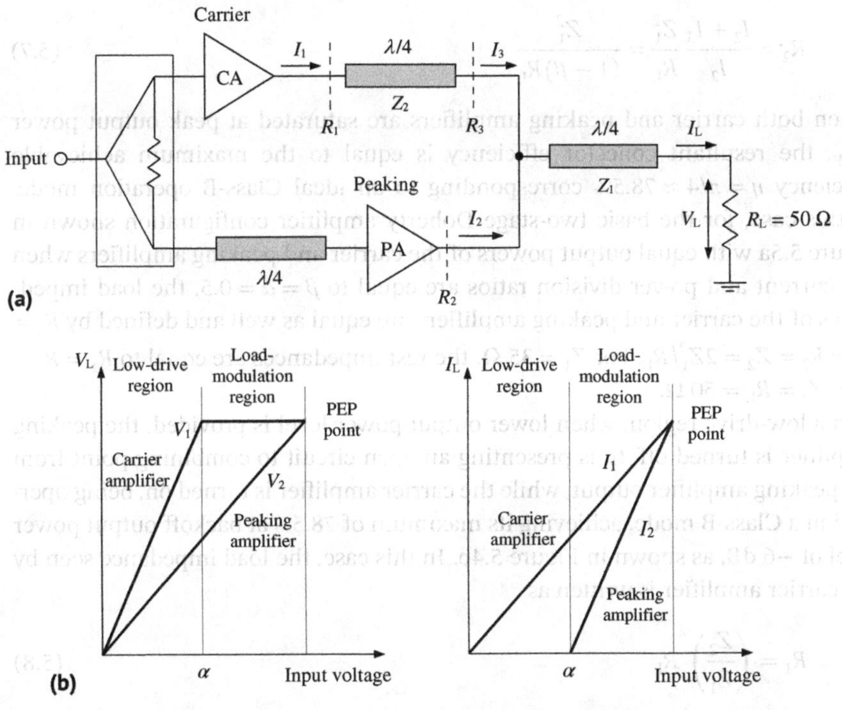

Figure 5.5 Conventional Doherty amplifier architecture and its operation principle.

respectively. Considering the condition of power conservation when the output matching quarterwave transmission line is lossless,

$$I_3 = I_1\sqrt{\frac{R_1}{R_3}} \tag{5.4}$$

where R_1 is the impedance seen by the carrier amplifier output and R_2 is the impedance seen by the peaking amplifier output, the current division ratio β is obtained as

$$\beta = \frac{I_3}{I_2 + I_3} \tag{5.5}$$

The output power P_{out} of the Doherty amplifier represents the output power $P_1 = \beta P_{out}$ of the carrier (main) amplifier and the output power $P_2 = (1-\beta)P_{out}$, of the peaking (auxiliary) amplifier while the impedance at the output of the carrier amplifying path is defined by

$$R_3 = \frac{I_2 + I_3}{I_3}\frac{Z_1^2}{R_L} = \frac{Z_1^2}{\beta R_L} \tag{5.6}$$

and the impedance at the peaking amplifier output is

$$R_2 = \frac{I_2 + I_3}{I_2}\frac{Z_1^2}{R_L} = \frac{Z_1^2}{(1-\beta)R_L} \tag{5.7}$$

When both carrier and peaking amplifiers are saturated at peak output power P_{PEP}, the resultant collector efficiency is equal to the maximum achievable efficiency $\eta = \pi/4 \approx 78.5\%$ corresponding to an ideal Class-B operation mode. In this case, for the basic two-stage Doherty amplifier configuration shown in Figure 5.5a with equal output powers of the carrier and peaking amplifiers when the current and power division ratios are equal to $\beta = \alpha = 0.5$, the load impedances of the carrier and peaking amplifiers are equal as well and defined by $R_1 = R_3 = R_2 = Z_2 = 2Z_1^2/R_L$. For $Z_1 = 35\ \Omega$, the rest impedances are equal to $R_1 = R_3 = R_2 = Z_2 = R_L = 50\ \Omega$.

In a low-drive region, when lower output power level is provided, the peaking amplifier is turned off, thus presenting an open circuit to combining point from the peaking amplifier output, while the carrier amplifier is turned on, being operated in a Class-B mode, achieving its maximum of 78.5% at backoff output power level of −6 dB, as shown in Figure 5.4b. In this case, the load impedance seen by the carrier amplifier is written as

$$R_1 = \left(\frac{Z_2}{Z_1}\right)^2 R_L \tag{5.8}$$

and, for general case of $Z_1 = 35.35\ \Omega$ and $Z_2 = R_L = 50\ \Omega$, the impedance at the output of the carrier device is equal to $R_1 = 2R_L = 100\ \Omega$.

In a load-modulation region corresponding to medium output power levels, the carrier amplifier is operated at saturation, while the peaking amplifier is turned on and continues to operate in the linear region. The collector efficiency of the carrier amplifier remains at its maximum value, while the collector efficiency of the peaking amplifier gradually increases up to its maximum value corresponding to Class-B operation mode at peak output power P_{PEP}. Consequently, the Doherty amplifier achieves its maximum efficiency at both the transition −6 dB output power backoff point and the peak output power, remaining relatively high between them, as shown in Figure 5.4b.

For a high-quality signal transmission in a Doherty amplifier, it is necessary to obtain a linear dependence between the input driving voltage and the output collector (or drain) voltage. In this case, it is important to optimize the bias of the carrier device and the amplitude of the driving signal, whose careful selection may allow the power amplifier to operate with low distortion. Considering a low-power region, the linearity of the Doherty amplifier is fully defined by the carrier amplifier that should be highly linear even though the load impedance is high. When using LDMOSFET devices, linearity can be improved in a high-power region by the harmonic cancellation using appropriate gate bias voltages [15].

A microwave version of the Doherty amplifier was first presented in the mid-1990s, based on existing at that time microwave design techniques and implementing a few modifications [16]. In this case, an efficiency of 61% was achieved at a 1-dB gain compression point, and this level of efficiency was maintained through a 5.5-dB reduction in output power at an operating frequency of 1.37 GHz. A few years later, the possibility of achieving high-efficiency performance of a microwave monolithic Doherty amplifier was demonstrated in the *Ku*-band and *K*-band frequencies using pHEMT and InP DHBT technologies, respectively [17, 18]. The first fully integrated millimeter-wave Doherty amplifier with a chip size of 2 mm^2 operating in a frequency bandwidth of 38–46 GHz was fabricated based on a 0.15-μm GaAs HEMT process [19]. Higher average efficiencies of 55% and greater were achieved with high-voltage HBT and GaN HEMT technologies [20, 21].

5.1.3 Offset Lines

In order to provide the required open-circuit condition at the peaking amplifier output when the peaking device is turned off, the transistor output reactance due to internal device and external package parasitics should be compensated by the so-called offset lines with optimized electrical lengths [15, 22]. The basic schematic of a fully matched microwave Doherty amplifier with equal-length offset lines in the load network is shown in Figure 5.6. Here, the phase-compensating

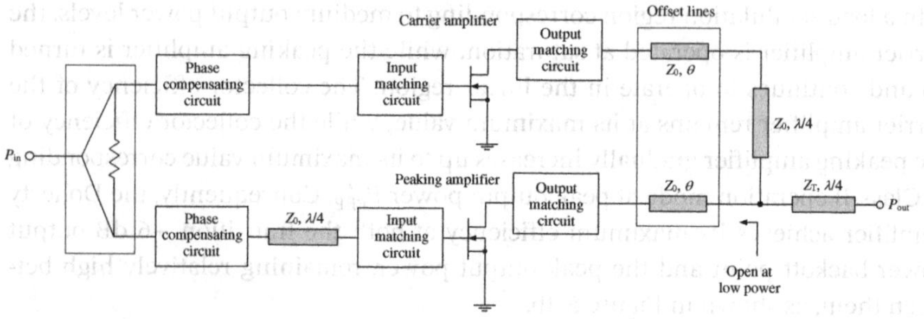

Figure 5.6 Block structure of transmission-line Doherty amplifier.

circuits in the input circuit which can be implemented using lumped elements or transmission lines are required to reduce the amplitude and phase variations over output power [22, 23]. However, usually it is sufficient to include an input phase-compensating transmission line at the input of the peaking amplifier to optimize linearity and efficiency for applications with a multicarrier input signal [24]. To further improve the Doherty amplifier performance, an adaptive power-dependent input power distribution between the carrier and peaking amplifiers can be used when more power is delivered to the carrier amplifier in the low-power region and to the peaking amplifier in the high-power region, which allows linearity to be improved from 5 to 7 dB and efficiency to be increased up to 5% over a wide range of output powers for WCDMA signals at 2.4 GHz [25].

For illustration, consider a design example with additional offset lines having the characteristic impedance of 50 Ω and different electrical lengths θ_c and θ_p connected at the outputs of the carrier and peaking amplifiers, respectively, as shown in Figure 5.7a. In a low-power region, setting the proper electrical length θ_p will cause the peaking amplifier to provide an open circuit, while setting the proper electrical length θ_c is required to double the impedance seen by the carrier device to 100 Ω because of an impedance-transforming property of a 50-Ω quarterwave transmission line in a carrier amplifying path. Here, it is assumed that the matching circuit in conjunction with the device output parasitics, such as shunt output capacitance and series bondwire inductance and offset line, provides the required impedance transformation to the optimum high impedance Z_{opt} seen by the carrier device output at 6-dB power backoff, as shown in Figure 5.7b. At the same time, the offset line in the peaking path is adjusted to transform the matching circuit impedance Z_{match_off} to a high impedance (ideally infinite or open circuit) so that it prevents power leakage to the peaking path when the peaking transistor is turned off. In practical design, effect of the saturation (or knee) voltage should be considered because of a nonzero value of the device

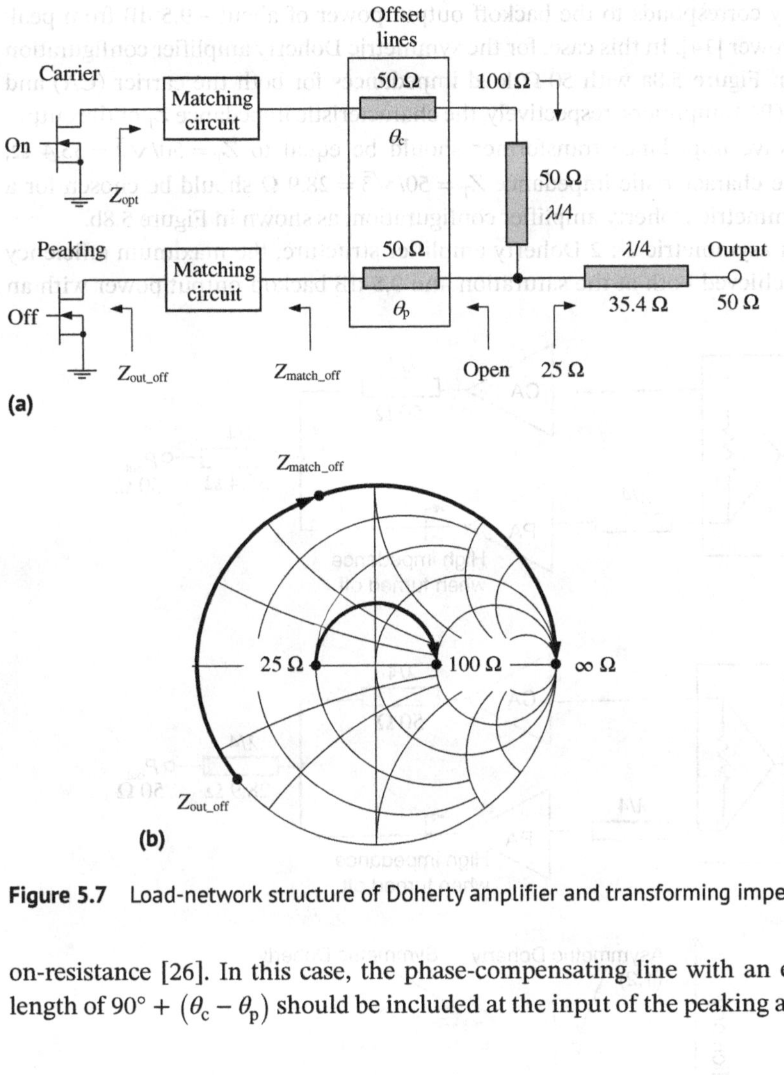

Figure 5.7 Load-network structure of Doherty amplifier and transforming impedances.

on-resistance [26]. In this case, the phase-compensating line with an electrical length of $90° + (\theta_c - \theta_p)$ should be included at the input of the peaking amplifier.

5.2 Asymmetric Doherty Amplifiers

There is a possibility to extend the high-efficiency region over a wider range of the backoff output powers if the carrier and peaking amplifiers are designed to operate with different output powers: smaller for the carrier (or main) amplifier and larger for the peaking (or auxiliary) amplifier. For example, for a 1 : 2 power division ratio when the output power from the peaking amplifier is two times larger than that of the carrier amplifier, the transition point with maximum drain (or collector) efficiency corresponds to the backoff output power of about −9.5 dB from peak output power [14]. In this case, for the symmetric Doherty amplifier configuration shown in Figure 5.8a with 50-Ω load impedances for both the carrier (CA) and peaking (PA) amplifiers, respectively, the characteristic impedance Z_T of the output quarterwave impedance transformer should be equal to $Z_T = 50/\sqrt{2} = 35.4\ \Omega$, while the characteristic impedance $Z_T = 50/\sqrt{3} = 28.9\ \Omega$ should be chosen for a 1 : 2 asymmetric Doherty amplifier configuration, as shown in Figure 5.8b.

For an asymmetric 1 : 2 Doherty amplifier structure, the maximum efficiency can be achieved both at the saturation and 9.5-dB backoff output power with an

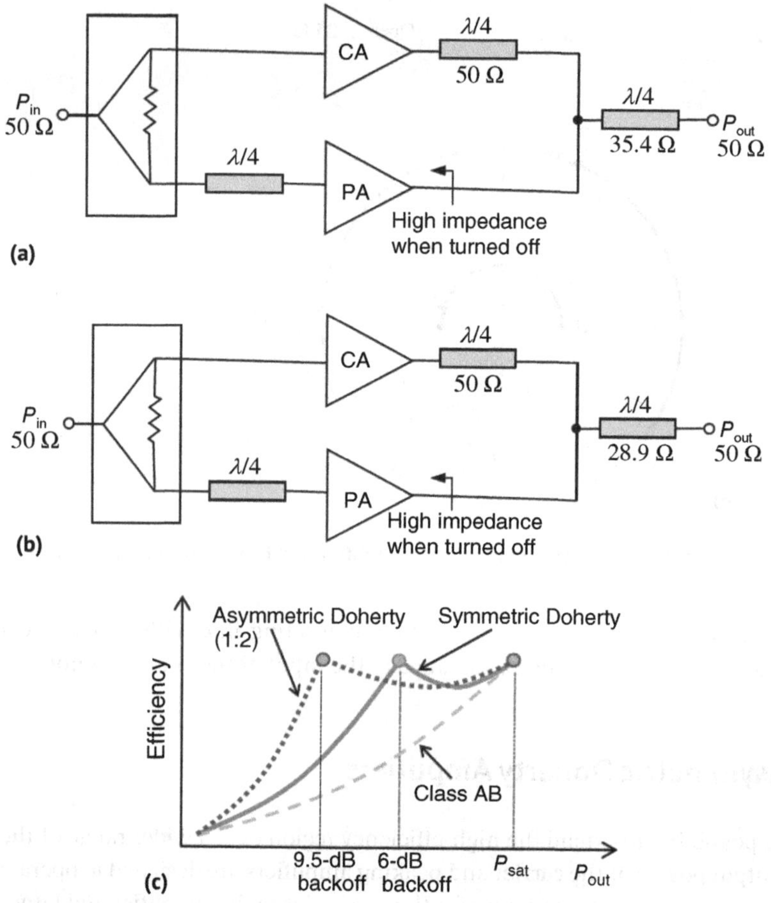

Figure 5.8 Block schematics of (a) symmetric and (b) asymmetric 1 : 2 Doherty amplifier and (c) their theoretical efficiencies.

enough high-efficiency level between these power levels, as shown in Figure 5.8c. At the same time, such an asymmetric Doherty structure is easier in implementation due to a much smaller size, compared to multistage Doherty configurations. The conventional Class-AB power amplifiers used for the same applications usually have the drain efficiencies about four times smaller at this backoff output power. To optimize the linearity and efficiency of an asymmetric Doherty amplifier, it is especially important to optimize the bias conditions for both the carrier and peaking amplifiers [27].

When a power division ratio is reduced to $\alpha = 0.25$, providing the output power from the peaking amplifier of three times greater than that of the carrier amplifier, the transition point with maximum drain efficiency will correspond to the backoff output power level of −12 dB from peak output power [14]. In this case, from consideration of their output powers, it follows that $R_1 = Z_2 = R_3 = 3R_2$, thus resulting in $I_2 = 3I_3$ and $\beta = 0.25$. From Eqs. (5.6) and (5.7), it follows that the output impedances R_2 and R_3 can be calculated through the load resistance R_L and characteristic impedance Z_1. For instance, by setting $Z_1 = 15\ \Omega$ and $R_L = 50\ \Omega$, the impedances Z_2 and R_1 are equal to $Z_2 = R_1 = 18\ \Omega$, and to $R_2 = R_1/3 = 6\ \Omega$. Because from Eqs. (5.6) and (5.7), it follows that

$$R_1 = \frac{Z_2^2}{\beta R_3} \tag{5.9}$$

hence, the output impedance R_1 at lower output power when the peaking amplifier is turned off is four times higher than that at peak output power where $R_1 = Z_2 = R_3$. As a result, the power ratio between the carrier and peaking amplifiers is defined as 1 : 3.

A monolithic Doherty amplifier using an extended technique based on GaAs HBT devices with total emitter areas of 840 and 3360 μm^2 (scaling ratio of 1 : 4) for the peaking and carrier amplifiers, respectively, was able to provide the power-added efficiencies of 45 and 23% at the maximum output power of 25 dBm and at 10-dB output power backoff, respectively [28].

In hybrid implementations using the packaged transistors, it is convenient to use the identical power amplifiers with the same-size devices where the carrier amplifier is connected in parallel to $(N-1)$ peaking amplifiers. In this case, the simplest hybrid approach is to combine all peaking amplifiers in a single peaking amplifier with a single transistor providing $(N-1)$ times higher output power larger than that in a carrier amplifier to compose an asymmetric two-way Doherty amplifier configuration [29]. Figure 5.9a shows the block schematic of an asymmetric N-way Doherty amplifier with a parallel connection of a carrier amplifier and $(N-1)$ identical peaking amplifiers. The ideal drain efficiencies of the N-way

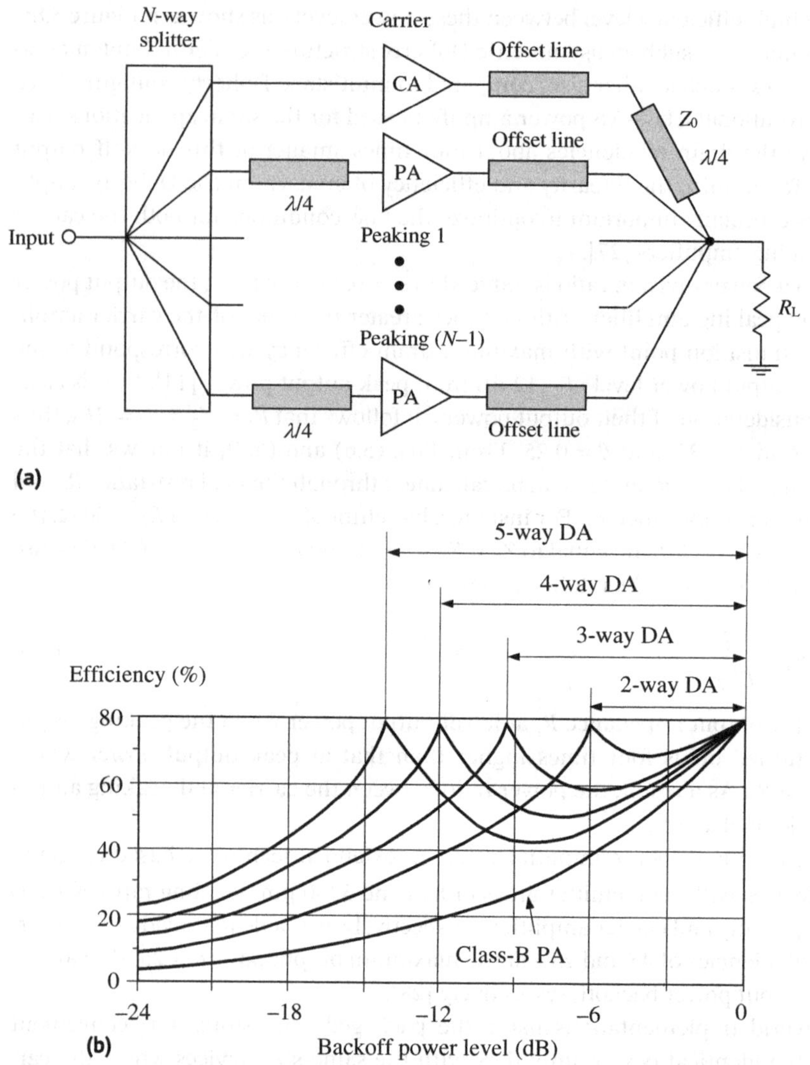

Figure 5.9 Doherty amplifier with asymmetric *N*-way structure and efficiencies.

Doherty amplifier (DA) structures with peak values at −6, −9.5, −12, and −14 dB output power backoff points according to $P_{\text{backoff}} = 20\log_{10}N$ for the two-, three-, four-, and five-way configurations, respectively, and for the conventional Class-B power amplifier are shown in Figure 5.9b.

Generally, the *N*-way Doherty amplifier is composed of the *N*-way power splitter, an identical fully matched carrier and $(N-1)$ peaking amplifiers, *N* offset lines, and an output combiner representing a quarterwave impedance transformer. An example of the asymmetric 1 : 2 Doherty amplifier is shown in Figure 5.9a, with 50-Ω load for the carrier amplifier, 25-Ω load for the peaking amplifier, and characteristic impedance $Z_0 = \sqrt{16.7 \times 50} = 28.9\ \Omega$ of the output quarterwave line [30]. In order to provide the operation with better linearity and produce more power, the asymmetric two-way and three-way Doherty amplifiers are designed with optimized individual bias conditions, uneven drive, and load matching for the carrier and peaking amplifiers [31, 32]. For example, a three-way 10-W GaN HEMT Doherty amplifier using Class-F load networks achieved a PAE of 45.9% for a single-carrier WCDMA signal with a PAR of 10 dB at an operating frequency of 2.14 GHz [33].

For a high-power amplifier with a very low output impedance, the width of the matching microstrip line becomes very wide being comparable to its length and the overall size of the matching circuit, including an offset line to create an open-circuit condition when the peaking amplifier is turned off and a quarterwave transforming microstrip line, becomes sufficiently large. In this case, it is very difficult to physically connect the output of the peaking amplifier directly to the carrier amplifier path. Besides, in order to provide a high-efficiency performance over a wide frequency bandwidth, an impedance conversion ratio should not be too high, resulting in an external circuit designed in a low-impedance system. Therefore, for convenience of its implementation, the classical *N*-way Doherty amplifier structure can be modified by including an additional half-wave line at the output of each peaking amplifier, as shown in Figure 5.10a for a 1 : 2 asymmetric Doherty amplifier [34, 35]. Adding a half-wave line at the peaking amplifier output can increase the flexibility of the circuit layout without losing the Doherty operation, though it may affect the performance when it is necessary to provide a very wide frequency bandwidth. To align the phases of both carrier and peaking amplifiers, an additional quarterwave line should be incorporated at the input side of the carrier amplifier.

For the packaged high-power transistors, when it is difficult to choose the proper power ratio between the transistors, it is convenient to use and combine the identical power amplifiers. For example, a 1 : 2 asymmetric two-way Doherty amplifier configuration can be transformed into a modified three-way asymmetric Doherty amplifier structure with a carrier amplifier (CA) and two identical peaking amplifiers (PA1 and PA2), all having equal-size devices, as shown in Figure 5.10b. Here, the half-wave transmission line in each peaking amplifier path can be split into two series quarterwave transmission lines with the corresponding characteristic impedances for the output combining and impedance matching.

The block schematic of a three-way asymmetric Doherty amplifier structure is shown in Figure 5.11a, where the output combiner provides both the power combining and impedance transformation to match with a 50-Ω load [34, 35].

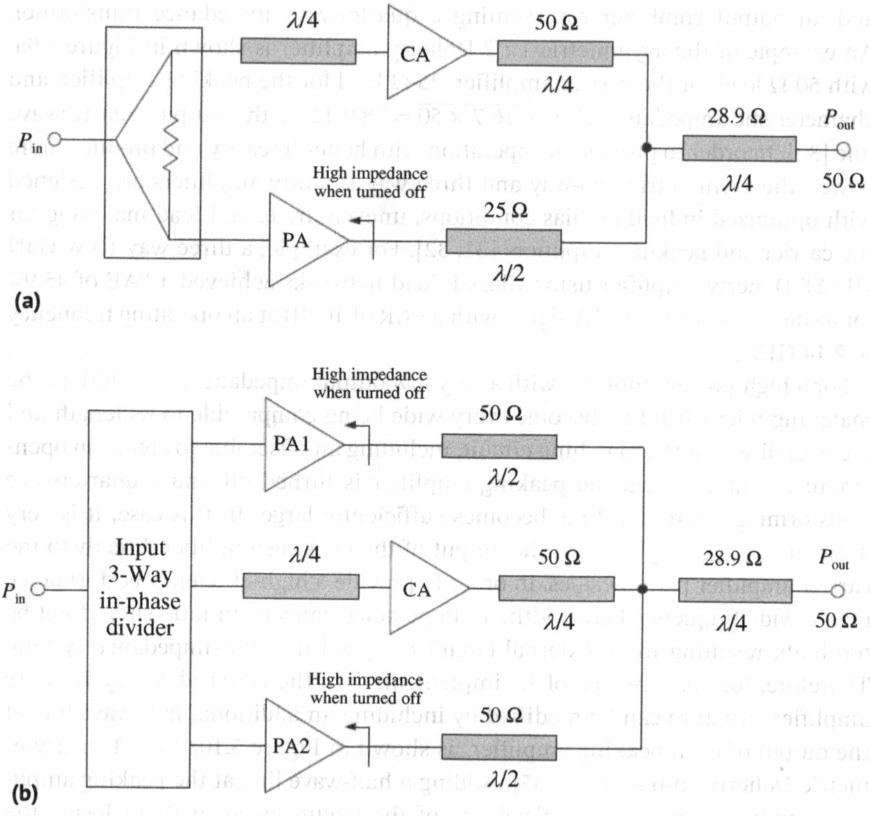

Figure 5.10 Modified 1 : 2 asymmetric Doherty amplifier structures.

Each amplifying path (carrier and two peaking) includes an equal-size packaged device, microstrip-line input and output matching circuits, and offset lines to provide the open-circuit conditions for the peaking amplifiers when they are turned off. Generally, various combinations of the characteristic impedances of the quarterwave microstrip lines in the output combiner can be used to optimize the Doherty amplifier performance.

A modified three-way Doherty amplifier based on three dual-path GaN HEMT devices in metal–ceramic flange packages, each including a pair of 180-W GaN HEMT dies with internal input matching, was implemented into the test board fabricated on a 20-mil RO4350 substrate. In this case, the output lead inductances of the packaged GaN HEMT devices were minimized for better performance prediction. As a result, the measured output power P_{1dB} of 60 dBm at 1-dB gain compression point and peak drain efficiency of 80% with a power gain of about 15 dB

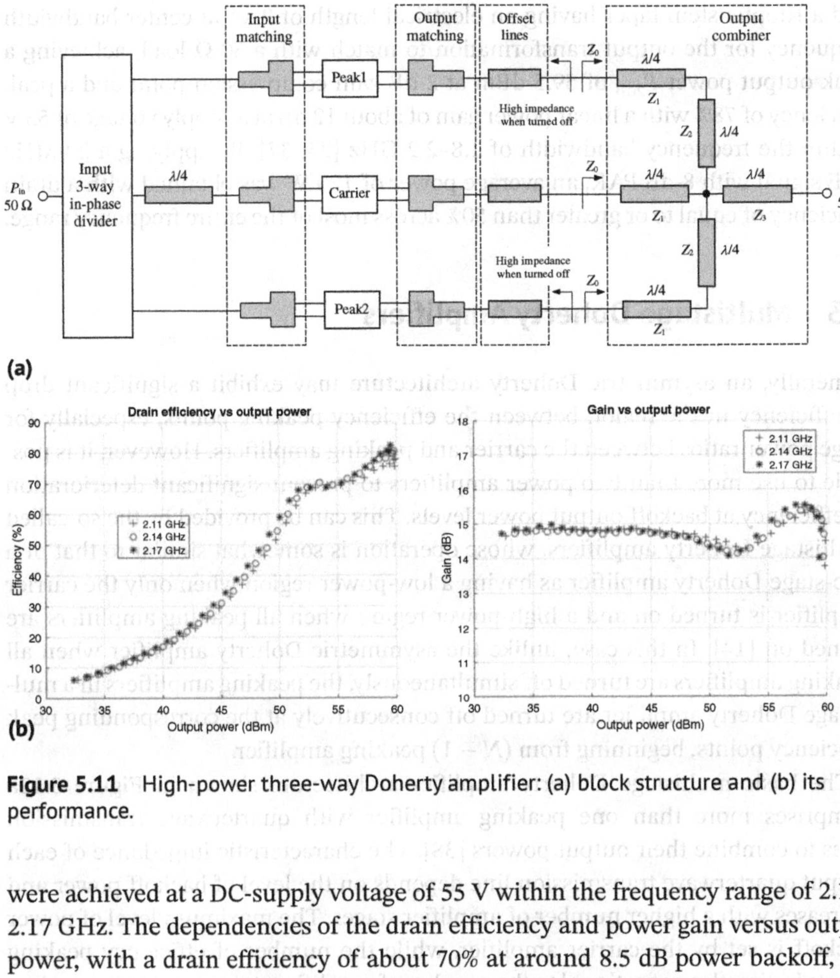

Figure 5.11 High-power three-way Doherty amplifier: (a) block structure and (b) its performance.

were achieved at a DC-supply voltage of 55 V within the frequency range of 2.11–2.17 GHz. The dependencies of the drain efficiency and power gain versus output power, with a drain efficiency of about 70% at around 8.5 dB power backoff, are shown in Figure 5.11b [34]. An average power of 52 dBm was obtained with a drain efficiency of about 65% for a single-carrier 20-MHz LTE signal with a PAR of 8 dB. In a linear operating region, a power gain of about 15 dB was achieved with a 2-dB flatness over the entire output power range up to 60 dBm when its value reduces by just 1 dB compared to its value in a linear region.

The operating frequency bandwidth of the Doherty amplifier can be extended by optimizing the circuit parameters and providing accurate load-pull testing. As a result, an average power of 85 W with a peak output power of 470–570 W and average efficiencies of 45–49% were achieved across the frequency bandwidth of 1.8–2.2 GHz for an asymmetric two-way Doherty amplifier [36]. The higher output power was achieved using the three dual-path GaN HEMT devices and a Klopfenstein taper having an electrical length of 270° at center bandwidth frequency for the output transformation to match with a 50-Ω load, achieving a peak output power P_{2dB} of 59.5 dBm at 2-dB gain compression point and a peak efficiency of 78% with a linear power gain of about 12 dB at a supply voltage of 55 V within the frequency bandwidth of 1.8–2.2 GHz [35, 37]. By applying a 20-MHz LTE signal with 8-dB PAR, an average power of 120 W was obtained with a drain efficiency of equal to or greater than 50% across most of the entire frequency range.

5.3 Multistage Doherty Amplifiers

Generally, an asymmetric Doherty architecture may exhibit a significant drop in efficiency in the region between the efficiency peaking points, especially for large power ratios between the carrier and peaking amplifiers. However, it is possible to use more than two power amplifiers to prevent significant deterioration of efficiency at backoff output power levels. This can be provided by the so-called multistage Doherty amplifiers, whose operation is somewhat similar to that of a two-stage Doherty amplifier as having a low-power region when only the carrier amplifier is turned on and a high-power region when all peaking amplifiers are turned on [14]. In this case, unlike the asymmetric Doherty amplifier when all peaking amplifiers are turned off simultaneously, the peaking amplifiers in a multistage Doherty amplifier are turned off consecutively at the corresponding peak efficiency points, beginning from $(N-1)$ peaking amplifier.

The basic multistage Doherty amplifier architecture shown in Figure 5.12a comprises more than one peaking amplifier with quarterwave transmission lines to combine their output powers [38]. The characteristic impedance of each output quarterwave transmission line depends on the level of backoff power and increases with a higher number of amplifier stages. The maximum level of power backoff is set by the carrier amplifier, while the number of efficiency peaking points is directly proportional to the number of amplifier stages used in the design. Figure 5.12b shows the simplified structure of a three-stage Doherty amplifier, with the characteristic impedances of the output impedance-transforming quarterwave transmission lines defined as $Z_1 = 30\ \Omega$ and $Z_2 = 120\ \Omega$ for $R_L = 15\ \Omega$ and the characteristic impedance of the output matching transmission line of $\sqrt{15 \times 50} = 27.4\ \Omega$.

Figure 5.13 shows the maximum theoretical drain efficiencies of the multistage Doherty amplifier (DA) structures for the two, three, and four stages, with maximum efficiencies at the transition points of −6, −12, and −18 dB backoff output powers, respectively. As shown in Figure 5.13, the multistage Doherty architecture provides higher efficiencies at backoff output powers between the

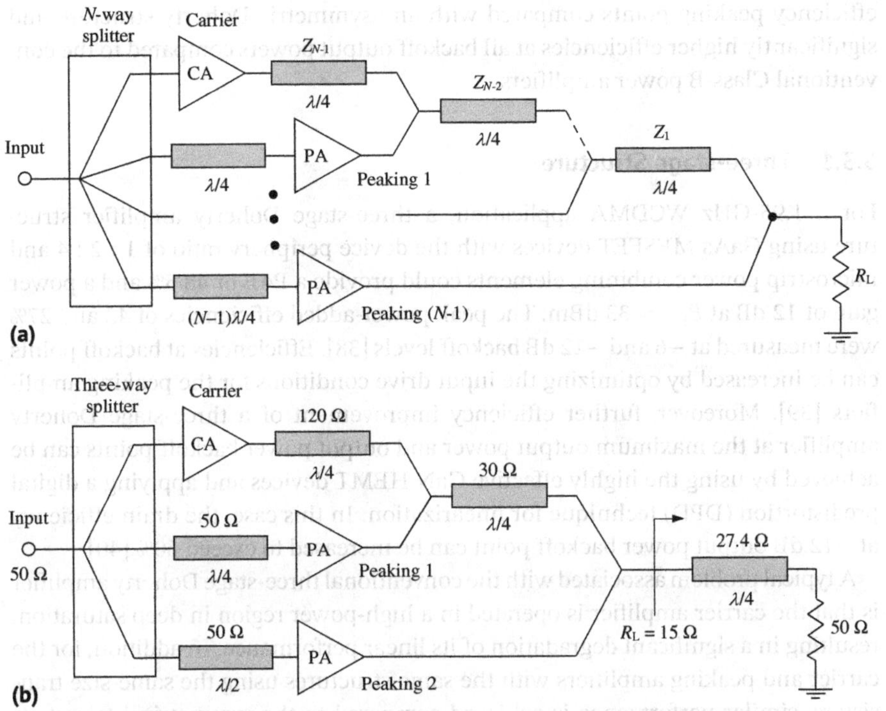

Figure 5.12 Multistage Doherty amplifier architectures.

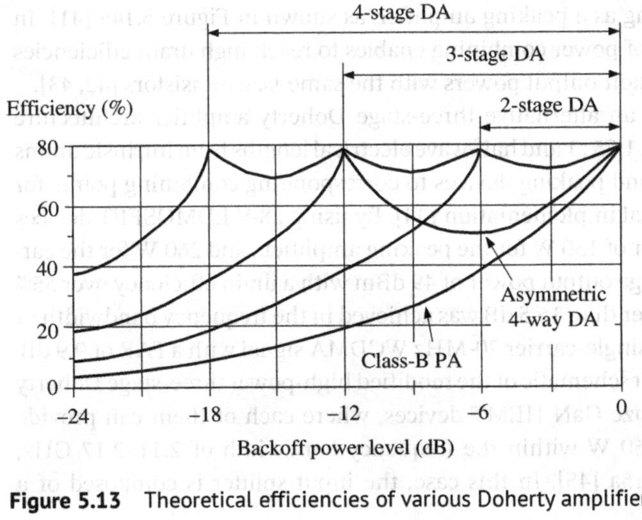

Figure 5.13 Theoretical efficiencies of various Doherty amplifier structures.

efficiency peaking points compared with an asymmetric Doherty structure and significantly higher efficiencies at all backoff output powers compared to the conventional Class-B power amplifiers.

5.3.1 Three-stage Structure

For a 1.95-GHz WCDMA application, a three-stage Doherty amplifier structure using GaAs MESFET devices with the device periphery ratio of 1 : 2 : 4 and microstrip power combining elements could provide a PAE of 48.5% and a power gain of 12 dB at P_{1dB} = 33 dBm. The peak power-added efficiencies of 42 and 27% were measured at −6 and −12 dB backoff levels [38]. Efficiencies at backoff points can be increased by optimizing the input drive conditions for the peaking amplifiers [39]. Moreover, further efficiency improvement of a three-stage Doherty amplifier at the maximum output power and output power backoff points can be achieved by using the highly effective GaN HEMT devices and applying a digital predistortion (DPD) technique for linearization. In this case, the drain efficiency at −12 dB output power backoff point can be increased to exceed 60% [40].

A typical problem associated with the conventional three-stage Doherty amplifier is that the carrier amplifier is operated in a high-power region in deep saturation, resulting in a significant degradation of its linear performance. In addition, for the carrier and peaking amplifiers with the same structures using the same-size transistors, similar performance is achieved compared to the symmetrical two-stage Doherty amplifier, with the efficiency peaking points at −3.5 and −6 dB backoff output powers. These problems can be partially solved by using a modified three-stage Doherty amplifier structure with a parallel combining of the carrier and a Doherty amplifier using as a peaking amplifier, as shown in Figure 5.14a [41]. In this case, a novel way of power combining enables to reach high drain efficiencies at −6 and −9.5 dB backoff output powers with the same-size transistors [42, 43].

Figure 5.14b shows an alternative three-stage Doherty amplifier architecture with a power ratio of 1 : 1.75 : 1 and half-wave electrical lengths from intrinsic drains of the carrier and second peaking devices to corresponding combining points for convenience of practical implementation [44]. By using 28-V LDMOSFET devices providing a peak power of 150 W for the peaking amplifiers and 260 W for the carrier amplifier, an average output power of 49 dBm with a drain efficiency over 55% and a power gain greater than 15.5 dB was achieved in the frequency bandwidth of 1805–1880 MHz for a single-carrier 20-MHz WCDMA signal with a PAR of 9.9 dB.

The simplified circuit schematic of the modified high-power three-stage Doherty amplifier with equal-size GaN HEMT devices, where each of them can provide an output power of 250 W within the frequency bandwidth of 2.11–2.17 GHz, is shown in Figure 5.15a [45]. In this case, the input splitter is composed of a

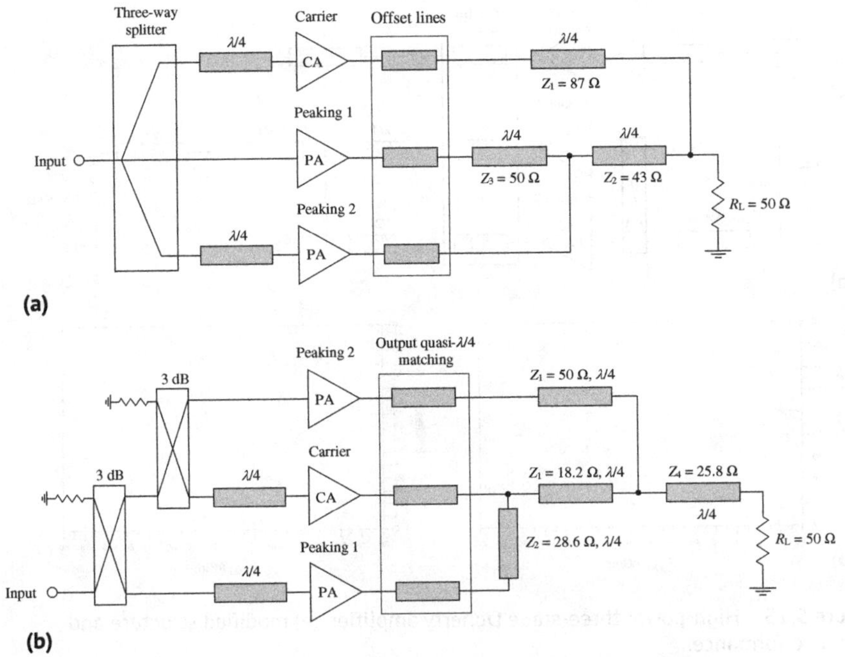

Figure 5.14 Three-stage Doherty amplifiers with modified structures.

two-way asymmetric 1 : 2 Wilkinson divider and a hybrid quadrature coupler to split the input power between the two peaking amplifiers, resulting in the presence of a half-wave transmission line at the output of the second peaking amplifier and a $3\lambda/4$ transmission line at the input of the carrier amplifier for proper phase compensation. Here, the short-length offset lines provide open-circuit conditions for the turned-off peaking device. Figure 5.15b shows the simulation results demonstrating the potential capability of achieving a very high drain efficiency over 70% at the peak output power greater than 59 dBm and 60% at 10-dB output power backoff, with a linear power gain above 14 dB and a gain flatness less than 3 dB.

5.3.2 Four-stage Architecture (Doherty-in-Doherty)

Figure 5.16 shows the theoretical drain efficiencies of the multistage (three- and four-stage) and four-way asymmetric Doherty amplifier (DA) structures for various power (or device size) ratios, with peak efficiencies ranging from −12 dB backoff output powers. As shown in Figure 5.16, the four-way or any asymmetric multiway Doherty amplifier structure provides much lower efficiency between the

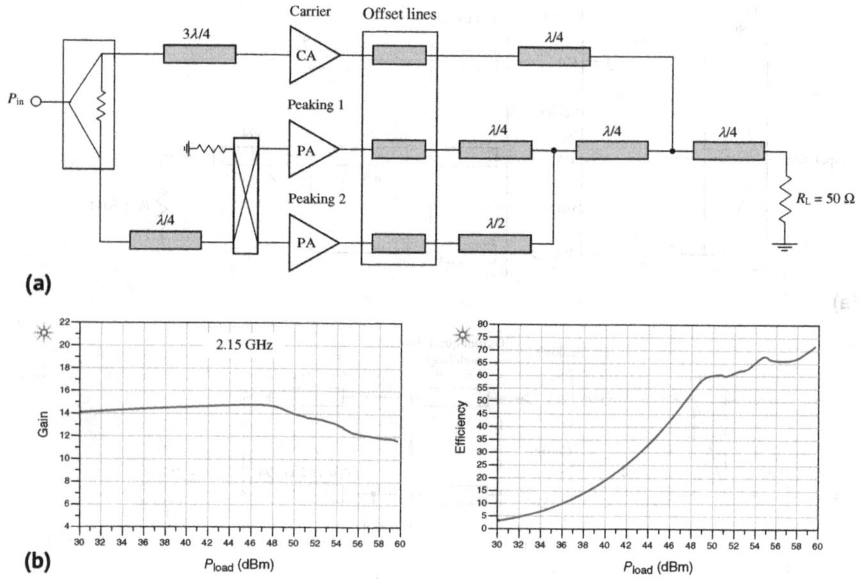

Figure 5.15 High-power three-stage Doherty amplifier: (a) modified structure and (b) its performance.

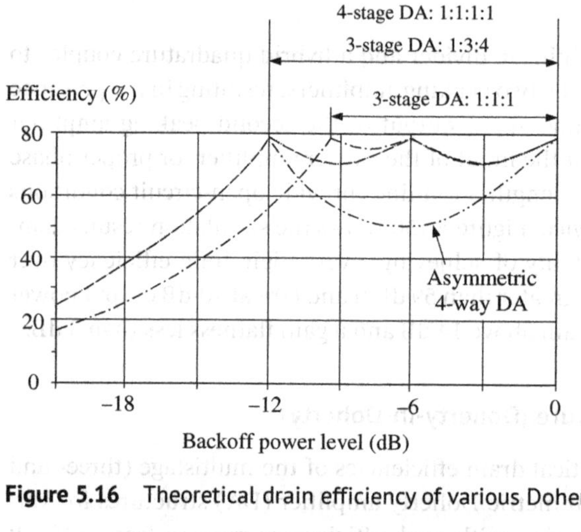

Figure 5.16 Theoretical drain efficiency of various Doherty amplifier structures.

corresponding peak efficiency points. However, for a multistage Doherty amplifier structure, the peak efficiency points at lower output power backoffs can be obtained using an optimum device size ratio. In this case, a peak efficiency at the lowest backoff point of −12 dB is achieved for a device periphery ratio of 1 : 3 : 4 in a three-stage Doherty amplifier, whereas the lowest output power backoff of about −9.5 dB corresponds to the peak drain efficiency for an equal-size device ratio of 1 : 1 : 1 in the modified three-stage Doherty amplifier, whose structure is shown in Figure 5.14a [38, 43].

In a classical four-stage Doherty amplifier with the corresponding peak efficiencies at −6, −12, and −18 dB backoff output power points, the maximum ratio between the characteristic impedances of the impedance-transforming quarter-wave transmission lines is equal to 16 [38]. In this case, it becomes difficult to correctly implement the corresponding microstrip lines with a high characteristic impedance ratio on a single substrate. However, in a modified four-stage Doherty amplifier structure with equal device size ratio, as shown in Figure 5.17, the two conventional two-stage Doherty amplifiers are properly combined in a four-stage Doherty amplifier structure (Doherty-in-Doherty), with the maximum impedance ratio of 50 Ω/25 Ω = 2 only [46]. Figure 5.16 shows the three efficiency peaking points corresponding to the performance of the modified four-stage Doherty amplifier with equal gate bias voltages for the second and third peaking amplifiers. It should be noted that this is a very practical version of a four-stage Doherty amplifier, when high output powers with high drain efficiency using three commercially available 90° hybrid couplers at the input and four quarter-wave microstrip lines at the output can be achieved.

The test board of the modified four-stage GaN HEMT Doherty amplifier based on four 25-W Cree CGH40025F devices, which is shown in Figure 5.18,

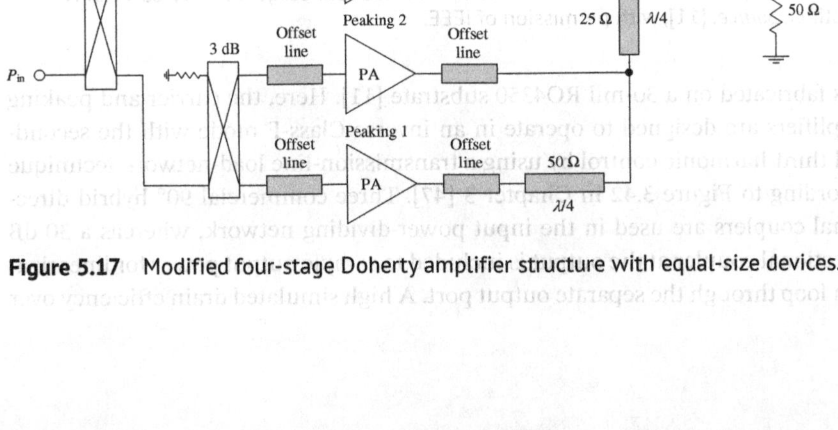

Figure 5.17 Modified four-stage Doherty amplifier structure with equal-size devices.

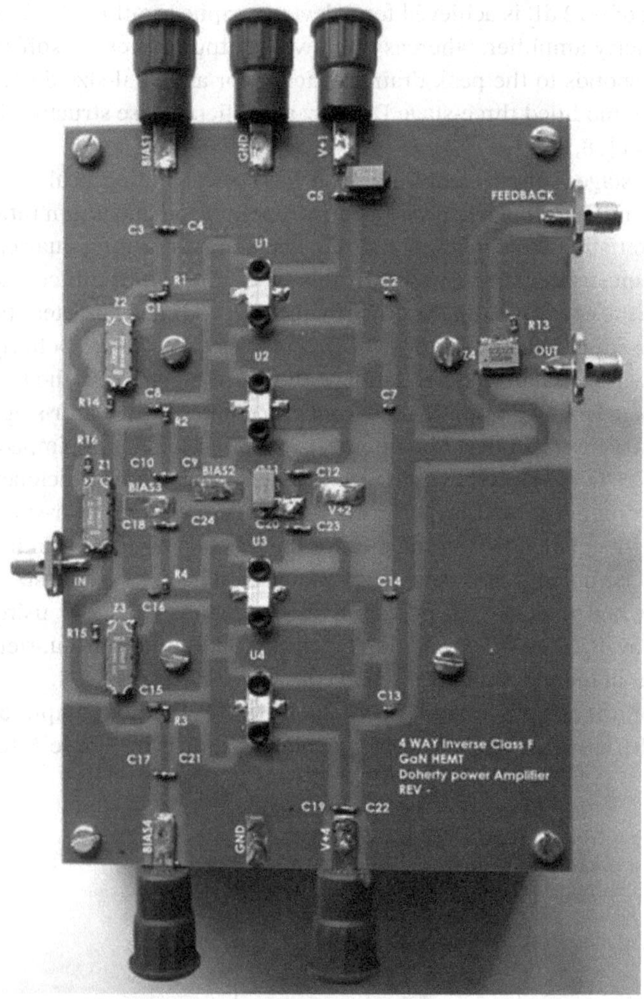

Figure 5.18 Test board of 2.14 GHz 100 W modified four-stage Doherty GaN HEMT amplifier. *Source*: [11], with permission of IEEE.

was fabricated on a 30-mil RO4350 substrate [11]. Here, the carrier and peaking amplifiers are designed to operate in an inverse Class-F mode with the second- and third-harmonic control by using a transmission-line load-network technique according to Figure 3.42 in Chapter 3 [47]. Three commercial 90° hybrid directional couplers are used in the input power-dividing network, whereas a 30-dB directional coupler at the output is included to sample output power for linearization loop through the separate output port. A high simulated drain efficiency over

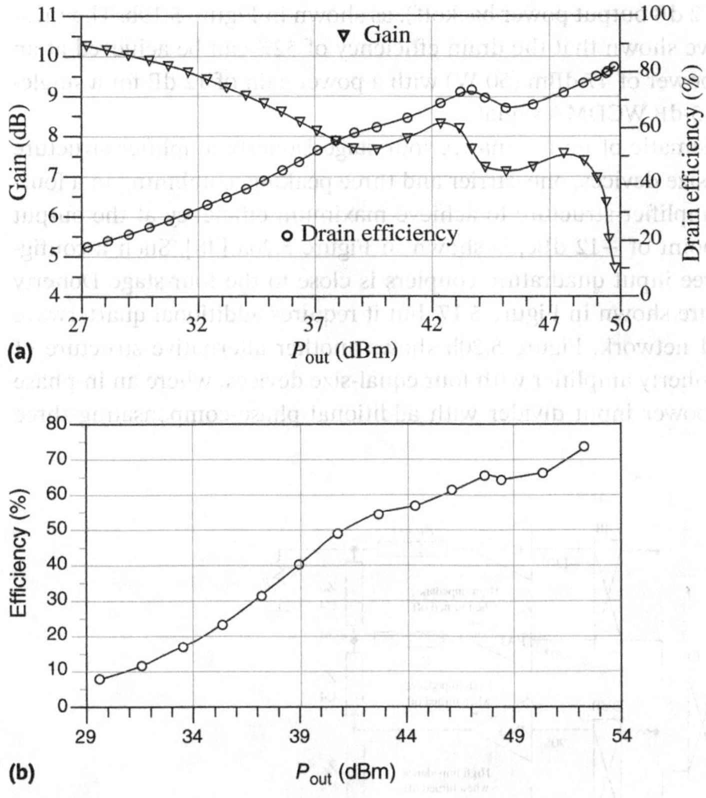

Figure 5.19 Comparative performance of modified (a) 100-W GaN HEMT and (b) 300-W LDMOSFET four-stage Doherty amplifier structures.

80% at saturation and greater than 60% across a 10-dB output power backoff range can be obtained with optimum gate bias voltages for each transistor, as shown in Figure 5.19a. A drain efficiency of 61% was measured at an average output power of 43 dBm (20 W) with an ACLR of −31 dBc for a single-carrier 2.14-GHz 6.5-dB WCDMA signal.

The modified four-stage LDMOSFET Doherty amplifier based on four 75-W NXP BLF6G22LS-75 devices was fabricated on the same 30-mil RO4350 substrate using the same Doherty amplifier architecture, as shown in Figure 5.17, where the carrier and peaking amplifiers are operated in an inverse Class-F mode with the second- and third-harmonic control each [47]. The input power splitter also includes three commercial 90° hybrid directional couplers. The simulated peak power of 52.7 dBm and a maximum drain efficiency of 72.5% were obtained at an operating frequency of 2.14 GHz, with a high efficiency exceeding 50% down to

41 dBm (about 12 dB output power backoff), as shown in Figure 5.19b. The measured results have shown that the drain efficiency of 52% can be achieved at an average output power of 47 dBm (50 W) with a power gain of 12 dB for a single-carrier 5-MHz 6.5-dB WCDMA signal.

The block schematic of an alternative four-stage Doherty amplifier structure with four equal-size devices, one carrier and three peaking, combining in a four-stage Doherty amplifier structure to achieve maximum efficiency at the output power backoff point of −12 dBc, is shown in Figure 5.20a [48]. Such a configuration with three input quadrature couplers is close to the four-stage Doherty amplifier structure shown in Figure 5.17, but it requires additional quarterwave lines in the load network. Figure 5.20b shows another alternative structure of the four-stage Doherty amplifier with four equal-size devices, where an in-phase four-way equal-power input divider with additional phase-compensating three

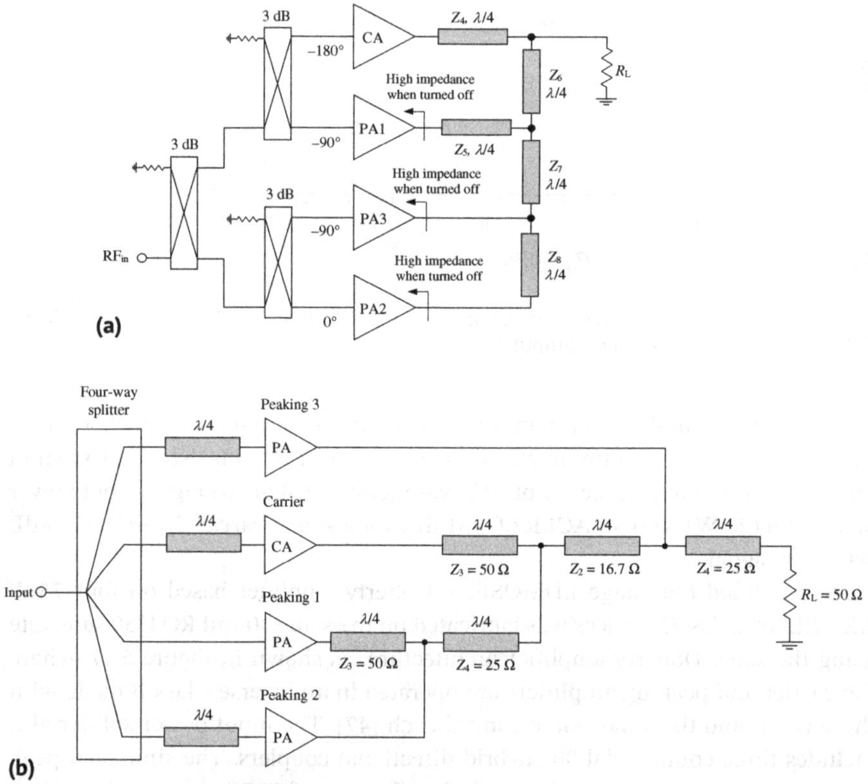

Figure 5.20 Modified four-stage Doherty amplifier architectures.

quarterwave transmission lines is used [49]. Here, the load network includes five quarterwave transmission lines with various characteristic impedances required to combine output powers from four devices and to match the resulting output impedance with 50-Ω load. By using two balanced 50-V LDMOSFET devices, an average output power of 50.2 dBm and a drain efficiency above 59% were measured across the frequency range of 925–960 MHz when the LTE and GSM signals are transmitted simultaneously.

5.4 Inverted Doherty Amplifiers

Figure 5.21 shows the block schematic of an inverted Doherty amplifier configuration where a quarterwave transmission line is included at the output of the peaking amplifier path. Here, the quarterwave transmission line does not operate as an impedance transformer but as an impedance inverter to achieve an open-circuit condition seen from the carrier amplifying path when the peaking device is turned off. Such an architecture is very useful in practical implementation when it is more convenient to provide a short circuit rather than an open circuit at the output of the peaking amplifier in a low-power region in view of the presence of the parasitic shunt drain–source capacitance and series bondwire and package lead inductance. The quarterwave impedance inverter which provides an open circuit at the conjunction point when the peaking device is turned off can be designed with much smaller size using an *LC* equivalent, which is very attractive for use in portable applications [50].

At a low-power level, the equivalent of a quarterwave transmission line combining the offset line, output matching circuit, and device parasitics is used to transform low output impedance (or a short circuit in an ideal case) after the offset line when the peaking device is turned off to a high impedance seen by the carrier amplifying path from the conjunction point. Similarly, considering the

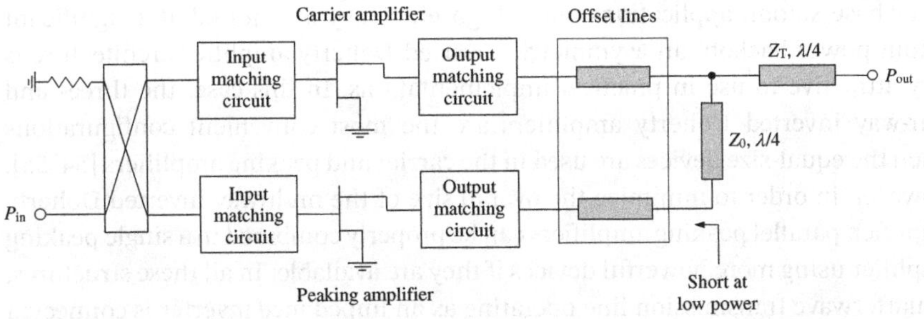

Figure 5.21 Block schematic of inverted Doherty amplifier.

device package parasitics of the carrier amplifier, an output matching circuit and a proper offset line are optimized to provide the proper output power from the carrier device at the required output power backoff [51]. At a high-power level, the load impedance seen from each amplifier with the same-size transistors is identical after the offset lines, followed by the output impedance-transforming quarterwave line to match a 50-Ω load. An average output power level of 46.3 dBm with a PAE of 32% was obtained for an inverted 2.14-GHz LDMOSFET Doherty amplifier with a four-carrier WCDMA signal, achieving a 9.5% improvement in efficiency and 1-dB improvement in the output power under the same ACLR conditions as for the balanced Class-AB power amplifier using the same devices [52]. An output power of 27 dBm with a 0.5-dB flatness and a PAE over 31% was achieved across the frequency bandwidth of 2.4–2.5 GHz with an ACLR better than −40 dBc [53].

Considering separately the load network shown in Figure 5.22a, where the peaking amplifier is turned off, allows a better understanding of the operation principle of an inverted Doherty amplifier. In a low-power region, the adjustment of electrical length θ of the offset line provides the short-circuited condition for the peaking amplifier at the input of the quarterwave inverter. At the same time, the offset line and matching circuit including the device parasitics such as lead inductance and output capacitance contributes to the required impedance transformation to achieve an optimum high-value impedance Z_{opt} seen by the carrier device output at 6-dB output power backoff, as shown in Figure 5.22b, where the optimum inductive impedance Z_{opt} at the fundamental frequency corresponds to a Class-E operation mode. The short circuit at the end of the quarterwave transmission line operating as an impedance inverter transforms into an open circuit at its input to prevent power leakage to the peaking amplifier path when the peaking transistor is turned off. In a high-power region, both carrier and peaking amplifiers designed to operate at a 50-Ω loads each are connected in parallel, and the output quarterwave transmission line with the characteristic impedance of 35.4 Ω transforms the resulting 25 Ω to the standard 50-Ω load.

For base station applications when high efficiency is required at a significant output power backoff, an asymmetric inverted Doherty amplifier architecture is very attractive to use in practical implementations. In this case, the three- and four-way inverted Doherty amplifiers are the most convenient configurations when the equal-size devices are used in the carrier and peaking amplifiers [54, 55]. However, in order to minimize the overall size of the multiway inverted Doherty amplifier, parallel peaking amplifiers can be properly combined in a single peaking amplifier using more powerful devices if they are available. In all these structures, a quarterwave transmission line operating as an impedance inverter is connected to the output of the corresponding peaking amplifier to achieve an open-circuit impedance condition seen by the carrier amplifier path in a low-power operation

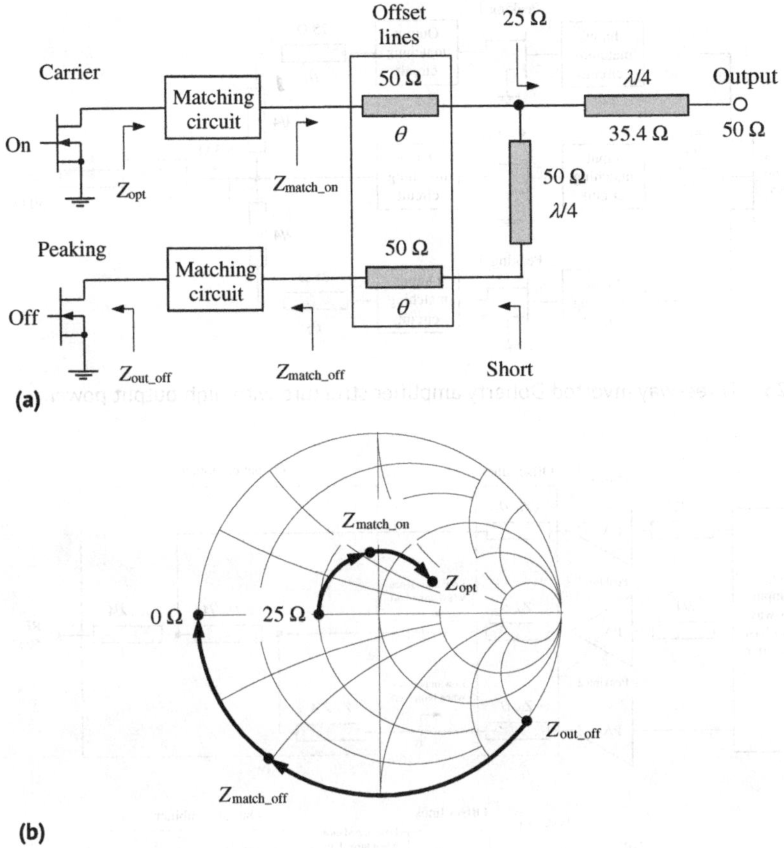

Figure 5.22 Load-network schematic and impedances.

mode. As an example, the block schematic of a three-way inverted Doherty amplifier with high output power is shown in Figure 5.23, where the carrier and two peaking amplifiers are matched to 25 Ω each at a high output power to achieve the saturated output power of 600 W and a peak drain efficiency greater than 57% across the frequency range of 1.9–2.2 GHz using three 50-V 200-W GaN HEMT devices [56]. Similarly, a three-way Doherty amplifier using three 28-V high-power LDMOSFET devices achieved greater than 45% efficiency with a power gain over 15 dB at an average output power of 48.7 dBm for a single-carrier 6.5-dB WCDMA signal over the frequency bandwidth of 2.11–2.17 GHz [57].

The block schematic of a three-stage inverted Doherty structure with equal-size devices is shown in Figure 5.24a, where the output power combiner contains only three quarterwave transmission lines. In order to provide low impedances at the

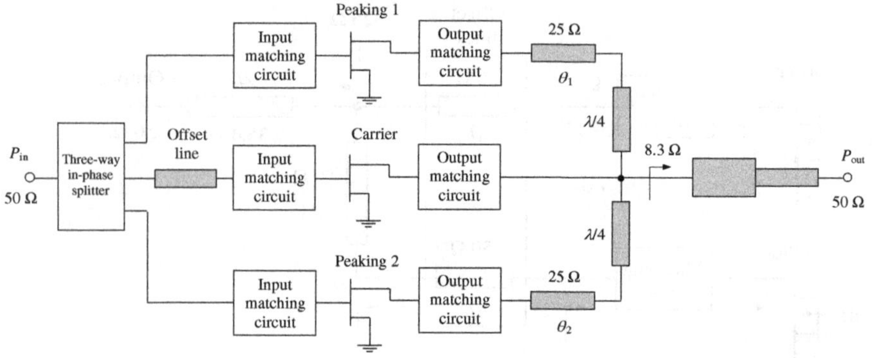

Figure 5.23 Three-way inverted Doherty amplifier structure with high output power.

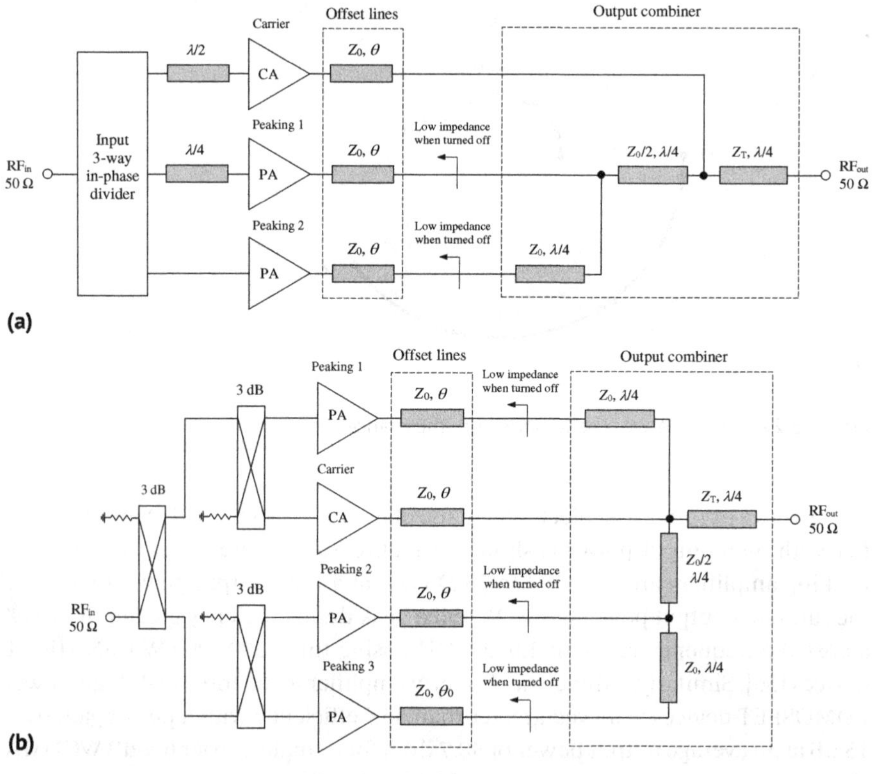

Figure 5.24 Three- and four-stage inverted Doherty amplifier structures.

outputs of the peaking amplifiers when they are turned off, the corresponding offset lines are necessary. In this case, an efficiency peak at the maximum power backoff is provided by the carrier amplifier, and an intermediate efficiency peak is achieved when both carrier and first peaking amplifiers are turned on. The proper impedance transformation with high impedance seen by the current source of the carrier transistor is provided by the output matching circuit of the carrier amplifier in conjunction with the properly tuned offset line. For the convenience of practical implementation, a half-wave transmission line in the carrier amplifying path may be transferred from the input divider to the output combiner.

The block schematic of a four-stage inverted Doherty amplifier with equal-size devices and three commercial 90° directional couplers at the input is shown in Figure 5.24b [58]. Such an inverted Doherty configuration can provide an efficiency peak at maximum backoff output power of −12 dBc, similar to the modified four-stage Doherty amplifier with equal-size devices shown in Figure 5.17. To achieve two additional efficiency peaks at power levels between the maximum power and output power backoff of −12 dB, it is necessary to provide the proper turning on of the second and third peaking amplifiers when a good compromise between efficiency and linearity can be achieved. In a 50-Ω environment, the characteristic impedance of the output transmission-line impedance transformer should be equal to $Z_T = 25\ \Omega$.

Figure 5.25 shows the simulated drain efficiencies of a high-power four-stage inverted Doherty amplifier based on the 250-W GaN HEMT devices which was designed first to achieve the maximum drain efficiency over 60% across the 10-dB power backoff region with a peak output power of more than 1 kW (60 dBm), as shown in Figure 5.25a and, secondly, to maximize the drain efficiency in the 10-dB power backoff region to achieve more than 50% across the frequency bandwidth of 1.8–2.2 GHz, as shown in Figure 5.25b. In both cases, it was assumed that the two dual-path devices (two 250-W dies in a single package) would be used, and the output matching circuit contains two consecutive quarterwave microstrip lines to cover a broader frequency bandwidth.

5.5 Three-way Doherty Amplifier with Enhanced Backoff Region

In an inverted Doherty amplifier architecture where the output impedance of the peaking amplifier becomes very small when it is turned off, it can be easier to provide a short circuit rather than an open circuit at the output of the peaking amplifier in a low-power region. In this case, the drain–source capacitance and parasitic series inductor compose the series resonant circuit with the resonant frequency

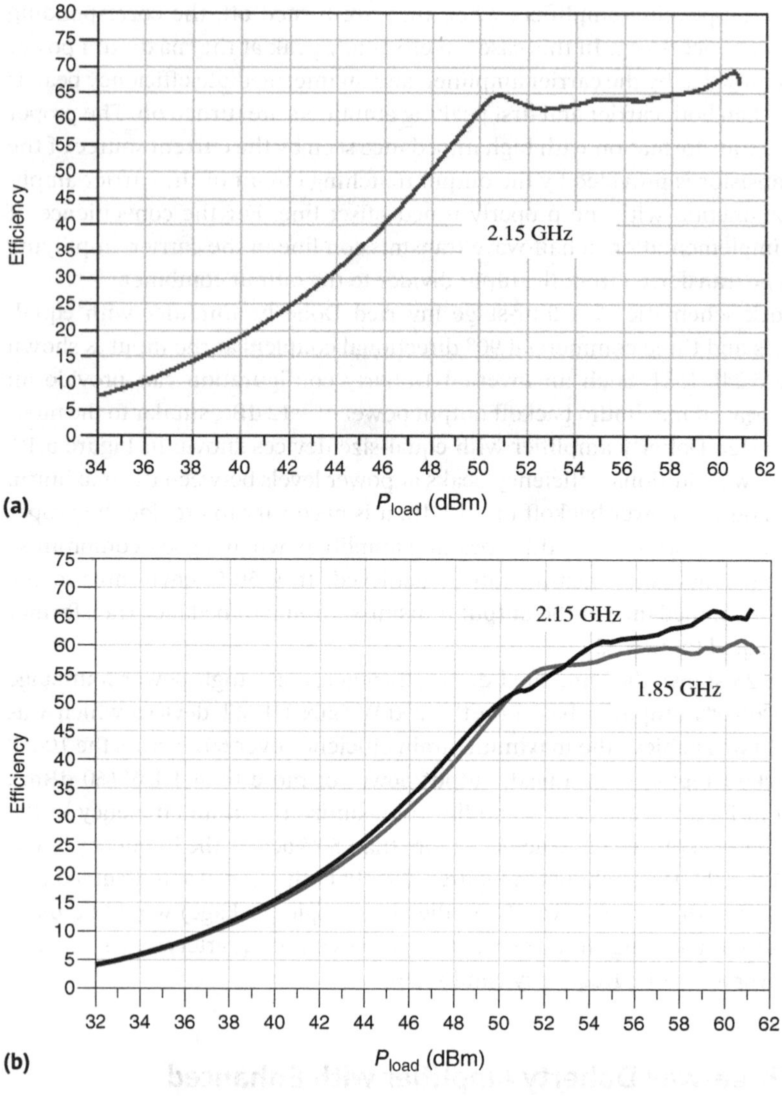

Figure 5.25 Simulated drain efficiency of four-stage inverted Doherty amplifier based on 250-W GaN HEMT devices.

close to the operating frequency. By adding an additional series offset line, the resonant point at the Smith chart can be properly tuned to the operating frequency. At the same time, the shunt drain–source capacitance and series (bondwire, package lead, and offset line) inductance represent a low-pass L-type matching section which increases the low impedance at the right-hand side of the offset line to the higher impedance seen by the carrier device current source.

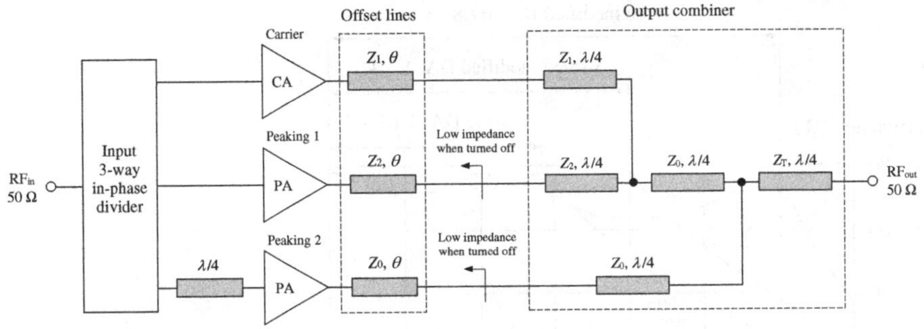

Figure 5.26 Three-stage inverted Doherty amplifier architecture with enhanced backoff region.

Figure 5.26 shows the circuit schematic of a modified inverted three-stage Doherty configuration with an extended high-efficiency backoff region [59]. Here, it is sufficient to use only one quarterwave line for in-phase power dividing at the input and five quarterwave lines in the output power combiner. In this case, it is very easy to vary the maximum efficiency backoff point by properly setting the device size of the carrier amplifier and the characteristic impedance Z_1 of the corresponding impedance-transforming quarterwave transmission line. For the device size ratio $1 : m_1 : m_2$, $Z_1 = m_2 Z_0$ and $Z_2 = (m_2/m_1)Z_0$, where m_1 is the size of the first peaking device and m_2 is the size of the second peaking device. When both peaking devices are turned off, the impedance at the right-hand side of the offset line in the carrier amplifying path, being equal to $(m_2^2/2)\,Z_0$, provides the peak efficiency at the lowest power backoff point. For example, for the device size ratio 1 : 3: 4, the lowest power backoff point with peak efficiency corresponds to −12 dBc, whereas for the device size ratio 1 : 7 : 8, the lowest power backoff point with peak efficiency reduces to −18 dBc, as shown in Figure 5.27.

For the device size ratio 1 : 1.7 : 1.7 in a three-stage inverted Doherty amplifier (DA), ideally, the first peak efficiency backoff point is obtained at −4.2 dBc, and the lowest power backoff point with peak efficiency is provided at −8.5 dBc, as shown in Figure 5.27. In a practical implementation with a 30-W carrier and two 50-W peaking amplifiers using LDMOSFET devices, the saturated output power of 50.5 dBm with a drain efficiency of 40.3% at 8.5-dB output power backoff was achieved at 2.14 GHz [60]. In this case, the output combiner includes six quarterwave transmission lines that makes the size of the final implementation too large, and the maximum efficiency output power backoff point is not very low to provide high operating efficiency for modern LTE communication signals with a peak-to-average power ratio (PAR) of 9 dB and higher.

Figure 5.28a shows the simulation setup of a modified inverted three-stage GaN HEMT Doherty amplifier with a device size ratio 1 : 7 : 8, corresponding to the

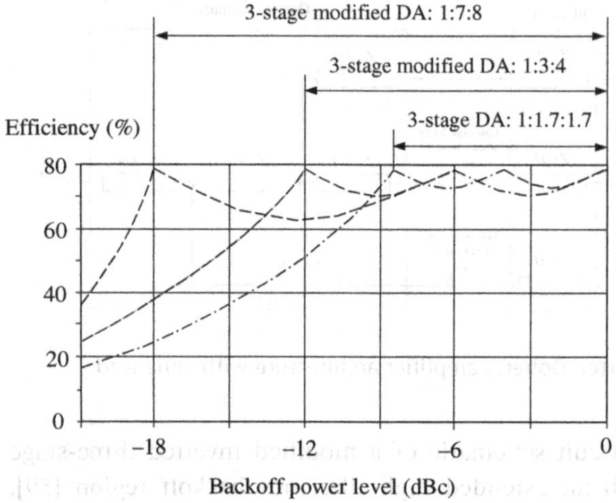

Figure 5.27 Theoretical efficiencies of inverted Doherty amplifier with different device size ratios.

device output powers of 30, 180, and 240 W, respectively, using a 30-mil RO4350 substrate and covering the frequency bandwidth of 1805–1880 MHz. In this case, the carrier (or main) device, which is directly connected to the gate of the first peaking device through the DC-blocking with capacitance, sees the load impedance greater than 170 Ω. The output matching a 50-Ω load is provided by the two consecutive quarterwave microstrip lines to satisfy the gain flatness requirements across the frequency bandwidth. As a result, the maximum drain efficiency exceeds 60% at −6 dB output power backoff and is greater than 52% at −14 dB output power backoff, with the saturated output power of 56 dBm at 1.85 GHz, as shown in Figure 5.28b, with a linear power gain of around 16 dB.

5.6 Integrated and Monolithic Doherty Amplifiers

At microwave and millimeter-wave frequencies, the transmission-line Doherty amplifier is implemented into the hybrid or monolithic microwave integrated circuit (MMIC) using different technologies depending on their implementations and operating frequencies. At the same time, the efforts to directly apply the Doherty technique to the design of power amplifier integrated circuits with a high level of integration at lower microwave frequencies usually face difficulties, because the physical size of the quarterwave transmission lines is too large in this case. For example, for an FR4 substrate with effective dielectric permittivity of $\varepsilon_r = 3.48$, the geometrical lengths of the quarterwave microstrip lines are 48, 19, and 8.7 mm at the operating frequencies of 900 MHz, 2.4 GHz, and 5.2 GHz, respectively.

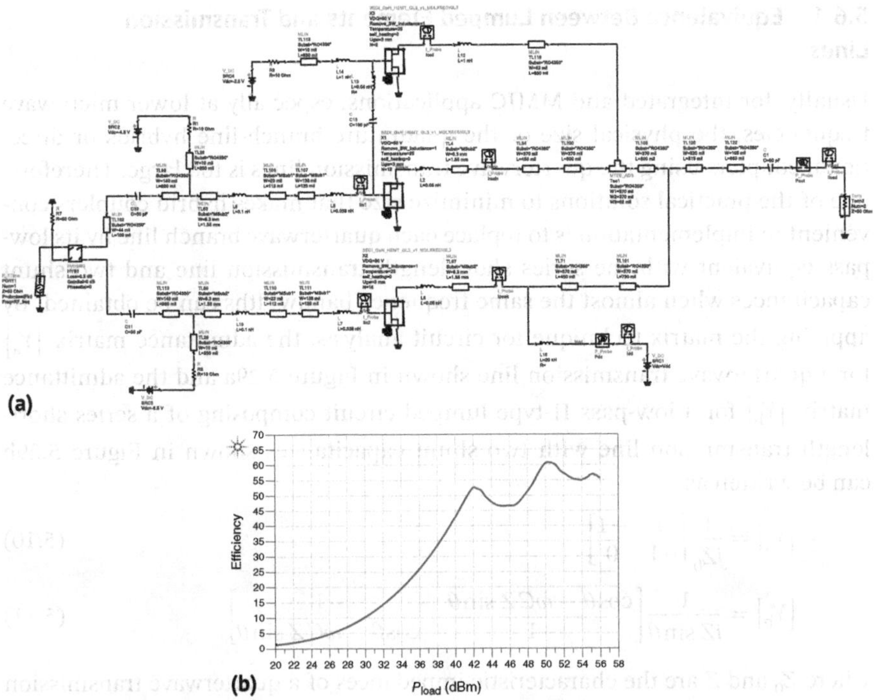

Figure 5.28 ADS schematic and performance of modified three-stage inverted Doherty amplifier with enhanced backoff region.

Therefore, one of the acceptable solutions for the fabrication of small-size Doherty amplifier hybrid integrated circuits or MMICs intended to operate in both handset transmitter or infrastructure multiple-input multiple-output (MIMO) systems represents a replacement of each quarterwave line in the input dividing, output combining, and output impedance-transforming circuits by its low-pass Π- or T-type lumped equivalent with the metal–insulator–metal (MIM) capacitors and bondwires or short-length high-impedance microstrip lines. Additionally, the small-size input or output second- or third-harmonic-control circuits can also be integrated to provide a high operating efficiency. To minimize the substrate losses and to provide a high level of integration in a CMOS process which can be used in handset applications, it is necessary to choose the substrates with a high resistivity and fully substitute the input divider, matching and harmonic-control circuits, as well as quarterwave impedance transformers, by their lumped equivalents. In integrated Doherty amplifiers intended to be used in MIMO applications, the quarterwave impedance transformer and load network can be implemented into a low-loss thin laminate substrate to maximize efficiency and minimize cost.

5.6.1 Equivalence Between Lumped Elements and Transmission Lines

Usually, for integrated and MMIC applications, especially at lower microwave frequencies, the physical size of the quadrature branch-line hybrids or directional couplers using the quarterwave transmission lines is too large. Therefore, one of the practical solutions to minimize size that makes hybrid couplers convenient in implementation is to replace each quarterwave branch line by its low-pass equivalent with the series short-length transmission line and two shunt capacitances when almost the same frequency bandwidths can be obtained. By applying the matrix technique for circuit analysis, the admittance matrix $[Y_a]$ for a quarterwave transmission line shown in Figure 5.29a and the admittance matrix $[Y_b]$ for a low-pass Π-type lumped circuit composing of a series short-length transmission line with two shunt capacitances shown in Figure 5.29b can be written as

$$[Y_a] = \frac{1}{jZ_0}\begin{bmatrix} 0 & -1 \\ -1 & 0 \end{bmatrix} \tag{5.10}$$

$$[Y_b] = \frac{1}{jZ\sin\theta}\begin{bmatrix} \cos\theta - \omega CZ\sin\theta & -1 \\ -1 & \cos\theta - \omega CZ\sin\theta \end{bmatrix} \tag{5.11}$$

where Z_0 and Z are the characteristic impedances of a quarterwave transmission line and a shortened transmission line, respectively, and θ is an electrical length of a shortened transmission line, and C is the shunt capacitance.

Equating the corresponding Y-parameters in Eqs. (5.10) and (5.11) results in simple analytical relationships between the lumped and transmission-line parameters as

$$Z = \frac{Z_0}{\sin\theta} \tag{5.12}$$

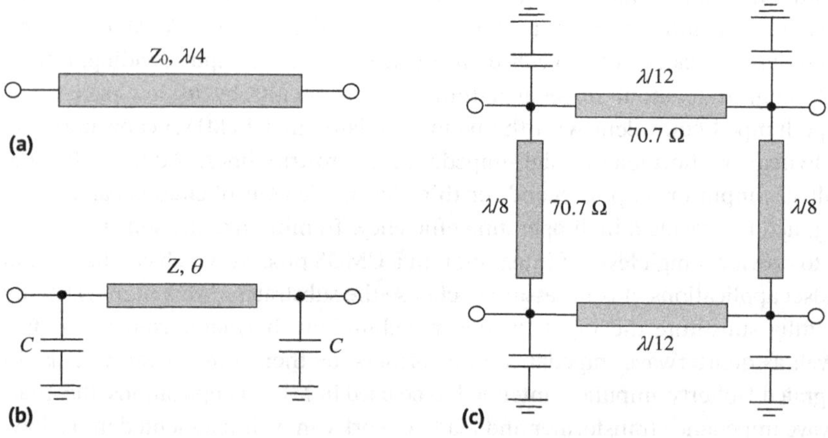

Figure 5.29 Reduced-size branch-line quadrature hybrid coupler.

$$C = \frac{\cos\theta}{\omega Z_0} \tag{5.13}$$

from which it follows that a much shorter transmission line with the increased characteristic impedance Z can be used when, for example, for $\theta = 45°$, the transmission-line characteristic impedance increases by a factor of $\sqrt{2}$.

Figure 5.29c shows the circuit schematic of a reduced-size branch-line quadrature hybrid coupler [61]. In comparison with the standard branch-line hybrid coupler including four quarterwave transmission lines with the characteristic impedances of its branch and main lines of Z_0 and $Z_0/\sqrt{2}$, respectively, the circuit parameters of the reduced-size branch-line hybrid coupler are obtained as

$$\theta_1 = \sin^{-1}\left(\frac{Z_0}{Z}\right) \tag{5.14}$$

$$\theta_2 = \sin^{-1}\left(\frac{Z_0}{Z\sqrt{2}}\right) \tag{5.15}$$

$$\omega C Z_0 = \sqrt{1 - \left(\frac{Z_0}{Z}\right)^2} + \sqrt{2 - \left(\frac{Z_0}{Z}\right)^2} \tag{5.16}$$

where θ_1 and θ_2 are the electrical lengths of the shunt branch line and series main lines, respectively. For $Z_0 = 50\ \Omega$, the characteristic impedance and electrical lengths of the shortened transmission lines are obtained as $Z = Z_0\sqrt{2}$, $\theta_1 = 45°$, and $\theta_2 = 30°$, which are shown in Figure 5.29c.

In monolithic implementations of the Doherty amplifier, the quarterwave transmission lines, phase compensation lines with different electrical lengths, and, in some cases, offset lines when a fully monolithic implementation is required need to be replaced by their lumped low-pass or high-pass Π- or T-type circuit equivalents. As an example, for a low-pass Π-type lumped equivalent, equating the *ABCD*-matrices of the transmission line shown in Figure 5.30a and low-pass Π-type lumped circuit, composing of a series inductance L and two shunt capacitances C, as shown in Figure 5.30b, results in

$$\begin{bmatrix} \cos\theta & jZ_0\sin\theta \\ j\sin\theta/Z_0 & \cos\theta \end{bmatrix} = \begin{bmatrix} 1-\omega^2 LC & j\omega L \\ j\omega C\left(2-\omega^2 LC\right) & 1-\omega^2 LC \end{bmatrix} \tag{5.17}$$

and the ratios between the circuit parameters corresponding to the transmission line and its low-pass Π-type lumped equivalent can be obtained as a function of electrical length θ by

$$\omega L = Z_0 \sin\theta \tag{5.18}$$

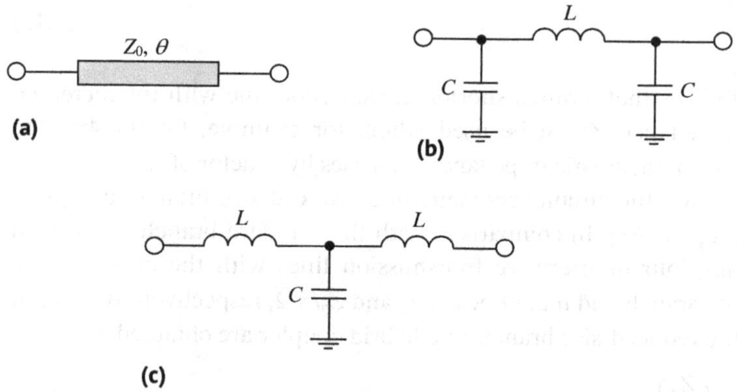

Figure 5.30 Transmission line and its lumped equivalents.

$$\omega C = \frac{1 - \cos\theta}{Z_0 \sin\theta} \tag{5.19}$$

where Z_0 is the characteristic impedance and θ is the electrical length of the transmission line.

Similarly, considering the *ABCD*-matrices of the transmission line with an electrical length of θ and its low-pass T-type lumped equivalent composing of a shunt capacitance C and two series inductances L, as shown in Figure 5.30c, and equating the appropriate elements of these matrices, the relationships between their elements are defined as

$$\omega L = \frac{Z_0(1 - \cos\theta)}{\sin\theta} \tag{5.20}$$

$$\omega C = \frac{\sin\theta}{Z_0} \tag{5.21}$$

For a quarterwave transmission line with $\theta = 90°$, Eqs. (5.18) and (5.19) as well as Eqs. (5.20) and (5.21) become identical, which can be written in a general form as

$$Z_0 \omega C = \frac{Z_0}{\omega L} = 1 \tag{5.22}$$

where Z_0 is the characteristic impedance of the quarterwave transmission line.

5.6.2 Integrated Doherty Amplifiers

To provide a high level of integration in handset or base station infrastructure transmitters, the input divider and input and interstage matching circuits of the Doherty amplifier are implemented with lumped elements using their monolithic implementation to minimize size and cost because the substrate loss in this case is not so important, what's more an additional resistive loss may contribute to improvement of the operation stability by increasing the amplifier stability factor. However, the load-network elements as the key elements for achieving a high efficiency of operation should be as lossless as possible, and usually they are implemented on a low-loss substrate using bondwire arrays or microstrip lines. Besides, the load-network parameters should be optimized to provide a high-impedance condition at the output of the peaking branch across the entire frequency bandwidth when the peaking amplifier is turned off. To simplify the load-network structure and minimize its size, the values of the shunt inductances are usually chosen to resonate with the corresponding shunt capacitances.

Figure 5.31 shows the circuit schematic of a GaAs HBT Doherty amplifier with fully integrated components into a single chip. In this case, the inductors are implemented using bondwires and slab inductors, the input dividing circuits are broadband based on low-Q matching networks, and the second- and third-harmonic impedances are controlled for high efficiency across the frequency bandwidth [62]. The open-circuit condition for the peaking device is provided by optimizing all elements of the load network, which include the drain bondwires and transistor output capacitance. As a result, such a lumped Doherty amplifier achieved a PAE over 27% and an average output power above 23.6 dBm across the frequency bandwidth of 2.2–2.8 GHz. A similar lumped Doherty amplifier provides a PAE above 30% at an average output power of 27.5 dBm over the frequency range of 1.6–2.1 GHz [63]. A hybrid design approach integrating MMIC and PCB technologies

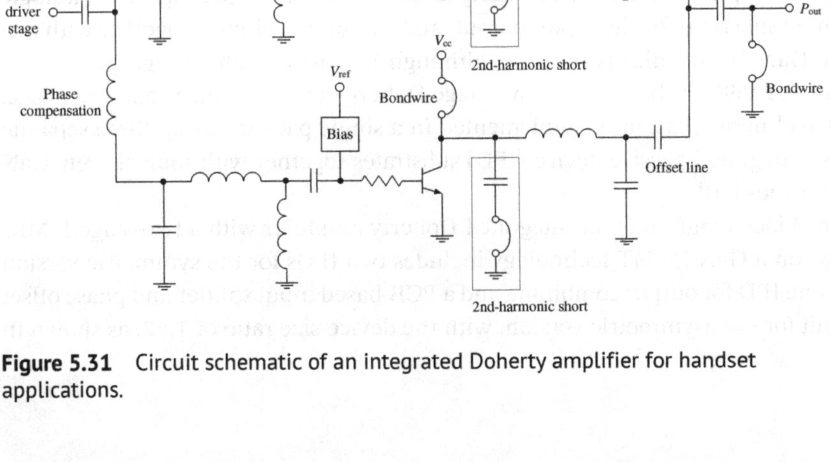

Figure 5.31 Circuit schematic of an integrated Doherty amplifier for handset applications.

where the entire input network is integrated within the MMIC, whereas the output network with harmonic control is implemented on a Rogers 4350 substrate, was used to cover the frequency bandwidth of 3.1–4.2 GHz, achieving a saturated output power over 33.8 dBm, a saturated PAE ranging from 39.6 to 44.9%, and a 6-dB power backoff PAE of 42.3–46.1% using a GaAs HBT process [64].

In order to implement a high-power integrated Doherty amplifier in a standard package, a quarterwave impedance-transforming transmission line should be equivalently replaced by a low-pass Π- or T-type lumped circuit. In this case, the input and output matching circuits are fabricated externally using a common laminate substrate. Figure 5.32a shows an example of the package integrated Doherty amplifier with a T-type input phase shifter in a peaking amplifying path and a T-type impedance transformer in a carrier amplifying path using series bondwires or bondwire arrays and shunt MIM capacitances [65]. If it is necessary to increase the overall output power, several Doherty amplifier device cells can be connected in parallel in a single device package. For example, by parallel combining the four 10-W MMIC Doherty amplifier cells, a drain efficiency of 39.8% at an average output power of 7.5 W for a two-carrier 2.14-GHz WCDMA signal with a PAR of 7.6 dB was achieved [66]. It is also convenient to use a low-pass Π-type lumped circuit where the drain–source capacitances represent the shunt capacitances and the bondwire array connected between the drains of the carrier and peaking devices represents a series inductance, as shown in Figure 5.32b [67]. By using a compact fixture of $50 \times 50\ \text{mm}^2$ to implement an integrated LDMOSFET Doherty amplifier with external input and output matching circuits, the peak output power of 50 dBm, a linear power gain of about 15.9 dB, and an average drain efficiency around 47% at 6-dB output power backoff were achieved for a 10-MHz 7.2-dB LTE advanced signal over the frequency range of 1.805–2.17 GHz [68].

For massive-MIMO and small-cell infrastructure applications in 4/5G and beyond communication systems, in order to further minimize the size and cost of the integrated Doherty amplifiers, higher integration capability is required. In this case, full implementation of a Doherty amplifier in a single package using a single ceramic substrate for the input, output, and combining circuits together with discrete GaN HEMT dies is possible, although its cost is high enough despite the small size [69]. In the case of a two-stage Doherty amplifier, the input, interstage, and load networks can be implemented in a single package using three separate GaAs integrated passive device (IPD) substrates together with four discrete GaN HEMT dies [70].

The block diagram of an integrated Doherty amplifier with a two-stage MMIC based on a GaN HEMT technology includes two IPDs for the symmetric version and one IPD for output combining and a PCB-based input splitter and phase offset circuit for the asymmetric version, with the device size ratio of 1 : 2, as shown in

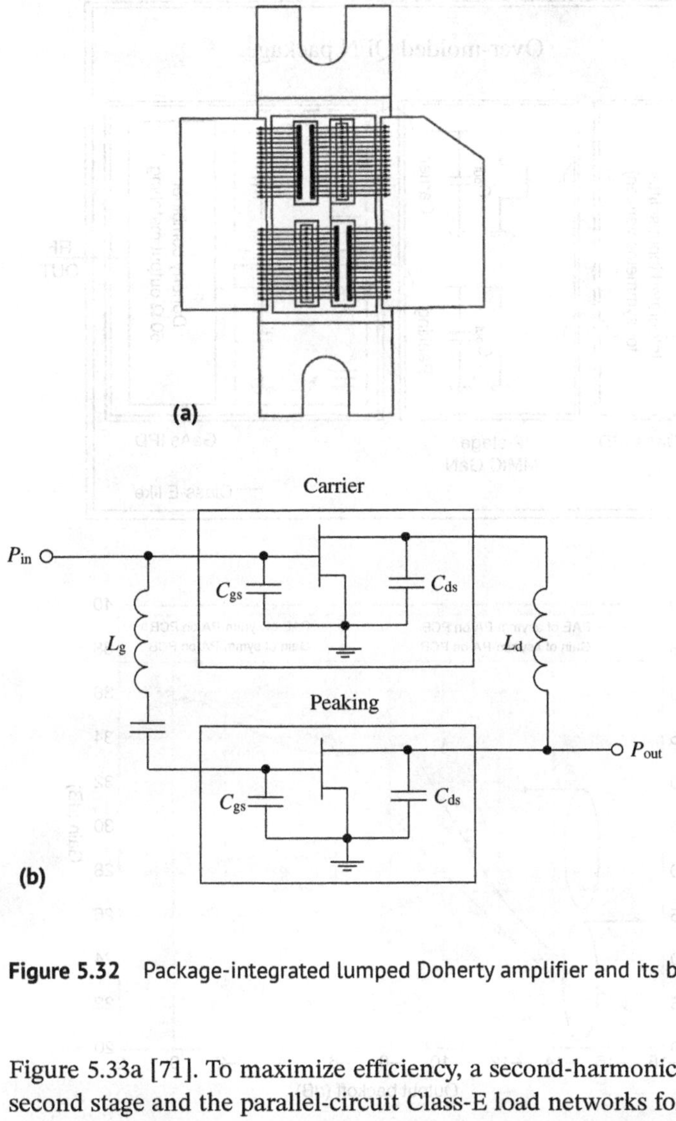

Figure 5.32 Package-integrated lumped Doherty amplifier and its basic schematic.

Figure 5.33a [71]. To maximize efficiency, a second-harmonic input short in the second stage and the parallel-circuit Class-E load networks for carrier and peaking power amplifiers are used in each version. When both 20-W asymmetric and 26-W symmetric Doherty amplifiers were implemented into separate small plastic packages of 7 × 7 mm^2 each, the saturated output power P_{3dB} of 44.3 dBm with a maximum PAE of 52% and a PAE of 44% at 6-dB power backoff for symmetric version, while the saturated output power P_{3dB} of 43 dBm with a maximum PAE of 49% and a PAE of 44% at 8-dB power backoff for the asymmetric version were achieved in a frequency range of 3.4–3.6 GHz, as shown in Figure 5.33b.

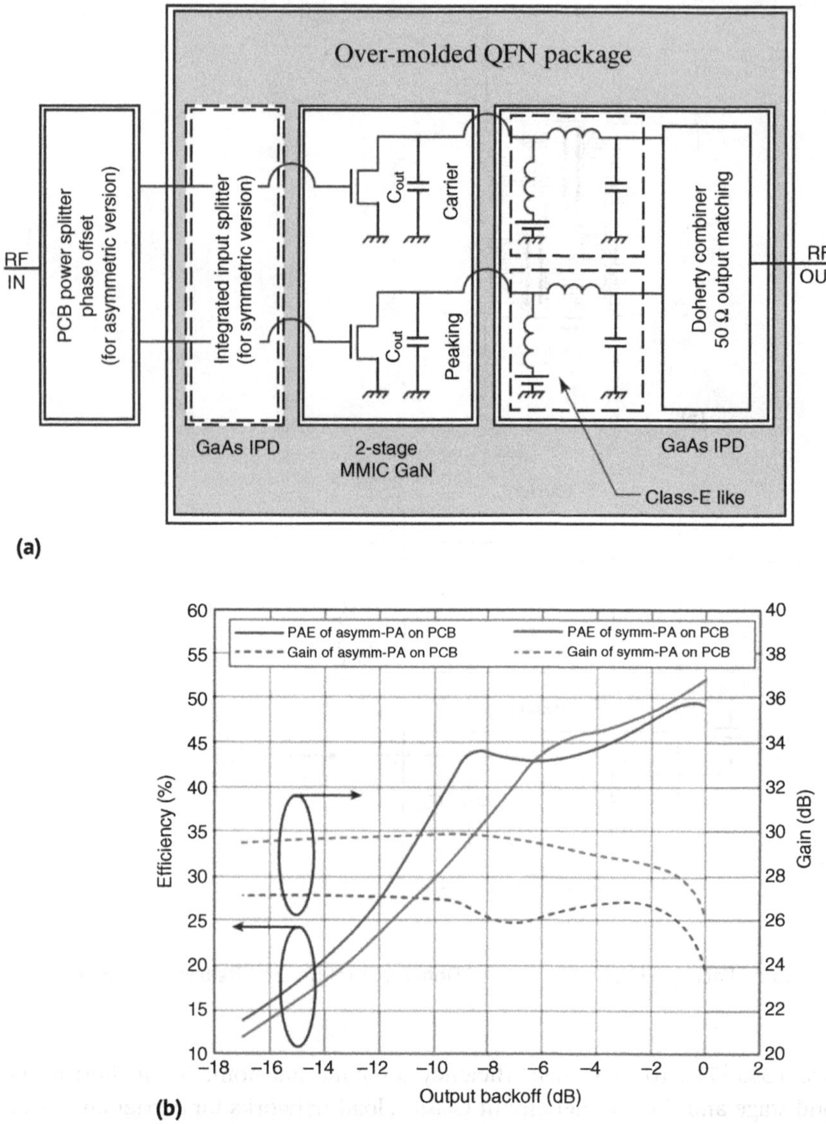

Figure 5.33 Package-integrated inverted Doherty amplifier: (a) block diagram and (b) its performance. *Source*: [71], with permission of EuMA.

An example of the possible implementation of the integrated inverted Doherty amplifier, whose circuit schematic is shown in Figure 5.34a, with separate carrier and peaking GaN HEMT devices, one IPD for the input circuit including a lumped Wilkinson power divider and phase shifter, and a PCB-based low-pass

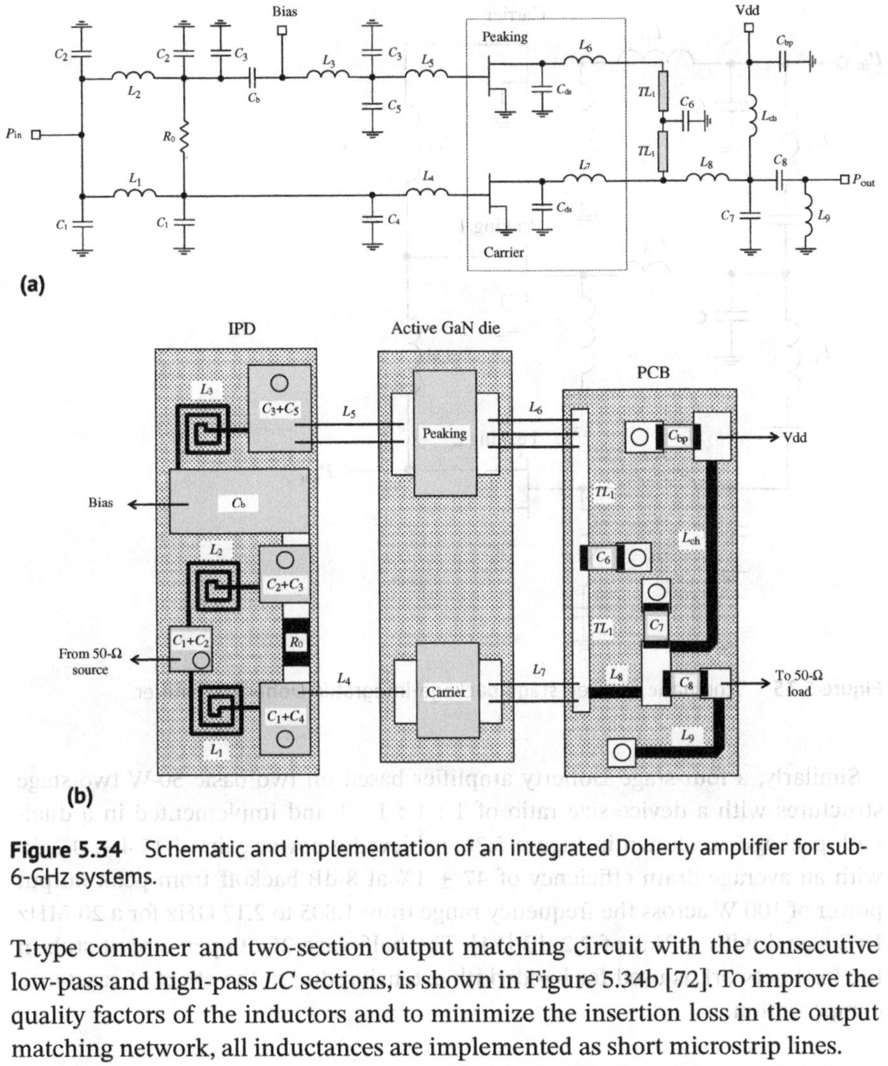

Figure 5.34 Schematic and implementation of an integrated Doherty amplifier for sub-6-GHz systems.

T-type combiner and two-section output matching circuit with the consecutive low-pass and high-pass *LC* sections, is shown in Figure 5.34b [72]. To improve the quality factors of the inductors and to minimize the insertion loss in the output matching network, all inductances are implemented as short microstrip lines.

The simplified circuit schematic of a three-stage package-integrated Doherty amplifier is shown in Figure 5.35, with a device size ratio of 1 : 4 : 4, an asymmetric input power divider, and a low-pass *LC* output combiner where the device gate–source and drain–source capacitances represent the shunt capacitances [73]. A peak output power of 47 dBm, a power gain of 12 ± 1 dB, and a PAE of 45–55% at backoff output power of −12 dBc were measured over the frequency bandwidth from 1.8 to 2.2 GHz using a compact package of 3 × 5 mm^2 with external input and output matching circuits, which is suitable for small-cell infrastructure applications.

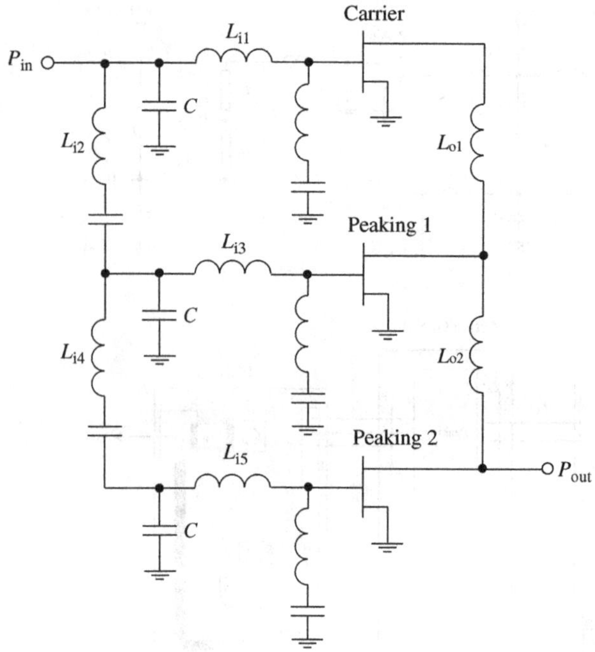

Figure 5.35 Schematic of three-stage package-integrated Doherty amplifier.

Similarly, a four-stage Doherty amplifier based on two basic 50-W two-stage structures with a device size ratio of 1 : 1 : 1 : 1 and implemented in a dual-path package, as shown in Figure 5.36, achieved a power gain of 13.4 ± 0.2 dB with an average drain efficiency of 47 ± 1% at 8-dB backoff from peak output power of 100 W across the frequency range from 1.805 to 2.17 GHz for a 20-MHz LTE signal with a PAR of 7.2 dB [74]. The half-wave 25-Ω open-circuit stub in the load network is used for bandwidth extension due to the effect of reactance compensation.

5.6.3 Monolithic Doherty Amplifiers

The transmission-line Doherty amplifier can be easily monolithically implemented using a GaAs or CMOS technology at microwave and millimeter-wave frequencies, although the CMOS process requires a high resistivity substrate to minimize insertion loss and to maximize high operating efficiency. However, generally, it is required to compromise between the cost and performance of the Doherty amplifier when choosing the implementation process, especially for higher-power and higher-frequency applications.

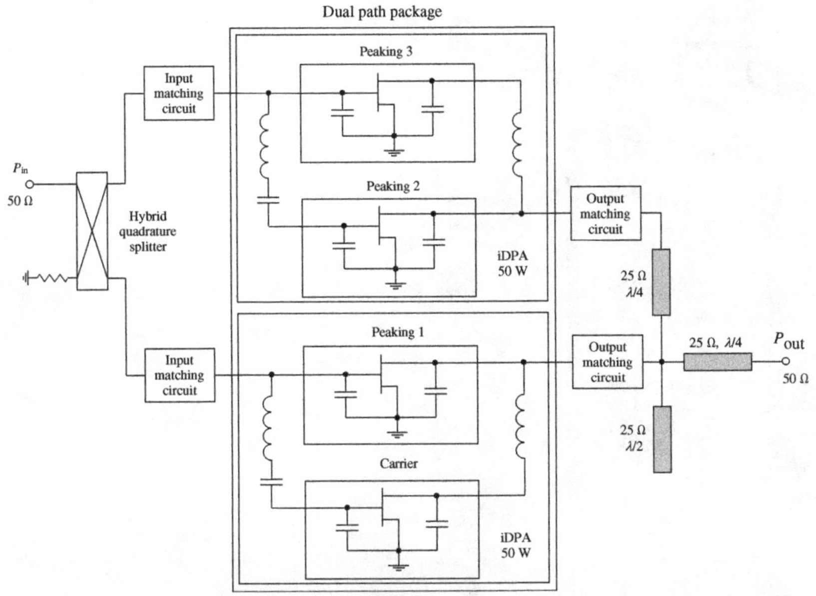

Figure 5.36 Block schematic of four-stage integrated Doherty amplifier.

The first implementation of a fully integrated *Ku*-band MMIC Doherty amplifier based on a 0.25-μm pHEMT technology, is shown in Figure 5.37, achieving a small-signal gain in excess of 8 dB, a two-tone PAE of 40%, and an output power of about 24 dBm at an operating frequency of 17 GHz [17]. Here, an input 3-dB hybrid splitter is implemented using a microstrip Lange coupler, the low-pass input matching circuits are used to maximize power gain, and shunt microstrip lines providing the drain bias also resonate the drain–source capacitances. The quarterwave impedance-transforming microstrip line is replaced by a capacitively loaded line with higher characteristic impedance in the form of a low-pass T-type equivalent. A year later, a single-tone PAE of 38.5% at P_{1dB} = 23.5 dBm and a PAE of 25% at 6-dB backed off compression were measured for a 20-GHz MMIC Doherty amplifier implemented in a 0.15-μm pHEMT process [75].

The test chip of a wideband monolithic 0.25-μm GaN HEMT asymmetric Doherty amplifier is shown in Figure 5.38a. Here, the input network includes a Wilkinson power divider, phase shifter, and two matching circuits using spiral inductances, whereas the load network represents a T-line impedance inverter based on microstrip lines with optimized characteristic impedances and electrical lengths [76]. A series resistor at the input of the carrier transistor and two shunt *LR* circuits at the inputs of the carrier and peaking transistors are included for stability of the operation. The drain efficiency above 35% at 9-dB power backoff

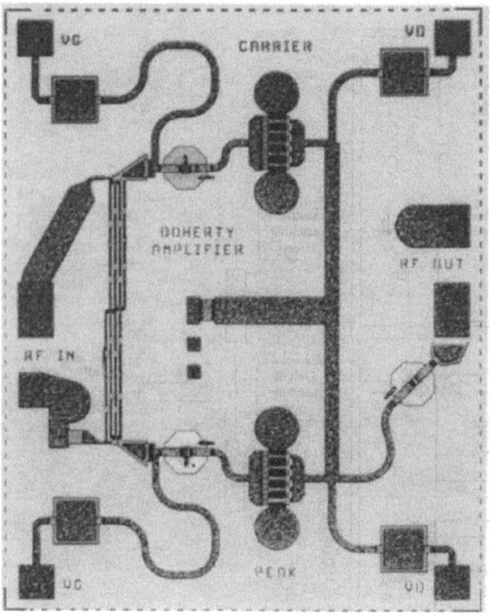

Figure 5.37 Photo of microwave Doherty amplifier MMIC. *Source*: [17], with permission of IEEE.

over the frequency range of 6.8–8 GHz, as shown in Figure 5.38b, and a maximum output power of 35 ± 0.5 dBm from 6.6 to 8.5 GHz were achieved. However, the linear power gain varies significantly across the entire frequency bandwidth, decreasing from 13 dB at 6.8 GHz to 6.5 dB at 8 GHz.

In some cases, the input divider can represent an unequal-power reactive divider instead of the directional coupler, as shown in Figure 5.39, which can improve linearity of a microwave Doherty amplifier [77]. In this case, the input divider changes the ratio of power delivered to the carrier and peaking devices due to an increased difference in the device gate–source capacitances with the increase of the input power under different bias conditions. The increased gate–source capacitance of the peaking amplifier reduces the load impedance presented to the input divider at the peaking amplifier port and therefore, as input power increases, more power is delivered to the peaking amplifier to relax saturated conditions for the carrier amplifier. Both the carrier and peaking amplifiers were matched to 100 Ω to provide parallel combining, thus simplifying the output network. The fabricated MMIC of a 10-GHz 20-V Doherty amplifier based on a 0.15-μm GaN HEMT technology achieved a peak PAE of 47% at the saturated output power over 36 dBm and a PAE of 41% at 6-dB output power backoff.

To increase the power gain of a Doherty amplifier at higher microwave frequencies, especially for *K*-band applications, the structure of a Doherty amplifier may include a high-gain driver stage in both carrier and peaking amplifier

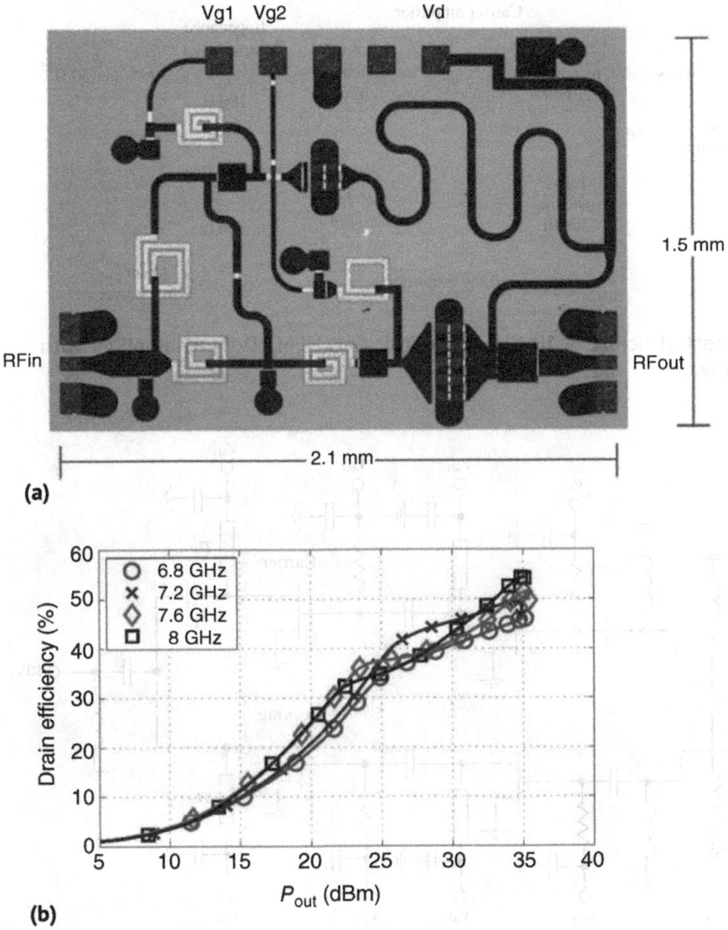

Figure 5.38 Wideband monolithic GaN HEMT Doherty amplifier: (a) test chip and (b) its performance. *Source*: [76], with permission of IEEE.

paths, providing potentially greater PAE compared to the approach with a single driver [78]. A two-stage symmetric Doherty amplifier was implemented using a 0.15-μm pHEMT process within the die size of 1.43 × 3 mm^2, achieving a PAE of 38% at the saturation power of 30.9 dBm and a PAE of 20% at 6dB output power backoff with a power gain of 12.5 dB at an operating frequency of 24 GHz. However, better performance in *K*-band for a two-stage Doherty amplifier can be achieved using a 0.15-μm GaN HEMT technology when a saturated power over 5 W, a PAE of 48% at P_{1dB} and a PAE of 25% at 8-dB power backoff, and a power gain of 16 dB were achieved [79]. The two-stage inverted asymmetric Doherty amplifier with the second-harmonic-control circuits located close to the gates of

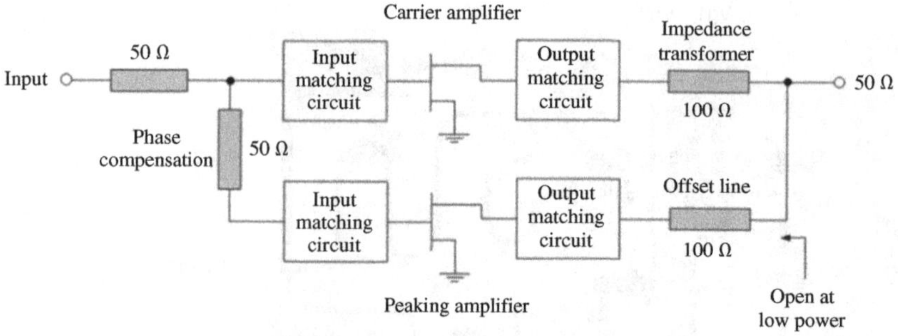

Figure 5.39 Block structure of 10-GHz monolithic GaN HEMT Doherty amplifier with input reactive divider.

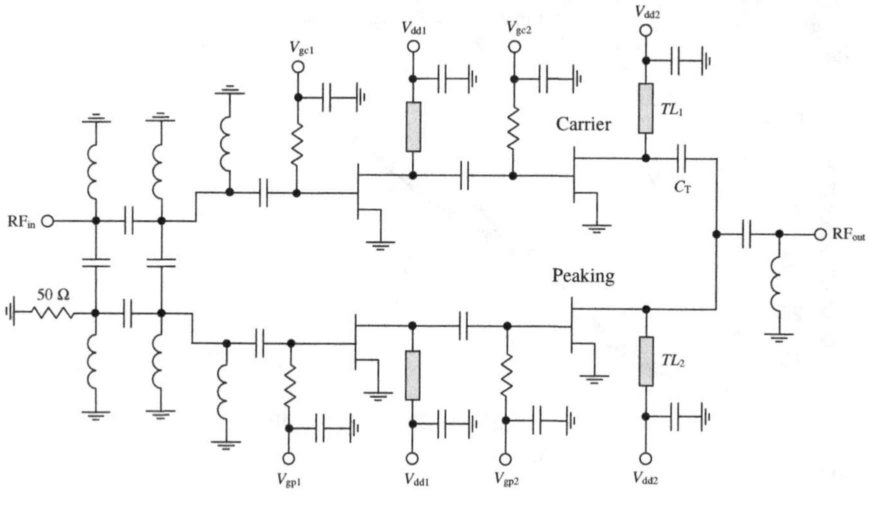

Figure 5.40 Circuit schematic of monolithic GaN HEMT Doherty amplifier for sub-6-GHz applications.

the final-stage devices achieved a peak PAE of 21.7% at the saturated power over 32 dBm across the frequency bandwidth of 24–28 GHz and a PAE over 20% at 6-dB output power backoff with a power gain of 15 dB at 26 GHz, implemented in a 0.15-μm GaN HEMT technology within a die size of 2.5 × 2 mm^2 [80].

Figure 5.40 shows the circuit schematic of a fully integrated *C*-band GaN HEMT Doherty amplifier with the driver stages in both carrier and peaking amplifiers, where a compact low-loss load network is used to provide high efficiency and to minimize chip size [81]. Here, a lumped branch-line coupler is used as an input power splitter, and simple L-type high-pass *LC* sections represent both input and interstage matching. In order to minimize the overall size of the Doherty amplifier, the quarterwave impedance transformer and output matching circuit are designed as a lumped high-pass Π-type equivalent circuit each, whereas the shunt inductances are replaced by the appropriate shunt microstrip lines to minimize power loss, resulting in a compact chip size of 3 × 2.5 mm^2. In this case, the saturated PAE of 61% and a PAE of 47% at 6-dB output power backoff with a power gain of around 15 dB were achieved over the frequency bandwidth of 4.4–5.1 GHz.

Figure 5.41a shows the circuit schematic of an asymmetric 1 : 2 monolithic two-stage 0.15-μm GaN HEMT Doherty amplifier with the asymmetric Wilkinson input divider, two driver devices of 4 × 75 and 8 × 75 μm, the carrier device and

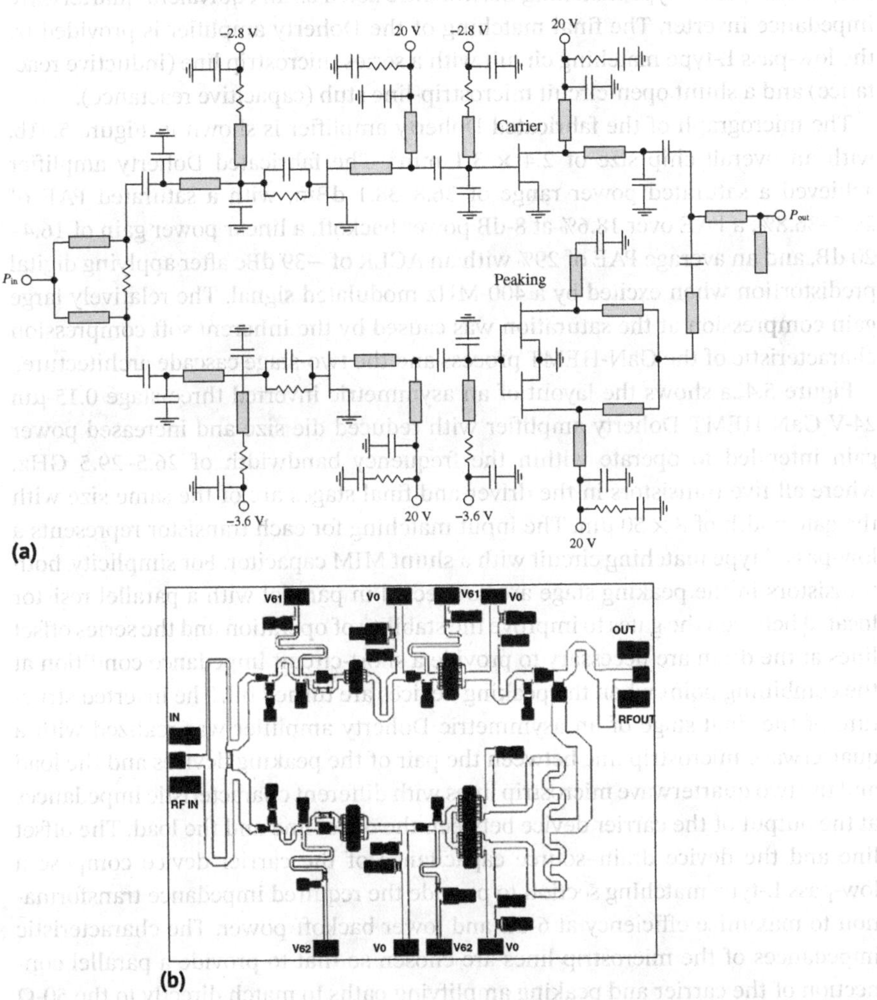

Figure 5.41 Circuit schematic and micrograph of monolithic millimeter-wave two-stage GaN HEMT Doherty amplifier.

two peaking devices of 8 × 75 μm each, operating in a frequency bandwidth from 24 to 28 GHz [82]. At the output of the peaking amplifying path of the load-modulation network, the three quarterwave microstrip lines are used providing an overall half-wavelength to create an open-circuit condition at the combining point seen by the carrier amplifying path when the peaking devices are turned off. Here, the drain–source capacitance C_{ds} of each peaking device is resonated out by the corresponding parallel inductance provided the parallel grounded microstrip lines connected to the DC-power supply. For the carrier branch, to realize wide-band load modulation, the output parasitics of the carrier device and the following output low-pass T-type matching network are acted as an equivalent quarterwave impedance inverter. The final matching of the Doherty amplifier is provided by the low-pass L-type matching circuit with a series microstrip line (inductive reactance) and a shunt open-circuit microstrip-line stub (capacitive reactance).

The micrograph of the fabricated Doherty amplifier is shown in Figure 5.41b, with an overall chip size of 2.4 × 3.1 mm^2. The fabricated Doherty amplifier achieved a saturated power range of 36.8–38.1 dBm, with a saturated PAE of 29.7–36.8%, a PAE over 18.6% at 8-dB power backoff, a linear power gain of 16.4–20 dB, and an average PAE of 29% with an ACLR of −39 dBc after applying digital predistortion when excited by a 400-MHz modulated signal. The relatively large gain compression at the saturation was caused by the inherent soft compression characteristic of the GaN-HEMT process and the two-stage cascade architecture.

Figure 5.42a shows the layout of an asymmetric inverted three-stage 0.15-μm 24-V GaN HEMT Doherty amplifier with reduced die size and increased power gain intended to operate within the frequency bandwidth of 26.5–29.5 GHz, where all five transistors in the driver and final stages are of the same size with the gate width of 8 × 50 μm. The input matching for each transistor represents a low-pass T-type matching circuit with a shunt MIM capacitor. For simplicity, both transistors in the peaking stage are connected in parallel with a parallel resistor located between the gates to improve the stability of operation and the series offset lines at the drain are necessary to provide a short-circuit impedance condition at the combining point when the peaking devices are turned off. The inverted structure of the final stage of an asymmetric Doherty amplifier was realized with a quarterwave microstrip line between the pair of the peaking devices and the load and the two quarterwave microstrip lines with different characteristic impedances at the output of the carrier device between the offset line and the load. The offset line and the device drain–source capacitance of the carrier device compose a low-pass L-type matching section to provide the required impedance transformation to maximize efficiency at 6-dB and lower backoff power. The characteristic impedances of the microstrip lines are chosen so that to provide a parallel connection of the carrier and peaking amplifying paths to match directly to the 50-Ω load. The layout of an asymmetric three-stage inverted Doherty amplifier is a very

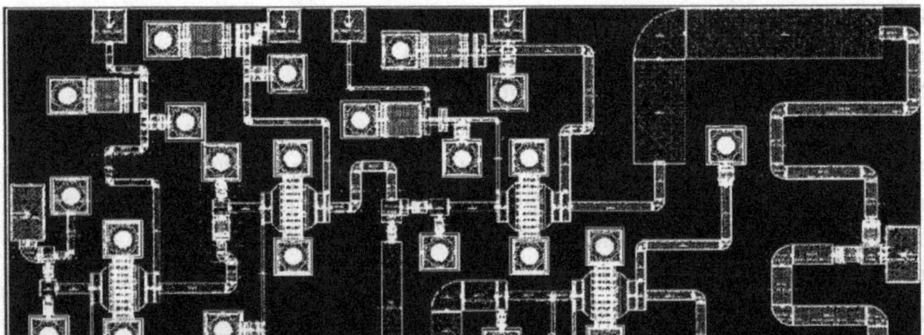

Figure 5.42 Layout and simulated performance of millimeter-wave three-stage inverted GaN HEMT Doherty amplifier.

compact with an overall chip size of 1.7 × 2.8 mm^2. The simulated performance of an asymmetric three-stage inverted Doherty amplifier is shown in Figure 5.42b, where a linear power gain in the range of 24–27 dB and a saturated output power of 35 dBm were obtained, with a PAE varying from 18 to 21% at 6-dB and around 15% at 8-dB backoff powers.

5.7 Multiband and Broadband Capability

For a multiband operation of the Doherty amplifier, generally all of its components should be properly designed to provide their broadband performance covering the required bands of operation without any tuning. For example, using broadband lossless or lossy input and interstage matching circuits in the carrier and peaking amplifiers to maximize output power flatness across the entire frequency range. The output matching circuit in this case can represent a multi-section low-pass or high-pass *LC* structure or cascaded transmission lines with different characteristic impedances instead of a single quarterwave transmission line. In order to improve overall efficiency over a wide frequency range, a broadband Class-E technique based on the reactance compensation technique using lumped or transmission-line elements can be used. In the simplest case, when the required frequency band is not so wide, the bandwidth extension can be achieved by optimizing the characteristic impedances of the quarterwave impedance transformer and quarterwave output matching line in a combining load network [83]. For a broader operation, the input splitter can represent a multi-section Wilkinson or branch-line power divider or a coupled-line directional coupler.

Generally, a simple two-stepped transmission-line impedance transformer with different characteristic impedance ratios and different electrical lengths of the transmission-line sections can provide sufficient bandwidth extension having a two-pole response [84]. As an example, a broadband output transformer for a dual-band Doherty amplifier can be based on the two cascaded quarterwave transmission lines, where the characteristic impedance of the first quarterwave transmission line is set to 30 Ω and the characteristic impedance of the second quarterwave transmission line is equal to 42 Ω, as shown in Figure 5.43a. In this case, the amplitude variations of ±0.5 Ω shown in Figure 5.43b by curve 1 and phase variations of ±1° shown in Figure 5.43c by curve 1 can be achieved across the frequency range from 2 to 2.8 GHz. For comparison, the narrowband amplitude and phase responses of a single quarterwave transmission line operating as a quarterwave transformer are shown in Figures 5.43b,c by curves 2, respectively. Moreover, as it follows from Figures 5.43b,c, an impedance transformer based on two cascaded quarterwave transmission lines provides the amplitude variations of ±1 Ω and phase variations of ±2° across the extended frequency bandwidth of 1.9–2.9 GHz.

5.7.1 Dual-band Parallel Doherty Architecture

In a classic two-stage Doherty amplifier with limited bandwidth capability, it is necessary to realize an impedance transformation from 25 to 100 Ω at the output of the carrier amplifier in a low-power region when the peaking amplifier is turned off, as shown in Figures 5.44a and 5.45a, that corresponds to a loaded quality factor $Q_L = \sqrt{100/25 - 1} = 1.73$ at 3-dB output power reduction level (approximation by a low-pass L-type impedance transformer near resonant frequency given by Eq. 2.18 in Chapter 2), whose value is sufficiently high for broadband operation. In this case, the bandwidth property can be improved in a low-power

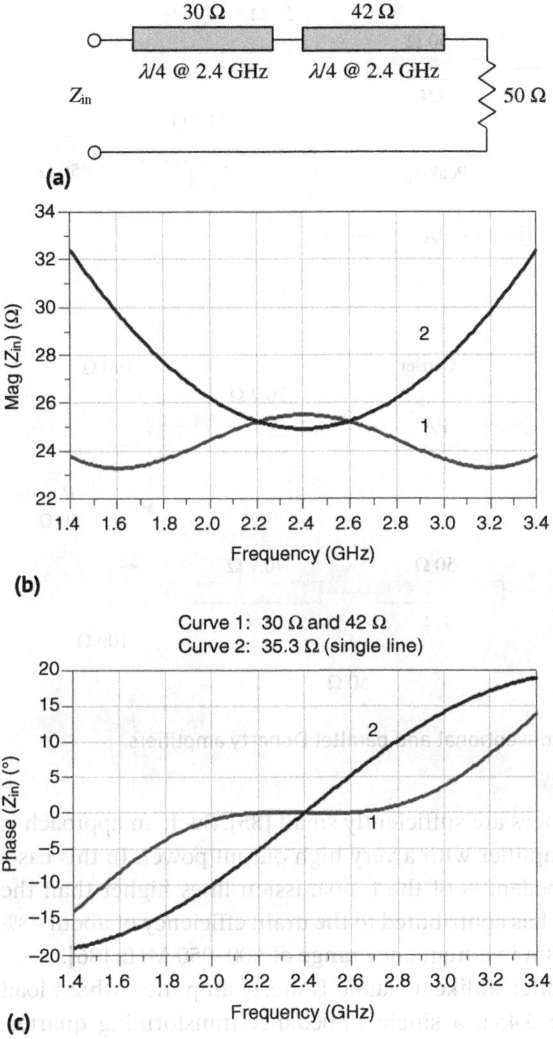

Figure 5.43 Transmission-line transformer with two cascaded transmission lines and its input impedances.

region using a parallel architecture of a classic Doherty amplifier with modified modulated load network. Figure 5.44b shows the block schematic of a parallel Doherty amplifier where the bandwidth property is improved by reducing the impedance transformation ratio by a factor of two, which is very important when

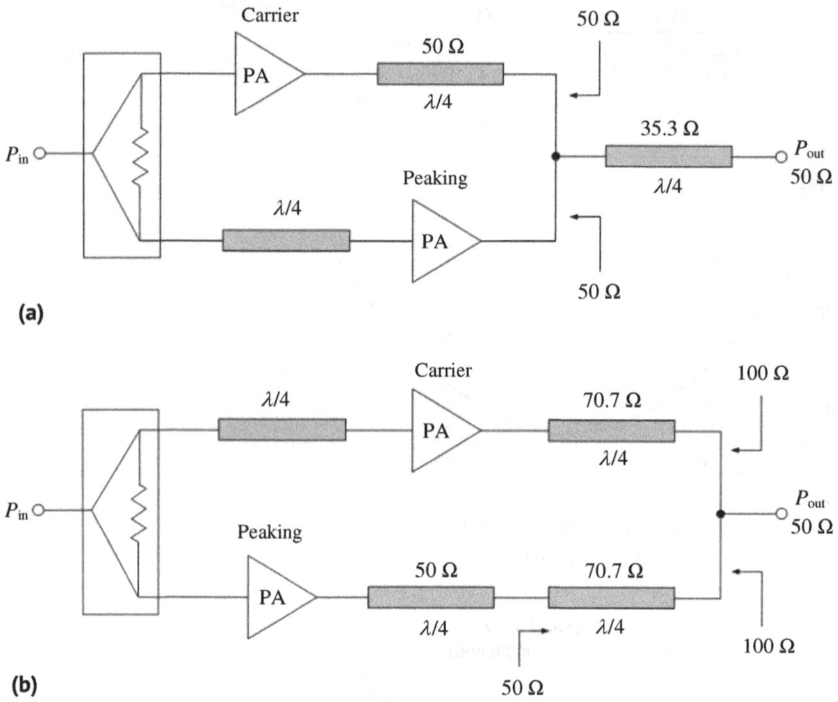

Figure 5.44 Block diagram of conventional and parallel Doherty amplifiers.

the transistor output impedances are sufficiently small [85]. Such an approach is very relevant to a Doherty amplifier with a very high output power. In this case, setting the characteristic impedances of the transmission lines higher than the output transistor impedances has contributed to the drain efficiency of about 48% and output powers over 49 dBm in a frequency range of 800–950 MHz [86].

In a parallel Doherty amplifier unlike a classic Doherty amplifier, whose load network is shown in Figure 5.45a, a single impedance-transforming quarterwave transmission line represents the load network for the carrier amplifier, the load network for the peaking amplifier includes a 50-Ω quarterwave transmission line and the impedance-transforming quarterwave transmission line. Each impedance-transforming quarterwave transmission line with a characteristic impedance of 70.7 Ω provides a 100-Ω impedance at its output with a 50-Ω impedance at its input seen by each amplifier output. Then, parallel combining of these two 100-Ω output impedances results in the required 50-Ω load impedance in a high-power region. At the same time, the quarterwave transmission line with the characteristic impedance of 70.7 Ω in a low-power region provides an impedance

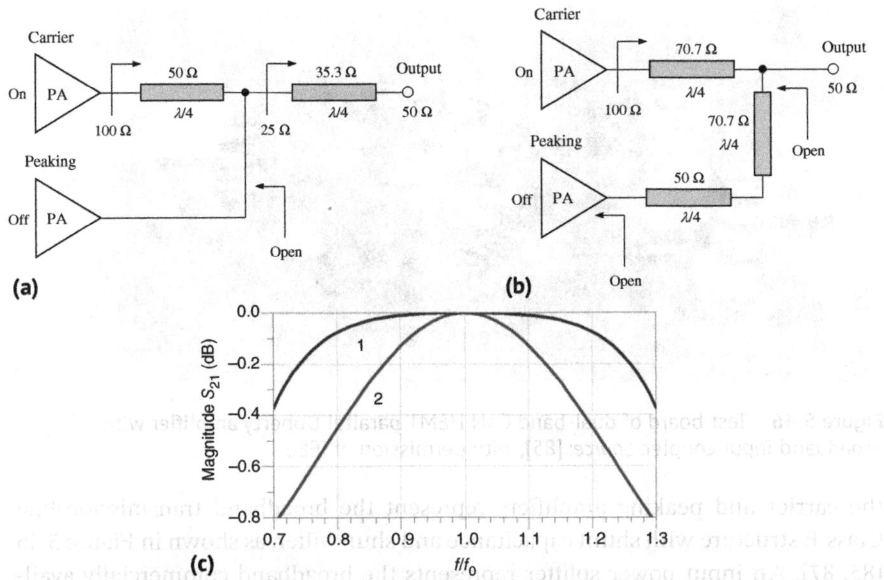

Figure 5.45 Load-network schematics of (a) conventional and (b) parallel structures, and (c) their bandwidth properties.

transformation of 50 to 100 Ω seen by the carrier amplifier output, as shown in Figure 5.45b.

Such a parallel architecture of a Doherty amplifier provides a loaded quality factor $Q_L = \sqrt{100/50 - 1} = 1$, contributing to 1.73 times wider operating frequency range, as shown in Figure 5.45c by curve 1 in contrast to the conventional structure (curve 2). At the same time, due to the series connection of two quarterwave transmission lines at the output of the peaking amplifier, an open circuit at the output of the peaking amplifying path is provided at the conjunction point when the peaking amplifier is turned off. For convenience of practical implementation, a commercially available broadband coupled-line 90° hybrid coupler can replace the input in-phase divider and phase-compensating transmission line. As shown in Figure 5.45c, a broadband operation within 25–30% near center bandwidth frequency with minimum output power variation can be provided by using a parallel Doherty architecture, covering, for example, the frequency bandwidths of 1.8–2.2 GHz or 2.1–2.7 GHz.

The small-size test board of a dual-band parallel Doherty amplifier using two 10-W GaN HEMT power transistors CGH40010P in metal–ceramic pill packages was fabricated on a 20-mil RO4360 substrate, where the load networks of both

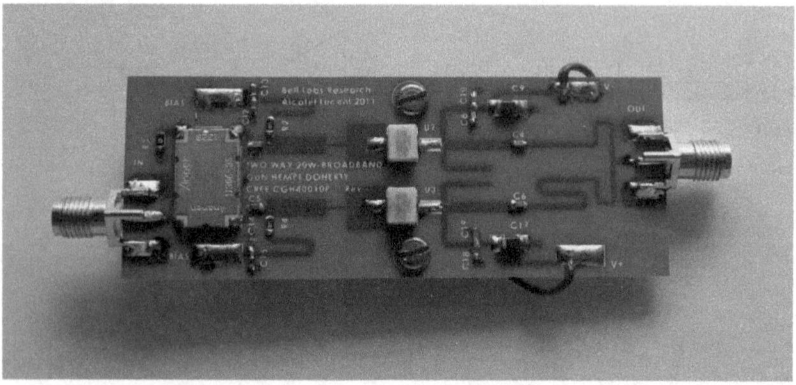

Figure 5.46 Test board of dual-band GaN HEMT parallel Doherty amplifier with broadband input coupler. *Source*: [85], with permission of IEEE.

the carrier and peaking amplifiers represent the broadband transmission-line Class-E structure with shunt capacitance and shunt filter, as shown in Figure 5.46 [85, 87]. An input power splitter represents the broadband commercially available coupled-line coupler with maximum phase balance of $\pm 5°$ and amplitude balance of ± 0.55 dB over the frequency range of 2–4 GHz. As a result, the drain efficiencies of 45% with a power gain of about 10 dB and of 40% with a power gain of about 11 dB were obtained at 2.14 and 2.655 GHz, respectively, at an average output power of 39 dBm for a single-carrier 5-MHz WCDMA signal. The load network of a dual-band parallel Doherty amplifier can be slightly simplified if two additional 35-Ω quarterwave lines, one in the carrier amplifying path and the other in the peaking amplifying path required match to the standard 50-Ω load, can be removed [88].

5.7.2 Tri-band Inverted Doherty Configuration

The inverted Doherty architecture is a very attractive configuration to maximize the operating frequency bandwidth because it allows minimizing the impedance transformation ratio for the impedance inverter, resulting in minimum matching sections between the carrier device current source and the combining point. In this case, the device output parasitic elements, such as the drain–source capacitance and drain series inductance, can represent the impedance inverter with the required impedance transformation ratio. As a result, the overall size of the Doherty amplifier can be reduced, and it is much more convenient to use the dual-path transistors.

The modified broadband load network of an inverted Doherty amplifier with two cascaded quarterwave transmission lines representing an output broadband impedance transformer is shown in Figure 5.47a. Here, each quarterwave

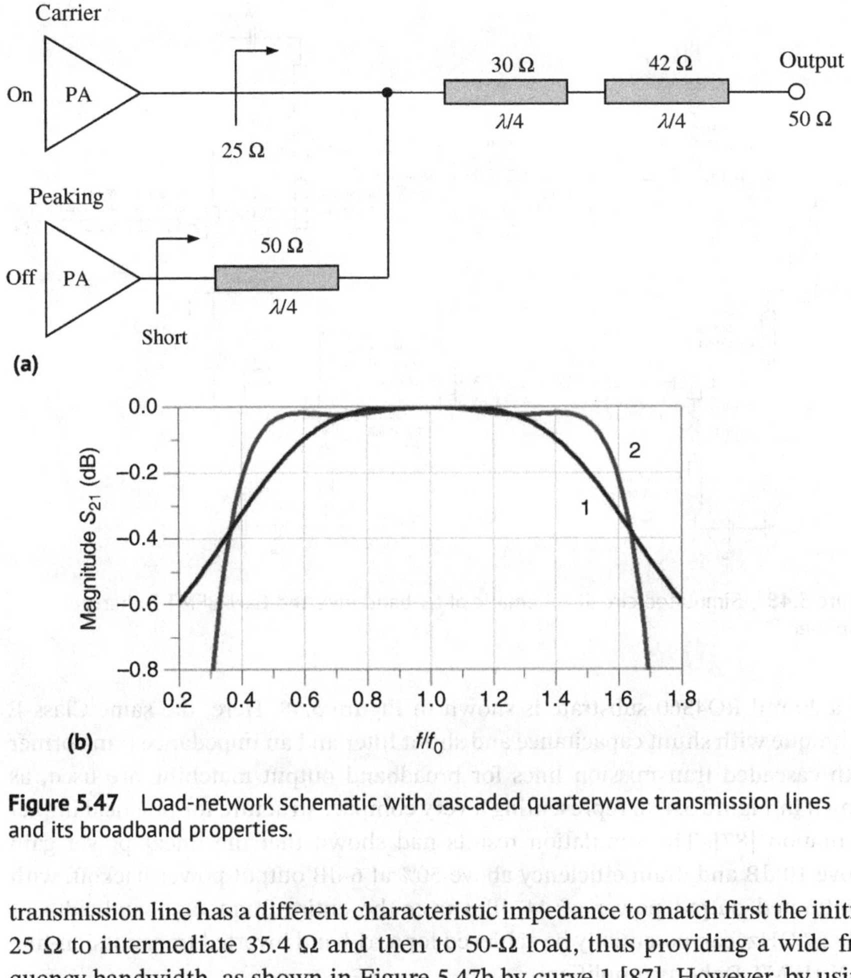

Figure 5.47 Load-network schematic with cascaded quarterwave transmission lines and its broadband properties.

transmission line has a different characteristic impedance to match first the initial 25 Ω to intermediate 35.4 Ω and then to 50-Ω load, thus providing a wide frequency bandwidth, as shown in Figure 5.47b by curve 1 [87]. However, by using a quarterwave transmission-line stub connected between the output of the peaking amplifier and the input of the cascaded transmission lines, which represents an open circuit at its output when the peaking amplifier is turned off, a flatter frequency response can be achieved in a broader frequency range, as shown in Figure 5.47b by curve 2, which may result in more than an octave frequency bandwidth in a low-power region. With such an approach to designing the broadband Doherty amplifier, an average drain efficiency close to 50% and an output power over 34 dBm were achieved across the frequency range of 3–3.6 GHz [89].

The simulated circuit schematic of a tri-band inverted GaN HEMT Doherty amplifier structure using the two CGH40010 GaN HEMT devices and fabricated

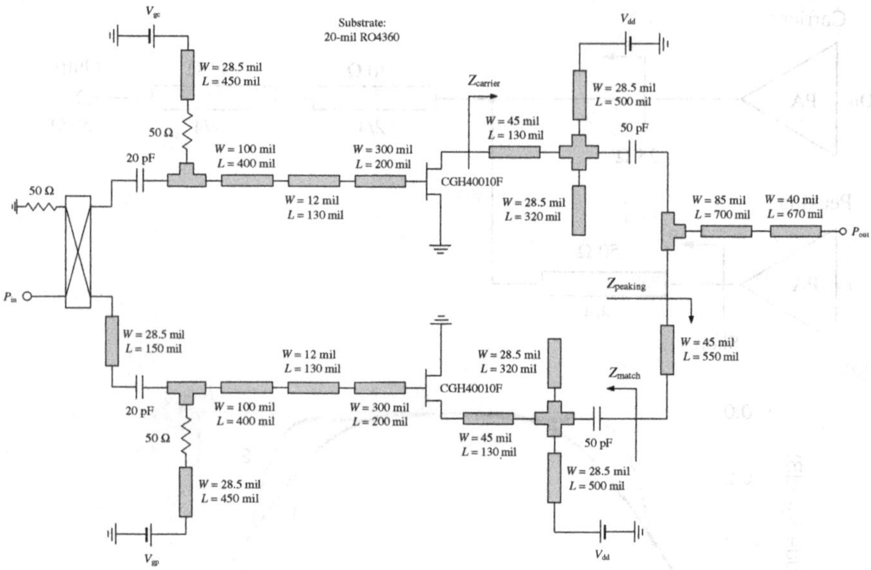

Figure 5.48 Simulated circuit schematic of tri-band inverted GaN HEMT Doherty amplifier.

on a 20-mil RO4360 substrate is shown in Figure 5.48. Here, the same Class-E technique with shunt capacitance and shunt filter and an impedance transformer with cascaded transmission lines for broadband output matching are used, as shown in Figure 5.47a, representing a very compact structure for practical implementation [87]. The simulation results had shown that the linear power gain above 10 dB and drain efficiency above 50% at 6-dB output power backoff, with a saturated output power of 44 dBm over the entire frequency bandwidth of 1.8–2.7 GHz, can potentially be achieved for a tri-band inverted transmission-line GaN HEMT Doherty amplifier.

The tri-band transmission-line inverted GaN HEMT Doherty amplifier was fabricated on a 20-mil RO4360 substrate with the input power divider, representing a broadband 3-dB hybrid coupler with maximum phase balance of ±5° and amplitude balance of ±0.5 dB over the frequency range of 690–2700 MHz. Figure 5.49 shows the test board of a tri-band inverted Doherty amplifier based on two 10-W GaN HEMT CGH40010P devices in metal–ceramic pill packages [87]. Here, the input matching circuits and load network are fully based on microstrip lines of different electrical lengths and characteristic impedances according to the simulation setup shown in Figure 5.48. In order to provide more predictable broadband performance of the tri-band inverted Doherty amplifier, the input and output lead inductances of the packaged GaN HEMT devices should be minimized. At the same

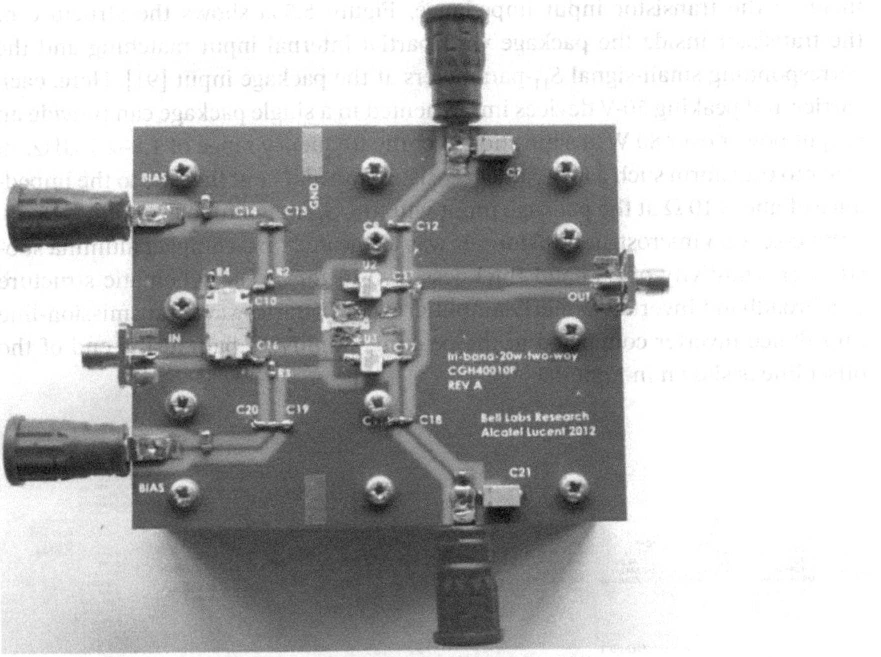

Figure 5.49 Test board of tri-band inverted GaN HEMT Doherty amplifier. *Source*: [87]. Summit Technical Media, LLC.

time, in order to maximize the power gain across the entire frequency bandwidth, especially at higher bandwidth frequencies, an additional tuning using copper foils was provided in the input matching circuits.

For a single-carrier 5 MHz WCDMA signal with a PAR of 6.5 dB, the drain efficiencies of 58, 50, and 42% at an average output power of 38 dBm with a power gain greater than 11 dB were achieved at the operating frequencies of 1.85, 2.15, and 2.65 GHz, respectively, with the ACLR (at 5-MHz offset) measured from −32 dBc at 1.85 GHz to −37 dBc at 2.65 GHz for the same bias conditions at all three operating frequencies. Close results with an output power of around 40 dBm and a drain efficiency of 45–55% were achieved over the frequency range of 1.7–2.6 GHz for an inverted GaN HEMT Doherty amplifier using a two-section Wilkinson divider at the input and open-circuit stubs to widen the frequency bandwidth in the load network [90].

In a high-power GaN HEMT device with a wider gate periphery, the increasing overall gate–source capacitance and decreasing gate resistance reduce the device input impedance to very low values, even less than 1 Ω. In this case, it is important to partly implement the input matching circuit inside the package to properly increase the transistor input impedance. Figure 5.50a shows the structure of the transistor inside the package with partial internal input matching and the corresponding small-signal S_{11}-parameters at the package input [91]. Here, each carrier and peaking 50-V devices implemented in a single package can provide an output power over 80 W at saturation over the frequency range of 1.8–2.7 GHz. In order to transform such a low-value transistor impedance at the gate to the impedance of about 10 Ω at the package input with the S_{11} magnitude less than −25 dB, a three-section microstrip transformer was fabricated on a compact alumina substrate (permittivity of 250 and thickness of 0.16 mm). The schematic structure of a broadband inverted Doherty amplifier with a quarterwave transmission-line impedance inverter connected to the peaking amplifying path at the end of the offset line is shown in Figure 5.50b.

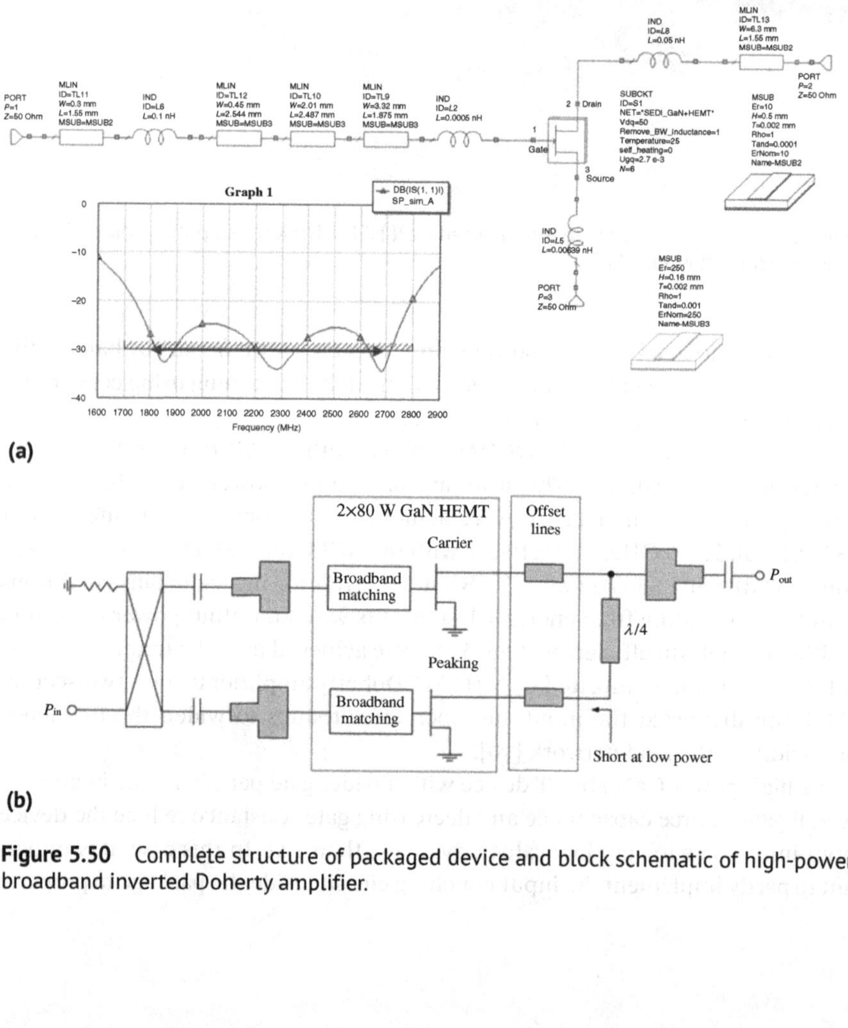

Figure 5.50 Complete structure of packaged device and block schematic of high-power broadband inverted Doherty amplifier.

The operation principle of an inverted symmetric Doherty amplifier can be demonstrated on the example of the basic load network shown in Figure 5.51a in a low-power region when the peaking amplifier is turned off. In this case, the electrical length θ of the offset line is adjusted to provide a short-circuit condition (0 Ω ideally) at its output, while a low-pass L-type matching circuit including output shunt capacitance, output series inductance, and an offset line provides an impedance transformation from 25 Ω to the high output impedance Z_{out} seen by the carrier device current source at 6-dB power backoff (100 Ω ideally), as shown in Figure 5.51b. The short circuit at the input of the quarterwave impedance inverter is transformed to the open circuit at its output seen from the carrier amplifying path to minimize power leakage to the peaking amplifying path.

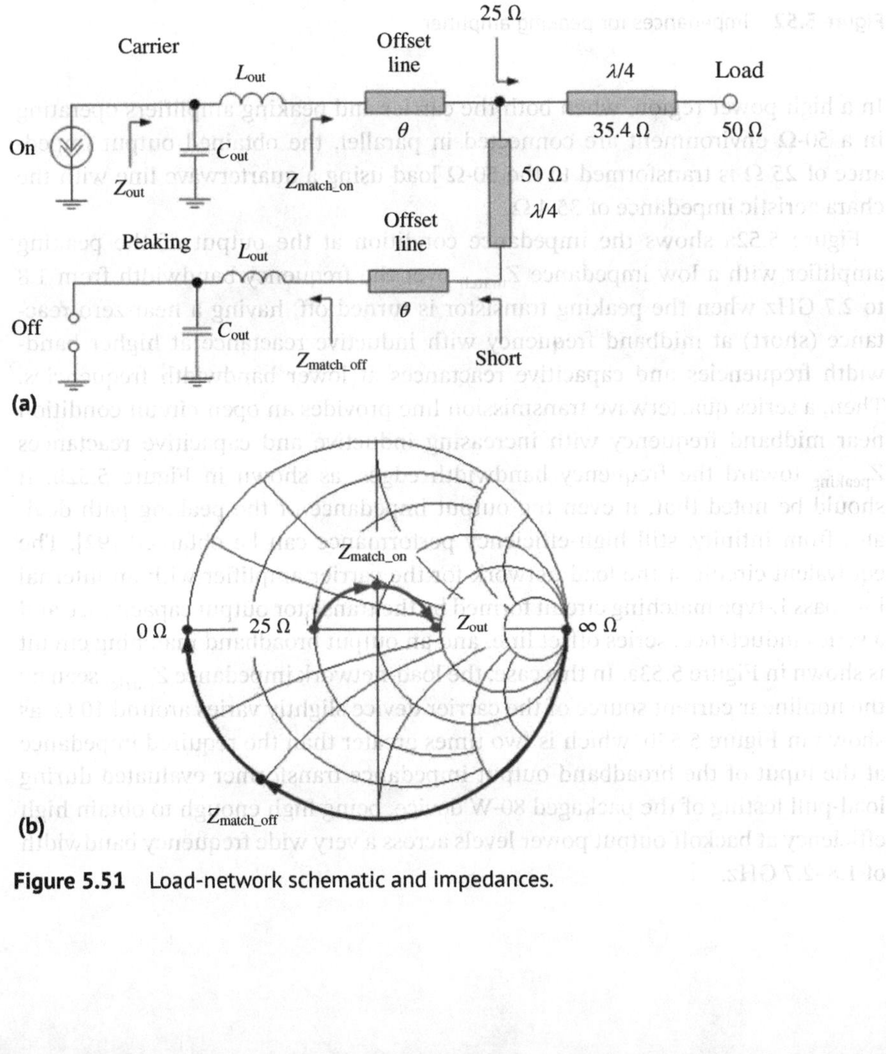

Figure 5.51 Load-network schematic and impedances.

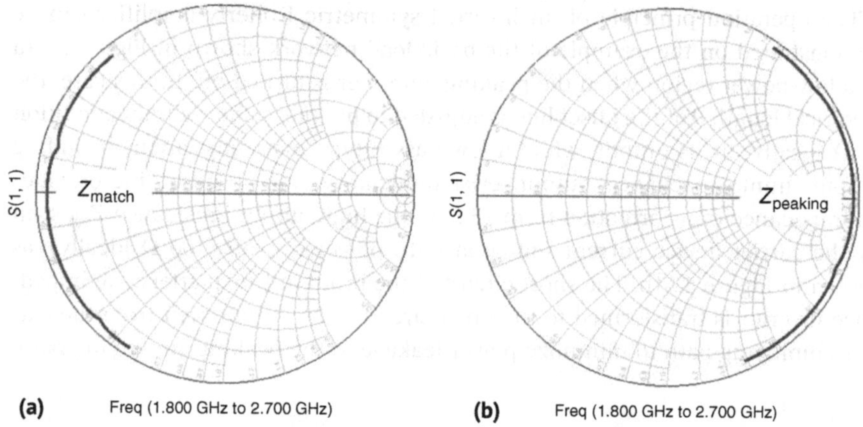

Figure 5.52 Impedances for peaking amplifier.

In a high-power region, when both the carrier and peaking amplifiers operating in a 50-Ω environment are connected in parallel, the obtained output impedance of 25 Ω is transformed to the 50-Ω load using a quarterwave line with the characteristic impedance of 35.4 Ω.

Figure 5.52a shows the impedance condition at the output of the peaking amplifier with a low impedance Z_{match} over the frequency bandwidth from 1.8 to 2.7 GHz when the peaking transistor is turned off, having a near-zero reactance (short) at midband frequency with inductive reactance at higher bandwidth frequencies and capacitive reactances at lower bandwidth frequencies. Then, a series quarterwave transmission line provides an open-circuit condition near midband frequency with increasing inductive and capacitive reactances Z_{peaking} toward the frequency bandwidth edges, as shown in Figure 5.52b. It should be noted that, if even the output impedance of the peaking path deviates from infinity, still high-efficiency performance can be obtained [92]. The equivalent circuit of the load network for the carrier amplifier with an internal low-pass L-type matching circuit formed by the transistor output capacitance and a series inductance, series offset line, and an output broadband matching circuit is shown in Figure 5.53a. In this case, the load-network impedance Z_{carrier} seen by the nonlinear current source of the carrier device slightly varies around 10 Ω, as shown in Figure 5.53b, which is two times greater than the required impedance at the input of the broadband output impedance transformer evaluated during load-pull testing of the packaged 80-W device, being high enough to obtain high efficiency at backoff output power levels across a very wide frequency bandwidth of 1.8–2.7 GHz.

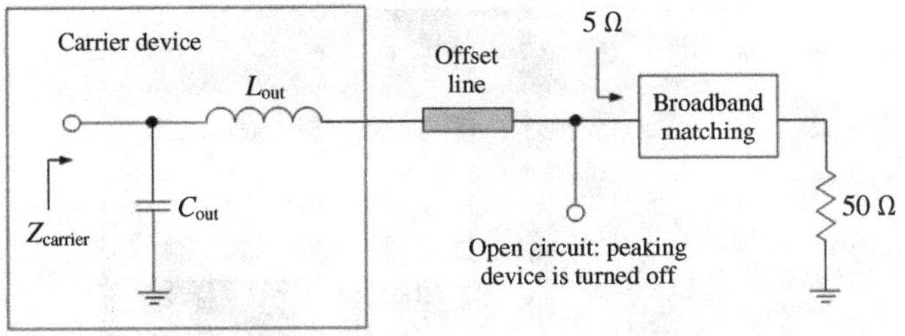

(a)

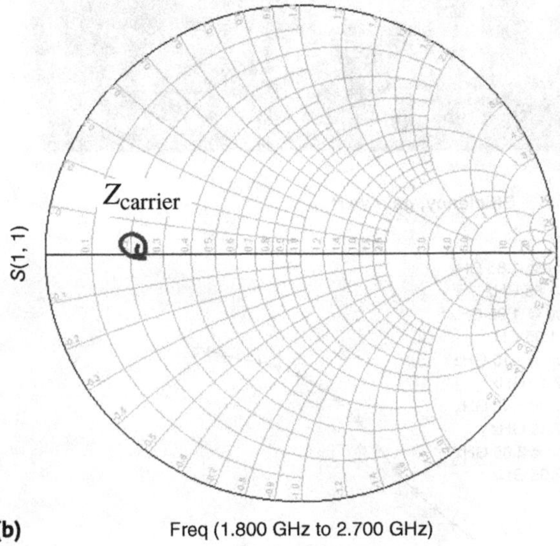

(b)

Figure 5.53 Matching network and load impedance for carrier amplifier.

Figure 5.54a shows the test board of a tri-band inverted microstrip Doherty amplifier based on the two 80-W GaN HEMT power transistors in a single package with internal three-section input matching being fabricated on a 20-mil RO4350 substrate [91]. Here, a commercially available broadband 90° hybrid coupler, which provides maximum phase balance of ±5° and amplitude balance of ±0.5 dB across the frequency range of 690–2700 MHz, represents an input splitter. The broadband output matching a 50-Ω load is provided by a three-section microstrip line with different characteristic impedances of its sections. To flatten the performance across the entire frequency range, the output lead inductances of the packaged GaN HEMT device should be minimized, and some tuning of the parameters of the output transmission-line impedance transformer using copper foil.

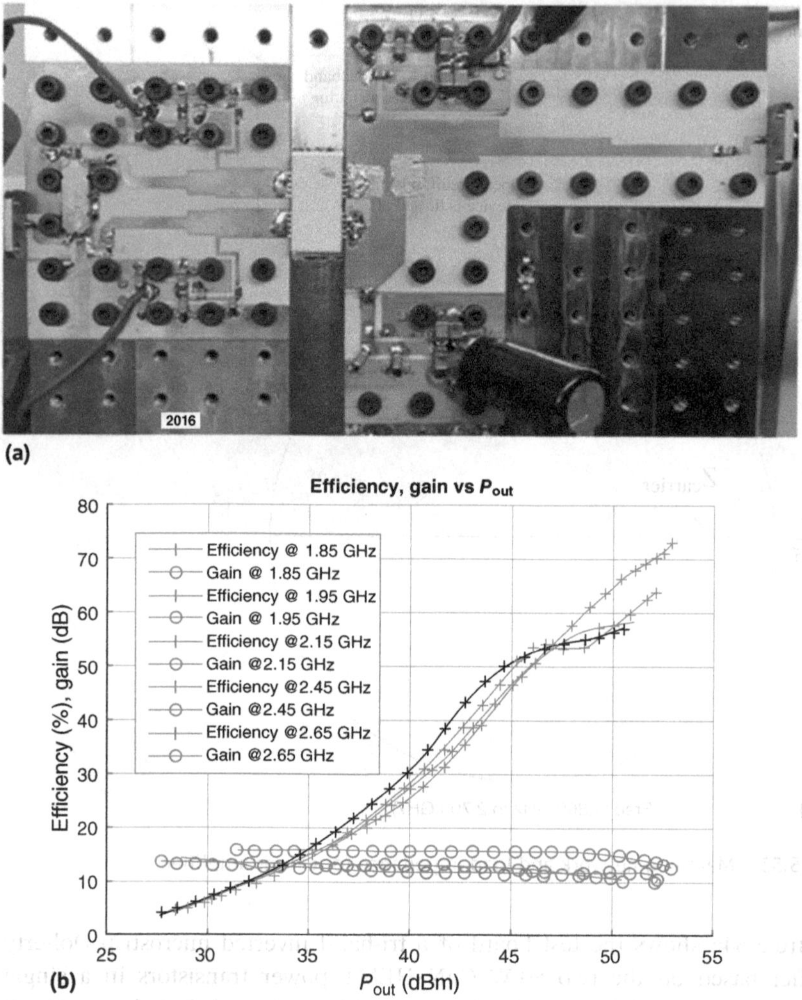

Figure 5.54 Broadband inverted Doherty amplifier using two 80 W GaN HEMT devices: (a) test board and (b) performance. *Source*: [91]. Microwave Journal.

Figure 5.54b shows the measured performance of a high-power broadband GaN HEMT inverted Doherty amplifier operating across the frequency bandwidth of 1.8–2.7 GHz, with a power gain greater than 9 dB, drain efficiencies over 55% at the output power at 3-dB gain compression point, and around 50% at 7-dB backoff output powers. Here, the maximum drain efficiency exceeds 70%

at lower bandwidth frequencies below 1.95 GHz. For concurrent transmission of a four-carrier GSM signal and 10-MHz LTE signal with a PAR of 8 dB, the drain efficiency of 51% at an average total output power of 45.5 dBm (18.2 W for GSM signal and 17 W for LTE signal) was achieved with the out-of-band intermodulation level lower than −70 dBc for a four-carrier GSM signal and ACLR lower than −57 dBc for a single-carrier 10-MHz LTE signal after dual-band DPD linearization. A broadband inverted Doherty amplifier using two GaN HEMT CGHV40100 devices with additional harmonic tuning designed with a similar approach achieved a drain efficiency of 40–50.2% at 6-dB backoff with a saturation output power of 52.7–54.3 dBm across the frequency bandwidth of 1.7–2.7 GHz, with a drain efficiency of about 45% at an average output power of 46 dBm for a single-carrier 10-MHz 7.5-dB LTE signal [93].

Figure 5.55a shows the test board of an inverted broadband Doherty amplifier structure based on two 120-W GaN HEMT devices in a single package with internal input matching, where a sufficiently wide quarterwave microstrip line is connected to the output of the peaking amplifier. Additionally, the external input and output stepped transmission lines both in the carrier and peaking amplifier paths and in the output broadband impedance-transforming network, were used to cover the frequency range of 1.8–2.7 GHz. The measured power gain and drain efficiency of such a tri-band transmission-line GaN HEMT inverted Doherty amplifier fabricated on a 20-mil RO4350 substrate across the frequency range of 1.8–2.7 GHz are shown in Figure 5.55b, with a linear power gain greater than 13 dB, the drain efficiencies over 65% at a saturation power (3-dB gain compression) of about 54 dBm and greater than 50% at 7-dB backoff output powers at a slightly higher DC-supply voltage of 55 V [94, 95].

In a particular case, the quarterwave microstrip line at the peaking amplifier output can be replaced by the compensation network with a series capacitor and a shunt microstrip stub to provide an open circuit seen from the carrier amplifying path when the peaking device is turned off that allows for the fabrication of a much more compact structure [96]. However, such an approach resulted in a lower drain efficiency of 48–54% at 6-dB power backoff from the saturation output powers of 53 and 54 dBm across the frequency bandwidth of 1.8–2.7 GHz. By using a group-delay control circuit at the input of the carrier amplifier, the frequency dispersion of load modulation can be minimized and a less dispersive performance at peak power can be achieved, resulting in 51–57% drain efficiency at 8-dB output power backoff with a small-signal gain of 14.4–17 dB over the frequency bandwidth of 1.8–2.7 GHz for a compact inverted Doherty amplifier with a 50-dBm peak power [97].

As shown in Figure 5.50a, the internal three-section microstrip transformer provides a flat amplitude response with more than 25-dB input return loss over the entire frequency range of 1.8–2.7 GHz. However, the transistor transconductance

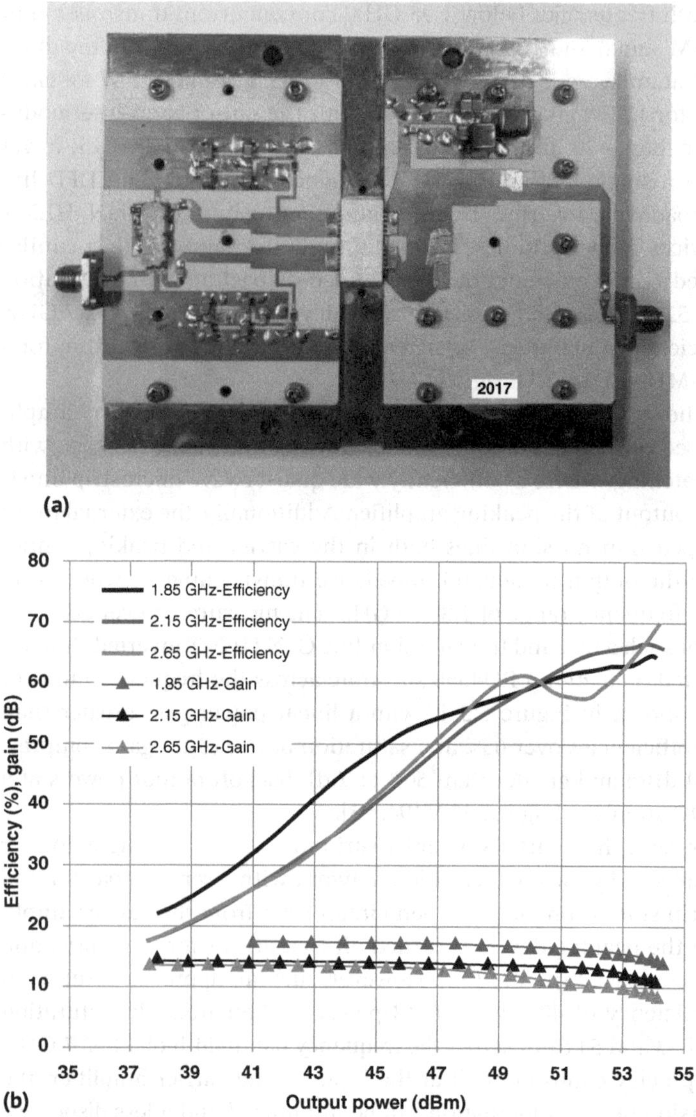

Figure 5.55 Broadband inverted Doherty amplifier using two 120-W GaN HEMT devices: (a) test board and (b) performance. *Source*: [95]. IEICE.

gradually reduces with increasing frequency, resulting in lowering the power gain across the operating frequency bandwidth, as shown in Figures 5.54b and 5.55b. Therefore, a flat power gain response can be achieved when the input matching network is designed with a prescribed amplitude response to compensate for the transistor gain roll-off at higher frequencies, which is relevant to any type of bipolar or field-effect transistors. As an alternative, the required prescribed amplitude response should be provided by preliminary stages of the multistage power amplifier. At the same time, using the broadband load network with both low-pass and high-pass filter sections contributes to an additional suppression of the low-frequency and high-frequency intermodulation and harmonic components.

5.8 Design Example: High-power 3.4–3.8 GHz Asymmetric Inverted GaN HEMT Doherty Amplifier for Base Station Applications

By using GaN HEMT technology and innovative Doherty architectures, the drain efficiencies of 50–60% for average output powers up to 120 W can be achieved that significantly reduces the cost, size, and power consumption of the cellular next-generation transmitters. The advantageous characteristics of the GaN HEMT device include high breakdown voltage, high current density, high transition frequency, low on-state resistance, and low parasitic capacitance, resulting in a high-power, wide-bandwidth, and high-efficiency operation. As an example, consider the design of an asymmetric 50-V GaN HEMT Doherty amplifier, which can provide a peak output power over 500 W in a frequency bandwidth of 3.4–3.8 GHz using a broadband Class-E approach to maximize efficiency across the operating bandwidth. In this case, the two high-power GaN HEMT devices which can be incorporated in a single package and can operate across the entire frequency bandwidth at output power levels of 220 W for the carrier amplifier and 380 W for peaking amplifier at a DC-supply voltage of 50 V are chosen.

Figure 5.56a shows the idealized ADS simulation setup of a 220-W carrier GaN HEMT amplifier with internal input single-section matching and output harmonic-control circuits where the bondwire arrays can be used instead of lumped inductors. Here, the external input matching is provided by two quarterwave lines with the corresponding characteristic impedances of 10 and 30 Ω, respectively, in order to match the impedance at the device package input of 5.5 Ω to the standard 50-Ω source, considering the package lead inductance. The load network represents a parallel-circuit Class-E structure with a low-Q series circuit composed of the bondwire and device lead inductance and a series capacitance of 8 pF to achieve a high efficiency across a wide frequency range. The optimum parameters of a parallel-circuit Class-E load network can be calculated using Eqs. (3.148)–(3.150) from Chapter 3, resulting in a 15-Ω load, parallel inductance of 0.5 nH, and shunt capacitance of 2 pF. At the same time, the device drain–source capacitance C_{ds} is substantially higher, being approximately equal to 1 pF per 20-W output power.

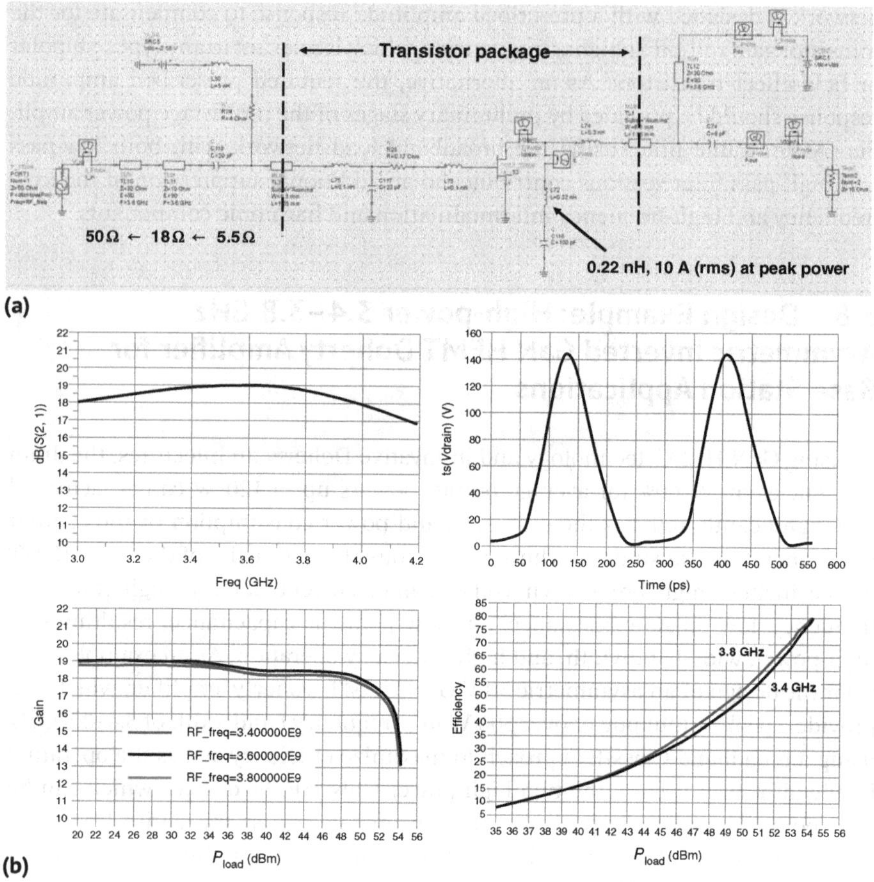

Figure 5.56 Simulation setup and performance of a 220-W Class-E GaN HEMT power amplifier.

However, the most important factor to approximate the high-efficiency Class-E operation is to provide an inductive impedance at the fundamental and capacitive reactances at the second- and higher-order harmonics.

Therefore, even though the shunt capacitance is too high, it would be enough to reduce the value of the parallel inductance to maximize efficiency which is demonstrated by the drain voltage waveform with a peak factor almost three times higher than the DC-supply voltage, as shown in Figure 5.56b. In this case, by decreasing the shunt inductance to 0.22 nH, a peak output power of 54 dBm, a drain efficiency close to 80%, and a linear power gain close to 19 dB across the entire frequency bandwidth of 3.4–3.8 GHz were simulated. To satisfy the fusing current requirements for parallel bondwires and minimize their number, the

2-mil bondwires should be chosen. In order to implement a 100-pF capacitance, required for DC isolation of the parallel bondwire array from the ground, a thin-film ceramic substrate with a high dielectric constant greater than 100 should be used. It should be noted that the presence of the shunt inductance directly connected to the device drain also contributes to the significant improvement of intermodulation distortion asymmetry in a video bandwidth by providing the baseband termination across the wide frequency range [98].

A similar design approach can be applied to a 380-W GaN HEMT power amplifier, which is intended to operate as a peaking amplifier in a 1 : 2 asymmetric Doherty amplifier structure. Figure 5.57a shows the idealized ADS simulation setup of a 380-W peaking GaN HEMT amplifier with internal input single-section

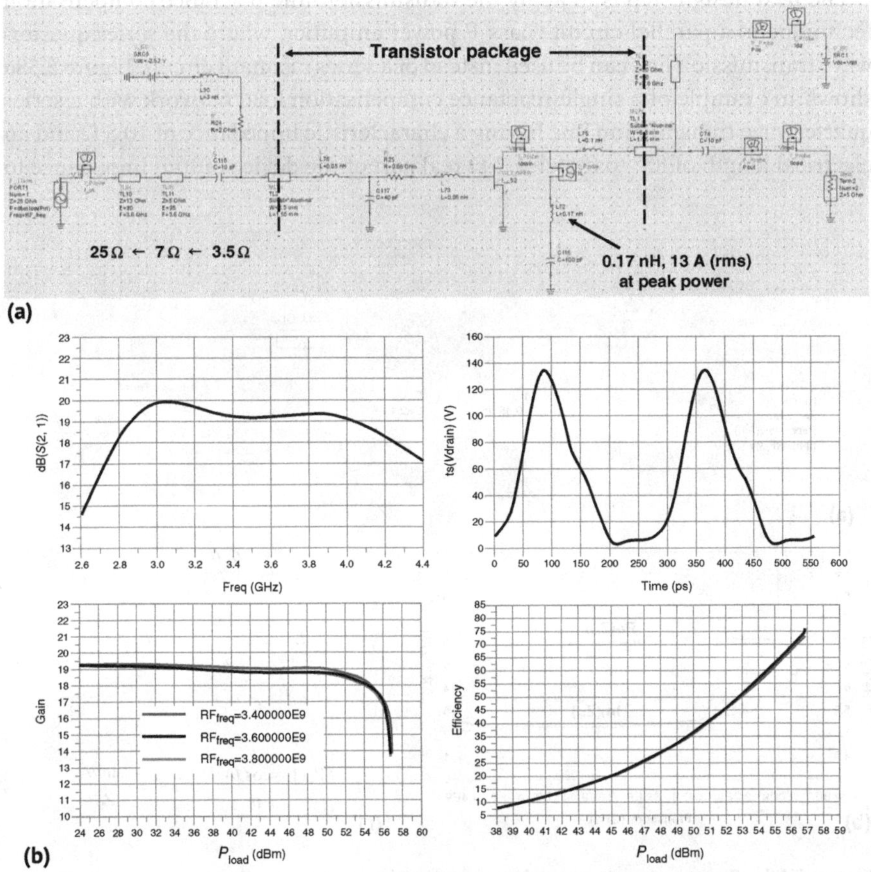

Figure 5.57 Simulation setup and performance of 380-W Class-E GaN HEMT power amplifier.

matching and output harmonic-control circuits where the bondwire arrays can be used instead of lumped inductors. Here, the external input matching is also provided by two quarterwave lines with the corresponding characteristic impedances of 5 and 13 Ω, respectively, to match the impedance at the device package input of 3.5 Ω to the 25-Ω source, considering the package lead inductance. The load network with the optimized parallel inductance of 0.17 nH and 5-Ω load represents a parallel-circuit Class-E structure with a low-Q series circuit composed of the bondwire and device lead inductance and a series capacitance of 10 pF to achieve a high efficiency across a wide frequency range. As a result, a peak power of about 57 dBm, a drain efficiency close to 75%, and a linear power gain close to 19 dB across the entire frequency bandwidth of 3.4–3.8 GHz were simulated, as shown in Figure 5.57b.

The next design step is to apply the transmission-line reactance compensation technique to a parallel-circuit Class-E power amplifier, where the series quarterwave transmission line can be used instead of a series resonant circuit. Figure 5.58a shows an example of a single-reactance compensation load network with a series quarterwave transmission line having a characteristic impedance of 10.8 Ω and an electrical length of 85° to match a 7-Ω real part of the device output impedance to

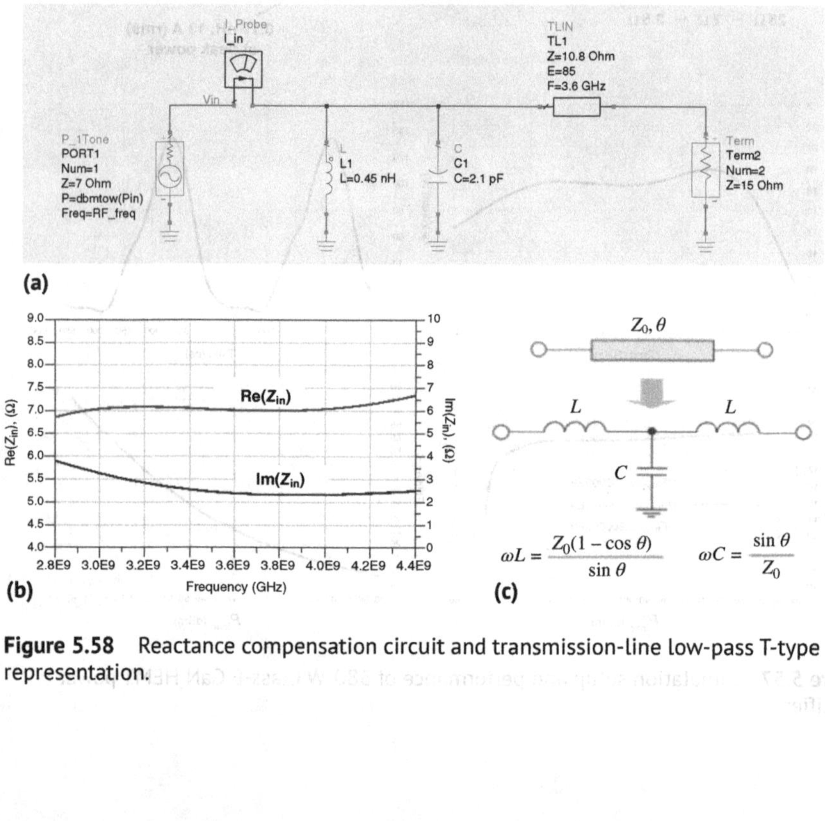

Figure 5.58 Reactance compensation circuit and transmission-line low-pass T-type representation.

a 15-Ω load at 3.6 GHz. The combination of a shunt LC circuit, where the shunt capacitance can be represented by the device drain–source capacitance and a series quarterwave transmission line provides minimum variations of the input resistance $\mathrm{Re}(Z_{\mathrm{in}}) = \mathrm{Re}(V_{\mathrm{in}}/I_{\mathrm{in}})$ of about 7 Ω and the input reactance $\mathrm{Im}(Z_{\mathrm{in}})$ of around 5.5 Ω in a very wide frequency range, as shown in Figure 5.58b. However, the value of the transmission-line characteristic impedance equal to 10 Ω or less can cause a problem in the practical implementation of the microstrip line when it becomes too wide. In this case, it can be replaced by a lumped low-pass T-type equivalent with a shunt capacitance and two series inductances, as shown in Figure 5.58c with their design equations obtained from the comparison of the corresponding matrix elements. Then, these lumped elements can be relocated inside the device package where the bondwire arrays can be used to implement both series inductances, while the ceramic capacitance with a high permittivity can be used to implement the shunt capacitance.

Figure 5.59a shows the idealized ADS simulation setup of a 220-W carrier GaN HEMT amplifier with an internal input matching section and a parallel-circuit Class-E reactance compensation circuit using a low-pass T-type lumped equivalent having a 15-Ω load resistance, and the bondwire arrays can be used

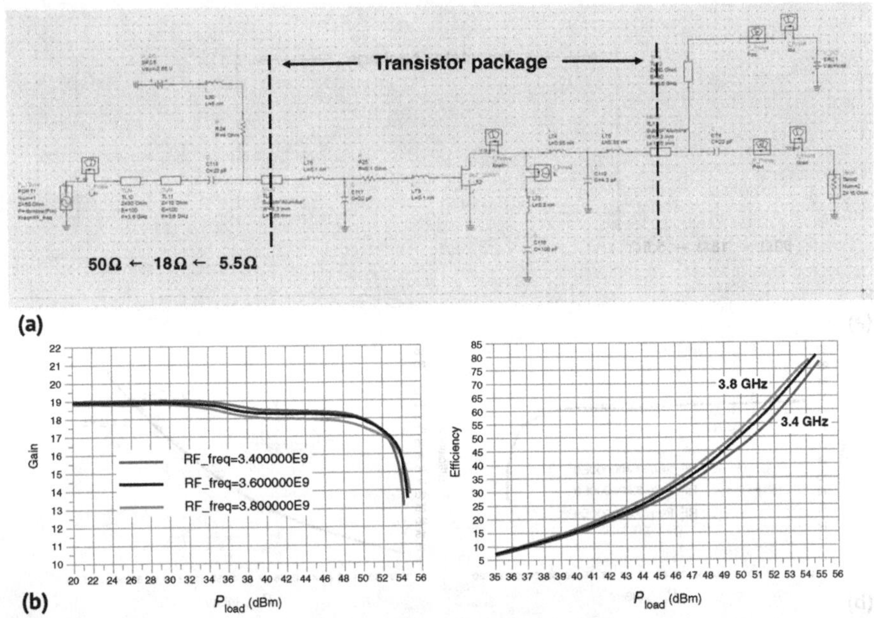

Figure 5.59 Simulation setup and performance of 220-W Class-E GaN HEMT power amplifier with partial internal matching.

instead of lumped inductors. The external parallel 30-Ω quarterwave line provides the isolation of the DC supply from the RF amplifying path. In order to optimize the amplifier performance over the entire operating frequency range, the parallel inductance was slightly increased to 0.3 nH. As a result, a peak power of about 54 dBm, a drain efficiency close to 80%, and a linear power gain close to 19 dB with minimum flatness across the frequency bandwidth of 3.4–3.8 GHz were simulated, as shown in Figure 5.59b. Similarly, for the 380-W peaking GaN HEMT amplifier with the idealized simulation setup shown in Figure 5.60a, a peak power close to 57 dBm, a drain efficiency over 75%, and a linear power gain over 19 dB with minimum flatness across the entire frequency bandwidth of 3.4–3.8 GHz were simulated, as shown in Figure 5.60b. In this case, the drain efficiency is slightly higher due to a smaller effect of the load on higher-order harmonics.

The circuit schematic of a load network of the peaking amplifier is shown in Figure 5.61a, where an additional 10-Ω transmission line with an electrical length of 87° (almost of a quarter wavelength) is added in series to provide an open-circuit condition near midband frequency with enough high inductive and capacitive reactances seen from the carrier amplifying path across the operating frequency bandwidth of 3.4–3.8 GHz, as shown in Figure 5.61b.

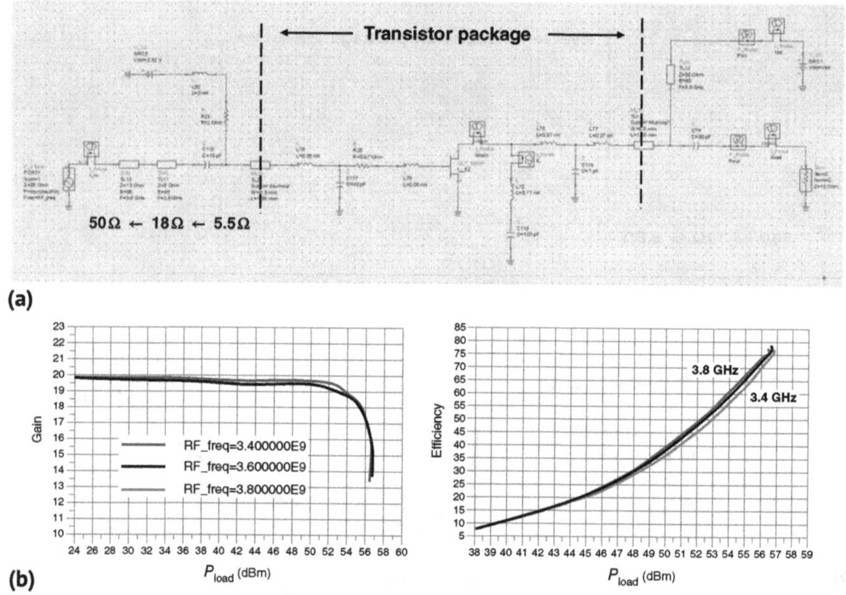

Figure 5.60 Simulation setup and performance of 380-W Class-E GaN HEMT power amplifier with partial internal matching.

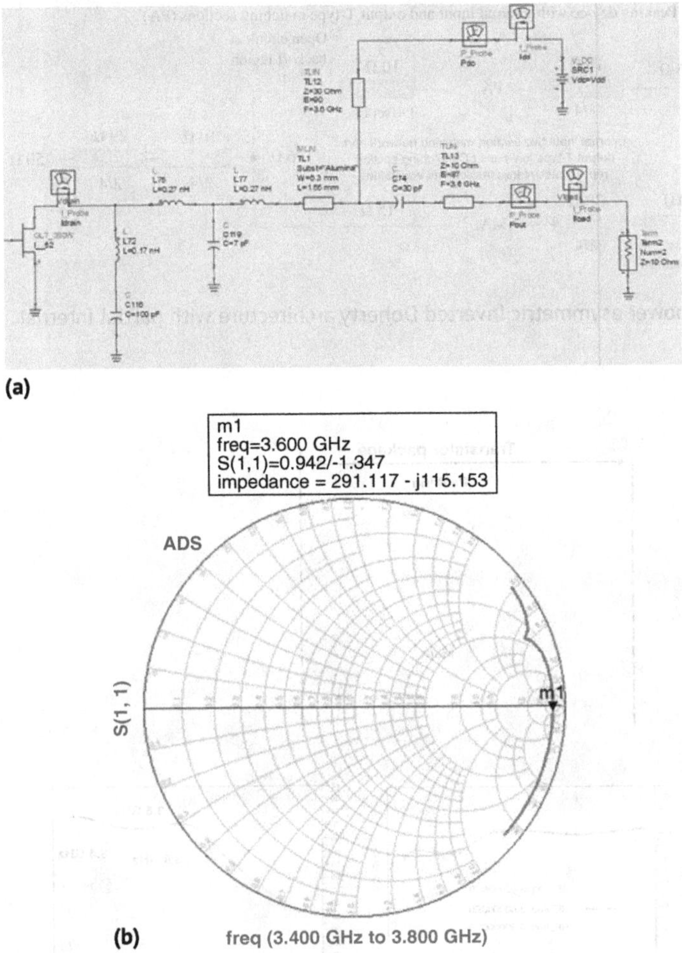

Figure 5.61 Peaking amplifier at power backoff mode: (a) load network and (b) output impedances.

Finally, both the carrier and peaking amplifiers can be combined into a Doherty amplifier structure. Figure 5.62 shows the block schematic of a high-power 1 : 2 asymmetric inverted Doherty amplifier where two quarterwave transmission lines with the characteristic impedances of 10 and 29 Ω provide a broadband impedance transformation in order to match the output impedance of 6 Ω at the connecting point to the standard 50-Ω load. A 1 : 2 asymmetric Wilkinson divider can be used at the input of the Doherty amplifier to properly split the input power,

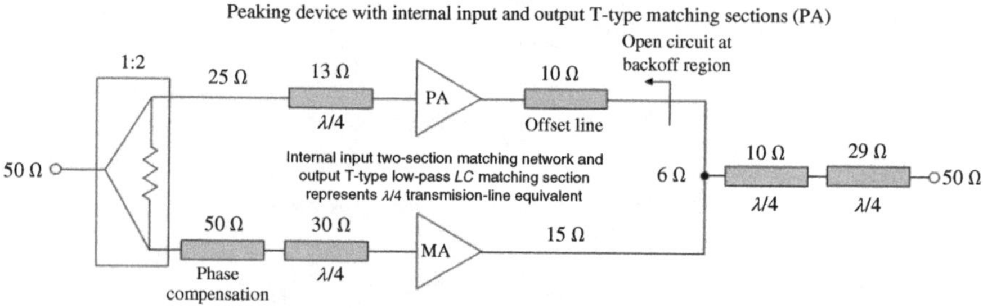

Figure 5.62 High-power asymmetric inverted Doherty architecture with partial internal matching.

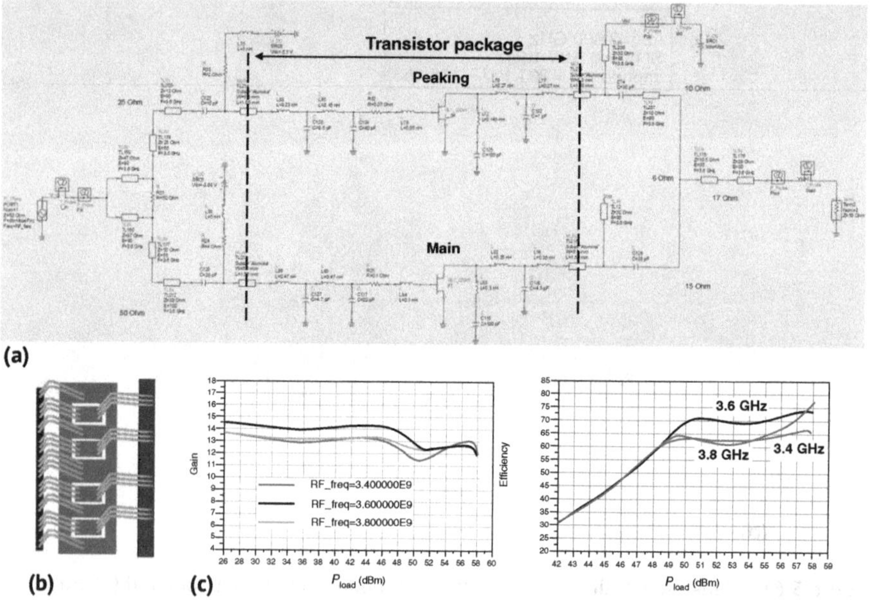

Figure 5.63 Simulation setup and performance of high-power 3.4–3.8 GHz asymmetric GaN HEMT Doherty amplifier for base station applications.

with the corresponding phase compensation line at the input of the main (carrier) amplifier. Here, it is assumed that the carrier and peaking devices in a single dual-path package contain the internal two-section matching network and output low-pass T-type matching section each.

Figure 5.63a shows the idealized simulation setup of a high-power 1 : 2 asymmetric inverted Doherty amplifier with two-section matching at the input of the carrier and peaking amplifiers to increase the impedances at both package inputs that simplifies the implementation of the input circuit using a standard PCB substrate. All internal inductors represent the bondwire arrays, assuming 1-mil bondwires at the input and 2-mil bondwires at the output. An implementation example of the output broadband Class-E reactance compensation circuit for the peaking amplifier is shown in Figure 5.63b, where both capacitances of 100 and 7 pF (four parallel pads) are fabricated on the ceramic substrate with dielectric constant of 150, and four pairs of series bondwire arrays with three bondwires in each array are used for symmetrical output power distribution. Figure 5.63c shows the simulated idealized performance with a linear power gain of around 13 dB, a peak drain efficiency over 65% at 58 dBm, and a drain efficiency over 60% at 9-dB power backoff with minimum flatness across the frequency range of 3.4–3.8 GHz.

5.9 Design Examples: Integrated Sub-6-GHz Two-stage GaN HEMT Doherty Amplifiers for Massive-MIMO Applications

In modern and next-generation communication systems employing the massive MIMO antenna arrays in their base stations to realize high data rates, modulation signals with a large PAR and multiple carrier aggregation, the power amplifiers are required to be efficient at larger backoff power levels from saturation with a high power gain over wide operating frequency ranges and without an increase in size and cost. In this case, an integrated high-efficiency two-stage Doherty amplifier structure in sub-6-GHz frequency bands can be designed based on GaN HEMT devices with high breakdown voltage and transition frequency using both monolithic integrated circuits and hybrid implementations on ceramic or thin laminate substrates.

Figure 5.64 shows the circuit schematic of a compact and fully integrated asymmetric two-stage GaN HEMT Doherty amplifier, where the Wilkinson power divider and phase offset line in the peaking path are equivalently replaced by the low-pass Π-type *LC* sections instead of the quarterwave lines [99]. Here, the input matching circuits for the carrier and peaking power amplifiers are fabricated on GaAs IPDs, whereas the input and interstage matching circuits in the driver stage, Wilkinson power divider, phase offset line, combining and output matching low-pass circuits are based on SMD (surface mount devices) components, resulting in a 10 × 6 mm^2 two-stage Doherty amplifier module using a multilayer epoxy substrate. The gate peripheries of the carrier and peaking GaN HEMT devices are 2.4 and 4 mm, respectively, with an asymmetry ratio of 1.67. The series bondwires connecting to the gates and drains of the carrier and peaking devices are included in the input and output matching circuits in the carrier and peaking amplifying paths. The shunt inductances and microstrip impedance-transforming line

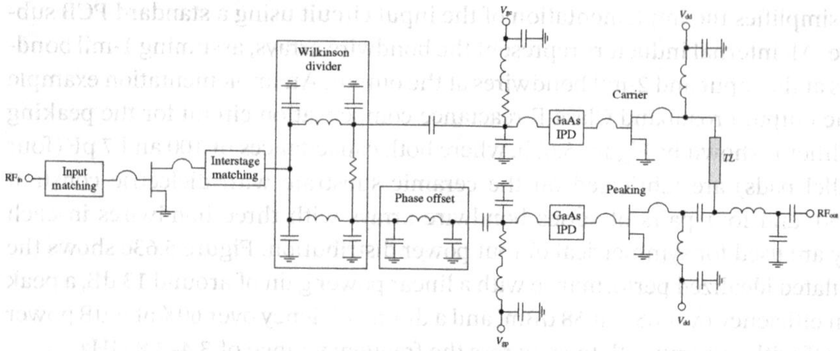

Figure 5.64 Circuit schematic of an integrated 3.4–3.8 GHz two-stage Doherty amplifier.

TL represent the combining network of the Doherty amplifier. The DC-supply voltage was set at 39 V, and the carrier (main) and driver amplifiers were biased in Class-AB, while the peaking amplifier was biased in Class-C to achieve a peak output power of 45.3 dBm. Over the frequency bandwidth of 3.4–3.8 GHz, a PAE of 42.9–47.8%, a power gain of 26.9–28.6 dB, and an average output power of 37.7–38.5 dBm at 8-dB output power backoff level were obtained with an ACLR of −50 dBc for a 10-carrier 20-MHz LTE signal after DPD linearization.

As an alternative, the input matching circuit and all interstage networks composing the Wilkinson power divider, phase compensation circuit, carrier, and peaking input matching circuits can be implemented on two IPDs inside a 7×7 mm^2 quad flat no-lead (QFN) package. As a result, in a frequency bandwidth of 3.4–3.8 GHz, such an integrated Doherty amplifier module achieved a peak output power between 47.5 and 47.9 dBm and a power gain between 26.8 and 28.4 dB with a PAE between 37.8 and 41.9% at an average output power of 39 dBm for a single-carrier 100-MHz LTE signal with a PAR of 9.3 [100].

The higher level of integration can be achieved if the output combining and matching circuits are also implemented on a separate IPD. Figure 5.65a shows the *LC* equivalent of the combining and output matching circuits where both the quarterwave impedance-transforming and output matching lines are replaced by a high-pass Π-type section with a series capacitor and two shunt inductances each, the series and parallel resonant circuits are added to the peaking path to extend the frequency bandwidth due to reactance compensation at the frequency band edges, and all three adjacent shunt inductances are combined into a single inductance L_{eq} [101]. The main and peaking amplifying paths of an asymmetric fully integrated two-way two-stage Doherty amplifier uses 10- and 20-finger 0.3-μm GaN HEMT devices, respectively. The four IPD circuits include an input, two

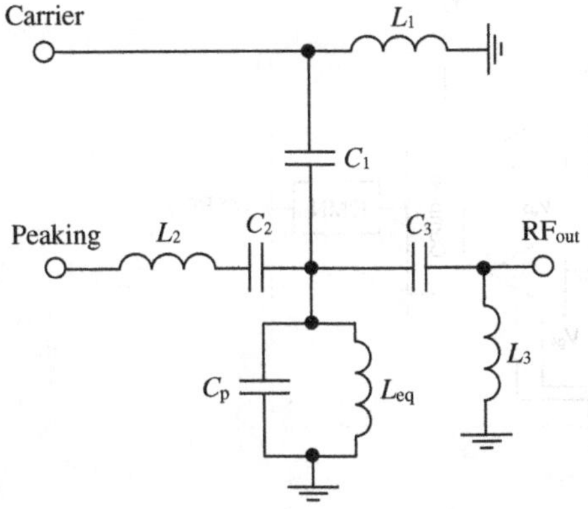

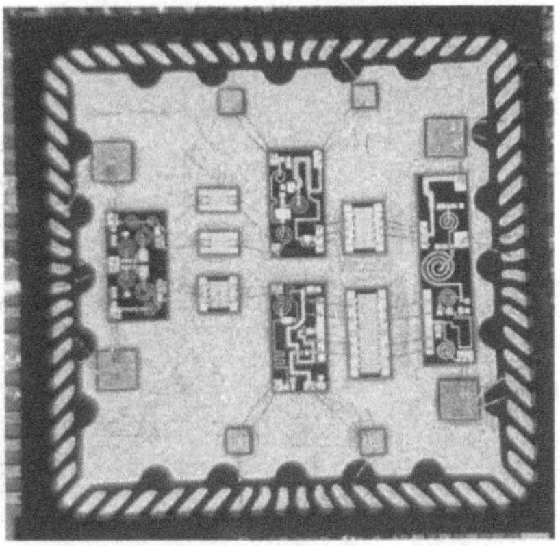

Figure 5.65 Fully integrated 3.4–3.8 GHz two-way two-stage Doherty amplifier. *Source*: [101]. IOP Publishing. CC BY 4.0.

interstage (one for the main device and the other for the peaking device), and output matching circuits, respectively. The GaN HEMT devices and IPDs on SiC substrates were assembled in a 10 × 10 mm^2 QFN package, as shown in Figure 5.65b. For a single-carrier 100-MHz LTE signal, a power gain higher than 22.3 dB and a PAE over 37.2% were measured across the frequency bandwidth of 3.4–3.8 GHz with a 50 V drain bias and at 9-dB output power backoff from a peak output power of 50 W.

There is also a possibility to improve the linearity of the inverted Doherty amplifier and its bandwidth performance by introducing separate driver stages into the main and peaking amplifying paths of the Doherty amplifier. Figure 5.66a shows the block schematic of an enhanced two-way two-stage inverted Doherty architecture where each final stage (main and peaking) includes its own driver stage [102]. Here, due to the separate input (IMN) and interstage (ISMN) matching networks in the main and peaking amplifying paths, the amplitude and phase responses of the Doherty amplifier can be properly adjusted to minimize their variations with the output power, thus resulting in better linearity performance, although making the Doherty configuration slightly more complicated. In this configuration, a gate width ratio of 1.8 between the carrier (main) and peaking devices in the final stage is used to maintain a high modulated efficiency at significant output power backoff levels. As a result, the total drain efficiency of 52.2–54.9% at an average output power of 39 dBm was obtained over the frequency bandwidth of 3.4–3.8 GHz, with a saturated output power (P_{5dB}) higher than 47.8 dBm and a linear power gain above 29.5 dB, as shown in Figure 5.66b.

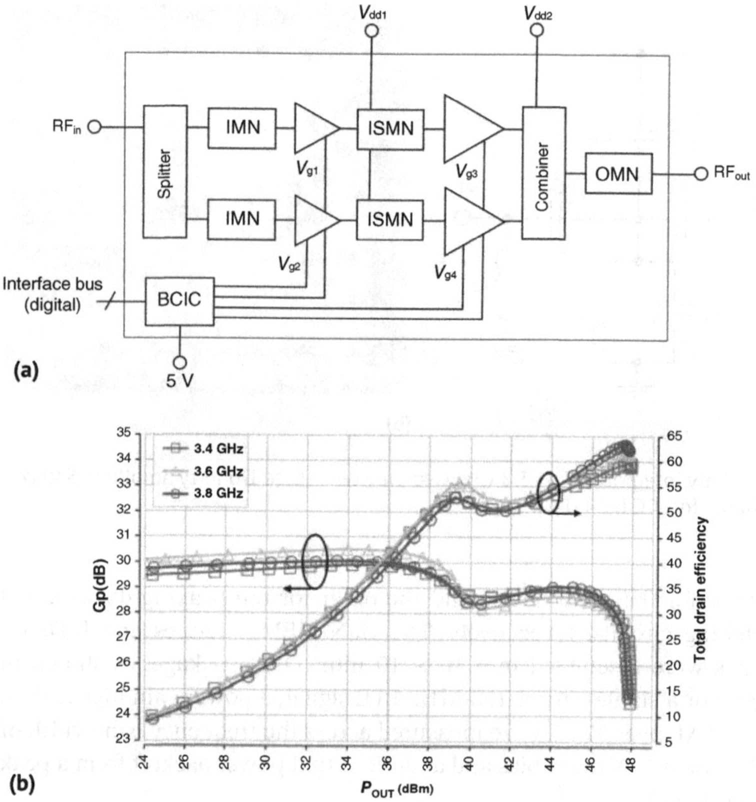

Figure 5.66 Simulation setup and performance of high-power 3.4–3.8 GHz asymmetric GaN HEMT Doherty amplifier for base station applications.

This design implemented on a 12 × 8 mm^2 land grid array (LGA) package allows obtaining the linearizability levels using DPD up to −50 dBc with a total drain efficiency of 46.7% for a dual-carrier 100-MHz LTE signal occupying the 400-MHz bandwidth.

An inverted Doherty amplifier architecture to achieve better broadband performance and a common interstage IPD for the carrier and peaking devices (in total, three IPDs) were chosen to design a fully integrated two-stage asymmetric Doherty amplifier to cover the frequency range of 4.5–5 GHz [103]. In this case, the two additional quarterwave lines are added in the peaking amplifying path, which introduce extra resonances to improve bandwidth, and all inductors in the matching networks are realized with microstrip lines to reduce the loss. The quarterwave lines can be replaced with a high-pass or low-pass lumped Π-type circuits, where it is necessary for implementation convenience. The main and peaking power amplifiers are designed to have an optimal load impedance of 100 Ω, so they can be directly connected to a 50-Ω load without the need for an output matching network. In the final stage of the Doherty amplifier, two GaN HEMT devices with a 4 : 5 gate width ratio are used to implement an asymmetric Doherty structure. A high-pass lumped Wilkinson power divider with an *LC*-ladder structure is used to minimize the insertion loss and to extend the frequency bandwidth. The entire module with three GaN HEMT devices and three GaAs IPDs placed on a four-layer substrate is assembled in a 10 × 6 mm^2 LGA package. As a result, a saturated output power of more than 45 dBm, a power gain of 27–28.2 dB (gain compression of less than 4 dB), and an average PAE of 36.3% for a single-carrier 100-MHz LTE signal with a PAR of 8.5 dB were measured in the frequency bandwidth of 4.5–5 GHz.

Figure 5.67 shows the idealized ADS simulation setup of an integrated asymmetric two-stage inverted Doherty amplifier, where a low-pass T-type equivalent is used to replace a quarterwave line at the output of the peaking amplifying path to minimize the overall size with 17- and 34-W GaN HEMT devices in the main and peaking amplifying paths, respectively, in order to provide a 1 : 2 asymmetry level to maximize efficiency at 8-dB power backoff in a frequency bandwidth of 3.7–4 GHz. In this case, the saturated power close to 47 dBm, a PAE over 47%, and a power gain greater than 31 dB were simulated at an output power of 39 dBm (8-dB power backoff), as shown in Figure 5.68a. In order to minimize the overall insertion loss in practical implementation, the driver IPD, common interstage IPD, and two IPDs for the carrier and peaking 50-V devices, where all quarter-wave transmission lines were replaced by their low-pass Π-type lumped equivalents with spiral inductors and MIM capacitances, were assembled inside the 5 × 5 mm^2 internal package, with an external DC-blocking chip capacitor. In this case, the saturated power of more than 46.8 dBm, a PAE over 45%, and a power

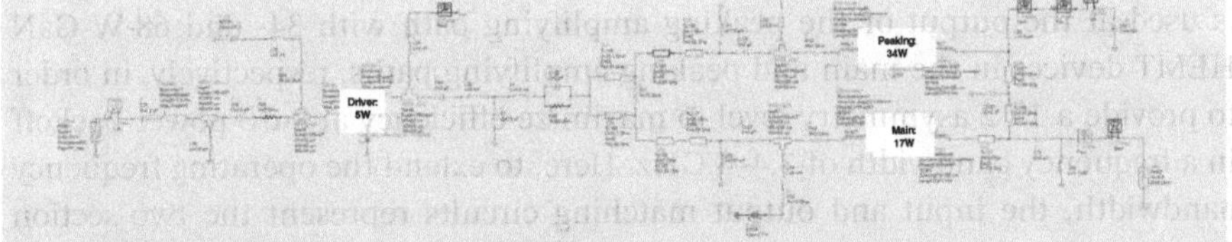

Figure 5.67 Simulation setup of integrated 3.7–4 GHz two-stage inverted GaN HEMT Doherty amplifier.

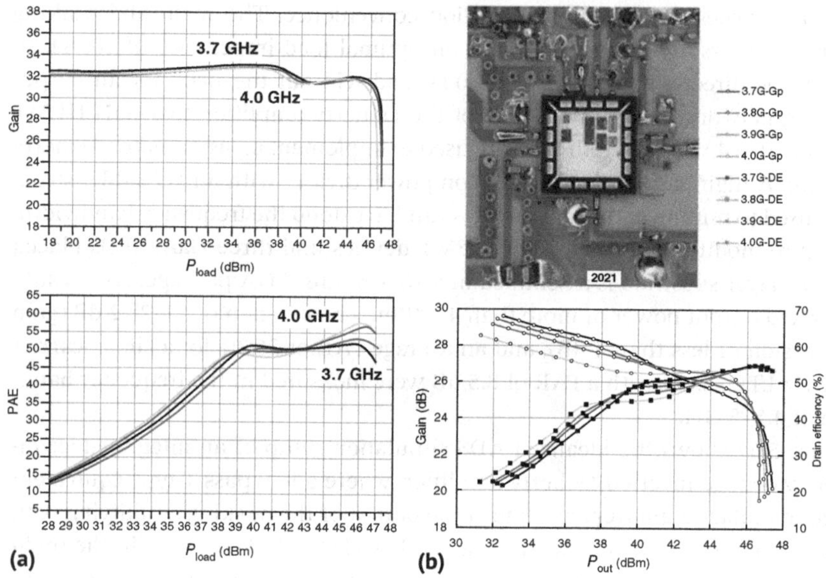

Figure 5.68 Performance and implementation of integrated 3.7–4 GHz two-stage inverted GaN HEMT Doherty amplifier.

gain greater than 26.8 dB were achieved at an output power of 39 dBm, as shown in Figure 5.68b. The power gain and its flatness can be improved by optimizing the gate bias voltages of the driver and carrier devices and using MIM capacitances (two in series to maximize the breakdown voltage) inside the interstage IPD instead of the external DC-blocking capacitor where it may be considered as a part of the interstage matching circuit.

A similar approach can be applied to the design of the Doherty amplifier with much wider operating frequency bandwidths, providing an increased output power and higher gain with minimum flatness over the entire frequency range. Figure 5.69 shows the idealized ADS simulation setup of an integrated asymmetric two-stage inverted Doherty amplifier where a low-pass T-type equivalent is used at the output of the peaking amplifying path with 34- and 68-W GaN HEMT devices in the main and peaking amplifying paths, respectively, in order to provide a 1 : 2 asymmetry level to maximize efficiency at 8-dB power backoff in a frequency bandwidth of 3.4–4 GHz. Here, to extend the operating frequency bandwidth, the input and output matching circuits represent the two-section structures with low-pass and high-pass L-type sections each. In practice, the input, interstage, and two input matching circuits for the carrier and peaking amplifiers are expected to be implemented on IPDs, while the low-loss laminate

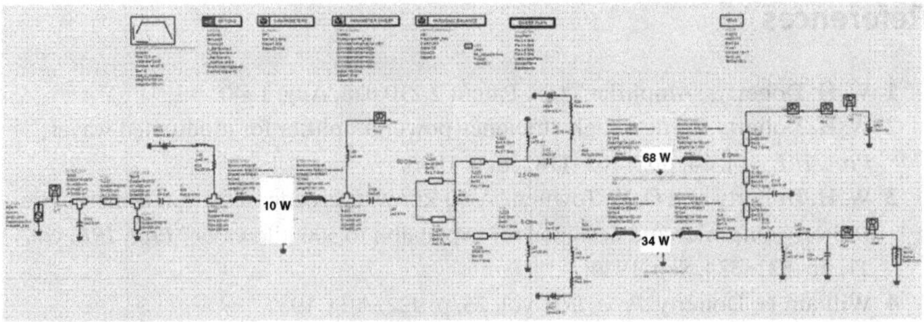

Figure 5.69 Simulation setup of integrated asymmetric 3.4–4 GHz two-stage Doherty amplifier.

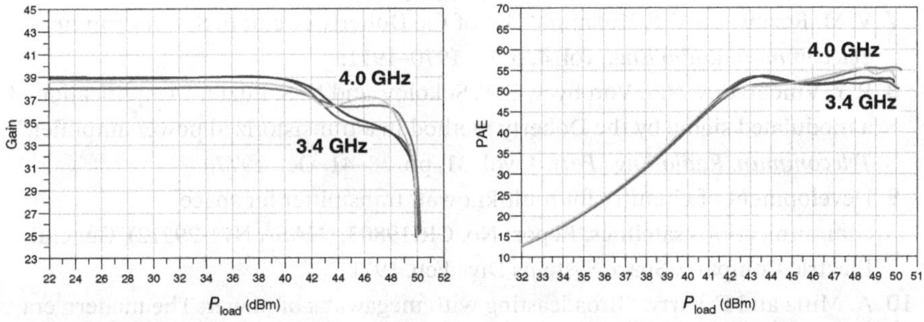

Figure 5.70 Performance of integrated asymmetric 3.4–4 GHz two-stage Doherty amplifier.

substrate is preferable to use to fabricate the combining and output matching circuits and possibly the input matching circuit. As a result, the saturated power of around 50 dBm, a PAE of about 50%, and a power gain greater than 37 dB were simulated at an output power of 42 dBm (8-dB power backoff), as shown in Figure 5.70.

In order to achieve high-efficiency, broadband, and compact fully integrated two-stage GaN HEMT Doherty amplifier module in a multichip configuration, an impedance inverter at the carrier device output in the form of a T-type transmission-line network with parasitic capacitance absorption can be used, resulting in a peak output power of 47.9 dBm, a PAE of 45.7%, and a power gain of 31.7 dB at 8-dB power backoff level at 3.9 GHz. The measurements from the experimental testing showed that the module achieved a PAE of 42–44.6%, a power gain of 28.6–31 dB, and an output power of 40.3–40.8 dBm in a frequency bandwidth of 3.4–4.1 GHz with an ACLR less than −47 dBc for a 20-MHz modulation signal with a PAR of 7.5 dB after DPD [104].

References

1 W. H. Doherty, "Amplifier," U.S. Patent 2,210,028, Aug. 1940.
2 W. H. Doherty, "A new high efficiency power amplifier for modulated waves," *Proc. IRE*, vol. 24, pp. 1163–1182, Sep. 1936.
3 W. H. Doherty and O. W. Towner, "A 50-kilowatt broadcast station utilizing the Doherty amplifier and designed for expansion to 500 kilowatts," *Proc. IRE*, vol. 27, pp. 531–534, Sep. 1939.
4 William H. Doherty, *Proc. IRE*, vol. 25, p. 922, Aug. 1937.
5 C. E. Smith, J. R. Hall, and J. O. Weldon, "Very high power long-wave broadcast station," *Proc. IRE*, vol. 42 pp. 1222–1235, Aug. 1954.
6 J. B. Sainton, "A 500 kilowatt medium frequency standard broadcast transmitter," *Cathode Press* (Machlett Company), vol. 22(4), pp. 22–29, 1965.
7 V. M. Rozov and V. F. Kuzmin, "Use of the Doherty circuit in SSB transmitters," *Telecommun. Radio Eng.*, vol. 4, p. 11, 1970–1971.
8 P. Y. Vinogradov, N. I. Vorobyev, E. P. Sokolov, and N. S. Fuzik, "Amplification of a modulated signal by the Doherty method in a transistorized power amplifier," *Telecommun. Radio Eng., Part. 1*, vol. 31, pp. 38–41, Oct. 1977.
9 Development of circuitry for multikilowatt transmitter for space communications satellites. Report No. CRI19803, (NASA N71-29212), General Electric Company, Space Systems Div., Feb. 1971.
10 A. Mina and F. Parry, "Broadcasting with megawatts of power: The modern era of efficient powerful transmitters," *IEEE Trans. Broadcast.*, vol. 35, pp. 121–130, June 1989.
11 A. Grebennikov and S. Bulja, "High-efficiency Doherty power amplifiers: Historical aspect and modern trends," *Proc. IEEE*, vol. 100, pp. 3190–3219, Dec. 2012.
12 R. Pengelly, C. Fager, and M. Ozen, "Doherty's legacy," *IEEE Microw. Mag.*, vol. 17, pp. 41–58, Feb. 2016.
13 G. A. Clark, "A comparison of current broadcast amplitude-modulation techniques," *IEEE Trans. Broadcast.*, vol. 21, pp. 25–31, June 1975.
14 F. H. Raab, "Efficiency of Doherty RF power-amplifier systems," *IEEE Trans. Broadcast.*, vol. 33, pp. 77–83, Sep. 1987.
15 B. Kim, J. Kim, I. Kim, and J. Cha, "The Doherty power amplifier," *IEEE Microw. Mag.*, vol. 7, pp. 42–50, Oct. 2006.
16 R. J. McMorrow, D. M. Upton, and P. R. Maloney, "The microwave Doherty amplifier," *1994 IEEE MTT-S Int. Microwave Symp. Dig.*, pp. 1653–1656, May 1994.
17 C. F. Campbell, "A fully integrated *Ku*-band Doherty amplifier MMIC," *IEEE Microw. Guided Wave Lett.*, vol. 9, pp. 114–116, Mar. 1999.

18 K. W. Kobayashi, A.K. Oki, A. Gutierrez-Aitken, P. Chin, L. Yang, E. Kaneshiro, P. C. Grossman, K. Sato, T. R. Block, H. C. Chen, and D. C. Streit, "An 18-21 GHz InP DHBT linear microwave Doherty amplifier," *2000 IEEE RFIC Symp. Dig.*, pp. 179–182, June 2000.

19 J. H. Tsai and T. W. Huang, "A 38-46 GHz MMIC Doherty amplifier using post-distortion linearization," *IEEE Microw. Wirel. Compon. Lett.*, vol. 17, pp. 388–390, May 2007.

20 C. Steinberser, T. Landon, C. Suckling, J. Nelson, J. Delaney, J. Hitt, L. Witkowski, G. Burgin, R. Hajji, and O. Krutko, "250 W HVHBT Doherty with 57% WCDMA efficiency linearized to −55 dBc for 2c11 6.5 dB PAR," *IEEE J. Solid-State Circuits*, vol. SC-43, pp. 2218–2228, Oct. 2008.

21 H. Deguchi, N. Ui, K. Ebihara, K. Inoue, N. Yoshimura, and H. Takahashi, "A 33 W GaN HEMT Doherty amplifier with 55% drain efficiency for 2.6 GHz base stations," *2009 IEEE MTT-S Int. Microwave Symp. Dig.*, pp. 1273–1276, June 2009.

22 Y. Yang, J. Yi, Y. Y. Woo, and B. Kim, "Optimum design for linearity and efficiency of a microwave Doherty amplifier using a new load matching technique," *Microw. J.*, vol. 44, pp. 20–36, Dec. 2001.

23 G. K. Wong, T. R. Shah, and K. Titizer, "Doherty power amplifier with phase compensation," U.S. Patent 7,295,074, Nov. 2007.

24 S. C. Jung, O. Hammi, and F. Ghannouchi, "Design optimization and DPD linearization of GaN-based unsymmetrical Doherty power amplifiers for 3G multicarrier applications," *IEEE Trans. Microw. Theory Tech.*, vol. MTT-57, pp. 2105–2113, Sep. 2009.

25 M. Nick and A. Mortazawi, "Adaptive input-power distribution in Doherty power amplifiers for linearity and efficiency enhancement," *IEEE Trans. Microw. Theory Tech.*, vol. MTT-58, pp. 2764–2771, Nov. 2010.

26 J. Moon, Ja. Kim, Ju. Kim, I. Kim, and B. Kim, "Efficiency enhancement of Doherty amplifier through mitigation of the knee voltage effect," *IEEE Trans. Microw. Theory Tech.*, vol. MTT-59, pp. 143–152, Jan. 2011.

27 J. Kim, B. Fehri, S. Boumaiza, and J. Wood, "Power efficiency and linearity enhancement using optimized asymmetrical Doherty power amplifiers," *IEEE Trans. Microw. Theory Tech.*, vol. MTT-59, pp. 425–434, Feb. 2011.

28 M. Iwamoto, A. Williams, P. F. Chen, A. G. Metzger, L. E. Larsson, and P. M. Asbeck, "An extended Doherty amplifier with high efficiency over a wide power range," *IEEE Trans. Microw. Theory Tech.*, vol. MTT-49, pp. 2472–2479, Dec. 2001.

29 Y. Yang, J. Cha, B. Shin, and B. Kim, "A fully matched N-way Doherty amplifier with optimized linearity," *IEEE Trans. Microw. Theory Tech.*, vol. MTT-51, pp. 986–993, Mar. 2003.

30 Y. Takayama, T. Harada, T. Fujita, and K. Maenaka, "Design method of microwave Doherty power amplifiers and its application to Si power MOSFET amplifiers," *Electron. Commun. Jpn. (Part II: Electron.)*, vol. 88, pp. 9–17, Apr. 2005.

31 J. Kim, J. Cha, I. Kim, and B. Kim, "Optimum operation of asymmetrical-cells-based linear Doherty power amplifiers – uneven power drive and power matching," *IEEE Trans. Microw. Theory Tech.*, vol. MTT-53, pp. 1802–1809, May 2005.

32 I. Kim, J. Cha, S. Hong, J. Kim, Y. Y. Woo, C. S. Park, and B. Kim, "Highly linear three-way Doherty amplifier with uneven power drive for repeater system," *IEEE Microw. Wirel. Compon. Lett.*, vol. 16, pp. 176–178, Apr. 2006.

33 J. Moon, Ja. Kim, I. Kim, Ju. Kim, and B. Kim, "Highly efficient three-way saturated Doherty amplifier with digital feedback predistortion," *IEEE Microw. Wirel. Compon. Lett.*, vol. 18, pp. 539–541, Apr. 2008.

34 J. Wong, N. Watanabe, and A. Grebennikov, "Efficient GaN Doherty amplifier peaks at 1 kW from 2.11 to 2.17 GHz," *Microw. RF*, vol. 56, pp. 64–69, 149, May 2017.

35 N. Watanabe, J. Wong, A. Grebennikov, and A. Akiyama, "High-power high-efficiency asymmetric Doherty amplifiers for base station applications," *Proc. 2018 Asia-Pacific Microwave Conf.*, pp. 989–991, Nov. 2018.

36 R. Hajji, L. Ledezma, G. Burgin, J. Gengler, and T. Langdon, "85 W Pavr, 500 W peak power, 1.8-2.2 GHz wideband GaN Doherty power amplifier," *2017 IEEE MTT-S Int. Microwave Symp. Dig.*, pp. 1106–1109, June 2017.

37 J. Wong, N. Watanabe, and A. Grebennikov, "Doherty amplifier combines high power and efficiency," *Microw. RF*, vol. 63, pp. 52–54, 56, Oct. 2017.

38 N. Srirattana, A. Raghavan, D. Heo, P. E. Allen, and J. Laskar, "Analysis and design of a high-efficiency multistage Doherty power amplifier for wireless communications," *IEEE Trans. Microw. Theory Tech.*, vol. MTT-53, pp. 852–860, Mar. 2005.

39 W. C. E. Neo, J. Qureshi, M. J. Pelk, J. R. Gajadharsing, and L. C. N. de Vreede, "A mixed-signal approach towards linear and efficient N-way Doherty amplifiers," *IEEE Trans. Microw. Theory Tech.*, vol. MTT-55, pp. 866–879, May 2007.

40 M. J. Pelk, W. C. E. Neo, J. R. Gajadharsing, R. S. Pengelly, and L. C. N. de Vreede, "A high-efficiency 100-W GaN three-way Doherty amplifier for base-station applications," *IEEE Trans. Microw. Theory Tech.*, vol. MTT-56, pp. 1582–1591, July 2008.

41 J. Gajadharsing, W. C. E. Neo, M. Pelk, L. C. N. de Vreede, and J. Zhao J, "3-Way Doherty amplifier with minimum output network," U.S. Patent 8,022,760, Sep. 2011.

42 B. Kim, I. Kim, and J. Moon, "Advanced Doherty architecture," *IEEE Microw. Mag.*, vol. 11, pp. 72–86, May 2010.

43 I. Kim, J. Moon, S. Jee, and B. Kim, "Optimized design of a highly efficient three-stage Doherty PA using gate adaptation," *IEEE Trans. Microw. Theory Tech.*, vol. MTT-58, pp. 2562–2574, Oct. 2010.

44 J. He, T. Zhang, Y. Zhang, Y. Wang, B. Zhang, and J. Gajadharsing, "A 500-W high efficiency LDMOS classical three-way Doherty amplifier for base-station applications," *2016 IEEE MTT-S Int. Microwave Symp. Dig.*, 1–4, 2016.

45 A. Grebennikov, J. Wong, N. Watanabe, "Modified three-stage Doherty amplifier," U.S. Patent 10,601,375, Mar. 2020.

46 A. Grebennikov, "A high-efficiency 100-W four-stage Doherty GaN HEMT power amplifier module for WCDMA systems," *2011 IEEE MTT-S Int. Microwave Symp. Dig.*, 1–4, June 2011.

47 A. Grebennikov, "High-efficiency transmission-line inverse Class F power amplifiers for 2-GHz WCDMA systems," *Int. J. RF Microw. Comput.-Aided Eng.*, vol. 21, pp. 446–456, July 2011.

48 R. Gajadharsing, "N-way Doherty amplifier," U.S. Patent 8,928,402, Jan. 2015.

49 X. Jiang, T. Zhang, J. He, S. Loysel, B. Zhang, and J. Gajadharsing, "A 750-W high efficiency LDMOS new four-way Doherty amplifier for base-station applications," *2019 IEEE MTT-S Int. Microwave Symp. Dig.*, 1324–1327, June 2019.

50 R. F. Stengel, W. A. Gu, G. D. Leizerovich, and L. F. Cygan, "High efficiency power amplifier having reduced output matching networks for use in portable devices," U.S. Patent 6,262,629, July 2001.

51 G. Ahn, M. Kim, H. Park, S. Jung, J. Van, H. Cho, S. Kwon, J. Jeong, K. Lim, J. Y. Kim, S. C. Song, C. Park, and Y. Yang, "Design of a high-efficiency and high-power inverted Doherty amplifier," *IEEE Trans. Microw. Theory Tech.*, vol. MTT-55, pp. 1105–1111, June 2007.

52 S. Kwon, M. Kim, S. Jung, J. Jeong, K. Lim, J. Van, H. Cho, H. Kim, W. Nah, and Y. Yang, "Inverted-load network for high-power Doherty amplifier," *IEEE Microw. Mag.*, vol. 10, pp. 93–98, Jan. 2009.

53 S. Jin, J. Zhou, and L. Zhang, "A broadband inverted Doherty power amplifier for IEEE 802.11b/g WLAN applications," *Microw. Opt. Technol. Lett.*, vol. 53, pp. 636–639, Mar. 2011.

54 S. M. Wood, R. S. Pengelly, and M. Suto, "A high power, high efficiency UMTS amplifier using a novel Doherty configuration," *Proc. 2003 IEEE Radio and Wireless Symp.*, pp. 329–332, Aug. 2003.

55 R.S. Pengelly, "N-way RF power amplifier with increased backoff power and power-added efficiency," U.S. Patent 6,700,444, Mar. 2004.

56 J. Li, J. Liu, W. Chen, G. Su, and Z. Wang, "A 600 W broadband three-way Doherty power amplifier for multi-standard wireless communications," *2015 IEEE MTT-S Int. Microwave Symp. Dig.*, pp. 1–3, May 2015.

57 J. Staudinger, G. Bouissi, and J. Kinney, "High efficiency 450W asymmetric three-device Doherty amplifier with digital feedback predistortion," *2010 IEEE MTT-S Radio and Wireless Symp. Dig.*, pp. 116–119, Jan 2010.

58 A. Grebennikov, J. Wong, N. Watanabe, "Doherty amplifier," U.S. Patent 10,187,015, Jan. 2019.

59 A. Grebennikov, J. Wong, N. Watanabe, "Inverted three-stage Doherty amplifier," U.S. Patent 10,432,147, Oct. 2019.

60 M. W. Lee, S. H. Kam, Y. S. Lee, and Y. H. Jeong, "Design of highly efficient three-stage inverted Doherty power amplifier," *IEEE Microw. Wireless Compon. Lett.*, vol. 21, pp. 383–385, July 2011.

61 T. Hirota, A. Minakawa, and M. Muraguchi, "Reduced-size branch-line and rat-race hybrid for uniplanar MMIC's," *IEEE Trans. Microw. Theory Tech.*, vol. MTT-38, pp. 270–275, Mar. 1990.

62 D. Kang, D. Kim, and B. Kim, "Broadband HBT Doherty power amplifiers for handset applications," *IEEE Trans. Microw. Theory Tech.*, vol. MTT-58, pp. 4031–4039, Dec. 2010.

63 D. Kang, D. Kim, Y. Cho, B. Park, J. Kim, and B. Kim, "Design of bandwidth-enhanced Doherty power amplifiers for handset applications," *IEEE Trans. Microw. Theory Tech.*, vol. MTT-59, pp. 3473–3483, Dec. 2011.

64 S. Liu, J. Pang, R. Cao, Z. Dai, M. Li, S. Chen, and W. Shi, "Highly-efficiency broadband GaAs HBT Doherty power amplifier with harmonic control technique for 5G application," *IEEE Trans. Circuits Syst.– II: Express Briefs*, vol. CAS-II-72, pp. 713–717, May 2025.

65 I. I. Blednov, "High power Doherty amplifier," U.S. Patent 7,078,976, July 2006.

66 I. I. Blednov and J. van der Zanden, "High power LDMOS integrated Doherty amplifier for W-CDMA," *2006 IEEE RFIC Symp. Dig.*, pp. 1–4, June 2006.

67 I. Blednov, "Doherty amplifier," U.S. Patent 7,800,448, Sep. 2010.

68 X. Moronval, J. van der Zanden, and M. A. Ercoli, "A 100 W tri-band LDMOS integrated Doherty amplifier for LTE-advanced applications," *2015 IEEE MTT-S Int. Microwave Symp. Dig.*, pp. 1–3, May 2015.

69 S. Lee, B. Jun, Y. Jung, S. Choi, M. Han, H. Jung, K. Oh, J. Jin, I. Seo, H. Kim, Y. Kim, and S. Ha, "Recent progresses in R&D and production of GaN HEMT based power amplifiers," *2017 IEEE Int. Radio-Frequency Integration Technology Symp. Dig.*, pp. 83–86, Aug. 2017.

70 D. Gustafsson, K. Andersson, A. Leidenhed, M. Malmstrom, A. Rhodin, and T. Wegeland, "A packaged hybrid Doherty PA for microwave links," *Proc. 46th European Microwave Conf.*, pp. 1437–1440, 2016.

71 S. Maroldt and M. Ercoli, "3.5-GHz ultra-compact GaN Class-E integrated Doherty MMIC PA for 5G massive-MIMO base station applications," *Proc. 12th European Microwave Integrated Circuits Conf.*, pp. 196–199, 2017.

72 A. Grebennikov, "Integrated Doherty amplifier," U.S. Patent Appl. 2020/0321918A1, Oct. 2020.

73 I. Blednov, "Wideband 3-way Doherty RFIC with 12 dB back-off power range," *Proc. 11th European Microwave Integrated Circuits Conf.*, pp. 17–20, 2016.

74 X. Moronval and J. Gajadharsing, "A 100 W multi-band four-way integrated Doherty amplifier," *2016 IEEE MTT-S Int. Microwave Symp. Dig.*, pp. 1–3, May 2016.

75 C. P. McCarroll, G. D. Alley, S. Yates, and R. Matreci, "A 20 GHz Doherty power amplifier MMIC with high efficiency and low distortion designed for broad band digital communication systems," *2000 IEEE MTT-S Int. Microwave Symp. Dig.*, pp. 537–540, June 2000.

76 D. Gustafsson, J. C. Cahuana, D. Kuylenstierna, I. Angelov, N. Rorsman, and C. Fager, "A wideband and compact GaN MMIC Doherty amplifier for microwave link applications," *IEEE Trans. Microw. Theory Tech.*, vol. MTT-61, pp. 922–930, Feb. 2013.

77 M. Coffey, P. MomenRoodaki, A. Zai, and Z. Popovic, "A 4.2-W 10-GHz GaN MMIC Doherty power amplifier," *2015 IEEE Compound Semiconductor Integrated Circuit Symp. Dig.*, pp. 1–4, Oct. 2015.

78 R. Quaglia, V. Camarchia, T. Jiang, M. Pirola, S. D. Guerrieri, and B. Loran, "K-band GaAs MMIC Doherty power amplifier for microwave radio with optimized driver," *IEEE Trans. Microw. Theory Tech.*, vol. MTT-62, pp. 2518–2525, Nov. 2014.

79 C. Campbell, K. Tran, M. Y. Kao, and S. Nayak, "A K-Band 5 W Doherty amplifier MMIC utilizing 0.15 μm GaN on SiC HEMT technology," *2012 IEEE Compound Semiconductor Integrated Circuit Symp. Dig.*, pp. 1–4, Oct. 2012.

80 R. Guo, H. Tao, and B. Zhang, "A 26 GHz Doherty amplifier and a fully integrated 2 × 2 PA in 0.15 μm GaN HEMT process for heterogeneous integration and 5G," *2018 IEEE MTT-S Int. Wireless Symp. Dig.*, pp. 1–4, May 2018.

81 G. Lv, W. Chen, L. Chen, and Z. Feng, "A fully integrated C-band GaN MMIC Doherty power amplifier with high gain and high efficiency for 5G application," *2019 IEEE MTT-S Int. Microwave Symp. Dig.*, pp. 560–563, June 2019.

82 R. Liu, X. W. Zhu, J. Xia, P. Cheng, L. Zhang, C. Yu, W. Hong, and A. Zhu, "A 24-to-28-GHz asymmetric GaN MMIC Doherty power amplifier with 32% PAE at 8-dB back-off using optimal phase inverter," *IEEE Microw. Wirel. Technol. Lett.*, vol. 34, pp. 1174–1177, Oct. 2024.

83 K. Bathich, A. Z. Markos, and G. Boeck, "Frequency response analysis and bandwidth extension of the Doherty amplifier," *IEEE Trans. Microw. Theory Tech.*, vol. MTT-59, pp. 934–944, Apr. 2011.
84 C. Monzon, "A small dual-frequency transformer in two sections," *IEEE Trans. Microw. Theory Tech.*, vol. MTT-51, pp. 1157–1161, Apr. 2003.
85 A. Grebennikov and J. Wong, "A dual-band parallel Doherty power amplifier for wireless applications," *IEEE Trans. Microw. Theory Tech.*, vol. MTT-60, pp. 3214–3222, Oct. 2012.
86 D. Y. Wu, J. Annes, M. Bokatius, P. Hart, E. Krvavac, and G. Tucker, "A 350 W, 790 to 960 MHz wideband LDMOS Doherty amplifier using a modified combining scheme," *2014 IEEE MTT-S Int. Microwave Symp. Dig.*, pp. 1–4, June 2014.
87 A. Grebennikov, "Multiband Doherty amplifiers for wireless applications," *High Freq. Electron.*, vol. 13, pp. 30–46, May 2014.
88 A. Barakat, M. Thian, V. Fusco, S. Bulja, and L. Guan, "Toward a more generalized Doherty power amplifier design for broadband operation," *IEEE Trans. Microw. Theory Tech.*, vol. MTT-65, pp. 846–859, Mar. 2017.
89 Y. Komatsuzaki, K. Nakatani, S. Shinjo, S. Miwa, R. Ma, and K. Yamanaka, "3.0-3.6 GHz wideband, over 46% average efficiency GaN Doherty power amplifier with frequency dependency compensating circuits," *Proc. 2017 IEEE Topical Conf. RF/Microwave Power Amplifiers for Radio and Wireless Appl.*, pp. 22–24, Jan 2017.
90 J. Pang, S. He, C. Huang, Z. Dai, J. Peng, and F. You, "A post-matching Doherty power amplifier employing low-order impedance inverters for broadband applications," *IEEE Trans. Microw. Theory Tech.*, vol. MTT-63, pp. 4061–4071, Dec. 2015.
91 J. Wong, A. Grebennikov, N. Watanabe, and E. Mochida, "200 W high efficiency 1.8-2.7 GHz GaN HEMT Doherty amplifiers for cellular," *Microw. J.*, vol. 61, pp. 66–80, Nov. 2018.
92 W. Shi, S. He, F. You, H. Xie, G. Naah, Q. A. Liu, and Q. Li, "The influence of the output impedances of peaking power amplifier on broadband Doherty amplifiers," *IEEE Trans. Microw. Theory Tech.*, vol. MTT-65, pp. 3002–3013, Aug. 2017.
93 X. Chen, W. Chen, Q. Zhang, F. M. Ghannouchi, and Z. Feng, "A 200-watt broadband continuous-mode Doherty power amplifier for base-station applications," *2017 IEEE MTT-S Int. Microwave Symp. Dig.*, pp. 1–4, June 2017.
94 J. Wong, N. Watanabe, and A. Grebennikov, "High-power high-efficiency broadband GaN HEMT Doherty amplifiers for base station applications," *Proc. 2018 IEEE Topical Conf. RF/Microwave Power Amplifiers for Radio and Wireless Appl.*, pp. 16–19, Jan. 2018.

95 A. Grebennikov, J. Wong, and H. Degouchi, "High-power high-efficiency GaN HEMT Doherty amplifiers for base station Applications," *IEICE Trans. Electron.*, vol. 104-C, pp. 488–495, Oct. 2021.

96 H. Jang and R. Wilson, "A 225-watt, 1.8-2.7 GHz broadband Doherty power amplifier with zero-phase shift peaking amplifier," *2018 IEEE MTT-S Int. Microwave Symp. Dig.*, pp. 797–800, June 2018.

97 M. Cavarroc, A. Lamy, O. Lembeye, R. McLaren, C. Duvanaud, and S. Bachir, "Compact 40% fractional bandwidth Doherty PA with input group delay engineering," *IEEE Microw. Wirel. Technol. Lett.*, vol. 33, pp. 851–854, June 2023.

98 I. Takenaka, K. Ishikura, H. Takahashi, K. Hasegawa, K. Asano, and N. Iwata, "Improvement of intermodulation distortion asymmetry characteristics with wideband microwave signals in high power amplifiers," *IEEE Trans. Microw. Theory Tech.*, vol. MTT-56, pp. 1355–1363, June 2008.

99 S. Sakata, K. Kato, E. Teranishi, T. Sugitani, R. Ma, K. Chuang, Y. C. Wu, K. Fukunaga, Y. Komatsuzaki, K. Horiguchi, K. Yamanaka, and S. Shinjo, "A fully-integrated GaN Doherty power amplifier module with a compact frequency-dependent compensation circuit for 5G massive MIMO base stations," *2020 IEEE MTT-S Int. Microwave Symp. Dig.*, pp. 711–714, Aug. 2020.

100 S. Bae, K. Kwon, H. Jeon, Y. C. Choi, S. Bin, K. Bae, H. Kang, W. Choi, Y. Woo, and Y. Yang, "8-W 2-stage GaN Doherty power amplifier module on 7 × 7 QFN for the 5G N78 band," *Electronics*, vol. 14, p. 2398, June 2025.

101 S. Lee, J. Jin, I. Kim, H. Jung, S. Koo, and D. Ahn, "GaN based 2-stage wide band Doherty PA for 3.4-3.8 GHz using hybrid integration with IPDs on HPSI SiC substrate," *Proc. 2023 Int. Conf. Compound Semiconductor Manufacturing Technol.*, pp. 1–4, 2023.

102 P. Ferris, S. Maroldt, N. Monsauret, K. Houssein, C. Quindroit, and X. Moronval, "Wideband fully integrated GaN Doherty power amplifier module for 5G MIMO applications," *Proc. 19th European Microwave Integrated Circuits Conf.*, pp. 166–169, 2024.

103 F. Huang, G. Lv, H. Jeon, H. Wu, W. Chen, and Z. Feng, "A hybrid integrated and low-cost multi-chip broadband Doherty amplifier for 5G massive MIMO application," *Engineering*, vol. 38, pp. 223–232, July 2024.

104 S. Sakata, M. Taguchi, K. Kato, E. Teranishi, S. Honda, Y. Komatsuzaki, M. Tsuru, and K. Yamanaka, "A 3.4-4.1 GHz broadband GaN Doherty power amplifier module for 5G massive-MIMO base-stations," *Proc. 2022 Asia-Pacific Microwave Conf.*, pp. 500–502, Nov. 2022.

6

Mixed-mode and Alternative Efficiency Enhancement Techniques

In this chapter, the different alternative and mixed-mode configurations of high-efficiency power amplifiers and oscillators are described and analyzed. The switchmode Class-FE power amplifiers or Class-E/F_3 power amplifiers and oscillators can provide lower voltage peak factors when zero-voltage and zero-voltage-derivative conditions in a Class-E operation mode, required to eliminate discharge loss of the shunt capacitance, are accompanied by harmonic tuning using the resonant circuits tuned to selected harmonic components corresponding to Class-F or inverse Class-F operation mode. To increase efficiency in a power backoff range, the envelope tracking and switched-path techniques for cellphone applications and load-modulated balanced power amplifier architectures for base station applications are discussed using practical examples. To provide a compact and broadband design solution, an asymmetric three-way Doherty architecture with a three-coupled-line combiner can be used.

6.1 Class-FE Power Amplifiers

The rectangular collector voltage and half-sine current waveforms characterizing the ideal Class-F operation can be realized when the effect of the device collector capacitance is negligible. Otherwise, especially at high frequencies, when the effect of the collector capacitance results in a finite switching time when the collector voltage and collector current exist at the same time simultaneously, such a load network with shunt capacitance cannot provide the instantaneous transition from the device pinch-off mode to saturation mode. Therefore, during a finite time interval, the device operates in an active region with the reverse-biased collector–base junction. In order to minimize the power losses during transients, the optimum zero-voltage and zero-voltage-derivative Class-E conditions to eliminate any collector current jumps can be applied to a Class-F circuit by connecting an additional shunt capacitor and series inductor to provide soft-switching operation mode, which was called Class Φ [1]. In this case, a high drain efficiency exceeding 80% can be achieved with a Class-C biasing at 13.56 MHz and a voltage peak factor of about two that is much lower compared to a Class-E operation mode. Since these results were achieved experimentally, it is important to provide an analytical insight into such a mixed-mode operation, which can be called the Class-FE mode, with the derivation of the optimum load-network parameters and additional Class-FE power amplifier configurations [2]. It should be noted that Class FE means that the elements of a Class E with shunt capacitance are added to an initial Class-F load network to compensate for the parasitic collector capacitance and maximize efficiency at higher operating frequencies.

The collector rectangular voltage and half-sine current waveforms can be provided by a push–pull voltage-switching Class-D power amplifier with a series L_0C_0 filter tuned to the fundamental frequency, as shown in Figure 6.1a [3]. In this case, the bipolar transistors must be driven to provide an alternate voltage switching between the on-state and off-state modes. However, by using a quarterwave transmission line corresponding to a Class-F power amplifier configuration with a parallel quarterwave transmission line, as shown in Figure 6.1b, the same collector waveforms with a voltage peak factor of two can easily be achieved due to the symmetrizing action of the transmission line operating as a voltage inverter, considering a traveling wave along its length. However, in such a Class-F configuration, the presence of the parasitic collector capacitance degrades the power amplifier's performance due to the capacitor charging and discharging processes, resulting in a finite switching time, especially at higher operating frequencies.

In order to eliminate the switching power losses and increase efficiency by compensating for the device output capacitances in a Class-D operation mode, it is necessary to introduce a dead time during the period when one device has already turned off, but the other has not turned on yet. In this case, an inductive load network should be used to charge and discharge the shunt capacitances. At the same time, to reduce the switching loss during off-to-on transition to zero, the Class-E switching conditions of zero voltage at zero-voltage slope should be provided at the end of the switching period using an additional series inductance L with an optimum value in the load network, as shown in Figure 6.2a for a Class-DE power amplifier. Similarly, a quarterwave transmission line can be used as a voltage inverter resulting in a Class-FE power amplifier by combining a Class-F operation mode with Class-E switching conditions, as shown in Figure 6.2b.

The optimum parameters of a Class-FE power amplifier can be determined based on its steady-state collector (or drain) voltage and current waveforms [2].

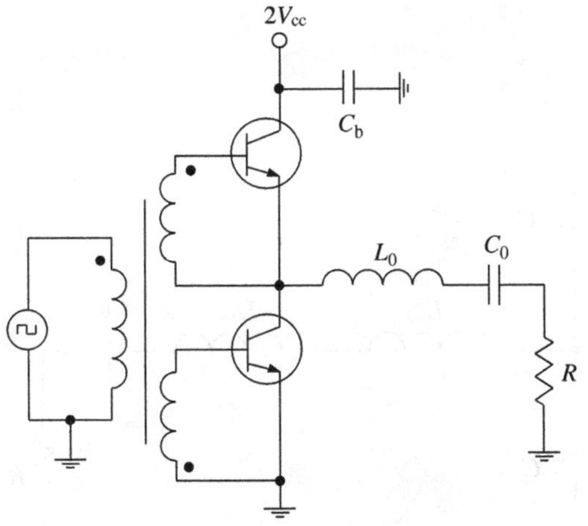

(a)

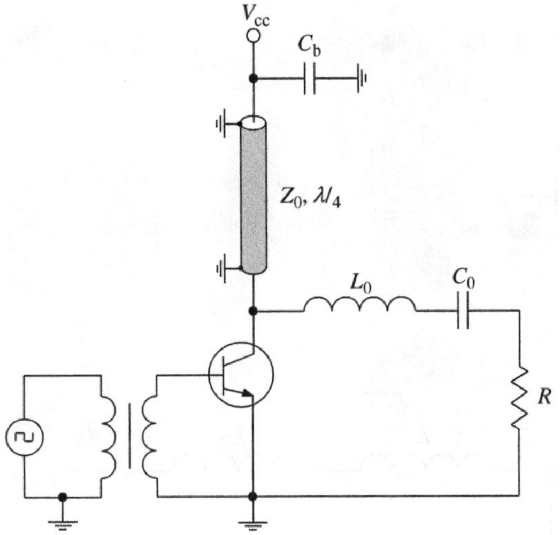

(b)

Figure 6.1 Schematics of voltage-switching Class-D and Class-F power amplifiers.

Figure 6.3 shows the two equivalent Class-FE switching circuits that occur during a switching cycle, the load network of which consists of an RF grounded parallel quarterwave transmission line, a shunt capacitor C, a series inductance L, a series fundamentally tuned L_0C_0 circuit, and a load resistor R. The active device is considered an ideal switch that is driven so as to provide the device switching between its on-state and off-state operation conditions.

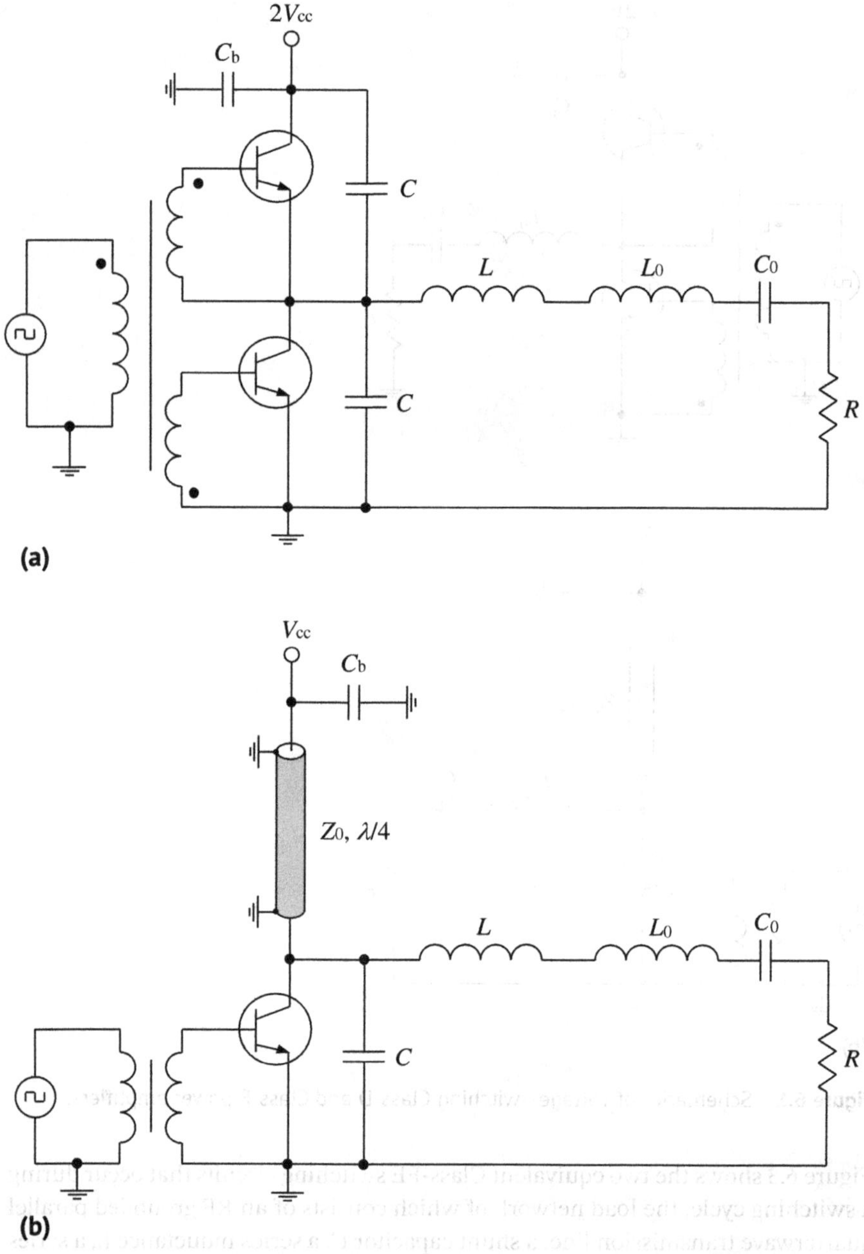

Figure 6.2 Schematics of voltage-switching Class-DE and Class-FE power amplifiers.

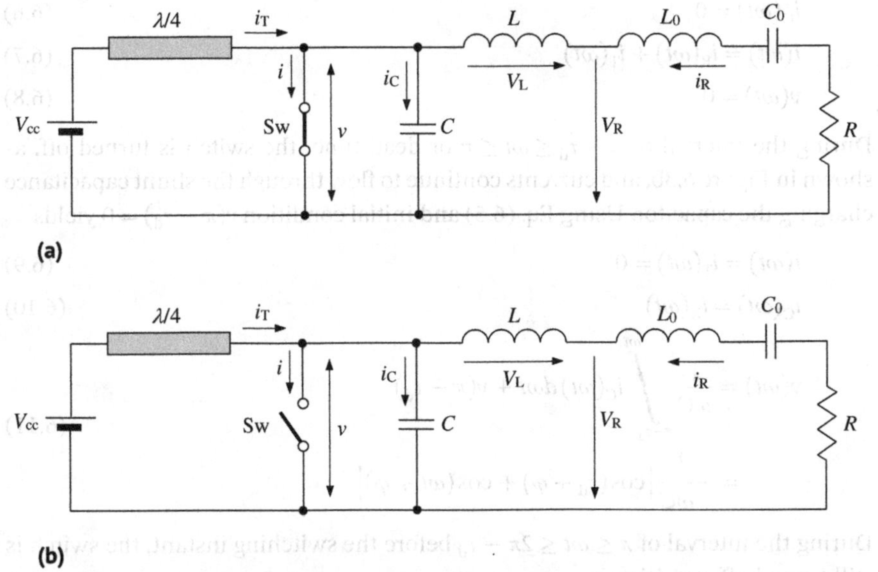

Figure 6.3 Equivalent circuits of Class-FE power amplifier.

To simplify an analysis of a Class-FE power amplifier, similar idealized assumptions as in a Class-E mode can be used, with the Class-E switching conditions for the transistor switch written as

$$v(\omega t)\big|_{\omega t=2\pi} = 0 \tag{6.1}$$

$$\left.\frac{dv(\omega t)}{d\omega t}\right|_{\omega t=2\pi} = 0 \tag{6.2}$$

where v is the voltage across the switch, and sinusoidal output current flowing through the load is

$$i_R(\omega t) = I_R \sin(\omega t + \varphi) \tag{6.3}$$

where I_R is the current amplitude and φ is the initial phase shift.

The basic Kirchhoff's equations characterizing the electrical behavior of the equivalent circuits of a Class-FE power amplifier are

$$i(\omega t) + i_C(\omega t) = i_T(\omega t) + i_R(\omega t) \tag{6.4}$$

$$i_C(\omega t) = \omega C \frac{dv(\omega t)}{d\omega t} \tag{6.5}$$

During the interval of $0 \le \omega t \le \pi - \tau_d$ before the switching instant, where τ_d is the dead time, the switch is turned on conducting the load current, as shown in Figure 6.3a, resulting in the current and voltage conditions

$$i_C(\omega t) = 0 \tag{6.6}$$

$$i(\omega t) = i_R(\omega t) + i_T(\omega t) \tag{6.7}$$

$$v(\omega t) = 0 \tag{6.8}$$

During the interval of $\pi - \tau_d \leq \omega t \leq \pi$ or dead time, the switch is turned off, as shown in Figure 6.3b, and currents continue to flow through the shunt capacitance charging the capacitor. Using Eq. (6.5) and initial condition $v(\pi - \tau_d) = 0$ yields

$$i(\omega t) = i_T(\omega t) = 0 \tag{6.9}$$

$$i_C(\omega t) = i_R(\omega t) \tag{6.10}$$

$$\begin{aligned} v(\omega t) &= \frac{1}{\omega C} \int_{\pi-\tau_d}^{\omega t} i_C(\omega t)\, d\omega t + v(\pi - \tau_d) \\ &= -\frac{I_R}{\omega C}\left[\cos(\tau_d + \varphi) + \cos(\omega t + \varphi)\right] \end{aligned} \tag{6.11}$$

During the interval of $\pi \leq \omega t \leq 2\pi - \tau_d$ before the switching instant, the switch is still turned off, resulting in

$$i_T(\omega t) = -i_R(\omega t) \tag{6.12}$$

$$i(\omega t) = i_C(\omega t) = 0 \tag{6.13}$$

$$v(\omega t) = 2V_{cc} \tag{6.14}$$

where V_{cc} is a DC-supply voltage.

During the interval of $2\pi - \tau_d \leq \omega t \leq 2\pi$ or dead time, the switch is turned off, and currents flow through the shunt capacitance discharging the capacitor. Using Eq. (6.5) and initial condition $v(2\pi - \tau_d) = 2V_{cc}$ yields

$$i(\omega t) = i_T(\omega t) = 0 \tag{6.15}$$

$$i_C(\omega t) = i_R(\omega t) \tag{6.16}$$

$$\begin{aligned} v(\omega t) &= \frac{1}{\omega C} \int_{2\pi-\tau_d}^{\omega t} i_C(\omega t) d\omega t + v(2\pi - \tau_d) \\ &= \frac{I_R}{\omega C}\left[\cos(\tau_d + \varphi) - \cos(\omega t + \varphi)\right] + 2V_{cc} \end{aligned} \tag{6.17}$$

Equation (6.16) can also be written in a differential form using Eqs. (6.3) and (6.5) as

$$\frac{dv(\omega t)}{d\omega t} = \frac{I_R}{\omega C}\sin(\omega t + \varphi) \tag{6.18}$$

Applying the switching condition given in Eq. (6.2) to Eq. (6.18) results in

$$\sin(2\pi + \varphi) = 0 \tag{6.19}$$

from which it follows that the initial phase φ can be set to zero.

Figure 6.4 shows the ideal current and voltage waveforms of a Class-FE power amplifier in an idealized optimum operation mode during the entire interval $0 \le \omega t \le 2\pi$ corresponding to its equivalent circuits shown in Figure 6.3. The quarterwave transmission line provides a half-wave symmetry, which means that the transmission line attempts to do the same work in the first and second halves of the cycle. Therefore, the transmission-line half-repeating current must necessarily instantaneously fall to zero at $\pi - \tau_d$ and $2\pi - \tau_d$, being zero during both collector voltage edges. Note that, for the sinusoidal driving signals, the transistor must be biased in a Class-C mode to provide a duty ratio D less than 50% ($D < 0.5$), because the period of time when the device is turned off must exceed the period of time when the device is turned on. As it follows from Figure 6.4, the maximum collector voltage peak factor of a Class-FE power amplifier does not exceed the factor of two for any saturation periods.

From the boundary condition $v(\pi) = 2V_{cc}$ or $v(2\pi) = 0$, by using the corresponding Eq. (6.11) or Eq. (6.17), it follows that

$$\frac{\omega C V_{cc}}{I_R} = \frac{1 - \cos\tau_d}{2} \tag{6.20}$$

The fundamental-frequency component of the voltage $v(\omega t)$ across the switch can be represented by the two quadrature components shown in Figure 6.3, whose amplitudes can be found using Fourier formulas by

$$V_R = -\frac{1}{\pi}\int_0^{2\pi} v(\omega t)\sin\omega t\, d\omega t = \frac{2(1 + \cos\tau_d)}{\pi} V_{cc} = I_R R \tag{6.21}$$

$$V_L = -\frac{1}{\pi}\int_0^{2\pi} v(\omega t)\cos\omega t\, d\omega t = \frac{\tau_d - \sin\tau_d\cos\tau_d}{\pi}\frac{I_R}{\omega C}. \tag{6.22}$$

In an idealized Class-FE operation mode with a collector (or drain) efficiency of 100%, there is no nonzero voltage and current simultaneously and the DC power P_0 and fundamental-frequency output power P_{out} are equal. Consequently, the load current amplitude I_R as a function of the dead time τ_d can be obtained from Eq. (6.21) as

$$I_R = \frac{\pi}{1 + \cos\tau_d} I_0 \tag{6.23}$$

As a result, the optimum normalized shunt capacitance C and series inductance L as functions of a dead time τ_d can be calculated using Eq. (6.20) from

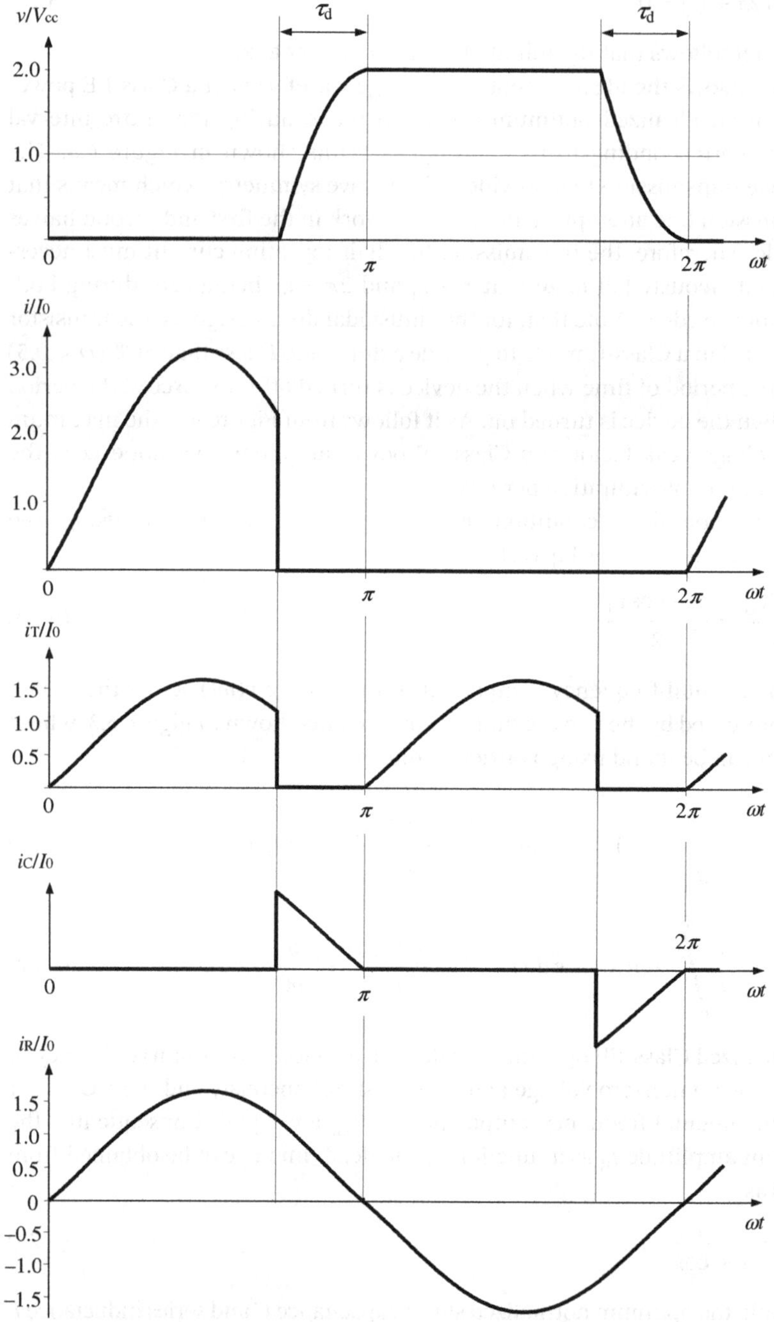

Figure 6.4 Waveforms for optimum Class-FE operation mode.

$$\omega CR = \frac{\sin^2\tau_d}{\pi} \tag{6.24}$$

$$\frac{\omega L}{R} = \frac{V_L}{V_R} = \frac{\tau_d - \sin\tau_d \cos\tau_d}{\sin^2\tau_d} \tag{6.25}$$

whereas the optimum load resistance R can be obtained using Eq. (6.21) as

$$R = \frac{1}{2}\frac{V_R^2}{P_{out}} = \frac{2(1+\cos\tau_d)^2}{\pi^2}\frac{V_{cc}^2}{P_{out}} \tag{6.26}$$

The maximum fundamental-frequency output power P_{out} delivered to the load as a function of the dead time τ_d and capacitance C can be obtained from Eqs. (6.24) and (6.26) as

$$P_{out} = \frac{2}{\pi}\left(\frac{1+\cos\tau_d}{\sin\tau_d}\right)^2 \omega C V_{cc}^2 \tag{6.27}$$

Since the device output (collector or drain–source) capacitance C_{out} represents a limiting factor, the maximum operating frequency f_{max} in an optimum Class-FE mode derived from Eq. (6.27) is written as

$$f_{max} = 0.25\left(\frac{\sin\tau_d}{1+\cos\tau_d}\right)^2 \frac{P_{out}}{C_{out}V_{cc}^2} \tag{6.28}$$

The dependence of the normalized maximum operating frequency $(f_{max} C_{out}V_{cc}^2)/P_{out}$ on the dead time τ_d is shown in Figure 6.5. From Eq. (6.28) and Figure 6.5, it follows that the maximum operating frequency f_{max} increases with smaller dead times according to

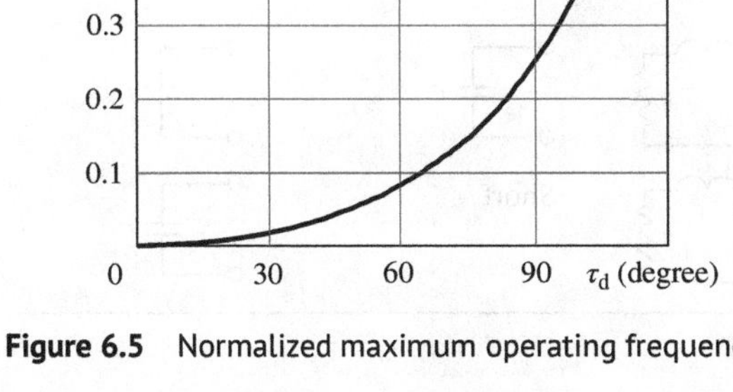

Figure 6.5 Normalized maximum operating frequency versus dead time.

$$f_{max} = 0.043\frac{P_{out}}{C_{out}V_{cc}^2} \tag{6.29}$$

for a $\tau_d = 45°$, which is close to the maximum operating frequency of a Class E with shunt capacitance and series filter with a 50% duty ratio.

Table 6.1 shows the load-network schematics seen from the device collector (or drain) at the fundamental and higher-order harmonics for different operation modes. For example, a purely resistive impedance at the fundamental is required for a Class-F operation mode with a quarterwave transmission line, with short-circuit conditions at even harmonics and open-circuit conditions at the third- and higher-order odd harmonics. Unlike regular Class-F mode, an inductive impedance at the fundamental is necessary for Class-E operation mode with shunt capacitance and series filter, with the capacitive reactance at any other harmonic component. Consequently, a Class-FE operation mode with a quarterwave transmission line will require the harmonic impedance conditions typical for both Class F with a quarterwave transmission line and Class E with a shunt capacitance and series filter.

The simulation setup for an ideal Class-FE operation mode in the frequency domain with the pulsed source, parallel quarterwave line, shunt capacitance C, series inductance L, and series LC filter tuned to the fundamental is shown in Figure 6.6, which provides a much faster simulation procedure than that in the time domain. Although the number of harmonic components is limited, with the harmonic number of 100 in this case, the simulation waveforms and numerical results for the optimum load-network parameters are sufficiently accurate.

If efficiency is chosen as an optimization parameter, the optimization procedure with an increased number of iterations for better calculation accuracy can be applied with respect to efficiency, too. Figure 6.7 illustrates the normalized switch

Table 6.1 Load-network schematics at fundamental and harmonics.

High-efficiency mode	f_0 (Fundamental)	$2nf_0$ (Even harmonics)	$(2n+1)f_0$ (Odd harmonics)
Class F with quarterwave line	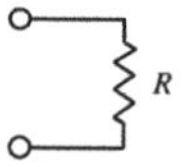	Short	Open
Class E with shunt capacitance and series filter	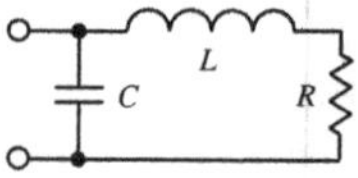	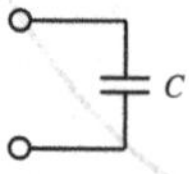	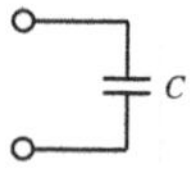
Class FE with quarterwave line	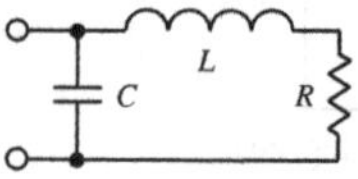	Short	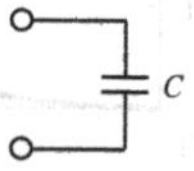

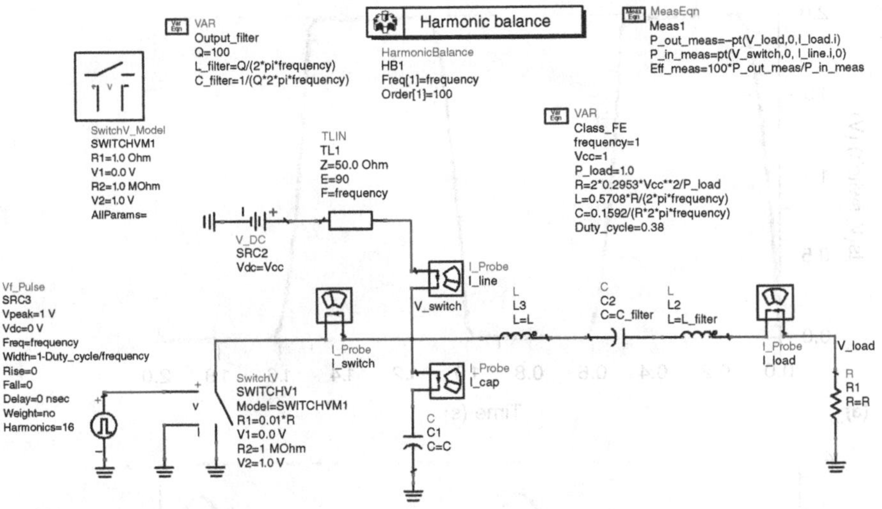

Figure 6.6 Simulation setup for idealized Class-FE mode in the frequency domain.

(a) voltage and (b) current waveforms obtained for an idealized Class-FE load network with optimum parameters given by Eqs. (6.24)–(6.26) having a dead time $\tau_d = 45°$ or duty ratio (or cycle) $D = 0.375$, which is defined as the ratio of the switch-on time to one period of the switching frequency. In this case, the smooth transitions between the switch-on and switch-off conditions and vice versa take place. For $r_{sat}/R = 0.01$ and lossless circuit elements, the overall circuit efficiency is equal to 98.2%, with the normalized switch voltage and current waveforms very close to the same theoretical waveforms shown in Figure 6.4.

Alternative configuration a Class-FE power amplifier with a series quarterwave transmission line is shown in Figure 6.8, which consists of a shunt capacitor C, a series quarterwave transmission line loaded by a parallel resonant L_0C_0 circuit tuned to the fundamental frequency, a shunt capacitor C_L, and a load resistor R_L. Here, it is assumed that the parallel L_0C_0 filter is ideal having infinite impedance at the fundamental frequency and zero impedances at the second- and higher-order harmonics. The operation principle of such a Class-FE power amplifier with a series quarterwave line is similar to a Class-FE power amplifier with a parallel quarterwave line when the short circuit at the output of the quarterwave transmission line produces a short circuit at its input for even harmonics and an open circuit at its input for odd harmonics, with a resistive load at the fundamental frequency.

As opposed to the parallel quarterwave transmission line, a series quarterwave transmission line can provide an impedance transformation at the fundamental

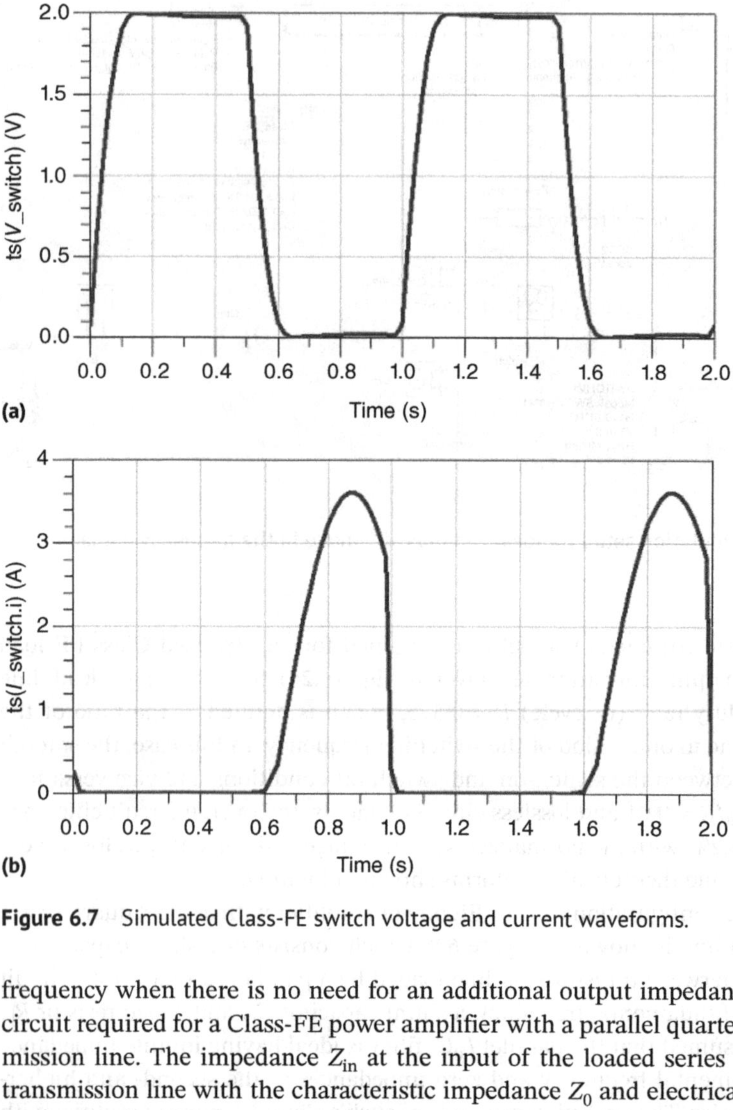

Figure 6.7 Simulated Class-FE switch voltage and current waveforms.

frequency when there is no need for an additional output impedance matching circuit required for a Class-FE power amplifier with a parallel quarterwave transmission line. The impedance Z_{in} at the input of the loaded series quarterwave transmission line with the characteristic impedance Z_0 and electrical length θ at the fundamental frequency can be written as

$$Z_{\text{in}} = Z_0 \frac{Z_L + jZ_0 \tan\theta}{Z_0 + jZ_L \tan\theta}\bigg|_{\theta=90^\circ} = \frac{Z_0^2}{Z_L} \tag{6.30}$$

where

$$Z_L = \frac{R_L}{1 + j\omega C_L R_L} \tag{6.31}$$

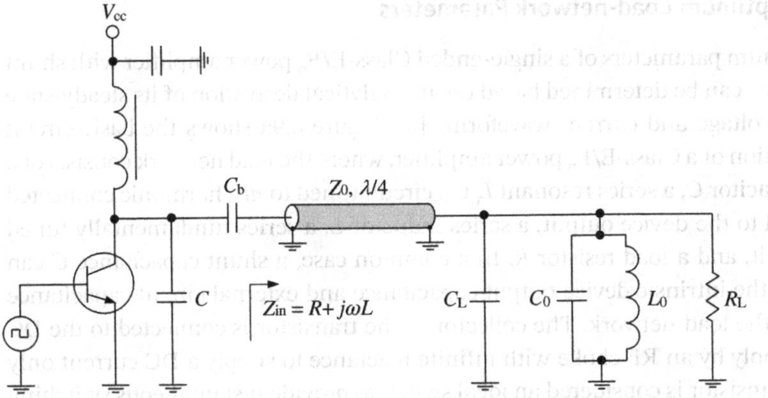

Figure 6.8 Schematic of Class-FE power amplifier with series quarterwave transmission line.

Then, by substituting Eq. (6.31) into Eq. (6.30) and separating Eq. (6.30) into real and imaginary parts, the load resistance R_L and shunt capacitance C_L can be expressed through the optimum load resistance R and series inductance L corresponding to the Class-FE operation mode according to Eqs. (6.25) and (6.26) as a function of the transmission-line characteristic impedance Z_0 as

$$R_L = \frac{Z_0^2}{R} \tag{6.32}$$

$$C_L = \frac{L}{Z_0^2} \tag{6.33}$$

6.2 Class-E/F$_3$ Power Amplifiers

Both Class-E and Class-F operation modes have their advantages and disadvantages with respect to each other. For example, Class E can provide high operation efficiency with a simple load network but with a high voltage peak factor, whereas Class F requires a separate resonant circuit at each harmonic for an efficiency improvement, resulting in a significant circuit complexity and limited switching speed, although with the low voltage peak factor. Class E can incorporate the device output capacitance into the load-network topology, although a large value of the shunt capacitance may become a limiting factor at high frequencies for high-power applications. For this case, only a simple Class-F topology with control of a few harmonics, mainly the second harmonic, can be realized with a limited efficiency.

6.2.1 Optimum Load-network Parameters

The optimum parameters of a single-ended Class-E/F$_n$ power amplifier with shunt capacitance can be determined based on an analytical derivation of its steady-state collector voltage and current waveforms [4]. Figure 6.9a shows the basic circuit configuration of a Class-E/F$_n$ power amplifier, where the load network consists of a shunt capacitor C, a series resonant L_nC_n, circuit tuned to nth harmonic connected in parallel to the device output, a series inductor L, a series fundamentally tuned L_0C_0 circuit, and a load resistor R. In a common case, a shunt capacitance C can represent the intrinsic device output capacitance and external circuit capacitance added by the load network. The collector of the transistor is connected to the DC power supply by an RF choke with infinite reactance to supply a DC current only and the transistor is considered an ideal switch to provide instantaneous switching between its on-state and off-state operation conditions.

To simplify the analysis of a Class-E/F$_n$ power amplifier, the corresponding equivalent circuit of which is shown in Figure 6.9b, the same idealized assumptions which were used to analyze a Class-E or Class-FE power amplifier are considered, with the optimum Class-E switching conditions being defined by Eqs. (6.1) and (6.2), where v is the voltage across the switch. It should be noted

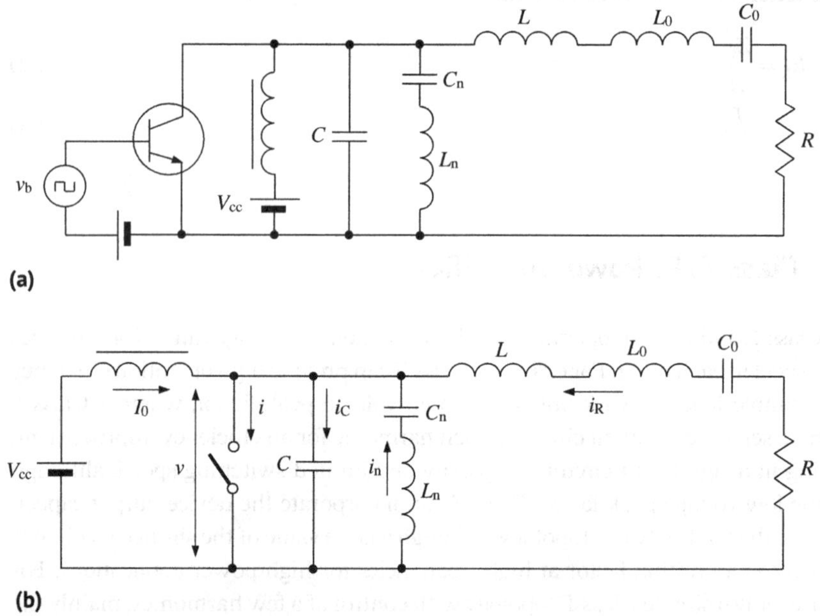

Figure 6.9 Basic circuits of a single-ended Class-E/F$_n$ power amplifier.

that these optimal Class-E switching conditions do not correspond to minimum dissipated power losses when real transistor with a nonzero value of its saturation resistance is used.

The output current i_R flowing through the load and nth-harmonic current i_n are assumed to be sinusoidal:

$$i_R(\omega t) = I_R \sin(\omega t + \varphi) \tag{6.34}$$

$$i_n(\omega t) = I_n \sin n\omega t \tag{6.35}$$

where I_R is the fundamental-frequency current amplitude, I_n is the nth-harmonic current amplitude, φ is the initial phase shift, and zero initial phase is considered for nth-harmonic current.

When the switch is turned on for $0 \le \omega t \le \pi$, the current on the capacitor C is zero:

$$i_C(\omega t) = \omega C \frac{dv(\omega t)}{d\omega t} = 0 \tag{6.36}$$

resulting in

$$i(\omega t) = I_0 + I_n \sin n\omega t + I_R \sin(\omega t + \varphi) \tag{6.37}$$

where I_0 is the DC-supply current.

Since $i(0) = 0$, Eq. (6.37) reduces to

$$1 + \frac{I_R}{I_0} \sin\varphi = 0 \tag{6.38}$$

When the switch is turned off for $\pi \le \omega t < 2\pi$, the current on the switch is zero and the current on the capacitor C can be written as

$$i_C(\omega t) = I_0 + I_n \sin n\omega t + I_R \sin(\omega t + \varphi) \tag{6.39}$$

which produces the voltage across the switch by charging the capacitor according to

$$\begin{aligned} v(\omega t) &= \frac{1}{\omega C} \int_{\pi}^{\omega t} i_C(\omega t) d\omega t \\ &= \frac{1}{\omega C} \left\{ (\omega t - \pi) I_0 - \frac{I_n}{n}(1 + \cos n\omega t) - I_R \left[\cos(\omega t + \varphi) + \cos\varphi \right] \right\} \end{aligned} \tag{6.40}$$

For a zero-voltage switching condition at $\omega t = 2\pi$, Eq. (6.40) results in

$$\frac{\pi}{2} - \frac{I_n}{nI_0} - \frac{I_R}{I_0} \cos\varphi = 0 \tag{6.41}$$

with the relative amplitude of the *n*th-harmonic component I_n/I_0 obtained for any *n* by

$$I_n = \frac{1}{\pi}\int_0^{2\pi} i(\omega t)\sin n\omega t\, d\omega t = \frac{2I_0}{n\pi}(1 - \cos n\pi) \tag{6.42}$$

which is reduced for odd *n* to

$$\frac{I_n}{I_0} = \frac{4}{n\pi} \tag{6.43}$$

The initial phase angle φ can directly be obtained for odd *n* by substituting Eqs. (6.38) and (6.43) into Eq. (6.41) as

$$\varphi = \tan^{-1}\left(\frac{2\pi n^2}{8 - \pi^2 n^2}\right) \tag{6.44}$$

and the DC-supply voltage V_{cc} can be determined by applying a Fourier-series expansion to Eq. (6.40) as

$$V_{cc} = \frac{1}{2\pi}\int_0^{2\pi} v(\omega t)d\omega t = \frac{I_0}{2\pi\omega C}\left[\frac{\pi^2}{2} - \frac{\pi}{n}\frac{I_n}{I_0} - \frac{I_R}{I_0}(2\sin\varphi + \pi\cos\varphi)\right] \tag{6.45}$$

For an idealized collector efficiency of 100%, the DC power $P_0 = I_0 V_{cc}$ and fundamental-frequency output power $P_{out} = V_R^2/2R$ at the load are equal and written as

$$I_0 V_{cc} = \frac{V_R^2}{2R} \tag{6.46}$$

where $V_R = I_R R$ is the voltage amplitude on the load resistor *R*.

In order to define the Class-E/F$_n$ load-network parameters, the amplitude of the two quadrature components of the fundamental-frequency voltage $v_1(\omega t)$ across the switch shown in Figure 6.10 can be calculated using Fourier formulas and Eq. (6.40) as

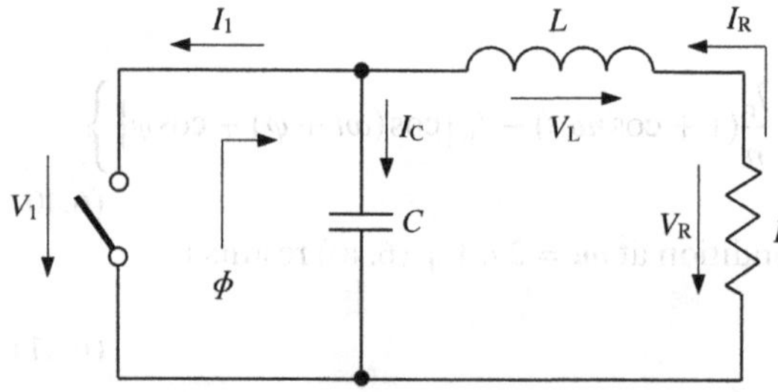

Figure 6.10 Equivalent Class-E/F$_n$ load network at fundamental frequency.

$$
\begin{aligned}
V_{\mathrm{R}} &= -\frac{1}{\pi}\int_{0}^{2\pi} v(\omega t)\sin(\omega t+\varphi)d\omega t \\
&= -\frac{2I_0}{\pi\omega C}\left[\sin\varphi + \cos\varphi\left(-\frac{\pi}{2}+\frac{1}{n}\frac{I_n}{I_0}+\frac{I_{\mathrm{R}}}{I_0}\cos\varphi\right)\right]
\end{aligned}
\tag{6.47}
$$

$$
\begin{aligned}
V_{\mathrm{L}} &= -\frac{1}{\pi}\int_{0}^{2\pi} v(\omega t)\cos(\omega t+\varphi)d\omega t \\
&= -\frac{I_0}{\pi\omega C}\left[\sin\varphi\left(\pi-\frac{2}{n}\frac{I_{\mathrm{n}}}{I_0}\right)+2\cos\varphi-\frac{I_{\mathrm{R}}}{I_0}\left(\frac{\pi}{2}+\sin 2\varphi\right)\right]
\end{aligned}
\tag{6.48}
$$

As a result, for a third-harmonic tuning when $n = 3$,

$$\varphi = \tan^{-1}\left(\frac{18\pi}{8-9\pi^2}\right) = -34.98^{\circ} \tag{6.49}$$

$$\frac{I_{\mathrm{R}}}{I_0} = -\frac{1}{\sin\varphi} = 1.7444 \tag{6.50}$$

$$\frac{I_{\mathrm{n}}}{I_0} = \frac{4}{3\pi} = 0.4244 \tag{6.51}$$

The normalized collector voltage and current waveforms for an idealized optimum Class-E/F$_3$ operation mode in the interval $0 \le \omega t \le 2\pi$ are shown in Figure 6.11, where there is no nonzero switch voltage and current simultaneously, resulting in an idealized collector efficiency of 100% given by Eq. (6.46).

The optimum normalized series inductance L and shunt capacitance C are calculated as

$$\frac{\omega L}{R} = \frac{V_{\mathrm{L}}}{V_{\mathrm{R}}} = 0.961 \tag{6.52}$$

$$\omega CR = \frac{\omega C V_{\mathrm{R}}}{I_0}\bigg/\frac{I_{\mathrm{R}}}{I_0} = 0.209 \tag{6.53}$$

whereas the optimum load resistance R using Eqs. (6.46) and (6.50) is defined by

$$R = \frac{1}{2}\frac{V_{\mathrm{R}}^2}{P_{\mathrm{out}}} = \frac{2}{}\frac{V_{\mathrm{cc}}^2}{P_{\mathrm{out}}}\bigg/\left(\frac{I_{\mathrm{R}}}{I_0}\right)^2 = 0.657\frac{V_{\mathrm{cc}}^2}{P_{\mathrm{out}}} \tag{6.54}$$

From Eqs. (6.52)–(6.54), it follows that the load resistance R is greater by 14% in an idealized optimum Class-E/F$_3$ mode compared with a Class-E mode with shunt capacitance and series filter, with close values for the shunt capacitance C and series inductance L for both classes.

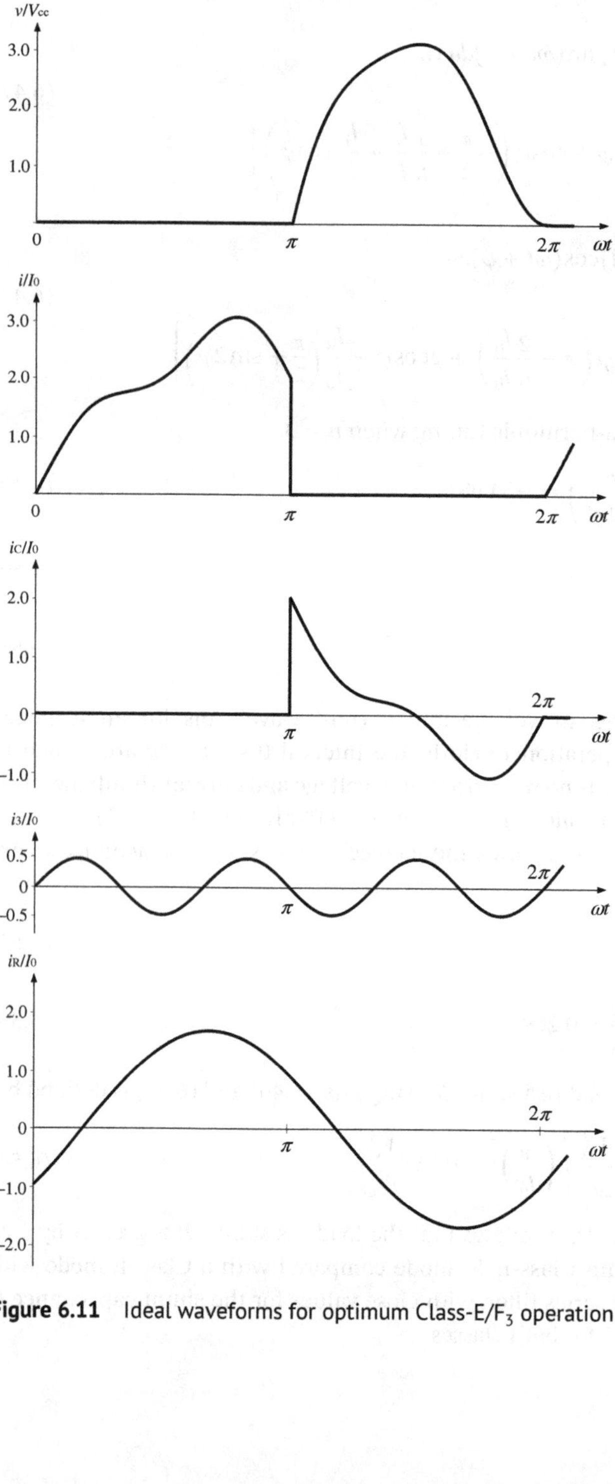

Figure 6.11 Ideal waveforms for optimum Class-E/F_3 operation mode.

The phase angle ϕ of the load network at the fundamental seen by the switch, as shown in Figure 6.10, and required is derived for an idealized optimum Class-E/F_3 operation mode through the load-network parameters using Eqs. (6.52) and (6.53) as

$$\phi = \tan^{-1}\left(\frac{\omega L}{R}\right) - \tan^{-1}\left(\frac{\omega CR}{1 - \frac{\omega L}{R}\omega CR}\right) = 29.2^\circ \tag{6.55}$$

From Eqs. (6.37) and (6.40), the peak collector voltage V_{max} and current I_{max} are obtained using Eqs. (6.45) and (6.49)–(6.51) as

$$\frac{V_{max}}{V_{cc}} = 3.142 \tag{6.56}$$

$$\frac{I_{max}}{I_0} = 3.056 \tag{6.57}$$

with the voltage peak factor lower by 13.4% compared to a Class-E operation mode with shunt capacitance and series filter.

The maximum operating frequency f_{max} in an idealized optimum Class-E/F_3 operation mode in view of the transistor output capacitance C_{out} using Eqs. (6.53) and (6.54) is written as

$$f_{max} = 0.0506\frac{P_{out}}{C_{out}\,V_{cc}^2} \tag{6.58}$$

corresponding to the maximum operating frequency of a Class-E operation mode with shunt capacitance and series filter.

Table 6.2 shows the load-network schematics seen from the device collector (or drain) at the fundamental and higher-order harmonics for different operation modes. For example, a purely resistive impedance at the fundamental is necessary for an inverse Class-F operation mode, with open-circuit conditions at even harmonics, and short-circuit conditions at odd harmonics. Unlike an inverse Class-F mode, an inductive impedance is required at the fundamental of the Class-E operation mode with shunt capacitance and series filter and capacitive reactance at any other harmonics. Consequently, an idealized Class-E/F_3 mode will provide the harmonic impedance conditions typical for both Class E with shunt capacitance and inverse Class F with third-harmonic control.

6.2.2 ADS Simulation Setup

The simulation setup for an ideal Class-E/F_3 operation mode in the time domain with a pulsed source is shown Figure 6.12, where a voltage-controlled switch with off-resistance of 1 MΩ and a small finite on-resistance represents the transistor.

Table 6.2 Load-network schematics at fundamental and harmonics.

High-efficiency mode	f_0 (Fundamental)	$2nf_0$ (Even harmonics)	$(2n+1)f_0$ (Odd harmonics)
Inverse Class F	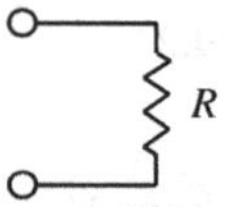	Open	Short
Class E with shunt capacitance and series filter	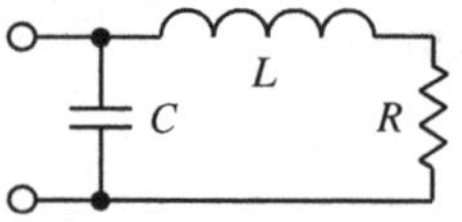	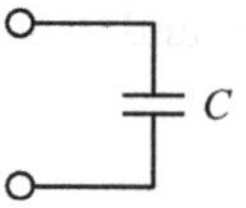	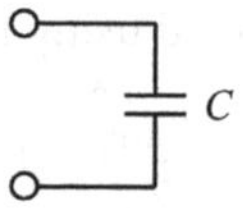
Class E/F$_3$	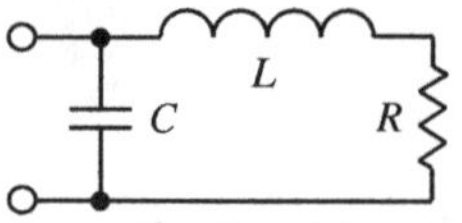	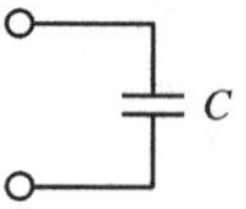	Short

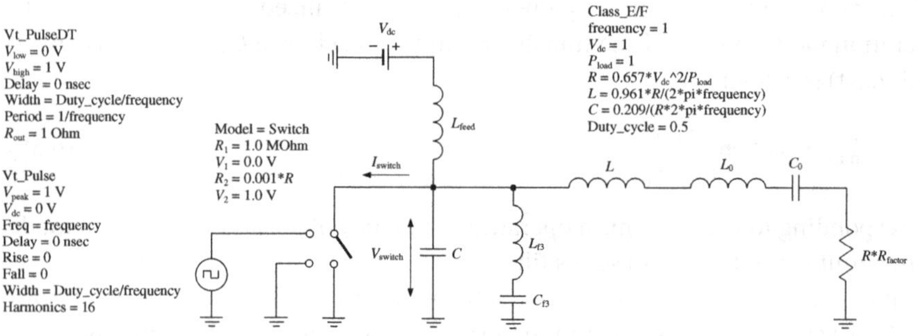

Figure 6.12 Simulation setup for idealized Class-E/F$_3$ mode in the time domain.

A discrete time pulse source is used, in contrast with a standard pulse source, providing the condition when there is no timing jitter in the pulse edges due to the waveform being sampled asynchronously by a fixed time interval simulation. As to the simulation time, it is much faster than the period of a square wave.

For a circuit simulation in the time domain, the transient simulator is added to the simulation setup, and the stop time of 80 seconds is chosen for a normalized operating frequency being sufficient to reach a steady-state mode. The loaded quality factor Q_L of the series resonant circuit is set to 10 to approximate a sinusoidal current flowing to the load. The measure equations to calculate efficiency include zero-voltage and zero-voltage-derivative Class-E switching conditions. Instead of the infinite RF choke inductance L_{feed}, its value was chosen to be 10 times greater than the value of a series inductance L. In this case, the value

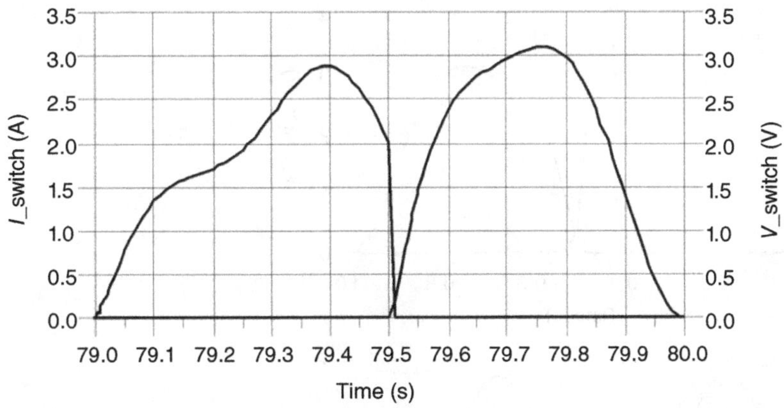

Figure 6.13 Simulated switch voltage and current waveforms in time domain.

of a load resistance increases by 22% ($R_{factor} = 1.22$) compared to its nominal value in an idealized optimum case to optimize the switch voltage and current waveforms shown in Figure 6.13, resulting in an efficiency of 99.8%.

The simulation setup shown in Figure 6.12 can also be used to obtain the idealized optimum Class-E/F$_3$ operation mode with infinite RF choke in the frequency domain with a voltage input source. If efficiency is chosen as an optimization parameter, the optimization procedure with increased number of iterations for better calculation accuracy can be used. The normalized (a) switch voltage and current waveforms and (b) sinusoidal third-harmonic and fundamental-frequency waveforms corresponding to the idealized optimum of the Class-E/F$_3$ load network with $R_{factor} = 1$ are shown in Figure 6.14, where the smooth transitions between the switching conditions occur. For $r_{sat}/R = 0.001$, where r_{sat} is the switch saturation resistance, and lossless load-network inductances and capacitances, the efficiency achieved 98.2%, resulting in the simulated normalized switch voltage and current waveforms similar to the same theoretical waveforms shown in Figure 6.11.

6.2.3 Class E/F$_3$ with Lumped Elements

For a lumped load network in a Class-E/F$_3$ operation mode, an additional network providing an open circuit at the second harmonic and a short circuit at the third harmonic should be used and the optimum parameters obtained for the shunt capacitance, series inductance, and load resistance at the fundamental should be calculated from Eqs. (6.52)–(6.54), respectively. A lumped Class-E/F$_3$ power amplifier based on this load-network technique and using a MOSFET device IRF510 achieved an output power 36.8 dBm and a drain efficiency of 96% at an operating frequency of 4 MHz [5]. It should be noted that the maximum switch

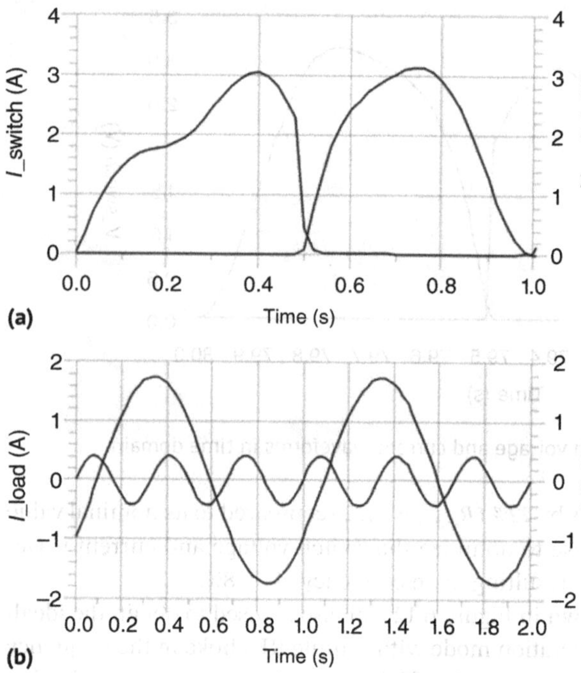

Figure 6.14 Simulated voltage and current waveforms in the frequency domain.

voltage is affected by the nonlinearity of the shunt capacitance, when V_{max}/V_{dd} is increased from 3 to 3.8 with increasing V_{dd} up to 100 V [6]. For such a Class-E/F_3 power amplifier, high efficiencies of about 95% can be achieved in a nonoptimal operation with zero-voltage and nonzero voltage-derivative switching conditions when the switch turns on with declining voltage having negative derivative [7].

Figure 6.15 shows the Class-E/F_3 lumped network consisting of the two cascaded parallel resonant circuits (L_1C_1 and L_2C_2), tuned to the fundamental frequency ($\omega = \omega_0$) and second harmonic ($\omega = 2\omega_0$), respectively, to provide the open-circuit conditions for Z_{net} at these frequencies, and a series inductance L_3. In this case, the network impedance Z_{net} at the input of this lumped network is written as

$$Z_{net}(\omega) = \frac{j\omega L_1}{1 - \omega^2 L_1 C_1} + j\omega L_3 + \frac{j\omega L_2}{1 - \omega^2 L_2 C_2} \tag{6.59}$$

In order to provide a short-circuit condition at the third harmonic ($\omega = 3\omega_0$), it needs to rewrite Eq. (6.59) as

$$Z_{net}(3\omega_0) = \frac{j3\omega_0 L_1}{1 - 9\omega_0^2 L_1 C_1} + j3\omega_0 L_3 + \frac{j3\omega_0 L_2}{1 - 9\omega_0^2 L_2 C_2} = 0 \tag{6.60}$$

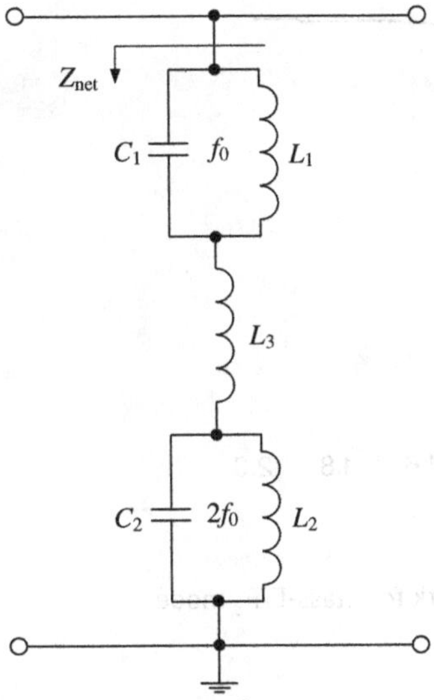

Figure 6.15 Lumped network for second- and third-harmonic tuning.

As a result, the relationships between the lumped network elements for specified values of the inductances L_1 and L_2 can be obtained by

$$C_1 = \frac{1}{\omega_0^2 L_1} \tag{6.61}$$

$$C_2 = \frac{1}{4\omega_0^2 L_2} \tag{6.62}$$

$$L_3 = 0.125L_1 + 0.8L_2 \tag{6.63}$$

Figure 6.16 shows the simulated frequency response for the lumped network shown in Figure 6.15, where its parameters were calculated by Eqs. (6.61)–(6.63) for $f_0 = 430$ MHz as $L_1 = 60$ nH, and $L_2 = 12.5$ nH as $L_3 = 17.5$ nH, $C_1 = 2.3$ pF, and $C_2 = 2.7$ pF with the quality factor Q for all three inductors set to 50. It is important to choose the inductances L_1 and L_2 enough far from both first (at fundamental) and second (at second harmonic) poles for a small low impedance (first zero) between them, as shown in Figure 6.16.

Figure 6.17 shows the simulation setup for a lumped Class-E/F_3 power amplifier based on a 28-V 10-W GaN HEMT power transistor CGH40010 operating at 430 MHz [4]. Here, the required optimum load resistance defined by Eq. (6.54) is close to 50 Ω, therefore it is enough to use a series high-Q_L resonant circuit with an optimum series inductance defined by Eq. (6.52). Since the device output capacitance of about 1.3 pF is slightly less than required by Eq. (6.53), an additional

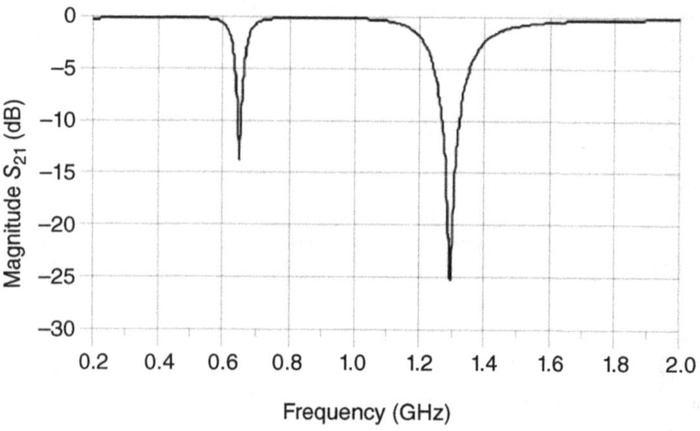

Figure 6.16 Frequency response of the lumped network for Class-E/F_3 mode.

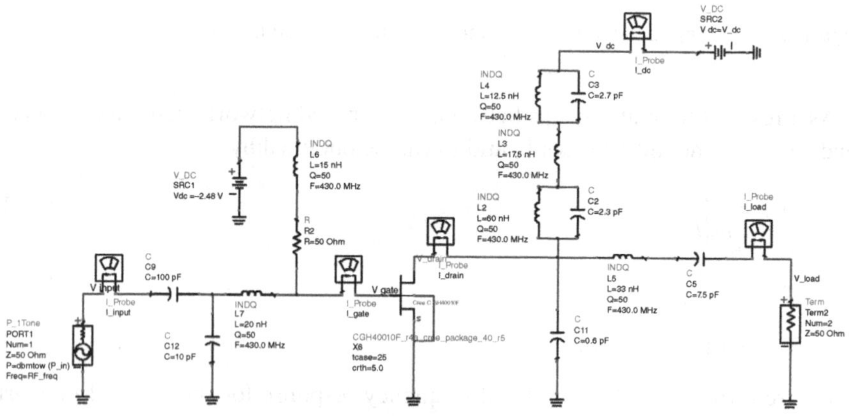

Figure 6.17 Simulation setup for lumped Class-E/F_3 GaN HEMT power amplifier.

external shunt capacitance of 0.6 pF is added. Initially, the power amplifier circuit was simulated for infinite inductor quality factors Q in the load network to understand the maximum achievable performance, resulting in an output power of 40.1 dBm, a drain efficiency of 87%, and a PAE of 84.5% with a power gain of about 15 dB, as shown in Figure 6.18a. However, for more practical values of $Q = 50$, an output power reduces to 39.3 dBm with a drain efficiency of 77.5% and a PAE of 74.9%. Figure 6.18b shows the simulated voltage drain waveform, which is very close to the theoretical one shown in Figure 6.11.

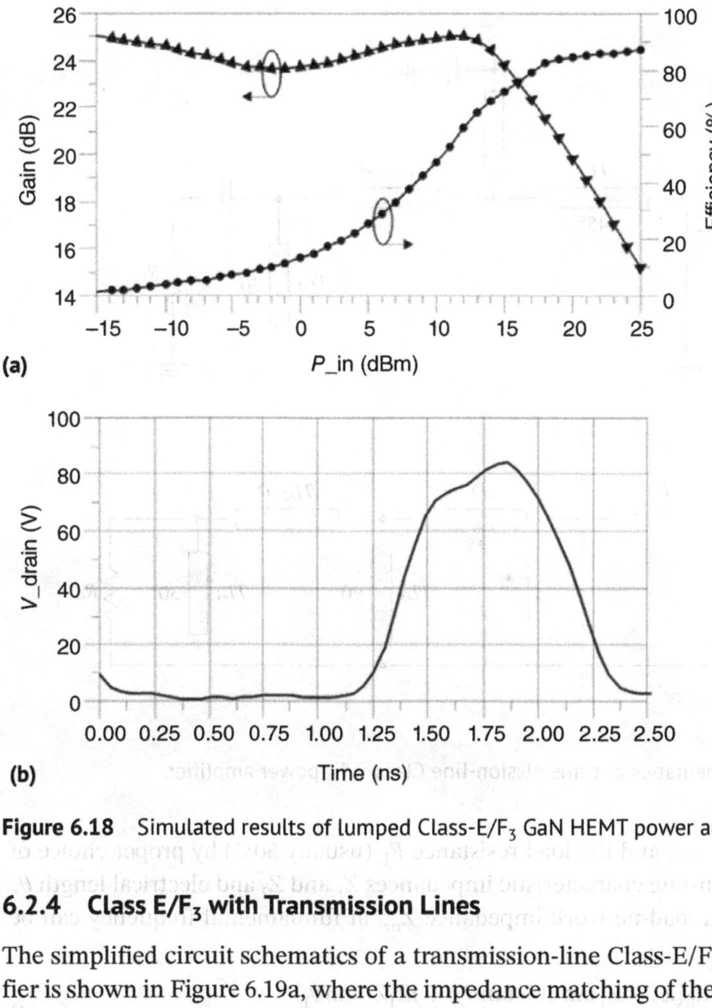

Figure 6.18 Simulated results of lumped Class-E/F_3 GaN HEMT power amplifier.

6.2.4 Class E/F_3 with Transmission Lines

The simplified circuit schematics of a transmission-line Class-E/F_3 power amplifier is shown in Figure 6.19a, where the impedance matching of the optimum load resistance R with the load impedance $R_L = 50\,\Omega$ is provided by a transmission-line network and the transistor output is represented by the multi-harmonic current source, shunt drain–source capacitance C and series bondwire inductance L, as shown in Figure 6.19b [4]. To approximate the Class-E/F_3 operation mode, the equivalent load networks seen by the device multi-harmonic source at the fundamental, second, and third harmonics can be considered separately. The load network at the fundamental frequency is shown in Figure 6.20a, where the combined series transmission line $TL_1 + TL_2$ with the same characteristic impedance Z_1 and an open-circuit capacitive stub TL_4 having an electrical length of 30° provide an impedance matching between the optimum Class-E/F_3 load

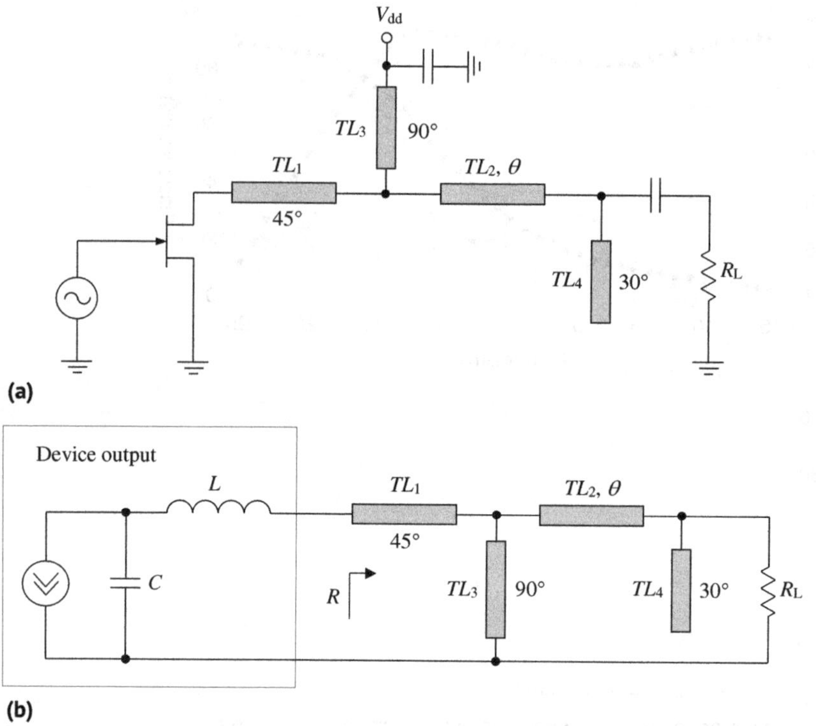

Figure 6.19 Schematics of transmission-line Class-E/F_3 power amplifier.

impedance $Z_{net} = R$ and the load resistance R_L (usually 50 Ω) by proper choice of the transmission-line characteristic impedances Z_1 and Z_2 and electrical length θ.

The optimum load-network impedance Z_{net} at fundamental frequency can be written as

$$Z_{net} = Z_1 \frac{R_L(Z_2 - Z_1 \tan 30° \ \tan\theta_1) + jZ_1 Z_2 \tan\theta_1}{Z_1 Z_2 + j(Z_1 \tan 30° + Z_2 \tan\theta_1) R_L} \tag{6.64}$$

where $\theta_1 = 45° + \theta$, Z_1 is the characteristic impedance of the series transmission line, and Z_2 is the characteristic impedance of the open-circuit stub with an electrical length of 30°.

Separating Eq. (6.64) into real and imaginary parts results in the system of two equations with two unknown parameters written as

$$(Z_1 \tan 30° + Z_2 \tan\theta_1)^2 R_L^2 R - (Z_1 Z_2)^2 \left[R_L\left(1 + \tan^2\theta_1\right) - R\right] = 0 \tag{6.65}$$

$$(Z_1 Z_2)^2 \tan\theta_1 + R_L^2 (Z_1 \tan 30° + Z_2 \tan\theta_1)(Z_2 - Z_1 \tan 30° \ \tan\theta_1) = 0 \tag{6.66}$$

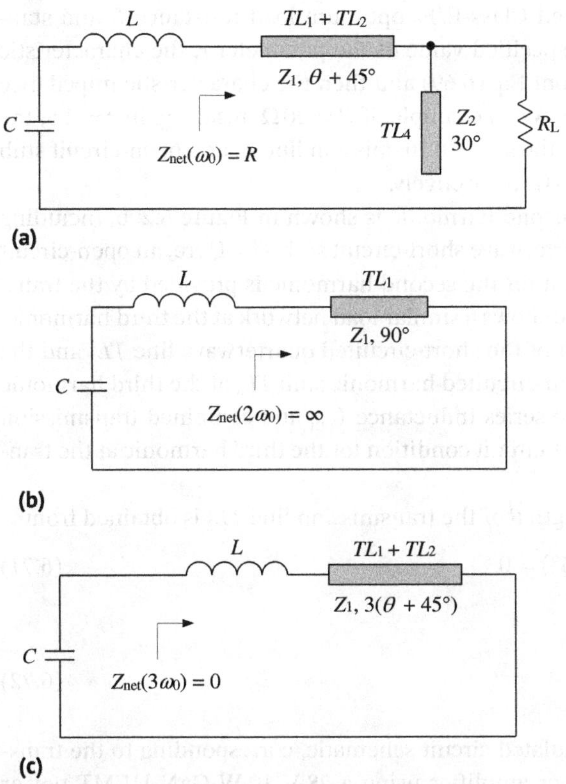

Figure 6.20 Load networks seen by the device output at three harmonics.

from which the characteristic impedances Z_1 and Z_2 can be explicitly calculated.

This system of two transcendental equations can be directly solved as a function of the load parameter $r = R_L/R$, resulting in

$$\frac{Z_1}{R_L} = \frac{\sqrt{r\left(1 + \tan^2\theta_1\right) - 1}}{r \tan\theta_1} \tag{6.67}$$

$$\frac{Z_1}{Z_2} = \frac{r - 1}{r \tan 30^\circ \ \tan\theta_1} \tag{6.68}$$

where both equations can be simplified for $\theta = 15^\circ$ and $\theta_1 = 45^\circ + \theta = 60^\circ$ to

$$\frac{Z_1}{R_L} = \frac{\sqrt{4r - 1}}{\sqrt{3}\, r} \tag{6.69}$$

$$\frac{Z_1}{Z_2} = \frac{r - 1}{r} \tag{6.70}$$

As a result, for the required Class-E/F_3 optimum load resistance R and standard load $R_L = 50\,\Omega$ for the specified value of the parameter r, the characteristic impedance Z_1 is obtained from Eq. (6.69) and then the characteristic impedance Z_2 is calculated by Eq. (6.70). As an example, if $R = 20\,\Omega$, resulting in $r = 2.5$, the characteristic impedances of the series transmission line Z_1 and open-circuit stub Z_2 become equal to 35 and 58 Ω, respectively.

The load network at the second harmonic is shown in Figure 6.20b, including the shorting effect of the quarterwave short-circuit stub TL_3. Here, an open-circuit condition at the device output for the second harmonic is provided by the transmission line TL_1. Figure 6.20c shows a similar load network at the third harmonic, where the open-circuit effect of the short-circuited quarterwave line TL_3 and the short-circuit effect of the open-circuited harmonic stub TL_4 at the third harmonic are included. As a result, the series inductance L_{out} and combined transmission line $TL_1 + TL_2$ provide a short-circuit condition for the third harmonic at the transistor output.

The required electrical length θ of the transmission line TL_2 is obtained from

$$3\omega_0 L + Z_1 \tan 3(\theta + 45°) = 0 \tag{6.71}$$

and can be calculated by

$$\theta = \frac{\pi}{12} - \frac{1}{3}\tan^{-1}\frac{3\omega_0 L}{Z_1} \tag{6.72}$$

Figure 6.21 shows the simulated circuit schematic, corresponding to the transmission-line Class-E/F_3 power amplifier using a 28-V 10-W GaN HEMT power transistor CGH40010 and a transmission-line load network fabricated on a 30-mil RO4350 substrate shown in Figure 6.19a [4]. The required conjugate matching with the standard 50-Ω source is provided by the input matching circuit consisting of the series low-impedance microstrip line and shunt microstrip stub. Here, since the device output capacitance C_{out} and series inductance L_{out} formed by the drain bondwires and package lead do not match the required values of the optimum shunt capacitance C and series inductance L corresponding to the idealized optimum Class-E/F_3 operation mode, the load network can be easily modified by optimizing the parameters of the series transmission line. For practical implementation, it needs to add the stripline models for the device input and output package leads to the simulation setup to include their parasitic series inductances, the values of which strongly depend on the assembling procedure. The maximum output power of 40.8 dBm, power gain of 12.8 dB (linear gain of about 18 dB), drain efficiency of 82.5%, and PAE of 78.2% were simulated at an operating frequency of 2.14 GHz with a DC-supply voltage of 28 V and a quiescent current of 40 mA.

The transmission-line GaN HEMT Class E/F_3 power amplifier was fabricated on a 30-mil RO4350 substrate. Figure 6.22 shows the test board of such a power amplifier using a 28-V 10-W GaN HEMT power transistor CGH40010P in a metal–ceramic pill package. The input matching circuit, output load network, and gate and drain bias circuits (having bypass capacitors on their ends) are fully based on microstrip lines of different electrical lengths and characteristic impedances according to the simulation setup shown in Figure 6.21. Special care should be taken in order to minimize the output lead inductance of the packaged GaN HEMT device, which can significantly affect the power amplifier characteristics such as output power and efficiency. The measured results demonstrate an output power of 40 dBm

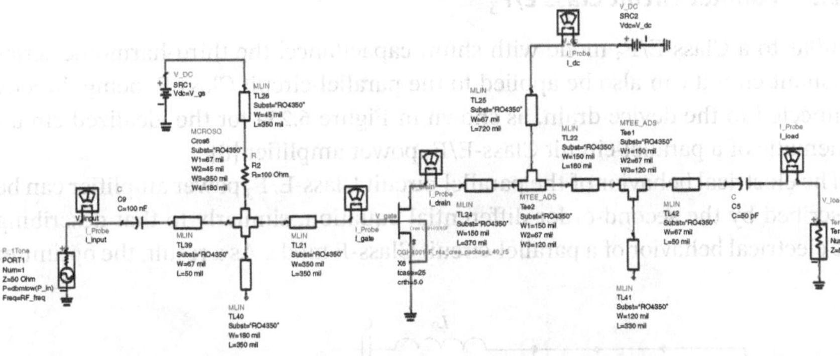

Figure 6.21 Circuit schematic of transmission-line Class-E/F_3 GaN HEMT power amplifier.

Figure 6.22 Test board of transmission-line Class-E/F_3 GaN HEMT power amplifier. *Source:* [4], with permission of IEEE.

with a drain efficiency of 76%, a PAE of 73.1%, and a power gain of 14.3 dB at an operating frequency of 2.14 GHz (gate bias voltage $V_g = -2.95$ V, quiescent current $I_q = 50$ mA, and drain supply voltage $V_{dd} = 30$ V), achieved without any tuning of the transmission-line input matching circuit and load network. It should be noted that the DC-supply voltage was increased up to 30 V in order to achieve an output power of 10 W, taking into account that the device typical breakdown voltage is equal to 100 V. A slightly greater quiescent current of 50 mA was set to increase the power gain at low input powers. The second and third harmonic components were measured at the levels of −35 and −25 dBc, respectively.

6.2.5 Parallel-circuit Class E/F_3

Similar to a Class-E/F_3 mode with shunt capacitance, the third-harmonic series resonant circuit can also be applied to the parallel-circuit Class E, being directly connected to the device drain, as shown in Figure 6.23a for the idealized circuit schematic of a parallel-circuit Class-E/F_3 power amplifier [8].

The electrical behavior of the parallel-circuit Class-E/F_3 power amplifier can be described by the second-order differential equation, similarly to that describing the electrical behavior of a parallel-circuit Class-E mode. As a result, the optimum

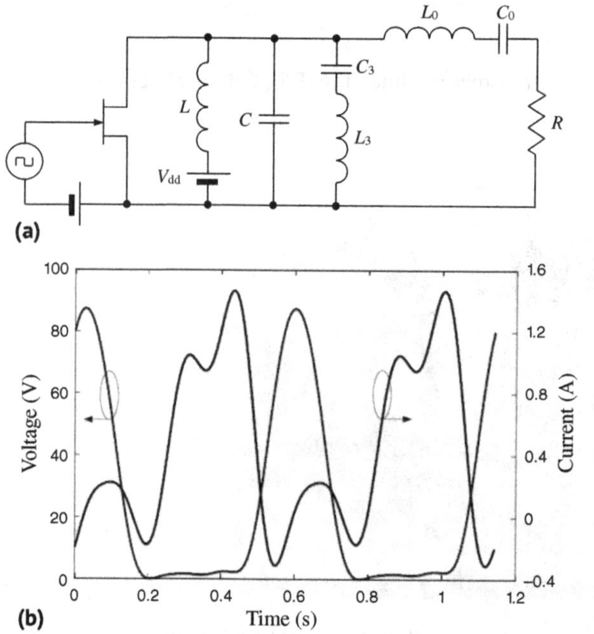

Figure 6.23 Circuit schematic and drain waveforms of parallel-circuit Class-E/F_3 power amplifier.

normalized parallel inductance L, shunt capacitance C, and load resistance R of the parallel-circuit Class-E/F_3 mode can be calculated according to

$$\frac{\omega L}{R} = 0.975 \tag{6.73}$$

$$\omega CR = 0.551 \tag{6.74}$$

$$R = 1.256\frac{V_{dd}^2}{P_{out}} \tag{6.75}$$

where V_{dd} is the drain (or collector) DC-supply voltage and P_{out} is the output power at the fundamental frequency.

From Eqs. (6.73)–(6.75), it follows that the normalized parallel inductance L is approximately the same in comparison with the Class-E/F_3 with shunt capacitance, whereas the normalized shunt capacitance C and load resistance R are significantly higher for the parallel-circuit Class-E/F_3.

The peak collector voltage V_{max} and current I_{max} are determined as

$$\frac{V_{max}}{V_{dd}} = 3.279 \tag{6.76}$$

$$\frac{I_{max}}{I_0} = 2.915 \tag{6.77}$$

that shows that the reduced voltage peak factor is close to that obtained for a Class E/F_3 with shunt capacitance, according to Eq. (6.56).

The maximum operating frequency f_{max} in a parallel-circuit Class-E/F_3 operation mode can be obtained from Eqs. (6.74) and (6.75) as

$$f_{max} = 0.0698\frac{P_{out}}{C_{out}\, V_{dd}^2} \tag{6.78}$$

which is slightly higher than the maximum operating frequency corresponding to Class E/F_3 with shunt capacitance, according to Eq. (6.58).

By providing an initial phase shift for the third harmonic and introducing an additional reactive element (inductance or capacitance) connected in series to the load resistance, the generalized case of the parallel-circuit Class-E/F_3 power amplifier with extended design flexibility can be obtained [9]. Figure 6.23b shows the intrinsic drain voltage and current waveforms of the transmission-line parallel-circuit Class-E/F_3 power amplifier, simulated at 1.75 GHz using the 10-W GaN HEMT device and 30-mil RO5880 substrate. Compared with the ideal voltage and current waveforms which are similar to those shown in Figure 6.11 for Class E/F_3 with shunt capacitance, the simulated results show very close drain waveforms. The small difference is due to the effect of the device saturation resistance, finite switching time, device package parasitic parameters, and the presence of the higher-order harmonic components. In this case, the output power of 40.4 dBm, drain efficiency of 83.9%, PAE of 77.2%, and power gain of 10.4 dB were measured at the operating frequency of 1.75 GHz. An additional second-harmonic suppression can be achieved by using a push–pull configuration of the parallel-circuit Class-E/F_3 power amplifier [10].

6.3 Class-E/F_3 Oscillators

Applying Class-F or inverse Class-F conditions for the second and third harmonic components to Class-E power oscillator provides an opportunity to reduce the maximum drain (or collector) voltage of the switching transistor that makes possible to increase the DC-supply voltage while still operating the transistors safely. Figure 6.24a shows the circuit schematic of a Class-E/F_3 MOSFET power oscillator

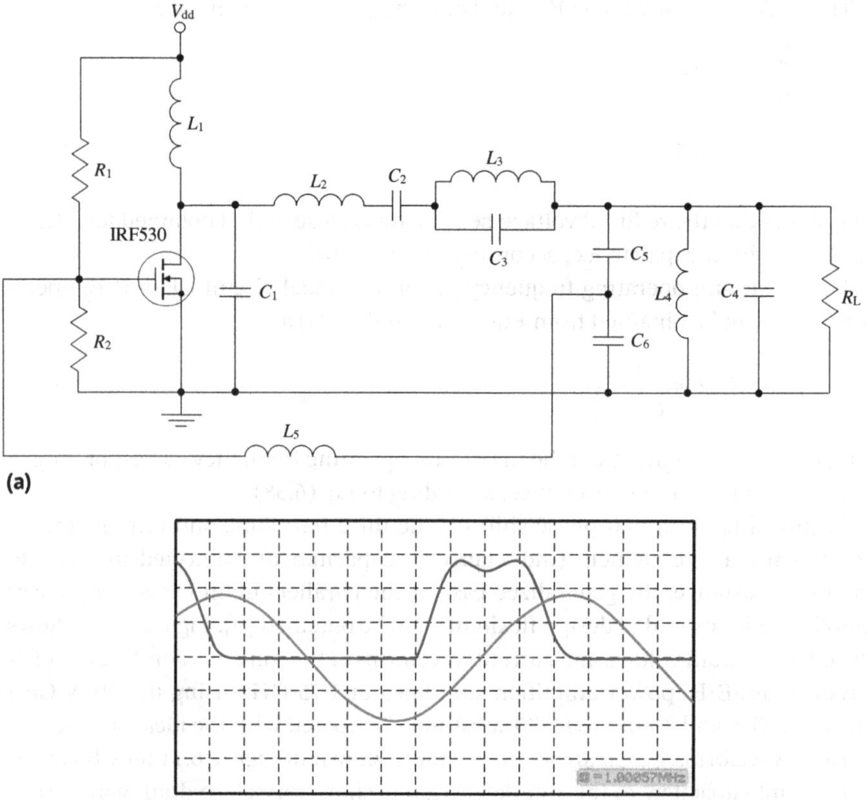

(a)

(b)

Figure 6.24 Circuit schematic and waveforms of Class-E/F_3 power oscillator with parallel resonator.

using a Class-E operation mode with shunt capacitance and shunt filter [11]. Here, additionally, the second-harmonic parallel L_3C_3 resonator connected in series to the load network and feedback network composed of the capacitors C_5 and C_6 and an inductor L_5 are included. The capacitor C_2 is used as a DC-blocking capacitor and due to its high capacitance has little effect on the circuit operation. In this case, the load impedance seen by the device drain must satisfy the Class-E condition (optimum inductive impedance) at the fundamental frequency, being very high at the second and near zero at the third harmonics, representing purely capacitive reactances at all higher-order harmonic components.

The required oscillation frequency of 1 MHz was achieved by tuning the feedback inductance L_5, whereas the drain voltage waveform corresponding to the Class-E/F_3 operation mode, as shown in Figure 6.24b, was obtained by varying the feedback capacitors C_5 and C_6, with the sinusoidal output voltage having extremely low content of the second- and higher-order harmonics suppressed by more than 40 dB. For the DC-supply voltage of 9 V, the drain voltage amplitude equals 28.5 V that translates to the drain peak voltage of 3.167, which is by 14% lower than that in Class E with shunt capacitance and shunt filter. The output power of 0.92 W and efficiency of 86% were achieved at 1-MHz oscillation frequency, with the output power variations from 77 mW to 2.64 W as a function of the DC-supply voltage. By varying the inductance in the feedback network, it was possible to tune the oscillation frequency within the frequency bandwidth of 0.94–1.04 MHz, providing an overall efficiency above 80%.

Figure 6.25a shows the circuit schematic of a Class-E/F_3 MOSFET power oscillator with a series third-harmonic resonator connected to the drain terminal [12]. Owing to the series resonant circuit tuned to the third harmonic, the drain voltage stress was reduced by about 22% compared to the corresponding Class-E power oscillator. The oscillator circuit also includes the two resistors R_1 and R_2 in order to provide gate voltage biasing, an RF choke, a shunt capacitor C, a series inductor L_1, a series capacitor C_0, and a feedback loop circuit composed of the capacitors C_{11} and C_{12} and an inductor L_2. Here, the inductor L_1 can be regarded as the series connection of an inductor L and an inductor L_0, which is a part of the series resonant circuit L_0C_0 tuned to the fundamental frequency. The experimental waveforms of the output sinusoidal voltage under the load quality factor of 12 and drain–source voltage with flattened top of a Class-E/F_3 power oscillator are shown in Figure 6.25b. The measured oscillation frequency was 801.6 kHz, with the output power of 0.9075 W and power conversion efficiency of 86.24%.

To minimize the sensitivity of the output power, efficiency, and oscillation frequency of a Class-E/F_3 power oscillator with series third-harmonic resonator to variation of the load network parameters, the feedback network based on a low-Q RC circuit can be used [13]. Such a feedback network is not a resonant circuit and provides a fairly constant phase shift in the vicinity of the oscillating frequency.

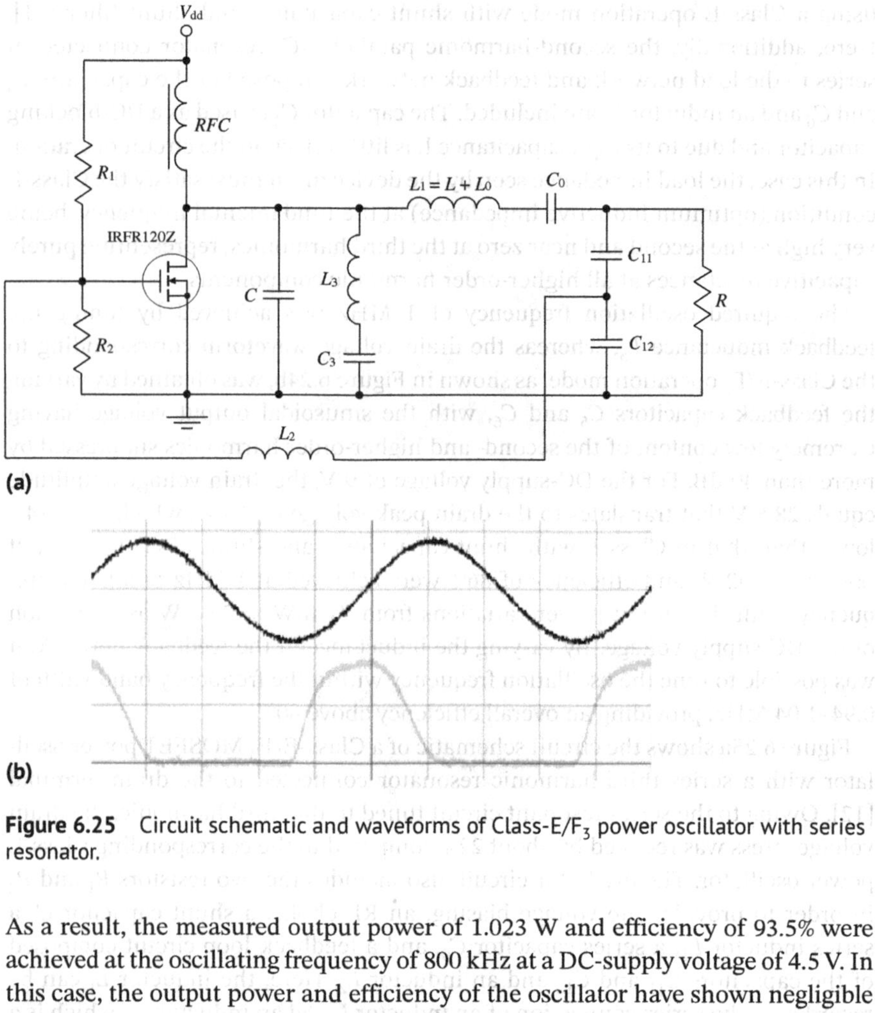

Figure 6.25 Circuit schematic and waveforms of Class-E/F_3 power oscillator with series resonator.

As a result, the measured output power of 1.023 W and efficiency of 93.5% were achieved at the oscillating frequency of 800 kHz at a DC-supply voltage of 4.5 V. In this case, the output power and efficiency of the oscillator have shown negligible sensitivities to the variations in the component values.

6.4 Switched-path and Load-modulated Balanced Power Amplifiers

Modern and future-generation wireless communications systems require that the power amplifier deliver a wide range of output powers with high efficiency and high linearity simultaneously. As a result, being designed for the highest power level with maximum available efficiency, the power amplifier tends to operate less efficiently at lower power levels, which leads to shortening the battery lifetime. In this case, it was suggested to provide switching between the power amplifier stages with their proper connection to the load [14]. In such a configurable power amplifier where each stage has its own output configuration, the power amplifier provides a variable output power level corresponding to the output powers of each stage and their combinations. It is assumed that each stage has its own matching to 50-Ω input and output impedances and requires switches between the stages. These requirements make it extremely difficult to implement this approach into a fully monolithic design, resulting in the increased size and cost of the entire power amplifier. The possibility to improve efficiency at a significant power backoff is to use a mechanism for switching the segments of the active device to reduce the quiescent current at low power mode and to switch the load network to provide different impedances at high and low output power modes [15]. However, the resulting improvement was not significant, and the PIN-diode was required in the load network.

In addition to high efficiency, it is very important to keep a high linearity at high output power with variation of the load impedance. Figure 6.26 shows the circuit schematic of the balanced power amplifier having a 90° phase shift due to 45° phase shifters in each amplifying path and impedance-transforming elements improving the linearity over different values of the load voltage standing wave ratio (VSWR) compared to the conventional single-ended prototype [16]. This is because each of the two transistors in a balanced power amplifier sees opposite impedances: high impedance and low impedance, respectively. Since the linear requirements remain for low impedance with some increase in the output power, only half an output signal has worse linearity performance compared to a conventional single-ended case where this is a concern for a total output signal. In the case of the wide frequency bandwidth requirements, the Wilkinson power combiner and two 45° phase shifters can be replaced by a broadband quadrature 90° hybrid coupler. However, in both cases, there is not any special mechanism for efficiency improvement at the power backoff conditions.

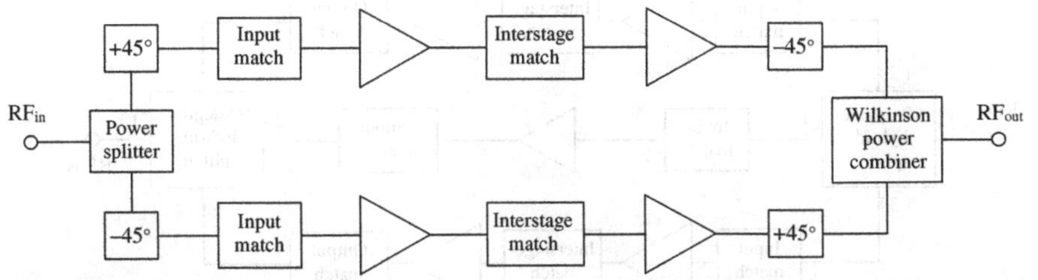

Figure 6.26 Block diagram of a balanced power amplifier with improved linearity against load variations.

In order to increase efficiency in a three-stage power amplifier, the direct bypass connection can be made between two isolated ports of the 90° hybrids in a final balanced amplifying stage, as shown in Figure 6.27 [17]. In this case, both high power paths are switched off, appearing as reflective impedances to the signal and, as a result, the power from the second stage flows directly to the load, thus improving efficiency at backoff. However, the two additional switches are required to connect both isolated ports in a low-power mode and to connect each isolated port to the ballast resistor in a high-power mode. The quadrature hybrid can generally be implemented with any lumped or transmission-line elements.

The block diagram of a two-stage balanced switched-path power amplifier having high efficiency at different output power levels is shown in Figure 6.28 [18]. It incorporates an additional low-power single-stage amplifying path connected between the isolated ports of a 90° input hybrid power splitter and a 90° output hybrid power

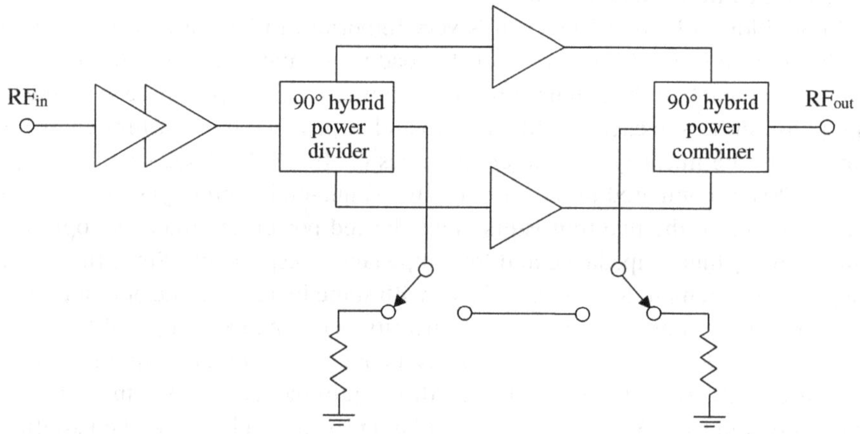

Figure 6.27 Block diagram of a balanced power amplifier with high efficiency at backoff.

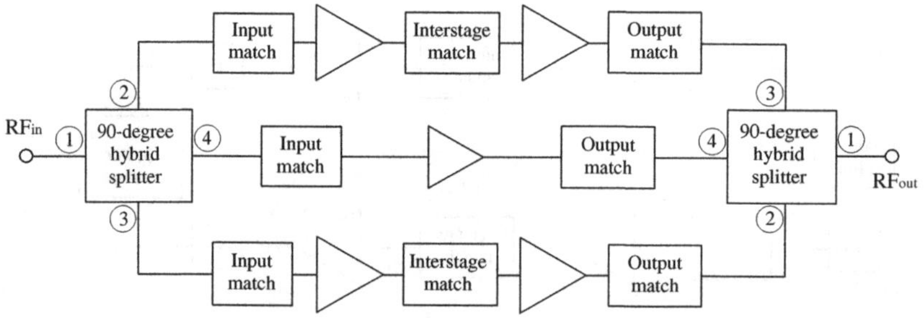

Figure 6.28 Block diagram of a balanced switched-path power amplifier.

combiner. In a high-power mode, the bias control unit provides proper biasing of the transistors in both balanced amplifying paths and switches off the biasing in a low-power amplifying path. In this case, the output powers from both balanced amplifying paths combine in the output hybrid combiner and the low-power amplifying path is isolated from them, thus providing a high efficiency at a high output power. For the matched hybrid branch-line or coupled-line combiner, power entering an input port 1 is divided between the output ports 2 and 3 with a 90° phase shift between these outputs and no power is delivered to the isolated port 4.

A quadrature 90° hybrid has an important advantage compared to in-phase splitters that at equal values of the reflection coefficients from loads connected to the output ports 2 and 3, there is no reflection wave at the input port 1 and, consequently, the input VSWR of a quadrature hybrid does not depend on the equal load mismatch level. In this case, all reflected power is dissipated in a 50-Ω ballast resistor connected to the isolated port 4. A similar process occurs in a quadrature 90° combiner when the reflected power from the two inputs will flow to the load. As a result, the isolated port of the input hybrid becomes the input for a low-power amplifying path, whereas the isolated port of the output hybrid operates as its output 50-Ω port. Hence, high efficiency at lower output power levels can be achieved and no need to include any additional switches or PIN-diodes to isolate the high-power balanced paths and a low-power path. Additionally, the adaptive biasing can be applied to the transistors in a low-power path to minimize the quiescent current at very low output powers, thereby contributing to increasing efficiency by two to three times compared to the conventional balanced architecture.

The two switched-path balanced power amplifiers following the above-described concept were fabricated. The first one covers the frequency band for WCDMA850 applications, the second one was designed to cover WCDMA1900/2100 bands. Both power amplifiers were implemented using a 0.35-μm SiGe HBT process intended for handset transmitter applications, with a typical collector-base breakdown voltage of 20 V for a high-voltage SiGe HBT device. The tested MMIC die of approximately 1.97 mm^2 in size for the low-band WCDMA850 design, including the two-stage high-power and low-power amplifying paths with the current-mirror biasing, input and interstage matching circuits, and electrostatic discharge (ESD) protection circuits, is shown in Figure 6.29. The emitter areas of the first and second stages of each high-power path are 496 and 4464 μm^2, respectively. The interstage matching of the high-power amplifying paths represents a two-section high-pass structure (series capacitor and shunt inductor to ground) to prevent the potential low-frequency oscillations, where the transistors provide higher gain. To reduce the chip size, the required shunt inductances were provided by the bondwires connected to the PCB.

Figure 6.30 shows the achieved maximum output power of 31.5 dBm with a maximum PAE of 58% for a low-band WCDMA850 balanced switched-path SiGe

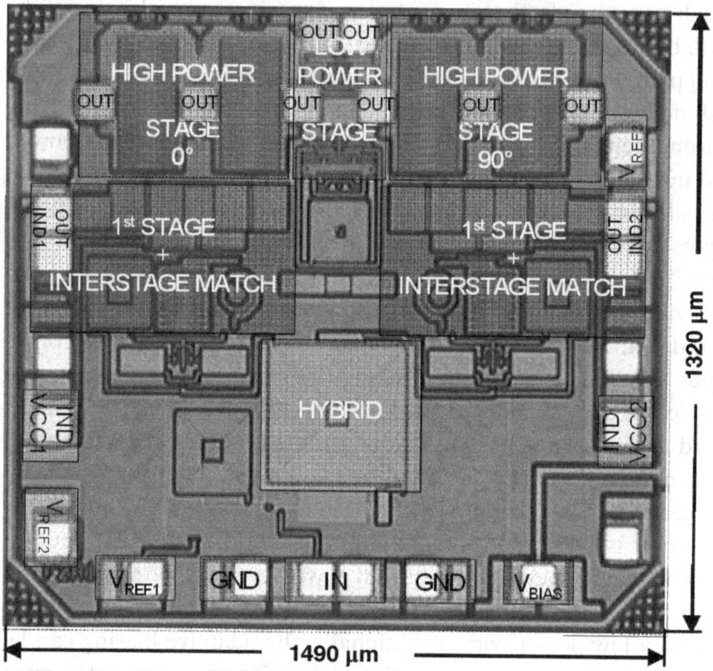

Figure 6.29 Photo of fabricated low-band balanced switched-path SiGe HBT power amplifier MMIC. *Source:* [18], with permission of EuMA.

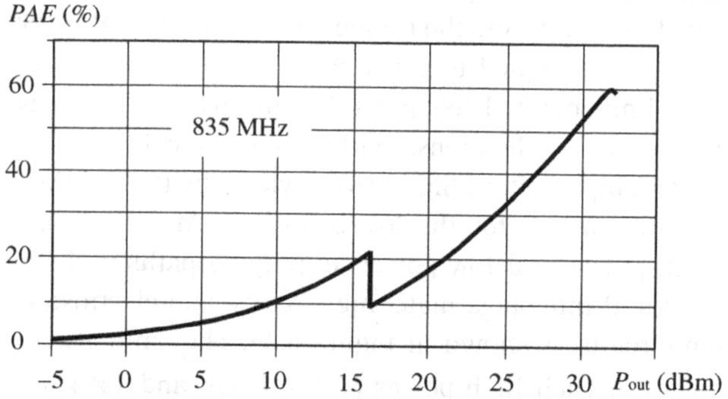

Figure 6.30 Measured power-added efficiency versus output power of low-band balanced switched-path SiGe HBT power amplifier.

HBT power amplifier. For an average WCDMA power of 28.5 dBm, efficiency remains still high, with the measured value of 41.5%. At 15-dB power backoff, efficiency is increased by a factor of two by turning off the bias current for the high-power amplifying path and switching on the low-power amplifying path

Table 6.3 Measured performance of high-band balanced switched-path SiGe HBT power amplifier.

Frequency	P_{out} (dBm)	PAE (%)	Gain (dB)	ACLR (dBc)
1.85 GHz	33	50	20	—
	30	39	26	−35
	16	20	22	−35
	12	14	18	−37
	6	7	12	−40

transistors. In this configuration, the bias current is only 14.5 mA and the saturated output power of 18 dBm appears with a maximum PAE of 30%. Table 6.3 shows the performance of a high-band WCDMA1900 balanced switched-path SiGe HBT power amplifier, with a PAE of 50% at the saturated output power of 33 dBm and an average PAE of 20% at the backoff power of 16 dBm.

A coupler-based load-modulated balanced amplifier (LMBA) architecture is shown in Figure 6.31a, where the primary drive signal is equally split between the first and second primary amplifiers through the hybrid quadrature splitter. The outputs of the primary amplifiers are connected to ports 1 and 2 of the output combiner, represented by a hybrid quadrature coupler [19, 20]. An additional auxiliary drive signal, amplified by an auxiliary amplifier, is injected into port 3 and the load is connected to port 4 of the output quadrature coupler, respectively. Due to the property of the quadrature splitter, there is a 90° phase difference between the signal input to the first and second primary amplifiers. As a result, the signals from the two primary amplifiers arrive at the load with the same phase and add constructively, whereas the signals arrive at the auxiliary port with 180° out of phase and thus add destructively. In this case, the auxiliary amplifier is disabled when the primary amplifiers operate below the saturation point. As the primary amplifiers approach saturation, the auxiliary amplifier is enabled and generates a reflection signal that is equally split and travels backward into the output ports of the first and second primary amplifiers. If the auxiliary drive signal is −180° out of phase with the original primary drive signal, the phase of these reflection signals at ports 1 and 2 will be both 180° offset from the phase of the output signals generated by the respective primary amplifiers. As a result, the injected reflection signals appear to the primary amplifiers as a load impedance mismatch and thus can effectively reduce the load impedance of the primary amplifiers. The primary amplifiers can therefore increase their output current and deliver more power to the load. So, this architecture represents a kind of load modulation scheme for the balanced power amplifier, similar to a Doherty amplifier, although the

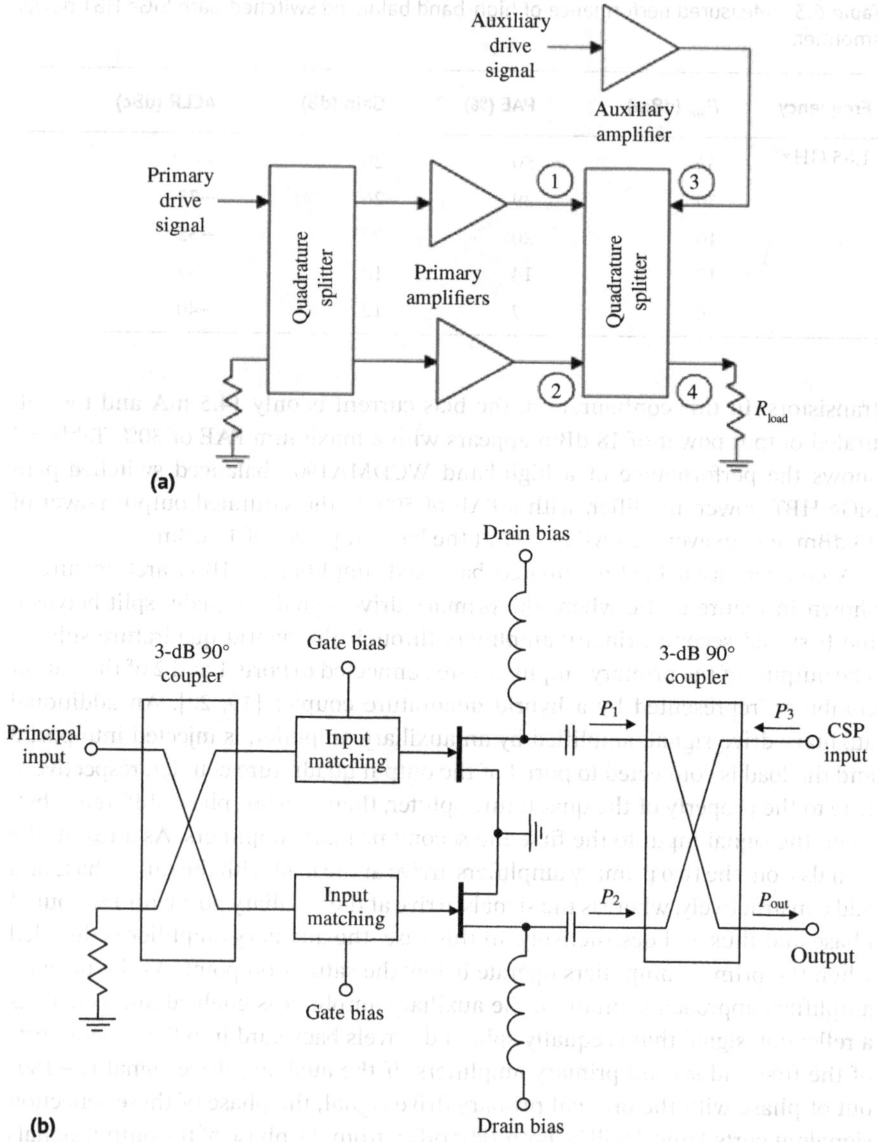

Figure 6.31 Simplified schematics of LMBAs.

main operating principle is based on the power reflection mode. In this implementation, the balanced amplifier is used as a primary amplifier and the injected signal is called the control signal power (CSP), whereas the entire configuration is named a LMBA.

From an analysis of the LMBA electrical behavior, it follows that the impedances presented at port 1 and port 2 change with the injected current at port 3, which means the load of the balanced amplifier can be "modulated" by adjusting the control signal. In this case, an efficiency enhancement can be achieved by dynamically adjusting the external control signal instead of using conventional passive networks. Ideally, this can remove the inherent bandwidth limitation of the Doherty amplifiers, where a bandwidth-limiting network based on an impedance inverter is required, and the frequency bandwidth of the LMBA fully depends on the frequency bandwidth of the hybrid quadrature coupler. Another feature of the LMBA is that if the coupler is lossless, the power injected at the CSP port can be fully recovered at the final output, thus positively contributing to the total output power, independent of the load modulation it is imposing. In the first prototype, the LMBA concept was pushed to its extreme: the packaged high-frequency 10-W GaN HEMT transistors were used directly in the balanced amplifier without any passive output matching, as shown in Figure 6.31b, where the LMBA could achieve a drain efficiency better than 50% at a 6-dB backoff power range over the frequency bandwidth of 0.8–2 GHz, with a 2-W phase-controlled signal [21]. Similar topology but this time with die transistors and substrate-integrated quadrature couplers was used to achieve a drain efficiency above 55% at 7-dB power backoff across the frequency bandwidth of 4.5–7.45 GHz, with a 1-W phase-controlled signal [22].

In Figure 6.31, the control signal is shown as an externally derived source, although it has to be frequency and phase matched to the main signal, the part of which has to be derived, directly or indirectly, from the main signal. The simplest way of achieving this is to split the input signal and apply suitable amplitude and phase controls to the coupled output to feed it to the output coupler through the CSP amplifier, as shown in Figure 6.32a. In this case, a drain efficiency above 29% at 6-dB power backoff was achieved across the wide frequency range of 1.8–3.8 GHz, with an output power above 42 dBm [23]. Ideally, the amplitude and phase controls in the CSP loop would be adjusted to optimize power and efficiency under modulated signal conditions. However, in practice, some compromise has to be made when an improved performance can be achieved by suitable synthesis of a phasing network in the control signal loop, as shown in Figure 6.32a, when the drain efficiency was improved up to 54% at 6-dB power backoff at 2.4 GHz [24].

Figure 6.32b shows the inverted LMBA architecture, where the CSP amplifier is used as the main amplifier and the balanced pair is used as the auxiliary amplifier [25, 26]. In this case, when the output power is lower than the output power backoff level, the main amplifier is turned on, whereas the auxiliary balanced amplifier is turned off, and the output power of the entire LMBA structure

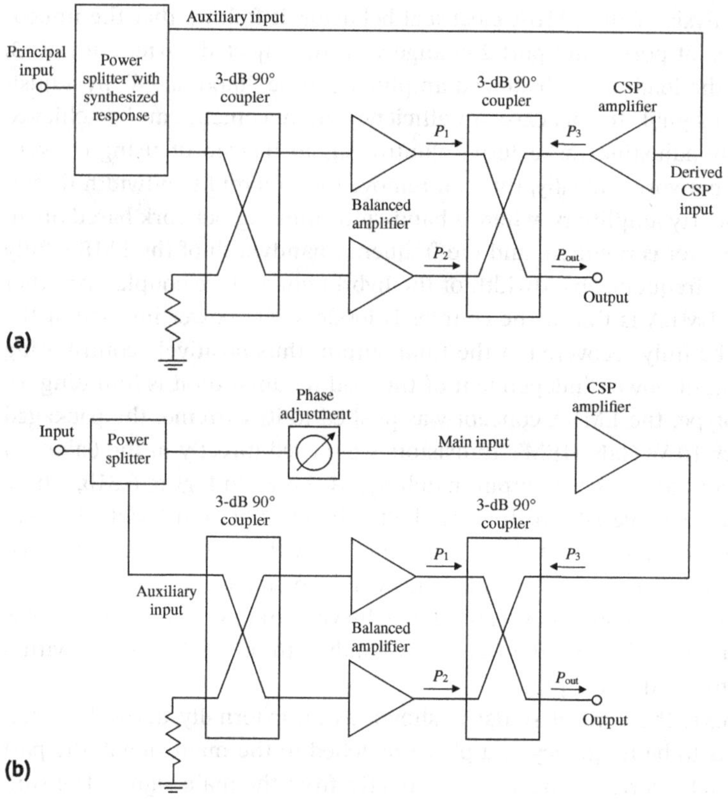

Figure 6.32 Block schematics of LMBAs.

is provided by only the CSP amplifier. When the output power is higher than the output power backoff level, the balanced amplifiers are turned on, and the CSP amplifier becomes saturated. When the output power is further increased, an additional power increase is mainly provided by the balanced amplifier due to the saturation status of the CSP amplifier. The most important feature of this inverted architecture is that there is no load modulation effect at the main CSP amplifier output, which means that CSP output power will not increase with the input power after reaching the output backoff power level. Theoretically, the additional power increase of the LMBA is entirely provided by the balanced amplifier pair due to the saturation status of the main amplifier. In this case, a high-efficiency performance can be achieved at a significant power backoff, for example, reaching a drain efficiency of 49% at 12-dB power backoff at 3.3 GHz and over 43% at 10-dB power backoff across the frequency bandwidth of 3.05–3.55 GHz [25]. At the same time, an ultrawideband performance can also be achieved because of removal of the carrier load modulation effect when a drain efficiency of 47–58% at 10-dB backoff was achieved across the frequency bandwidth of 1.5–2.7 GHz with a peak output power of 42.5 dBm and a drain efficiency of 40–64% at 10-dB backoff was obtained over the frequency bandwidth of 0.55–2.2 GHz using a 6-W GaN transistor CGH40006P and two 10-W GaN transistors CGH40010F [26, 27].

However, the main drawback of the LMBA is its too big size for lower frequency applications due to the significant size of both couplers and for lower power handset applications where a very small size of the monolithic integrated circuits is required and too low power in existing implementations for higher power base station applications. Besides, its efficiency level is not high enough for practical use, especially for frequency bandwidths close to an octave and wider. Therefore, an asymmetric Doherty amplifier looks like the best compromise between the output power, efficiency, size, and cost for both handset and base station applications when the frequency bandwidth is lower than an octave band.

6.5 Three-way Doherty Amplifier with Three-coupled-line Combiner

An asymmetric Doherty amplifier represents a very convenient implementation and a very efficient in operation architecture, which can be used in both handset and base station applications, especially in sub-6-GHz and millimeter-wave frequency bandwidths. Figure 6.33a shows the simplified three-way structure of an asymmetric Doherty amplifier with two peaking (PA) amplifiers and one carrier (CA) amplifier, which is located between them, and all having the same output powers. The main drawback of this structure is that all three power amplifiers are connected in parallel, resulting in a sufficiently low output impedance when it is necessary to provide a high output power and limited operating frequency bandwidth, less than an octave, because of the presence of a quarterwave line at the output of the carrier (or main) amplifier. The output impedance can be increased, and the frequency bandwidth can be widened by using a three-coupled-line combiner at the output, combining these three power amplifiers, as shown in Figure 6.33b, which represents an extended version of a two-coupled-line combiner, which combines the two power amplifiers and provides a very wide operating frequency bandwidth [28]. It should be noted that using a coupled-line structure at the output can provide the lower sensitivity to the load variations in view of a high load VSWR, because the carrier and peaking amplifiers see different load impedances (for example, higher impedance for the peaking amplifiers and lower impedance for the carrier amplifier and vice versa) in the case of mismatched load.

Figure 6.34 shows the block schematic of an asymmetric Doherty amplifier with the three-way branch-line splitter on the input side and the three-coupled-line

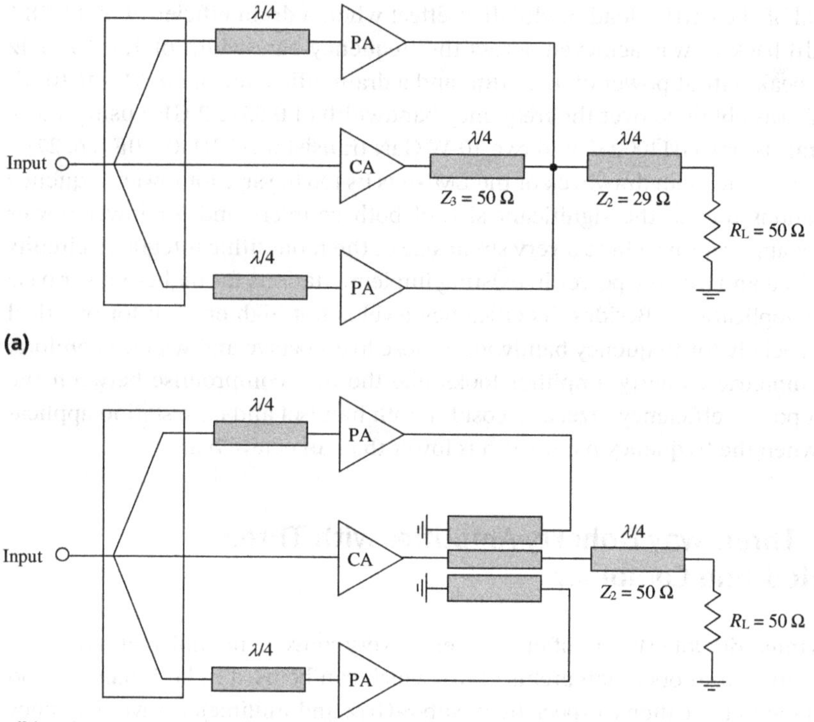

Figure 6.33 Three-way asymmetric Doherty amplifier architectures.

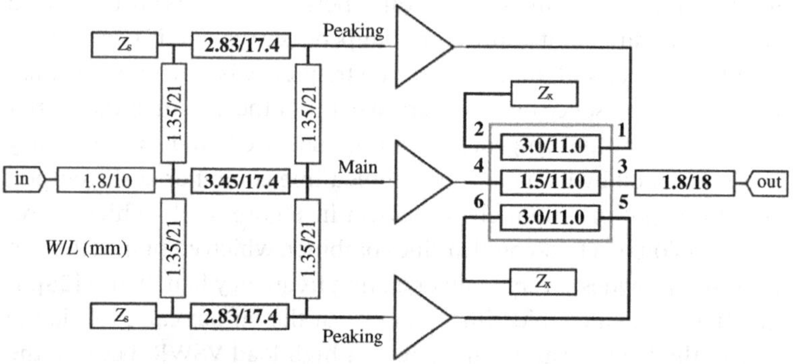

Figure 6.34 Block schematic of asymmetric Doherty amplifier with three-coupled-line combiner.

combiner on the output side, which provides an output power of 30 W at an operating frequency of at 2.4 GHz using three identical 10-W GaN HEMT power amplifiers based on the CGH40010F devices and a multilayer substrate to implement the three-coupled-line coupler using the microstrip lines fabricated at different metal layers underneath each other [29]. A multilayer substrate structure is used for stronger coupling at 2.4 GHz, with the microstrip line preferred for higher frequencies, especially in millimeter-wave and MMIC designs, due to easier tuning of the coupling coefficient and technological capabilities. By adjusting the impedance Z_s (initial value of 50 Ω) at the isolated ports of the branch-line input splitter, the split ratio between the peaking and main power amplifiers can be modified without reducing the port isolation or altering phase balance. An external impedance Z_x connected to each of the isolated ports 2 and 6 of the three-coupled-line combiner to compensate for the effect of the off-state impedances Z_{os} of the peaking amplifiers connected to ports 1 and 5 significantly influences the circuit performance and controls the output phase difference between the peaking and main amplifiers, particularly in the low-power region. This impedance can be realized using a short- or open-circuit microstrip-line stub.

The equation for the load impedance Z_m seen by the main device at the power backoff region when the peaking device is turned off can be written as

$$Z_m = Z_0 \frac{Z_0^2 + Z_x Z_0 + 2Z_{os} Z_x}{2Z_0^2 + Z_{os} Z_0 + Z_{os} Z_x} \tag{6.79}$$

which should be equal to $2Z_0$, thus resulting in the external impedance Z_x in

$$Z_x = 3Z_0 + 2Z_{os} \tag{6.80}$$

where Z_0 is the characteristic impedance of the Doherty amplifier [29].

The layout of the three-coupled-line coupler, which was simulated and optimized using CST Studio, is shown in Figure 6.35a. For practical implementation, a 32-mil RO4003C substrate with the dielectric permittivity of 3.38 was used, where the separate PCB pieces were assembled by stacking and screwing them together. The simulated and measured S-parameters of the three-coupled-line combiner are shown in Figure 6.35b, demonstrating the reliable performance with minimum flatness variations across the wide frequency bandwidth of 2–3 GHz. In this case, the optimal load impedance seen by the main amplifier equals 100 Ω at 10-dB power backoff and 30 Ω at peak power, whereas the optimal impedance for each peaking amplifier is equal to 35 Ω. For concept verification purposes, the Class-F power amplifiers with microstrip-line input and output matching and harmonic-control circuits having a narrowband amplitude and phase response were chosen. However, they can be replaced by the broadband parallel-circuit Class-E power amplifiers with an input three-coupled-line splitter to provide a high-efficiency performance over a very wide frequency bandwidth.

Figure 6.36 shows the PCB implementation of an asymmetric three-way Doherty amplifier with three-coupled-line combiner, which is built on a Rogers RO4003C substrate with a thickness of 0.8 mm. In order to minimize overall size and cost, the three-coupled-line coupler is constructed from eight individual board pieces, as shown on the right-hand side of Figure 6.36, all fabricated from

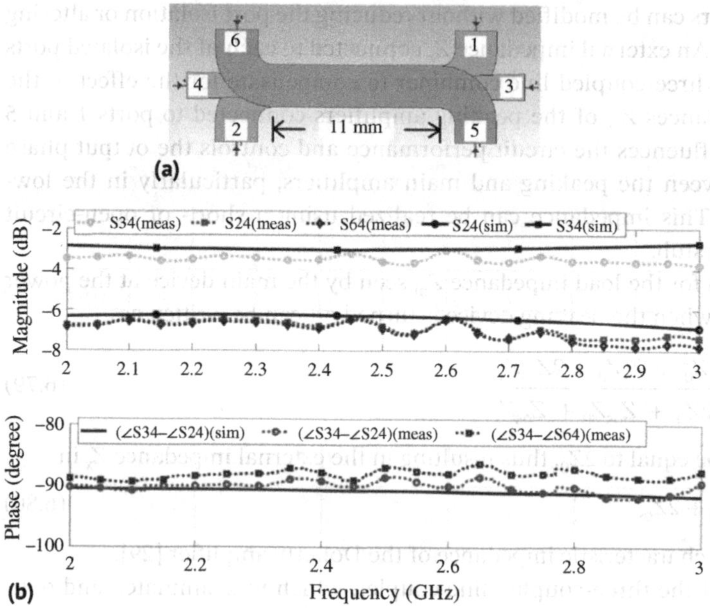

Figure 6.35 Measured and simulated *S*-parameters of a three-coupled-line combiner.

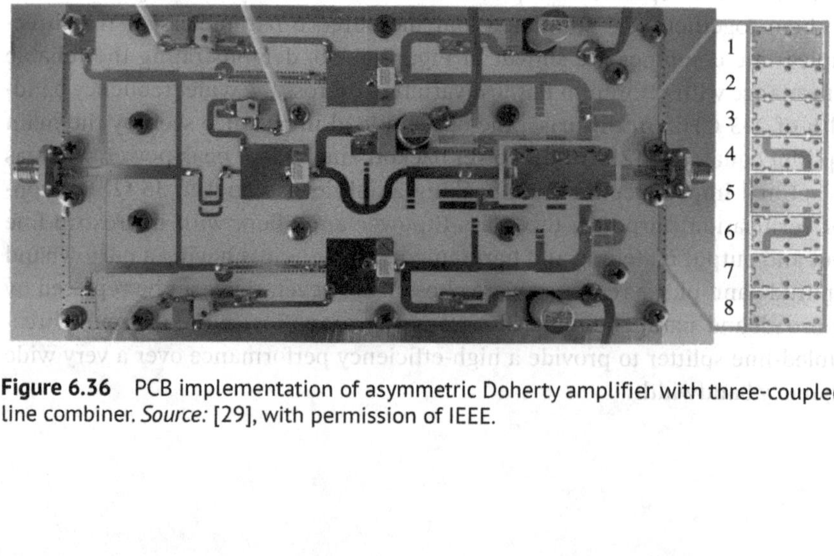

Figure 6.36 PCB implementation of asymmetric Doherty amplifier with three-coupled-line combiner. *Source:* [29], with permission of IEEE.

a single RO4003C sheet. The piece 5, serving as the center line, is designed for independent testing of the three-coupled-line coupler and doubles as the main PCB. Pieces 1–4 are mounted on top of the main PCB, while pieces 6–8 are positioned beneath it. After proper alignment, all pieces are secured with screws. Multiple vias are used to interconnect the layers, with soldering the completed post-assembly for enhanced stability and performance. As shown in Figure 6.36, some trimming options, such as additional paths to change the length and width of microstrip lines, are also available on the PCB, but in practice there was no need to use them.

The simulated and measured total system power gain and efficiency of the asymmetric three-way Doherty amplifier with three-coupled-line combiner at an operating frequency of 2.4 GHz are shown in Figure 6.37a, demonstrating their close proximity to each other over the entire range of the output powers. The measured results show that the drain efficiency remains above 50% across the entire 10-dB power backoff range, exceeding 60% within the 8-dB power backoff

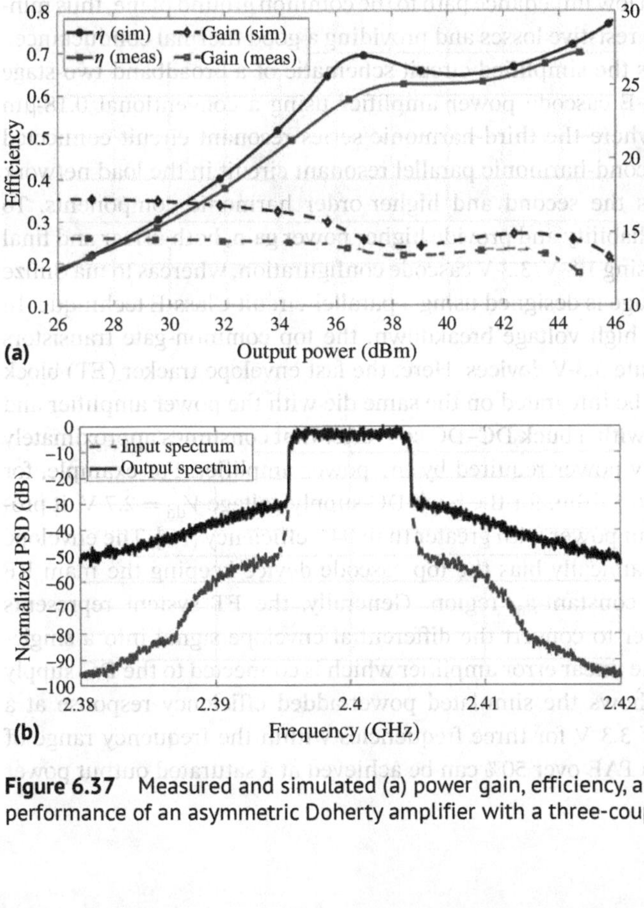

Figure 6.37 Measured and simulated (a) power gain, efficiency, and (b) linearity performance of an asymmetric Doherty amplifier with a three-coupled-line combiner.

range and reaching 70% at a peak output power of 45 dBm. For a 10-MHz LTE signal with a PAR of 10 dB, such an asymmetric Doherty with three-coupled-line combiner achieved the measured efficiency of 51% with an average output power of 35.5 dBm. In this case, an ACLR of −35 dBc was obtained without any digital predistortion, as shown in Figure 6.37b, using an input driver amplifier based on two SHF0589 GaAs transistors with a 3-W linear output power having the corresponding input spectrum.

6.6 Envelope Tracking for Cellphone Applications

Modern CMOS power amplifier technology is another possibility for designing broadband, lower-cost power amplifiers with a sufficiently high efficiency for 5G/6G and beyond handset transmitters. This is due to the current availability of the silicon substrates with a high resistivity of above 1 kΩ cm and silicon vias (TSVs) connecting the front-side metallization of the die to the backside ground metal and providing a low impedance path to the common ground plane, thus minimizing the substrate resistive losses and providing a good thermal conductance.

Figure 6.38a shows the simplified circuit schematic of a broadband two-stage parallel-circuit Class-E cascode power amplifier using a conventional 0.18-μm CMOS technology, where the third-harmonic series resonant circuit connected to the ground and second-harmonic parallel resonant circuit in the load network are used to suppress the second and higher-order harmonic components. To increase operation reliability and provide higher power gain, both driver and final stages are designed using 1.8-V/3.3-V cascode configuration, whereas to maximize efficiency the final stage is designed using a parallel-circuit Class-E technique. In order to prevent the high voltage breakdown, the top common-gate transistors represent the thick-gate 3.3-V devices. Here, the fast envelope tracker (ET) block is the one which can be integrated on the same die with the power amplifier and can operate together with a buck DC–DC converter that consumes approximately 20% of the DC-supply power required by the power amplifier. For example, for an output power of 26.5 dBm, for the buck DC-supply voltage $V_{dd} = 2.7$ V, it provides 80% of the output power with greater than 94% efficiency [30]. The envelope signal is used to dynamically bias the top cascode device keeping the main RF transistor in almost constant-g_m region. Generally, the ET system represents a differential amplifier to convert the differential envelope signal into a single-ended one, feeding the linear error amplifier which is connected to the DC-supply path. Figure 6.38b shows the simulated power-added efficiency response at a DC-supply voltage of 3.3 V for three frequencies within the frequency range of 2.3–2.7 GHz, where a PAE over 50% can be achieved at a saturated output power of around 31 dBm.

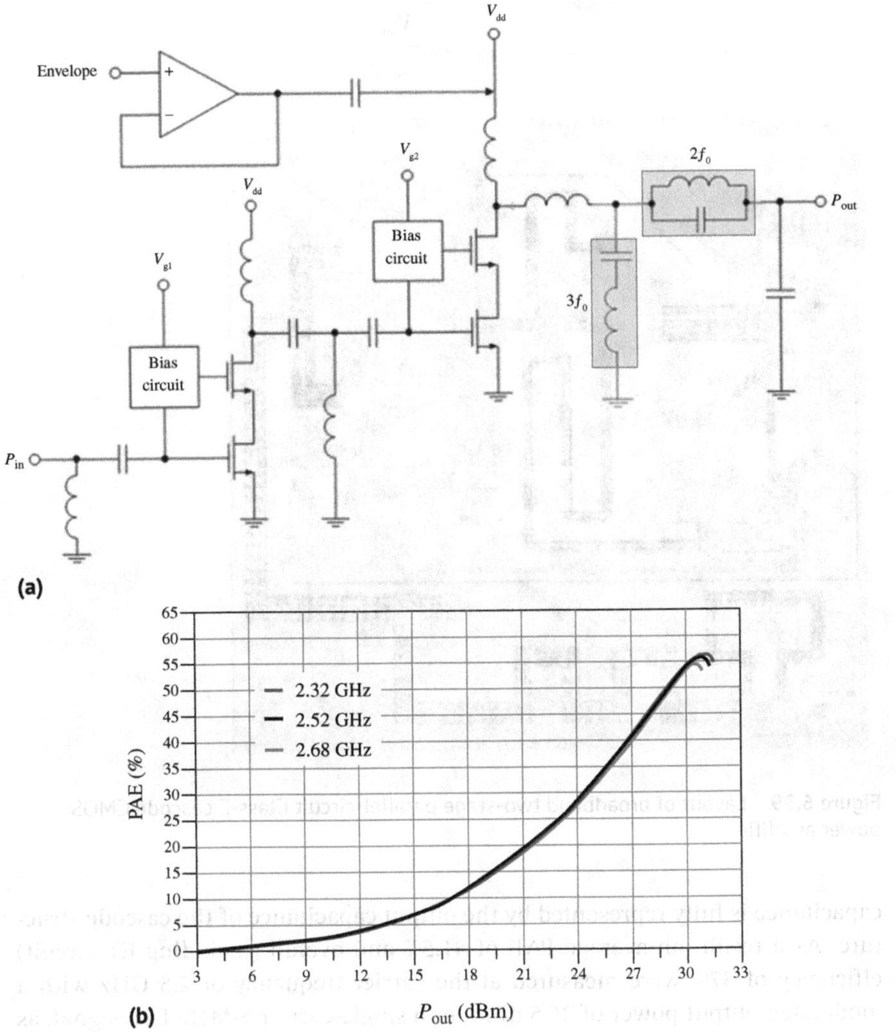

Figure 6.38 Circuit schematic and efficiency of broadband two-stage parallel-circuit Class-E cascode CMOS power amplifier.

Figure 6.39 shows the chip implementation of the broadband two-stage parallel-circuit Class-E cascode CMOS power amplifier with an overall chip size of 1.7×1.6 mm^2, where the meandered microstrip lines in the load network are used instead of the spiral inductors to minimize power losses and increase efficiency [30, 31]. Here, the required parallel-circuit Class-E inductance is provided by two parallel bondwires in a DC-supply path, and a shunt

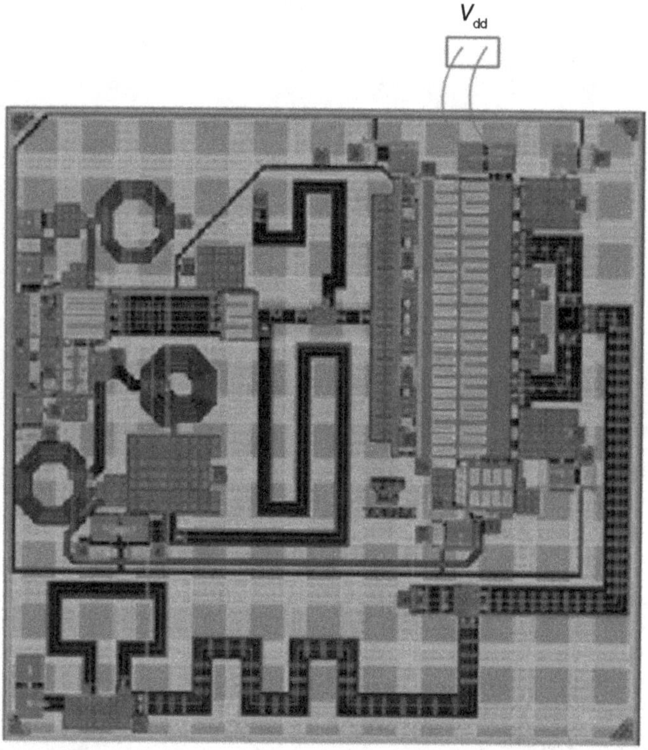

Figure 6.39 Layout of broadband two-stage parallel-circuit Class-E cascode CMOS power amplifier.

capacitance is fully represented by the output capacitance of the cascode structure. As a result, an average PAE of 41.5% and overall (including ET circuit) efficiency of 37% were measured at the carrier frequency of 2.5 GHz with a modulated output power of 26.5 dBm for a single-carrier 5-MHz LTE signal, as shown in Figure 6.40. By proper setting the required time delay between the ET and power amplifier paths for envelope signal, a significant linearity improvement was achieved, from initial level of −23.7 dBc to better than −36 dBc for $ACLR_1$ (improvement by more than 12 dB) and from −43.5 to −50.5 dBc for $ACLR_2$ (improvement by 7 dB).

The most widely used architecture for ET systems combines the DC-supply voltage provided by the DC–DC converter with the fast variable-envelope signal provided by an error amplifier. The error amplifier can generally be DC or AC connected to the output and represents a part of the fast tracker. For an ET AC coupling system, the use of a DC tracking loop is required, which provides tracking at

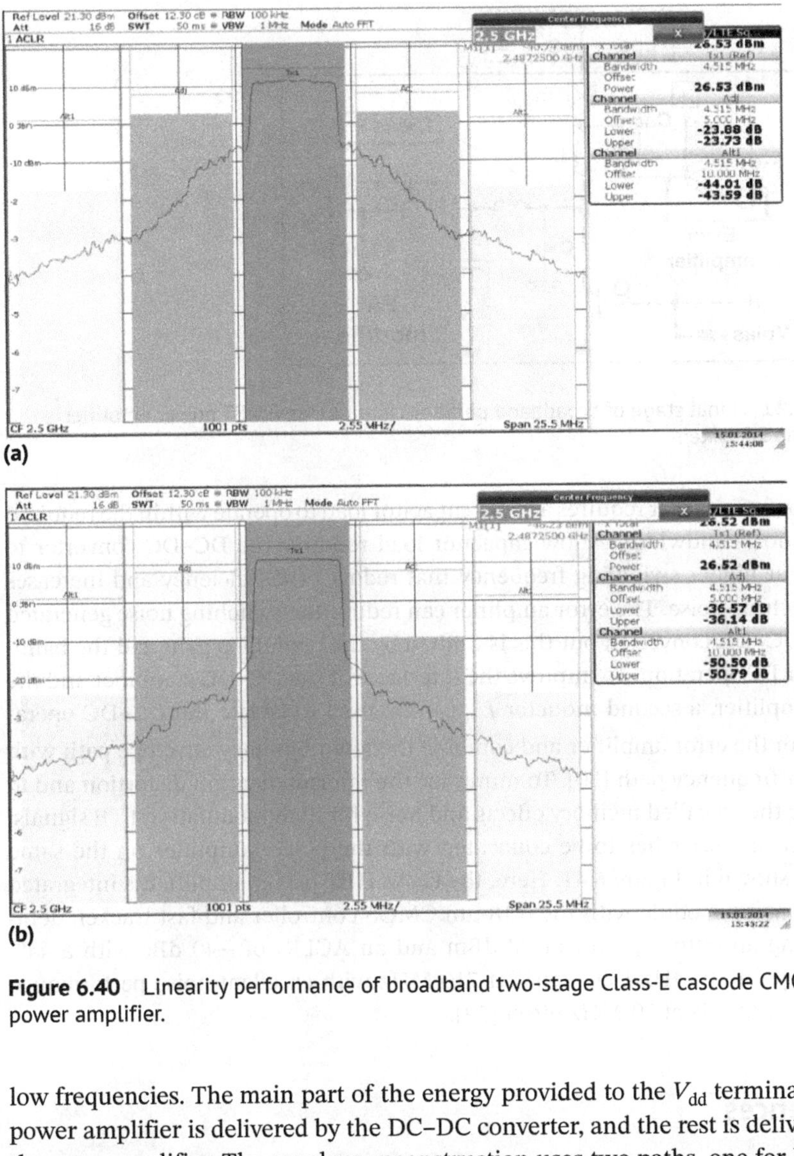

Figure 6.40 Linearity performance of broadband two-stage Class-E cascode CMOS power amplifier.

low frequencies. The main part of the energy provided to the V_{dd} terminal of the power amplifier is delivered by the DC–DC converter, and the rest is delivered by the error amplifier. The envelope reconstruction uses two paths, one for low frequency (DC–DC converter) and another for high frequency (error amplifier), and they don't have perfect alignment in time and frequency. For this reason, there is no flat gain over frequency of the envelope signal for configurations with both DC and AC ET coupling. During ET operation, there is an interaction between the error amplifier and the DC–DC switching.

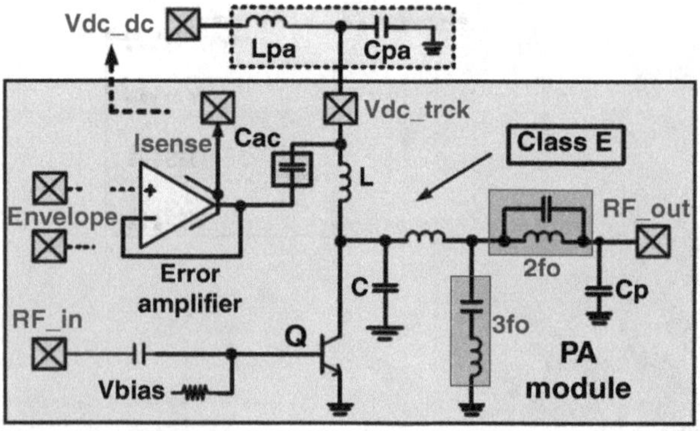

Figure 6.41 Final stage of broadband parallel-circuit Class-E HBT power amplifier within the ET system.

The error amplifier requires a lower capacitor load to operate within the required modulation bandwidth. A low capacitor load requires the DC–DC converter to operate at higher switching frequency that reduces the efficiency and increases the switching noise. The error amplifier can reduce the switching noise generated by the DC–DC converter, but this is limited by the open loop gain and the bandwidth of ET operation. To improve the interface for the DC–DC combiner and the error amplifier, a second inductor L_{pa} can be used to isolate the DC–DC operation from the error amplifier and combine the low-frequency envelope path with the high-frequency path [30]. To minimize the intermodulation distortion and to improve the so-called memory effects and noise for high modulation LTE signals, the error amplifier has to be collocated with the power amplifier on the same chip, as shown in Figure 6.41. Here, the GaAs HBT power amplifier is integrated into the same module with the 0.18-μm CMOS controller and fast tracker, demonstrating an output power of 28 dBm and an $ACLR_1$ of −40 dBc with a 44% efficiency for 10-MHz LTE signal at 710 MHz with excellent noise performance of −21.9 dBm/Hz at 30-MHz offset [32].

References

1 J. W. Phinney, D. J. Perreault, J. H. Lang, "Radio-frequency inverters with transmission-line input networks," *IEEE Trans. Power Electron.*, vol. PE-22, pp. 1154–1161, July 2007.

2 A. Grebennikov, "High-efficiency Class FE tuned power amplifiers," *IEEE Trans. Circuits Syst. - I: Regul. Pap.*, vol. CAS-I-55, pp. 3284–3292, Nov. 2008.

3 H. L. Krauss, C. W. Bostian, and F. H. Raab, *Solid State Radio Engineering*, Wiley, 1980.
4 A. Grebennikov, "High-efficiency Class E/F lumped and transmission-line power amplifiers," *IEEE Trans. Microw. Theory Tech.*, vol. MTT-59, pp. 1579–1588, June 2011.
5 A. Sheikhi, M. Hayati, and A. Grebennikov, "A design methodology of Class-E/F_3 power amplifier considering linear external and nonlinear drain-source capacitance," *IEEE Trans. Microw. Theory Tech.*, vol. MTT-65, pp. 548–554, Feb. 2017.
6 M. Hayati, A. Sheikhi, and A. Grebennikov, "Effect of nonlinearity of parasitic capacitance on analysis and design of Class E/F_3 power amplifier," *IEEE Trans. Power Electron.*, vol. PE-30, pp. 4404–4411, Aug. 2015.
7 M. Hayati, A. Sheikhi, and A. Grebennikov, "Design and analysis of Class E/F_3 power amplifier with nonlinear shunt capacitance at nonoptimum operation," *IEEE Trans. Power Electron.*, vol. PE-30, pp. 727–734, Feb. 2015.
8 C. Liu and Q. F. Cheng, "Analysis and design of high-efficiency parallel-circuit Class-E/F power amplifier," *IEEE Trans. Microw. Theory Tech.*, vol. MTT-67, pp. 2382–2392, June 2019.
9 Z. Yang, M. Li, Z. Dai, C. Xu, Y. Jin, T. Li, and F. Tang, "A generalized high-efficiency broadband Class-E/F_3 power amplifier based on design space expanding of load network," *IEEE Trans. Microw. Theory Tech.*, vol. MTT-68, pp. 3732–3744, Sep. 2020.
10 H. Lu, J. Jiang, and H. Zhang, "A high-efficiency push-pull parallel-circuit Class-E/F_3 power amplifier for harmonic suppression," *IEEE Microw. Wirel. Technol. Lett.*, vol. 34, pp. 1170–1173, Oct. 2024.
11 V. G. Krizhanovski, D. Chernov, and A. Grebennikov, "Low-voltage Class E/F_3 high frequency oscillator," *Proc. 14th Int. Conf. Advanced Trends in Radioelectronics, Telecommunications and Computer Eng.*, pp. 607–611, 2018.
12 T. Inaba and H. Koizumi, "Class E/F_3 tuned power oscillator," *IEEE Trans. Power Electronics*, vol. PE-33, pp. 1420–1427, Feb. 2018.
13 S. Pezeshkpour and M. M. Ahmadi, "A self-tuned Class-E/F_3 power oscillator," *IEEE Trans. Power Electron.*, vol. PE-39, pp. 11398–11408, Sep. 2024.
14 J. Staudinger, "Applying switched gain stage concepts to improve efficiency and linearity of mobile CDMA power amplification," *Microw. J.*, vol. 43, pp. 152–162, Sep. 2000.
15 A. V. Grebennikov, H. Jaeger, and E. P. Heaney, "High-efficiency cellular phone transmitters with reduced power consumption," *Proc. European Conf. Wireless Technology*, pp. 115–118, 2002.
16 S. W. Chang, H. J. Finlay, N. S. Cheng, and B. S. Park, "Load variation tolerant radio frequency (RF) amplifier," U.S. Patent 6954623, Oct. 2005.

17 J. Klaren, C. J. Persico, S. Walter, and P. L. Chan, "Balanced power amplifier with a bypass structure," U.S. Patent 6806768, Oct. 2004.

18 A. Grebennikov, B. Zogl, H. Hermann, C. Roth, and W. Thomann, "High-efficiency balanced switched-path monolithic SiGe HBT power amplifiers for wireless applications," *Proc. 37th European Microwave Conf.*, pp. 1189–1192, 2007.

19 R. Qualia, J. Pang, S. C. Cripps, and A. Zhu, "Load-modulated balanced amplifier," *IEEE Microw. Mag.*, vol. 23, pp. 60–70, Dec. 2022.

20 P. W. Dent, "Doherty amplifier," U.S. Patent Appl. 2005/0134377A1, Dec. 2003.

21 D. J. Shepphard, J. Powell, and S. C. Cripps, "An efficient broadband reconfigurable power amplifier using active load modulation," *IEEE Microw. Wirel. Compon. Lett.*, vol. 26, pp. 443–445, Jun. 2016.

22 D. J. Shepphard, J. Powell, and S. C. Cripps, "A broadband reconfigurable load modulated balanced amplifier (LMBA)," *2017 IEEE Int. Microwave Symp. Dig.*, pp. 947–949, June 2017.

23 P. H. Pednekar, E. Berry, and T. W. Barton, "RF-input load modulated balanced amplifier with octave bandwidth," *IEEE Trans. Microw. Theory Techn.*, vol. 65, pp. 5181–5191, Dec. 2017.

24 P. H. Pednekar, W. Hallberg, C. Fager, and T. W. Barton, "Analysis and design of a Doherty-like RF-input load modulated balanced amplifier," *IEEE Trans. Microw. Theory Techn.*, vol. 66, pp. 5322–5335, Dec. 2018.

25 J. Pang, Y. Li, M. Li, Y. Zhang, X. Y. Zhou, Z. Dai, and A. Zhu, "Analysis and design of highly efficient wideband RF-input sequential load modulated balanced power amplifier," *IEEE Trans. Microw. Theory Techn.*, vol. 68, pp. 1741–1753, May 2020.

26 Y. Cao and K. Chen, "Pseudo-Doherty load-modulated balanced amplifier with wide bandwidth and extended power back-off range," *IEEE Trans. Microw. Theory Techn.*, vol. 68, pp. 3172–3183, July 2020.

27 Y. Cao and K. Chen, "Dual-octave-bandwidth RF-input pseudo-Doherty load modulated balanced amplifier with ≥10-dB power back-off range," *2020 IEEE Int. Microwave Symp. Dig.*, pp. 703–706, Aug. 2020.

28 Y. Chen, W. Choi, J. Shin, H. Jeon, S. Bae, K. Bae, J. Song, Y. Ju, H. Oh, H. Kang, Y. Woo, K. Y. Lee, K. C. Hwang, and Y. Young, "Comprehensive analysis of coupled transmission lines for broadband high-efficiency Doherty power amplifiers," *IEEE Trans. Microw. Theory Tech.*, vol. 73, pp. 4937–4953, Aug. 2025.

29 R. M. Salehabad and A. Grebennikov, "Generalized active N-port network analysis for load modulation: Doherty power amplifier incorporating tri-coupled-line combiner," *IEEE Trans. Circuits Syst. - I: Regul. Pap.*, vol. CAS-I-72, pp. 6936–6947, Nov. 2025.

30 F. Balteanu, "Single die broadband CMOS power amplifier and tracker with 37% overall efficiency for TDD/FDD LTE applications," *2016 IEEE RFIC Symp. Dig.*, pp. 322–323, May 2016.

31 F. Balteanu, "Linear front-end module for 4G/5G LTE advanced applications,". *Proc. 48th European Microwave Conf.*, pp. 251–254, 2018.

32 F. Balteanu, H. Modi, S. Khesbak, S. Drogi, and P. DiCarlo, "Envelope tracking LTE multimode power amplifier with 44% overall efficiency," *Proc. 2017 Asia-Pacific Microwave Conf.*, pp. 37–40, Nov. 2017.

Index

Note: Page numbers in *italics* and **bold** refers to figures and tables respectively.

h

i

n

o

Printed in the USA/Agawam, MA
October 7, 2026

911622.006